To my parents,
Celia and Konstantin,
and Slava.

The Soviet Economy

1970-1990

A Statistical Analysis

Dmitri Steinberg

International Trade Press

International Trade Press

©1990 International Trade Press
2 Townsend Street, #2-304
San Francisco, CA 94107
1-800-359-6031

First Edition, December 1990

Library of Congress Cataloging-in Publication Data

The Soviet Economy 1970-1990: A Statistical Analysis
Dmitri Steinberg

Includes bibliographical references and index.
1. Soviet Union--Economic conditions--1985- --Econometric models.
2. Soviet Union--Economic conditions--1975-1985- --Econometric models.
3. Soviet Union--Economic conditions--1965-1975- --Econometric models.
4. Soviet Union--Economic conditions--1985- --Statistics.
5. Soviet Union--Economic conditions--1975-1985--Statistics.
6. Soviet Union--Economic conditions--1965-1975--Statistics.
I. Title.
HC336.26.S74   1990   330.947'085--dc20
90-41575 CIP

ISBN 0-926476-03-3 $115.00

Publisher: Vance T. Petrunoff
Associate Publisher: Christine V. Judge

All material presented in this document is compiled from reliable sources, but accuracy cannot be guaranteed.

Table of Contents

Abbreviations

AFC	Adjusted Factor Cost
AFCS	Adjusted Factor Cost Standard
CIA	U.S. Central Intelligence Agency
CMEA	Council of Mutual Economic Assistance
DIA	U.S. Defense Intelligence Agency
GDP	Gross Domestic Product
GFI	Gross Fixed Investment
GNP	Gross National Product
Gosbank	State Bank
Goskomstat	U.S.S.R. State Committee on Statistics
Gosplan	U.S.S.R. State Committee on Planning
GSP	Gross Social Product
GVO	Gross Value of Output
HC&E	Housing, Communal and Everyday Services
IDS	Intelligent Decision Systems
I-O	Input-Output
KGB	Committee on State Security
MBMW	Machine Building, Metal Works and Capital Repair of Machinery
M & E	Machinery and Equipment
MVD	Ministry of Internal Affairs
Narkhoz	Narodnoe Khozyaistvo SSSR (Statistical Manual)
NFI	Net Fixed Investment
NMP	Net Material Product
NS	National Security
O&M	Operations and Maintenance
PNI	Produced National Income
R&D	Research and Development
SME	Soviet Military Expenditures
T&C	Transportation and Communications
T&D	Trade and Distribution
UB	Unified Balance Table of the U.S.S.R. Economy
UN	United Nations
UNI	Utilized National Income
VS	Vestnik Statistiki (Goskomstat Monthly Journal)
VT	Vneshnyaya Torgovlya SSSR (Foreign Trade Statistical Manual)

List of Charts

List of Tables

Soviet Data Base

I. Allocation of Resources

Soviet Data Base

II. Growth Rates and Annual Price Changes

Introduction

INTRODUCTION

What is the real size and structure of the USSR economy, how has it grown in the 1970s and 1980s, and to what extent is it militarized? During the past two decades, most specialists and government officials have answered these questions by citing estimates published in CIA annual reports to the U.S. Congress. These estimates are widely believed to be reliable because they are based on tested methods which have been periodically enhanced by CIA analysts and several prominent academic economists. However, as the Soviet crisis began to unfold, a growing number of specialists began to express doubts about the accuracy of some major CIA estimation results.

This erosion of confidence was initially triggered by impressionistic accounts indicating that the real size of the USSR economy is much smaller and the total defense burden much larger than CIA reports have indicated. During 1989-1990, this evidence was corroborated by unexpected findings obtained by prominent Soviet economists which indicate that the USSR national economy grew little in the past decade, that its size comprises no more than 15-20 percent of the U.S. level and that the total defense burden is no less than 20-25 percent of GNP. These results were in stark contrast with CIA estimates which suggest that average growth rates exceeded 2 percent a year, that the size of the USSR economy is 50 percent of the U.S. level, and that the defense burden is 15-17 percent of GNP.

Has the CIA provided the U.S. Congress with a misleading assessment of the USSR economy? Or did Soviet economists derive their sensational results using faulty statistical methods that cannot be verified with acceptable tools of statistical analysis?

In 1990, the Goskomstat--the main Soviet statistical agency--released a considerable amount of previously unpublished data which provided fresh insights into the size, structure and growth of the USSR economy. These data are unprecedented in scope and coverage, making it possible to simplify a notoriously cumbersome procedure for compiling an independent set of Soviet GNP accounts. In addition, the new data provide a unique opportunity--probably the first in decades-- for testing the accuracy of the CIA estimates.

In this study, I analyze the new Soviet economic data and then attempt to substantiate a number of revisions in the CIA methodology and estimates. My analysis is based on an alternative approach to collecting and systematizing Soviet economic statistics, an approach specifically designed to deal with difficulties often encountered in interpreting published Soviet data. In short, I integrated all available Soviet data on production, input-output, national income and financial flows as well as on capital and labor resources. Afterward, I converted integrated Soviet national accounts into a GNP format.

I compiled Soviet GNP accounts in established (prevailing) prices that are actually used in financial transactions as well as in factor cost prices that take into account the real cost of labor and capital employed in production and service sectors. Factor cost prices provide the best available approximation of price weights which serve as measures of the relative contribution of each sector to the national economic growth.

To compile Soviet national accounts in constant prices, I combined data on price weights with independently estimated price deflators. The latter were estimated for given sectors by comparing data on the gross output evaluated in established prices and in terms of physical quantities. Whenever data on physical quantities was not available I designed various synthetic measures as substitutes. I thus avoided using officially published price indexes which, as known, are not well designed for measuring hidden inflation in manufacturing and construction sectors. I believe that the proposed estimates of Soviet economic growth fully take into account annual changes in most hidden price increases which have puzzled specialists for decades.

The fact that it is difficult to analyze the USSR economy because of large uncertainties in interpreting published data must seem perplexing to specialists on market economies. Indeed, it is an established tradition for governments to inform their population and the outside world about their nations' economic activities by publishing national accounts that are designed according to the UN format.

National accounts contain aggregate as well as detailed statistics on the structure and growth of GNP--the most widely accepted aggregate measure of all legal economic activities. National accounts thus serve as a reliable source of information which can be used to trace the past and current development of the economy and to make predictions about its future prospects.

As opposed to other governments, the Soviet party-state has made little effort to conform to the above tradition. In addition to refusing a GNP-type format, it has strived to conceal important production and financial facts without which one is unable to have a full and undistorted view of the Soviet economic life. To monopolize all information flows in the country, the Soviet party-state was first in the world to publish not one but two sets of national accounts. Even though the two sets are essentially compiled from one pool of detailed information collected by various Soviet statistical agencies, they serve quite different purposes.

The first set--called "the national economic balance" (NEB)--serves as a private source of information for top policy makers. Secret NEB tables are known to contain the aggregated data on all national economic activities, including those performed in sensitive military, financial and foreign sectors. Not a single NEB table has ever been released to the public in the unexpurged form. Each table apparently contains sensitive data whose publication could allegedly damage Soviet national security interests.

The second set, which is commonly known as the official statistics, is compiled in the form of poorly organized excerpts from NEB tables. The official statistics has a purpose to inform the public about carefully selected socio-economic facts that cannot possibly damage Soviet national security interests. The official statistics are designed to conceal not only sensitive data but also numerous, seemingly "harmless" facts on the allocation of resources, price trends and finances.

Although it is unknown why particular facts are omitted from publication, the preoccupation with secrecy that is deeply embedded in Russian culture is often cited as the main reason. Party and state bureaucrats also may not want to divulge facts in their possession because this may result in the dercease of their monopolistic power and privilege in the society. This explains why even Gorbachev has experienced difficulties with receiving reliable information on the budgetary deficit, defense and KGB budgets, foreign debt, secret privileges of party and government bureaucrats and numerous other sensitive data that undermine ideological claims of the Soviet party-state.

It is also possible that the infamous bureauctatic environment of gathering statistics has served as a convenient smokescreen for concealing carefully selected secrets. What supports this possibility is that concealment policy has been so efficient that economists around the world are still unable to resolve decades-old controversies about the real size, structure and growth of the USSR economy. Discrepancies between various proposed estimates have been of such an enormous size that observers have wondered whether the purpose of the official statistics is to confuse rather than to inform.

When Soviet post-war statistical publications were designed in the mid-1950s their authors--top experts on national accounts--were confronted with a difficult dilemma. On the one hand, they were supposed to publish information on national and sectoral trends without revealing secrets. On the other hand, the had to be careful not to violate the internal consistency of NEB tables.

Soviet experts were able to solve their dilemma using the arcane Marxist concept of productive labor. According to this concept, national income is generated by material production sectors. This created a convenient opportunity to conceal data on so-called "nonproductive" sectors involved in sensitive national security activities. These include defense, space, research, police and secret banking activities.

Soviet experts also took advantage of the fact that defense production activities had been until recently heavily subsidized. The evidence discussed in this study indicates that defense subsidies are approximately the same in size as value added in secret industries that assemble and repair weapons intended for domestic use. This made it possible to lower the total weapon procurement bill without changing estimates of published totals for the gross and net material product.

Western specialists on the Soviet economy have long realized that in order to make sense out of the official statistics they must treat all published data as extracts from original NEB tables. Instructions on how to compile most NEB tables have always been readily available in published Soviet planning manuals, journal articles and in UN publications written by Soviet experts. A careful reading of these instructions provides the only opportunity for specialists working outside Soviet government agencies to reconstruct NEB tables using the official statistics.

There is enough information for reconstructing most segments of NEB tables pertaining to the civilian economy. Estimates for national security activities then can be derived as the difference between national economic aggregates and the derived aggregates for the civilian economy. For example, wages of secret national security sectors can be estimated as the difference between total wages and wages of those employees who are reported in the official statistics on labor.

Overall, the process of reconstructing original NEB tables entails a four-stage procedure: 1) identifying numerous "harmless" facts which are withheld from publication, 2) designing algebraic equations with which to estimate these facts using published data as known variables, 3) recognizing individual components of published national economic aggregates where Soviet economic secrets are hidden, and 4) making residual estimates of these secrets.

The same procedure is followed in this study in estimating Soviet economic secrets contained in seven key NEB tables which are compiled by the Goskomstat--the main Soviet statistical agency. These tables contain data on: supply and demand for goods and services: production, distribution and end use of national income; input-output (i-o) flows; foreign trade activities; household budgets; financial accounts of public sectors; and resources of labor and capital.

Known difficulties experienced in reconstructing NEB tables suggest that someone in the Soviet government anticipated future attempts to use the official statistics to compile NEB tables well in advance. Over the years the Goskomstat has been allowed to publish a limited amount of the NEB data, thus frustrating numerous attempts to uncover major Soviet economic secrets.[1]

Given the dearth of the available NEB data, it is not surprising that most specialists have based their analysis on GNP accounts compiled by the CIA. Unfortunately, it is not always recognized that the GNP format is much less efficient than the NEB format in extracting data on Soviet economic secrets from the official statistics. Secrets are hidden as components of national aggregates that are compiled according to the NEB format. Most of these components cannot be even found in GNP accounts because they are estimated with unique Soviet methods of national accounting. This makes it difficult for researchers who are unfamiliar with NEB tables to focus on those segments of Soviet national accounts where economic secrets are hidden.

For example, after compiling Soviet household accounts CIA analysts have discovered a huge gap between outlays and revenues. Unable to explain this gap, they have referred to it as a "statistical discrepancy." In reality, this gap represents wages of secret sectors that are omitted from the official statistics on labor force and incomes.

Similarly, the precision of estimates of Soviet GNP depends to a large extent on estimates of total gross investment. Published data on gross investment are always presented in comparable prices. Since the combined investment price index is never published, it is difficult to convert published data on investment components into current prices.

[1] For major publications on the subject refer to D. Gallik, B. Kostinsky and V. Treml, Input-Output Structure of the Soviet Economy: 1972, Foreign Economic Report, No.18, Washington, D.C.: U.S. Department of Commerce, 1983; A. Becker, Soviet National Income, 1958-1964. Berkeley: University of California Press, 1969; W. Lee, The Estimation of Soviet Defense Expenditures, 1955-1975, An Unconventional Approach, New York: Praeger, 1977; I. Birman, Secret Incomes of the Soviet State Budget. The Hague: Martinus Nijhofff Publishers, 1981; and S. Rapawy, "Labor Force and Employment in the USSR," The U.S. Congress, Joint Economic Committee, Gorbachev's Economic Plans, Study Papers, Volume 1, Washington, D.C.: GPO, 1987.

Some of these components simply cannot be estimated using a standard GNP methodology.

Instead of the GNP format, I propose to use a format of integrated input-output and financial accounts. I believe that the advantage of using integrated accounts is two-fold: it enables one to check the accuracy of the same indicators compiled with several estimation methods, and it can be used to analyze the full impact of financial policies on production, consumption and investment trends.

To avoid any future misunderstanding I would like to emphasize that no integrated system of Soviet national accounts indeed exists. In fact, as the Goskomstat Chairman Kirichenko emphasized, one of the major tasks facing his organization is to make several sets of Soviet statistics which are compiled with various national accounting methods compatible.[2] I believe that this compatibility can be achieved by analyzing how Soviet production, national income and financial methods differ in compiling particular economic indicators.

The idea of compiling integrated accounts was born after years of frustration in trying to make sense out of various unexplained residuals found in official statistical tables on the annual allocation of material, labor and financial resources. These residuals are defined as statistically large gaps between total resources that are available to all sectors of the economy and those resources which are allocated for specific civilian uses. Although some of these residuals have attracted attention on a number of occassions, no attempt has ever been made to analyze these residuals in relation to each other. As argued in this study, the approximate size and contents of these residuals are impossible to determine outside the framework of integrated accounts.

It is usually assumed that the Goskomstat has managed to conceal the entire defense production within published aggregates for production outputs and end uses of national income. As a result of this assumption, a number of specialists on the Soviet defense economy have repeatedly searched in vain for traces of large-scale armaments production in published accounts of machinery production and

investment. The acceptance of this assumption has led to the widespread confusion about the real size of the Soviet economy and its defense burden.

As evident from published CIA reports, after applying a standard GNP methodology one arrives at the size of the Soviet defense burden that is less than 8-9 percent. At the same time, the building bloc method of counting commissioned weapons and other defense activities leads to results that suggest that the Soviet defense burden is around 15 percent. Unable to deal with this discrepancy, some specialists offer the analysis of Soviet consumption, investment and defense trends whereby the sum of individual end uses far exceeds total GNP. As was discussed above, the defense sector can be fully integrated within Soviet national accounts by taking into account secret defense subsidies.

Estimates of Soviet GNP presented in this study can be viewed as the first available test for estimates contained in widely quoted annual CIA reports on the Soviet economy submitted to the U.S. Congress. While the proposed estimates generally confirm CIA ruble estimates of Soviet GNP and military expenditure, there are some notable discrepancies in estimates of the size, structure and growth of Soviet GNP.

I found that household consumption accounted for 47 percent of GNP, investment--31 percent, defense expenditure--18 percent, and civilian government--4 percent in 1985-1988. In published CIA reports, household consumption accounts for 55 percent, while investement is never divided into civilian and defense uses. I also found that during the past two decades the average annual rate of hidden inflation was 4 percent in consumer sectors, 3 percent in weapon production sectors and 4.5 percent in investment sectors. In comparison, CIA reports indicate that the rate of inflation was 2 percent in consumer and investment sectors and as much as 5.2 percent in weapon production sectors.

Due to different assessments of hidden inflation, my estimate of the Soviet civilian economic growth was smaller than that of the CIA by 0.5-0.6 percent a year. Similarly, my estimate of the Soviet defense economic growth was larger than that of the CIA by 0.3-0.4 percent a year. As a result of these discrepancies in inflation and growth rates, my estimate of the total Soviet GNP in dollars are

---

[2] V. Kirichenko, "Ochishchenie Statistiki," EKO, N2, 1990, p. 36.

notably smaller than those of the CIA. While the CIA estimates that the size of the Soviet economy is around half of the U.S. economy, preliminary estimates presented in this study indicate that the ratio is around 32-34 percent.

The existence of such disparate estimates when the Soviet society has become much more open is by itself remarkable; it serves as a poignant reminder of how little is known about the USSR economy after five years of perestroika, including three years of radical economic reforms. Given the important role played by the USSR in the world affairs and given its enormous military might, it seems that more effort should be made by Western government agencies and by the UN in the near future to compile a reliable set of integrated Soviet national accounts that will be acceptable to all specialists.

I present my research results in five chapters. In the opening chapter I review the CIA methodology and estimates on the size, structure and growth of the Soviet economy. In Chapter 2, I introduce the reader to Soviet production and financial methods of compiling national accounts. I argue that these methods provide a key to understanding the Soviet economic statistics.

In Chapter 3, I present my alternative approach to compiling Soviet national accounts which is based on integrating new Soviet data on production and financial flows. In Chapter 4, I convert Soviet national accounts into a GNP format, present GNP accounts by sector of origin and end use, analyze the real defense burden and the cuurrent financial

crisis. In Chapter 5, I present my alternative approach to estimating the real Soviet economic growth and price changes adjusted for hidden inflation. Afterward, I discuss my research results on the performance of Soviet civilian and defense sectors and the rate of inflation in different sectors of the economy.

The text of this study is accompanied by the data base on the annual allocation of resources and growth trends in the Soviet economy. The data base on growth and price trends is organized following a standard format of GNP accounts compiled in current and constant prices.

Altogether the presented data base consists of around 100 tables, each having on the average 57 rows. The detailed computerized documentation for each of the table will require a high level of effort and technical support which was not available at the time of preparing this study.

Therefore, readers will only find a short description of data source and methodology used in preparing the data base. Readers are urged to send inquiries to the author for the clarification of specific methodological issues, estimation procedures and sources of data not found in the present study.

The study contains a short description of activities performed by major Soviet economic sectors which was prepared by UN experts. This description will hopefully assist readers in better understanding the structure of economic sectors analyzed in the study.

Chapter 1

Sizing up the USSR Economy:
Review of the CIA Methodology and Estimates

CHAPTER 1

SIZING UP THE USSR ECONOMY:
REVIEW OF THE CIA METHODOLOGY AND ESTIMATES

This chapter has two objectives: 1) to review major tenets of the CIA methodology and estimates of Soviet GNP in dollars, established, AFC and constant prices; and 2) to propose improvements in this methodology.

The CIA Methodology and Estimates

The CIA methodology is designed to make periodic base-year estimates in established and AFC prices and annual estimates in constant prices. A recent CIA report on Soviet GNP in 1982 prices states that the CIA's estimates "are based on the same concept used in the U.S. and other Western countries: the total value of goods and services sold to final purchasers during a given period of time."[1] The major problem with the way the CIA applies this concept is that the CIA claims to have no reliable information on Soviet government purchases of defense goods and services which comprise a sizable part of GNP by end use.

CIA's Accounts in Established Prices

The CIA presents Soviet GNP accounts in current established prices--prevailing prices in the economy--in three major tables containing data on: 1) incomes of households and public sectors, 2) final purchases of all goods and services, and 3) the structure of value added in 32 major sectors of the economy, including civilian police and military personnel. These tables summarize more detailed tables on incomes and outlays of household and public sectors and on end uses of goods and services. In accordance with Professor Bergson's method, the control total for Soviet GNP is estimated as the sum of household outlays and incomes of the public sector.

Each of three tables contains large residuals which are referred to as unidentified incomes and outlays and statistical discrepancy. The household

income residual is derived as the difference between a) purchases of goods and services plus net savings and b) identified household incomes. The outlay residual of the public sector is derived as the difference between a) identified plus unidentified incomes and b) identified purchases. The unidentified income of the public sector is estimated as the state budget income residual.

The table on the structure of value added by sector of origin in established prices provides a starting point for compiling GNP accounts in AFC prices. Value added in this table is divided into the following components: state wages, social insurance deductions, other labor income (agriculture, private services and military subsistence), depreciation, profits, indirect taxes (turnover tax and various miscellaneous charges), subsidies, and other nonlabor income.

The comparison of this table with tables on household and public sector accounts indicates that the total for other nonlabor income is derived as business expenditures on services plus the household income residual. It also indicates that this total is distributed among sectors of origin arbitrarily. It is unclear why the household income residual is treated as part of nonlabor income.

The table on GNP by end use contains estimates for household consumption of goods and services, new fixed investment, capital repair, changes in inventories, government administrative services (GAS), R&D, net exports and other unidentified uses. The GAS consist of agricultural, state government, municipal and cultural services, forestry and civilian police. Unidentified end uses were estimated as the difference between total GNP and all identified end uses.

Consumption of goods is estimated using published data on retail trade and private agricultural production. Purchases of services by household and public sectors was estimated using published data on paid services for 1970 and service wages. The derivation of data on value added in most public and private services and on material purchases made by various services was primarily

---

[1] CIA, Measures of Soviet Gross National Product in 1982 Prices. Submitted to Joint Economic Committee of the U.S. Congress. Washington, D.C.: USGPO, 1990 (forthcoming), p. iii.

based on conjecture. In estimating fixed investment, the CIA made several assumptions which it also did not justify. For example, it assumed that 1982 prices on machinery were the same as 1984 comparable prices or that all deductions on capital repair were allocated to pay for performed work.

The CIA methodology has several peculiar features which set it apart from the methodology proposed by the Goskomstat and Gosplan--the two Soviet government agencies responsible for compiling Soviet national accounts:[2]

1. Net cash savings and purchases of producer goods and services appear to be excluded from the CIA's estimates of household outlays, thus resulting in the underestimation of the household income residual.

2. Household incomes and outlays include military subsistence, even though it constitutes state budgetary purchases of food and uniforms for the military.

3. The public sectors' incomes include hidden state budgetary revenues, which are essentially deficits, and purchases of business services, but exclude depreciation of capital operated by budget-supported sectors.

4. Housing subsidies are double counted as negative profits of other sectors.

5. Financing of broadcasting and some recreational services is counted as subsidies, even though these are government services.

6. Wages of seasonal agricultural workers are double counted as wages paid by collective farms and the state agricultural sector.

7. Net exports are estimated as the difference between exports and imports in foreign trade prices which differ significantly from established prices.

8. Consumption is overestimated by the amount of military purchases of consumer goods and services.

9. Additions to unfinished construction are double counted as components of both new fixed investment and working capital.

10. New fixed investment and hence total GNP are not reduced by losses of capital.

11. Forestry services are treated as part of end use, even though they offer intermediate inputs for wood production sectors.

12. Total GNP is increased by the amount of paid and imputed services of the second economy.

The CIA estimated that total Soviet GNP in established prices equaled 714 billion rubles (b.r.) in 1982. This exceeds Goskomstat's estimate by 21 b.r. Differences in methodology can explain most of this discrepancy. For example, purchases of business services plus budgetary deficits alone exceed 21 b.r. for this year. Other discrepancies in estimates resulting from the noted differences in methodology evidently cancel each other.

In CIA's accounts, unidentified outlays amounted to 55 b.r. Had the CIA decided to reduce Soviet GNP by CIA's own estimate of other budgetary revenues in the amount of 8 b.r., these outlays would equal 47 b.r.[3] In 1982, outlays on military R&D equaled around 12 b.r. Did Soviet military expenditure (SME) equal only 59 b.r. in 1982? SME (without military pensions and other transfer payments) derived by the CIA using a building bloc approach that is based on pricing individual defense goods and services equaled 103 b.r. in 1982. The discrepancy between the two CIA measures of SME thus amounted to as much as 44 (103-59) b.r. It remains unclear how this large amount can be possibly concealed in consumption and investment uses.

---

[2] For the detailed discussion of the Goskomstat-Gosplan methodology refer to Yu. Ivanov and B. Ryabushkin, "Integratsiya Balansa Narodnogo Khozyaistva i Sistemy Natsional'nykh Schetov." Vestnik Statistiki (VS), N9, 1989; Yu. Ivanov, B. Ryabushkin and M. Eidel'man, "Ischislenie Valovogo Natsional'nogo Produkta SSSR." (VS) N7, 1988; and Goskomstat SSSR and Gosplan SSSR, Metodika Ischisleniya Valovogo Natsional'nogo Produkta SSSR. Moscow, 31 March 1988.

[3] CIA, Measures.., op. cit., p. 106.

CIA's Accounts in AFC Prices

In converting Soviet GNP accounts from established to AFC prices, the CIA follows the so-called adjusted factor cost standard (AFCS) that was also originally developed by Professor Bergson. The principal idea underlying the AFCS is to replace all revenues of the public sector--profits, turnover taxes, subsidies, net budgetary revenues from foreign trade, and other nonlabor income--with the hypothetical return on capital which is uniform for all sectors. Compared to established prices estimated by the CIA, AFC prices also include depreciation of capital in free household services, science and housing sectors.

The CIA currently estimates this return as 12 percent of capital which equals the sum of the average annual value of depreciated fixed assets, inventories and unfinished construction. In estimating 1982 AFC prices, the CIA used published Soviet data on the value of fixed capital stock in 1973 prices as well as data on working capital and unfinished construction. Since all these data had not been available for individual service sectors, the CIA had to make assumptions about the availability of capital in these sectors.

It is usually assumed that AFC prices better reflect relative average resource costs, the production structure, patterns in the allocation of resources, and sectoral contribution to economic growth than producers' prices. What is not well recognized is that AFC prices have been accepted as reliable tools for measuring the production potential of the economy but not as tools for evaluating annual changes in consumer welfare. The reason for this is that in reality even theoretically justified AFC prices differ substantially from market-clearing prices (some combination of ex-village and black market prices) that determine consumers' expenditure patterns. As known, compared to AFC prices, market-clearing prices better capture severe structural disbalances in the Soviet economy that are evident in shortages of the ever growing number of traded consumer and producer goods.

As estimated by the CIA, the structure of Soviet GNP estimated in current AFC prices by sector of origin has remained essentially unchanged between 1970 and 1982. Increases in weights were observed for total industry--from 32 to 32.4 percent, for construction--from 7.3 to 7.8 percent, and for transportation--from 8.7 to 9.5 percent. Declines in weights were observed for agriculture--from 21.1 to 20.6 percent, for trade--from 7.3 to 6.5 percent, and for services from 20.5 to 20.1 percent.[4]

CIA's Accounts in Constant Prices

The CIA has compiled Soviet GNP accounts in current prices for 1970 and 1982, while for other years it presents estimates in constant 1982 AFC prices. The latter derives constant prices by multiplying base-year values and growth rates. The CIA estimates the growth of each GNP component by using the published Soviet data on quantities of output in physical units, values of output in comparable prices and the average annual employment in service sectors.

Data on value in comparable prices and on employment are used as a proxy for quantities of output in physical units. For machinery and apparel sectors, value data comprise around 40 and 30 percent of the total sample of output. In estimating the growth of value added in industrial and transportation sectors, the CIA uses Soviet data on gross output as a proxy. For agriculture, the CIA makes adjustments for waste and losses, while it uses data on quantities of produced construction materials to measure the growth of the construction sector. The CIA presents estimates for the growth of consumption of goods and services in established as well as AFC prices. No independent estimates of constant prices are made for imports and exports, nor for investment in machinery.

As estimated by the CIA, average growth rates of Soviet GNP in constant AFC prices by sector of origin declined from 4 percent a year in 1965-1975 to 2 percent in 1975-1985 and to 2.3 percent in 1986-1989. Since 1975 annual growth rates have averaged 2.4 percent in industry, transportation and services; 1.7 percent in construction, and 1.4 percent in agriculture. Consumption has grown by 2.5 percent, investment--by 3 percent, and government expenditures--by -0.2 percent. The shares of consumption and investment have remained 54-56 and 30-32 percent respectively,

---

[4] ibid, p. 9.

while the share of other unidentified end uses has surprisingly declined from 10 to 6 percent.[5]

CIA 's Dollar Estimates of Soviet GNP

To measure the relative size of the USSR economy, the CIA combines its dollar/ruble purchasing power parity ratios, which were derived for consumption, investment and defense uses, with its annual data on the growth of these major end uses. Parity ratios for consumption were obtained using a quite representative sample of 334 Soviet consumer goods and the detailed information on the size and quality of Soviet consumer services for 1976.[6] The same approach was applied to evaluate the comparative size of Soviet investment goods and services.[7] The CIA investment study entailed the comparison of 245 machinery and equipment items and 277 examples of construction works. Defense activities were re-evaluated several times in dollars, each "on the basis of a detailed identification and listing of Soviet forces and their supporting elements."[8] The CIA presents separate dollar estimates for weapons procurement, operating expenditures and military R&D works.

The CIA's estimation results indicate that dollar-ruble ratios derived using US and Soviet quantity-weighted averages were 1.67 and 2.48 respectively. These averages led to the average (geometric mean) ratio of 2.00, which implied that the size of the USSR economy was more than 60 percent of the U.S. level. Other average ratios were

as follows: consumption--1.8, investment--2.4 (including machinery--3.00 and construction--2.00), and defense--2.00.[9]

As reported by the CIA, the size of the USSR economy is currently above 50 percent of the U.S. level, thus suggesting that the average dollar-ruble ratio has grown since 1976 from 2.00 to around 2.7. The same increase is implied by the CIA's estimates of Soviet defense activities whose dollar value is currently the same as that of the US.[10] Then, by extrapolating the 1976 data, it follows that by 1989 average ratios for consumption and investment have increased to 2.3 and 2.8 respectively. This in turn suggests that consumption and investment currently constitute 31-32 and 100 percent of the U.S. level.

CIA's In-House Review

CIA reports acknowledge that limitations in the available data make it possible to make only approximate estimates of real growth. Estimation errors supposedly occur because:

- physical quantities often do not capture fully all changes in mix and quality
- the employment data for services can be used as a proxy for output as long as there are no large changes in labor productivity
- comparable Soviet prices are derived with methods which are not designed to account for various hidden price increases, and
- published Soviet output series exclude statistics on privately provided services which must be independently accounted for in estimating total GNP.

CIA reports also publicly acknowledge some additional problems with limitations in the available data that undermine the reliability of results. Foremost are difficulties encountered in integrating sector-of-origin and end-use estimates of the GNP growth. For example, for 1987 the end-use data suggest that the USSR economy declined by 0.5 percent, while the sector-of-origin data indicate the 1.5 percent growth.[11] For some years

[5] ibid, p. 83.

[6] Gertrude Shroeder and Imogene Edwards, Consumption in the USSR: An International Comparison. A study prepared for the use of the Joint Economic Committee, U.S. Congress, Washington, D.C.: USGPO, 1981.

[7] Imogene Edwards, Margaret Hughes and James Noren, "US and USSR: Comparisons of GNP," in U.S. Congress, Joint Economic Committee, Soviet Economy in a Time of Change, Volume I. Washington, D.C.: USGPO, 1979, pp. 369-401.

[8] CIA, Soviet and US Defense Activities, 1970-1979: A Dollar Cost Comparison. A Research Paper, SR 80-100005, January 1980, p. 1. For a more recent publication refer to CIA, A Guide to Monetary Measures of Soviet Defense Activities. A Reference Aid. SOV 87-10069. November 1987.

[9] Edwards, et. al., pp. 378-379.

[10] CIA, A Guide..., op. cit., pp. 13-14.

[11] See John S. Pitzer, "Alternative Methods of

the discrepancy between the two sets of estimates reached as much as 3 percent.

CIA analysts find several causes for this huge discrepancy. These include:

- the uncertainty about the nature of unidentified household incomes
- the uncertainty about the way major portions of defense expenditures are hidden with consumption and investment uses in Soviet GNP accounts
- the inability to employ the theoretically preferred double deflation method of converting both gross value of output and material inputs from current to constant prices, and
- the absence of constant-price time series for exports, imports, defense expenditures, and inventory changes.

Despite the acknowledged lack of internal consistency in the CIA's estimates, the CIA report concluded that[12]

the income and outlay accounts are essentially independent measures of each sectors's transactions, and both sets of accounts do seem to match fairly well. The principal problems have already been described and, while troubling, do not seem major faults. If any errors are concentrated in specific areas, any resulting analysis of those subject areas may be misleading, but the general scope and distribution of the incomes and outlays of both sectors appear broadly correct.

Similarly, all major tenets of the CIA methodology were recently upheld at the 1989 CIA conference with most Western specialists on Soviet national accounts in attendance.[13] The last CIA report submitted to the U.S. Congress similarly concluded that "our [CIA] estimates of total GNP growth are not far from the mark." The report continued that "growth is subject to sources of both

overestimation and underestimation, so errors in opposite directions offset each other..."[14]

The report further specified several major sources of bias in CIA growth estimates. It noted that reliance on quantity data in measuring growth of basic industrial sectors--metals, chemicals and construction materials--and housing and service sectors results in a downward bias. At the same time, reliance on value data in measuring the growth of machinery, apparel, furniture, repair and personal care results in an upward bias. On balance, CIA estimates of annual Soviet economic growth are believed to contain a margin of error not exceeding -0.1 to -0.3 percentage points.

According to a recent in-house review of CIA's work on the comparative size of Soviet consumption, the CIA's estimates contain an upward bias of around 10 percent.[15] This bias was primarily a result of difficulties encountered in evaluating the generally inferior quality of Soviet goods and services. There were also other reasons for the bias:[16]

the sample better represents the Soviet product mix; many products available in the U.S. are not to be found (or are extremely highly priced in the USSR; the product matches do not reflect the U.S. advantage in terms of style, design, color and attractiveness...Moreover, subsequent upward revisions in U.S. consumption for 1976 reduces the Soviet comparative standing.

Had the CIA accounted for the bias, its estimate of per capita consumption in the USSR would have dropped to 31 percent of the U.S. level in 1976 and to 28 percent in 1988.[17]

Independent Reviews

How accurate are the CIA in-house reviews?

---

14 CIA, Measures..., op. cit., p. 45.
15 Gertrude E. Schroeder, "Consumption in the USSR and the U.S.: A Western Perspective." A paper presented at AEI Conference on the Comparison of U.S. and Soviet Economies. Virginia, April 1990, p. 19.
16 ibid, pp. 18-19.
17 ibid, p. 20.

Valuing Soviet Gross National Product," in CIA, Measuring Soviet GNP: Problems and Solutions. A Conference Report. SOV 90-10038, September 1990, p. 25.
12 ibid, p. 6.
13 ibid, p. xxii-xxiv.

Independent reviews usually focus on four areas: the relative quality of Soviet goods and services, the reliability of comparable Soviet prices, the soundness of assumptions on which AFC prices are based, and the real scope of the second economy. Few researchers also gathered some evidence from alternative sources of information to demonstrate that the CIA has grossly overestimated the size of the USSR economy, particularly its consumer sector. However, no attempt was ever made to explore why CIA's accounts are internally inconsistent or why these accounts are not well designed for making comparisons between growth rates of Soviet civilian and defense economies. Such a comparison could throw light on the extent to which the overall Soviet economic growth has depended on increases in military expenditure.

The CIA's Dollar Estimates

Igor Birman, a known CIA critic in the U.S., once suggested a provocative short-cut for measuring the extent to which CIA's accounts are internally inconsistent. He compared the size of USSR and U.S. economies by using the commonly accepted data on the role played by the agricultural sector in both economies. With the U.S. producing more agricultural products than the USSR, the share of the agricultural sector in Soviet GNP is seven times as large as the same share in the U.S.

Indeed, according to CIA reports quoted above the agricultural sector accounts for 20.6 percent of Soviet GNP, while in the U.S. the share is below 3 percent. Does this mean that the USSR economy is only 14 percent of the U.S. level and not 50-53 percent as reported by the CIA? Or has the CIA overestimated the role played by the agricultural sector in the USSR economy? Even after accounting for large differences in the structure of the two economies, in levels of productivity and in uses of the agricultural output, one must still conclude that the inner inconsistency of CIA's accounts is on an intolerably large scale.

A Swedish economist Anders Aslund and a prominent Soviet economist Victor Belkin independently cited a wide range of interesting but unsystematically organized evidence to support their view that the size of the USSR economy may

well be around 20 percent of the U.S. level.[18] Compared to Aslund and Belkin, Birman made a systematic effort to review the CIA study on the comparison of Soviet U.S. consumption levels.

Birman paid particular attention to difficulties encountered by CIA analysts in evaluating the output of the trade sector as well as the comparative quality of Soviet consumer goods and services. He concluded that after accounting for errors made by the CIA and for the inferior quality of Soviet goods and services the comparative size of Soviet consumption should be reduced not by 10 percent as was recommended by CIA's in-house review but by at least one third.[19] At the same time, he admitted that problems with comparing quality parameters in market and non-market economies remain unsolved. In his more recent works, he expresses the view that when the existing problems are solved, then the comparative size of Soviet consumption will have to be reduced even further.

Although Birman presented a number of new and interesting ideas for improving the CIA's estimates, the lack of data on the relative quality of Soviet consumer goods makes it exceedingly difficult to apply his ideas in a systematic way. It is from this standpoint that his ideas appear unacceptable to UN and World Bank experts who conduct comparison studies. It appears that only after the Goskomstat will make an effort to compile the detailed data on quality will it be possible to settle the debate on the extent to which the CIA has overestimated the relative size of Soviet consumption.

In this respect, it is interesting to compare CIA's and Birman's estimates with the recently published Goskomstat's results of four comprehensive studies comparing U.S. and USSR consumption and investment levels with those of Hungary and Poland

[18] Anders Aslund, "How Small is Soviet National Income?" A paper presented at the First Biennial RAND-Hoover Symposium on the Defense Sector in the Soviet Economy, Stanford, March 1988; and Viktor Belkin, Market and Non-Market Sytems: Limits to Macroeconomic Comparability." A paper presented at the AEI Conference on the Comparison of U.S. and Soviet Economies. Virginia, April 1990.

[19] I. Birman, Ekonomika Nedostach. New York: Chalidze Publications, 1983, p. 392.

that were separately performed under the auspicies of the UN and CEMA.[20]

The average of these results indicates that the ratio between per capita consumption in the USSR and the U.S. was around 31-32 percent in 1980, which is quite close to the adjusted CIA estimate. This ratio, however, decreased dramatically to 26 percent in 1983 and to 24 percent in 1985. By continuing the trend, one arrives at a ratio of 22 percent for 1988-1990. If these results are reliable, then CIA's growth indexes for consumption have a strong upward bias. However, there is also another possibility that UN/CEMA results for 1980 are quite unreliable and that the ratio for this year was around 27-28 percent.

Average results of the noted studies indicate that in 1980-1985 the ratios between per capita investment and between GDP in two countries increased from 86 to 100 percent and from 40-41 percent to 36-37 percent respectively. In deriving the GDP ratio, the Goskomstat made an assumption that the size of government expenditures (GE) is approximately the same in the two countries since the total for GE primarily consists of SME. While this is the case for the USSR, in the U.S. military expenditures account for only 60-65 percent of GE. As a result of this serious (hopefully unintentional) oversight, the Goskomstat grossly overestimated the relative size of Soviet GDP.

After accounting for the civilian GE the ratio between per capita GDP of the two countries decreases from 36-37 to 34-35 percent for 1985. If one assumes that the parity between the two countries' gross investment and military expenditures remained unchanged in 1985-1988, then one can use data on relative consumption trends to conclude that per capita GDP for the USSR was around 31-32 percent of the U.S. level. For total GNP, the ratio appears to be 33-34 percent.

It must be noted that the parity for gross investment is derived by the Goskomstat using a

methodology which results in international economic comparisons that have an upward bias for the USSR. As is known, in estimating Soviet GNP in rubles, the Goskomstat fails to subtract such losses of capital as abandoned construction cites which are not available for end use. In addition, capital repair works account for 17-18 percent of Soviet gross fixed investment. In the U.S., like in other market economies, most capital repair works are treated as current expenditures and thus are removed from GNP altogether.

Problems also exist in treating all Soviet construction projects which are never finished as a GNP component. The value of unfinished construction works currently exceeds total capital investment. A number of unfinished construction projects will be eventually abandoned. It is clear that these projects should not have been started in the first place. Had Goskomstat's experts compared the value of completed construction projects as opposed to the entire volume of gross fixed investment, then they would have estimated the size of per capita investment in the USSR as no more than 75 percent of the U.S. level. In view of this fact, it can be tentatively assumed that the lower bound estimate for Soviet GNP is 31-32 percent of the U.S. level.

Undoubtedly, these and other results quoted above remain highly hypothetical because most aspects of Goskomstat's methodology for international comparisons have remained secret. It also must be noted that indirect comparisons lead to a significant margin of error, especially for a non-market economy. Only a direct comparison with a large market economy performed for 1988 or some later year would make it possible to provide an unequivocal answer about the relative size of the USSR economy today.

The CIA's Estimates in Established Prices

For years specialists on the USSR economy have paid little attention to Soviet GNP accounts in established prices. Starting with Bergson, the prevailing attitude has been that established prices distort the Soviet economic reality to such an extent that little effort should be made to apply these prices for analytical purposes. As a result, the CIA has focused its research activities on compiling

[20] V. Martynov, "SSSR i SSha po Materialam Mezhdunarodnykh Sopostavleniy OON i SEV (raschoty Goskomstata SSSR)." Vestnik Statistiki, N9, 1990, p. 15.

growth indexes and on improving methods for compiling AFC prices. During the last 20 years the CIA has undertaken only two relatively small-scale studies of Soviet GNP accounts in established prices.

During the last couple years a growing number of specialists have begun to realize, however, that a lot can be learned about the USSR economy by compiling a complete and internally consistent set of GNP accounts in established prices on the annual basis. Overall, there are at least three reasons why established prices have a much greater analytical value than had been realized. First, the Soviet financial crisis can be analyzed only in established prices which are actualy used in financial transactions. Second, it is impossible to determine AFC prices before learning about all inadequacies of established prices. And third, a reliable set of GNP accounts in established prices is needed to understand Gorbachev's allocative choices, including in the military area. It is for the same three reasons that Soviet GNP compiled by the CIA do not fully serve their purpose.

This is particularly evident in the way the CIA measures seven major items in Soviet household and public sectors' accounts. For the household sector, these items are: total wages of state employees, the income residual, and purchases of goods and services. For the public sector, these items are: business purchases of services, miscelaneous charges, government services, and total gross fixed investment.

1. In determining the state wage bill, the CIA has applied a simple but unreliable procedure using published Soviet data on the average wage rate and employment. This procedure, however, cannot be applied because the average wage rate is derived by Goskomstat officials using data on wages and employment for all state sectors, including secret sectors which are never published in the official statistics. Following the established convention, employees of the secret sector represent three groups: civilians, police and the military. A recent Goskomstat report revealed that the size of the previously unreported civilian group alone amounts to 4.3 million persons or 10 percent of the entire industrial labor force.[21]

Regardless of what type of industrial activities this secret group is engaged in, its wages must be treated as part of the total state wage bill. If this secret group is engaged in production and repair of machinery intended for military use, then its wages should be added to those of published machinery sectors, thus increasing the total value added of the entire machinery sector. As was noted above, CIA's accounts contain a household income residual which is not, however, identified as wages of unpublished sectors. For unknown reasons, this residual is distributed in CIA's accounts among published sectors as other nonlabor income. In this way, the CIA not only grossly underestimates the wage bill but also distorts the price structure in the USSR economy.

2. There are also problems with the way the household income residual is derived using the CIA methodology. In the absence of reliable data on total wages, the household income residual can be derived as the difference between monetary outlays and net savings on the one hand and published monetary incomes on the other. Instead of using this formula, the CIA decided to employ a procedure--first introduced by Bergson--which is based on combining two incompatible tasks.

The first task was to add total household outlays on consumption and investment as represented in GNP accounts, while the second task was to derive total household incomes as the sum of the above outlays and transfer payments. The incompatibility arises because some outlays--for example, purchases of producer goods--are excluded from GNP but are still part of household financial accounts.

GNP accounts also have no place for cash savings (net household savings kept at home), even though these savings constitute income received during a given year. As a result, the CIA underestimated the household income residual by as much as 50 percent. There is evidence indicating that cash savings have sky-rocketed during the last three years, further worsening the Soviet financial crisis. Cash savings thus can be no longer ignored as a minor item in Soviet household accounts.

3. There are additional problems with the way

---

[21] For the discuusion of the report refer to N.

Zhelnorova, "Statistika Stanovitsa Ob'yektivnoi." Arqumenty i Fakty, N4, 1990, p. 1.

the CIA estimates household purchases of consumer goods and services. Households make most purchases through the retail trade network. As known, this network also serves civilian institutions. What is not acknowledged is that retail trade purchases are made not only by civilian but also by military institutions. The same applies to purchases of services made outside the retail trade network. As can be judged from open publications, the CIA methodology is not designed to account for any such military purchases. Uncertainties also surround CIA's approximate estimates of the total volume of household services. For reasons that are unclear, the CIA decided not to use the available Soviet data on services and instead designed a questionable procedure based on continuing trends since 1970. The comparison with the Soviet data indicates that using this procedure leads to estimates with a notable downward bias.

4. To estimate business purchases of services the CIA uses a cumbersome procedure which is based to a large extent on guesswork. For example, charges for militarized guards are estimated at half the value of current outlays on civilin police. Data on these current outlays are all based on the questionable assumption that the size of the civilian police force is 67.6 percent of employment in so-called "other material production sectors."

As reported in Soviet statistical handbooks, these sectors are engaged in other material production activities. However, there is no evidence indicating that civilian police is treated in Soviet national accounts as material production labor. The easiest and most reliable way of estimating business purchases of services is to compile a table on sources of financing service activities in the USSR economy. After compiling such a table, these purchases can be estimated as the difference between the total volume of services and the sum of household and state government (budgetary) purchases of services. Following the UN methodology, Goskomstat officials exclude business purchases of services from value added. Why the CIA decided to treat these purchases as part of value added remains unclear.

5. What the CIA refers to as miscelaneous charges is known in Soviet national accounts as other revenues of enterprises and net revenues from foreign trade. While a large part of other revenues

is extracted to the state budget, some of these revenues remain on banking accounts of enterprises. The CIA procedure is to equate these revenues with most of the state budget income residual.[22]

The Goskomstat procedure is to estimate the difference between total nonlabor income and net profits (profits less charges against profits). This procedure is undoubtedly much more precise, especially considering that each year a larger share of other revenues remains on accounts of enterprises and that the Soviet government has finally admitted that deficits comprise most of the state budgetary income residual.

6. In Soviet national accounts, the concept "government services" applies to a broad range of household, scientific and administrative services which are financed by the state budget. Except for housing and entertainment, all these services receive most of their financial support directly from the state budget. In this sense, these services can be analyzed in the same way as government services that are found in market economies. Housing and entertainment industries suffer operational losses which are treated in Soviet national accounts as charges against profits.

There are two major problems with the way government services are presented in CIA's accounts. First, it is unclear why the CIA treats budgetary financing of housing, broadcasting and social services as subsidies that reduce total value added. Housing subsidies are already implicitly incorporated into data on profits of service sectors that were used by the CIA. At the same time, there was no need to treat broadcasting and social services as profit-seeking sectors. Second, the CIA ignored published Soviet budgetary and national income data on services and instead decided to make independent estimates using the wage data combined with conjecture about the structure of individual services. The apparent objective behind this exercise was to measure total public-sector outlays on services as components of GNP by end use. As I discussed above, GNP should include only those public-sector outlays on services that are

---

[22] The CIA procedure is to exclude allocations from depreciation deductions and 10 percent of the unexplained residual.

financed by the state budget.

7. Gross fixed investment is estimated in Soviet national accounts as the sum of net fixed investment (additions to fixed capital stock net of depreciation), depreciation deductions, assumed depreciation of capital operated by budgetary and private sectors, additions to unfinished construction, and losses of capital. In the absence of data on fixed investment in current prices, the above estimation procedure appears to be the only reliable one available. Instead of following this procedure, the CIA has designed its own procedure, which as was noted above, is based on several unjustified assumptions. It is difficult to accept this procedure not because CIA's assumptions lead to overestimating the investment share in Soviet GNP but because these assumptions make it impossible to integrate published data on capital flows in the USSR economy. Without having integrated capital accounts, users of the CIA's estimates can feel free to make endless conjectures about large amounts of military hardware that are concealed in published Soviet fixed investment series. However, after compiling integrated capital accounts it becomes clear that there is no room for military hardware in these investment series.

Review of the CIA's Estimates in AFC Prices

The unconditional acceptance of the AFCS over the last three decades has created the impression that AFC prices can solve every methodological problem created by established prices. As it turned out, this impression has been thrice misleading. First, the analysis based on AFC prices has distracted everyone's attention from solving longstanding problems with understanding the structure of the USSR economy that were outlined above. Second, it has been often forgotten that the lack of internal consistency in CIA's accounts in established prices are multiplied in various unexpected ways when these accounts are converted into AFC prices. And third, the euphoria about a theoretical superiority of AFC prices has lasted for so long that no one has bothered to test the crucial hypothesis on which the AFCS is based.

In brief, the hypothesis states that stocks of fixed capital, unfinished construction and inventories provide a much better measure of relative

profitability then profits and other revenues contained in established prices. One way to test the hypothesis is to compare the production capacity of investment goods--producer durables and construction works--that can be purchased with one ruble of capital investment in each sector of the economy. If the variance in production capacity is large--for example, it exceeds 20-25 percent for several sectors--than the hypothesis should be rejected as invalid.

A group of Soviet economists headed by Rogovsky has recently completed a large-scale study on a relative purchasing power of capital investment in major production and service sectors of the USSR economy.[23] Their study was based on a detailed comparison of Soviet producer durables and construction works with those that are sold on world markets. Soviet economists emphasize that their results are less reliable for housing and service sectors than for production sectors. The reason for this is that the accepted methodology for international comparisons is not well designed to account for the inferior quality of Soviet construction works performed in service sectors. The major results of their research are reproduced in the chart below.

As the chart vividly demonstrates, there is an unexpectedly large variance in the relative quality of capital investment in many sectors of the USSR economy. The quality of producer durables in machinery production and power sectors is twice as high as in transportation and agricultural sectors and five times as high as in service sectors. The quality of construction works in transportation and metallurgy sectors is six times as high as in coal mining and agriculture. Overall, the quality of capital investment in machinery production and non-ferrous metallurgy sectors is three times higher than in coal mining and agriculture, which helps explain why the latter sectors' activities have traditionally been non-profitable.

The results obtained by Soviet economists are quite stunning as they indicate that the AFCS has

[23] For their research results refer to Rogovsky, E.A., Rutkovskaya, E. A., and Polyakov, E. V., "Kachestvennaya Neodnorodnost' Investitsionnykh Resursov i Metody Yeyo Izmereniya," Ekonomika i Matematicheskie Metody, Vol. 26, N3, 1990, p. 438.

Chart 1: New Soviet Data on the Relative Purchasing Power of One Ruble of
Capital Investment in Major Sectors of the USSR Economy

	Weighted Average	Producer Durables	Construction Works
Ferrous Metals	1.04	0.74	1.32
Non-Ferrous Metals	1.61	1.21	1.94
Petroleum		0.79	
Coal	0.43	0.50	0.33
Gas		0.69	
Chemicals	1.38	1.32	1.44
Machine-Building	1.65	1.57	1.75
Power	1.06	1.48	0.88
Wood and Paper	1.46	1.39	1.52
Constr. Materials	1.45	1.13	1.72
Textiles and Apparel	1.56	1.46	1.72
Total Industry	1.15	1.26	1.07
Construction	1.27	1.08	1.72
Agriculture	0.52	0.82	0.33
Transport and Comm.	1.20	0.80	1.87
Trade and Distrib.	1.07	0.97	1.18
Total Production	1.00	1.00	1.00
Total Services	1.04	0.30	1.18
Total Economy	1.02	0.97	1.07

provided a misleading impression of the structure of the USSR economy. Specifically, the AFCS has led to the overestimation of the share of fuels and agricultural sectors and to the underestimation of machinery production, non-ferrous metallurgy, wood making, construction materials, textile and apparel sectors in Soviet GNP.

There are some additional problems with the way the AFCS has been recently applied. As was noted above, the CIA currently estimates AFC prices using the Goskomstat's data on the size of fixed capital stock evaluated in 1973 comparable prices and adjusted for depreciation. The fact that comparable prices are not designed to account for hidden inflation is well recognized. What is rarely acknowledged is that there is a significant variance in rates of hidden inflation, whereby the largest rates are observed in fuels, transportation, agricultural, and housing sectors.

A separate issue is whether it is necessary to make adjustments for depreciation. Stocks of fixed capital are least depreciated in agriculture (22

percent) and housing (18 percent). At the same time, for construction and industry, the share of depreciated capital amounts to 54 and 46 percent.[24] Why should the return on capital in housing be thrice as high as in construction because of adjustments for depreciation remains unclear.

Has the AFCS outlived its usefulness? Has the AFCS always tried to accomplish an impossible task? Or should the AFCS be further revised to account for differences in capital productivity? I believe that adjusting established prices is an endless exercise that the real 'evil' of established prices has less to do with profits than with wages. Once this will be realized, current debates will be eventually settled with a consensus that the AFCS was originally an imaginative theoretical construct that had no or little practical application for the non-market economy.

[24] Goskomstat SSSR, Osnovnye Pokazateli Balansa Narodnogo Khozyaistva SSSR i Soyuznykh Respublik, Moscow: IIT, 1990, p. 67.

prices has less to do with profits than with wages. Once this will be realized, current debates will be eventually settled with a consensus that the AFCS was originally an imaginative theoretical construct that had no or little practical application for the non-market economy.

A non-market economy is commonly associated today with deficits and privileges rated according to one's ability to have access to goods and services which are in short supply. A system of privileges is not distributed evenly among various sectors of the economy. Employees in such sectors as state administration and trade have monopolized access to most high quality goods and services, while employees in such sectors as agriculture, construction, non-priority industries and education enjoy little privileges. Somewhere between these two extremes, there are employees in priority industries and defense sectors who enjoy a number of privileges granted to them by the government administration sector.

It is clear that, given the existing distribution system, relative wage rates provide a distorted picture of the real demand for labor in the USSR economy. At issue is how to adjust wage rates so that it will be possible to measure the real relative cost of labor resources in all major sectors of the economy. I doubt that the issue can be definitely resolved even if there were detailed statistics on privileges and 'supplemental' incomes of trade and other sectors' workers who re-sell goods and services which are in short supply. The reason for my scepticism is that there is no standard measure in a non-market economy for evaluating market-clearing prices for most goods and services. Thus, any attempt to improve on the AFCS would lead to results with a margin of error that is impossible to evaluate with any precision.

Probably some combination of dollar and black market prices for Soviet goods and services could be used temporarily as a substitute for the AFCS until the market mechanism is introduced into the USSR economy. In the absence of a representative sample of these prices, one has no alternative to using an improved version of the AFCS that accounts for differences in capital productivity and supplemental wages. For example, after making these adjustments I estimated that the share of agriculture in Soviet GNP drops from 20.6 to 14.8 percent.

Finally, I would like to sum up the discussion about the AFCS by comparing the results on end uses of Soviet GNP in 1982 established and AFC prices obtained by the CIA. According to the latest CIA report, the share of R&D and other outlays in Soviet GNP comprised 15.9 percent in established prices and 11.4 in AFC prices.[25] The same share estimated in 1982 AFC prices dropped to 9.4 percent of Soviet GNP in 1985-1987.[26] Civilian R&D works (8-9 billion rubles in 1982) and additions to inventories (31 billion rubles as estimated by the CIA) comprise 5.6 percent in established prices and around 6 percent in AFC prices.[27] The share of defense outlays in 1982 Soviet GNP that was identified by the CIA then amounts to 10.3 percent in established prices and 5.4 percent in AFC prices. The share of defense outlays in Soviet GNP for 1985-1987 evaluated in 1982 AFC prices is below 5 percent.

The absurdity of these results is self-evident, especially considering that the CIA has made no effort to identify military expenditures in consumption and investment components of Soviet GNP. What seems totally unbelievable is that established prices are twice as high as AFC prices for military expenditures. How is this possible, given all the privileges enjoyed by the defense sector in the USSR economy?

Review of the CIA's Estimates of Growth

The CIA's estimates of Soviet economic growth have usually been criticized in the West on the ground that the CIA methodology is not designed to account fully for quality improvements, especially in investment and service sectors. Some CIA critics claim that indexes based on comparable Soviet prices are more reliable than indexes based on physical quantities of output.[28] These CIA critics cite no new evidence to support their claims.

25 CIA, Measures..., op. cit., p. 26.
26 ibid, p. 72.
27 ibid, p. 108.
28 See Michael Boretsky, "The Tenability of the CIA Estimates of Soviet Economic Growth," Journal of Comparative Economics, December 1987; Mark Prell, "The Role of the Service Sector in Soviet GNP and Productivity Estimates," Journal of Comparative Economics, September 1989

During the past two years comparable prices have been totally discredited even in the USSR as the Goskomstat began to introduce more refined methods that resemble those used by Western statistical agencies.

Other CIA critics propose to make adjustments for quality improvements by considering several alternative sources of information. Thus, Michael Alexeev suggested that the housing floorspace index should be adjusted for increases in the supply of amenities.[29] After making such an adjustment, he concluded that the CIA underestimates the real growth of Soviet housing services by 2 percent a year and that this growth averaged 5.7 percent a year in 1970-1987.[30]

I believe that one cannot use data on changes in amenities as the only measure of real changes in quality of housing services. The other three indicators are the volume of performed current and capital repair works per unit of housing space, the share of ovedepreciated housing units in the total housing stock, and changes in the quality of new construction. All these indicators have dived dramatically, especially since 1975, thus explaining why the overall Soviet housing stock has seriously deteriorated despite sizable increases in new construction. More research is required to determine the extent to which this deterioration has offset increases in amenities.

In my work on long-term Soviet economic growth I observed a) that the issue of real changes in quality of Soviet goods and services cannot be resolved because of the dearth of available data and b) that the issue can be circumvented in most cases using alternative methods of estimation.[31] Thus, I proposed to measure the growth of service sectors as the weighted average growth of labor, material and capital inputs into these sectors.

My proposal was based on the assumption that changes in labor productivity in service sectors depend on changes in the amount of supplies (food, medicine, uniforms, materials, etc.), equipment and available floorspace. I found that labor resources in service sectors grew faster than supplies and capital assets. As a result, my estimates of rates of growth were much smaller for most sectors than The CIA's estimates, especially for the 1980s.

My other proposed improvements were as follows:

1. Replace CIA's base-year weighted Laspeyres index with the Paasche index. The advantage of using the Paasche index is that it makes it possible a) to trace annual price changes in various sectors of the economy, including sectors that produce intermediate goods and services, and b) to employ the double deflation method of estimatiing the growth of value added by sector.

2. Employ the double deflation method of estimating value added in constant prices as the difference between gross value of output and intermediate purchases. The CIA assumes that value added grows in tandem with gross output is not supported by the available evidence. This unproven assumption makes it practically impossible to integrate the CIA's estimates of Soviet economic growth by sector of origin and end use. Therefore, it is not surprising that the CIA's growth estimates for consumption, investment and defense far exceed its estimates of the growth of GNP by sector of origin.

3. Design an alternative methodology for measuring the growth of output for production sectors for which no data on quantities of physical output are available. This concerns primarily machinery production, furniture and apparel sectors. For machinery production, I proposed to use various data on hidden inflation published by those Soviet authors who had access to the detailed Goskomstat production reports which are not available in the West.[32] For furniture and apparel

[29] See Michael Alexeev, "Soviet Residential Housing in the GNP Accounts," in CIA, Measuring Soviet GNP..., op. cit.

[30] ibid, p. 90.

[31] See Dmitri Steinberg, "Measures of Soviet Economic Growth, 1965-1985, Volume I: Discussion of Alternative Estimates and Methodology, and Volume II: IDS Data Base on the Soviet Economy (USSR Integrated Economic Accounts)," IDS Report. October 1989.

[32] It is interesting to note that those Soviet economists who had access to more detailed and complete machinery data found the largest size of hidden inflation. For example, see K. Val'tukh and B. Lavrovsky, "Proizvodstvennyi Apparat Strtany:

sectors, I proposed to use published data on production of wood products and textiles. My assumption was that the output of the furniture and apparel industry could not grow much faster than the amount of materials used in production.

4. Test the CIA methodology for measuring the growth of agricultural, construction, trade, transportation, other industries and services. The CIA methodology was successfully tested for agricultural and trade sectors. Various tests performed for other noted sectors resulted in estimates of growth rates which are much smaller than those derived by the CIA. For construction, I proposed to use published Soviet data on the completed floor space and other measures of physical output. This made it possible to account for hidden inflation in unfinished construction which has grown much faster than finished construction. I rejected the CIA procedure for using data on the production of construction materials because this procedure was not designed to deal with the problem of cost maximization that is prevalent in the Soviet construction industry whereby enterprises are interested in using more rather than less construction materials.

For transportation, I proposed to use the ton index as opposed to the ton-kilometer index which is known to be flawed because of inefficiencies in the distribution system and largely padded mileage reports. For other industries, I proposed to use published Soviet data on quantities of output. In comparison, the CIA assumed that other industries grew at the same rate as the entire industrial sector. The alternative methodology for estimating the growth of service sectors was already discussed above.

After making all these proposed adjustments I was able to compile an internally consistent set of Soviet national accounts where end-use estimates of growth were only slightly higher than sector-of-origin estimates. I also was able to derive two separate estimates for the growth of the civilian and

defense sectors of the USSR economy.[33] It turned out that during the past two decades the defense sector grew 2.8 percent a year, while the civilian sector grew only 1.6 percent a year with the average growth of 1.8 percent for the entire economy.

In comparison, CIA estimates indicate that growth rates for the entire economy averaged 2.8 percent, while those for the defense sector--2.2 percent. Using CIA's data on the size of the defense burden and the CIA's estimate of growth rates, it follows that the civilian economy grew by around 3 percent a year. I believe that end-use estimates of growth are more reliable than sector-of-origin estimates. However, the CIA quotes the latter as its final estimates, thus reporting that the USSR economy has grown by 2.1 percent. In this way, the CIA indirectly admits that it overestimates the average annual growth of civilian consumption and investment by 0.7 percent.

Conclusion

The above discussion pursued two related objectives. The first objective was to review the CIA methodology, paying particular attention to major assumptions on which the CIA's estimates of the size, structure and growth of the USSR economy are based. The second objective was to propose improvements in the CIA methodology and estimates.

It was demonstrated that CIA's accounts contain a number of internal inconsistencies which undermine the reliability of the CIA's estimates presented in established, AFC and constant prices. These inconsistencies are first of all the result of several questionable assumptions made by the CIA in compiling Soviet household and public sectors' accounts which make it impossible to integrate defense activities within the framework of GNP accounts. Some inconsistencies also result from a number of peculiarities in the CIA methodology of estimating the size and growth of Soviet GNP.

Ispol'zovanie i Rekonstruktsiya," Ekonomika i Organizatsiya Promyshlennogo Proizvodstva, N2, 1986, p. 29. They found that the machinery growth averaged only 4 percent a year in 1966-1980, which is smaller than Khanin's and The CIA's estimates by 0.6 and 1.5 percent respectively.

[33] The results were summarized in Dmitri Steinberg, "The Growth of Soviet GNP and Military Expenditures in 1970-1989: An Alternative Assessment," A paper presented at the Second Biennial RAND-Hoover Symposium on the Defense Sector in the Soviet Economy, Santa Monica, March 1990.

Although each of these peculiarities may not have a significant individual effect on the overall results, their cumulative effect leads to several quite controversial conclusions.

Thus, CIA's accounts contain a residual for military expenditures which comprises 10.4 percent in established prices and only 5.4 percent in AFC prices. The small size of the residual and the fact that estimates in established prices are so much higher then in AFC prices are serious causes for concern. The same applies to CIA's residual estimates of growth rates which were derived for the defense component of Soviet GNP. The negative growth of 3 percent a year for 1966-1985 is too unrealistic. There is something inherently wrong with the methodology that leads to such unrealistic results.

In the course of the above discussion I tried to determine particular problems that cause such results. Specifically, I argued that during the past three decades the CIA paid insufficient attention to the Soviet economic data in established prices. This often led to the misinterpretation of methods which are used by the Goskomstat to compile these data.

Without having a reliable set of GNP accounts in established prices, it was indeed impossible to compile reliable AFC prices. I also argued that the traditional CIA methodology for computing AFC prices is based on two crucial assumptions which are contradicted by the available evidence. In brief, I considered the issue of differences in capital productivity and real wage rates and pointed out why the Bergson-CIA methodology is not designed to deal with this issue.

In addition, I tried to demonstrate why any estimate based on Soviet comparable prices leads to the overestimation of Soviet economic growth. For this reason, I proposed various alternative procedures which avoid using comparable prices in measuring the growth of consumer industries, construction, transportation and government services. I argued that these procedures can lead to internally consistent results, providing they are employed in conjunction with the double deflation method of estimating value added.

Finally, I compared the CIA's estimates of the relative size of the USSR economy with recently published Goskomstat estimates. I concluded that the CIA's estimates contain a strong upward bias largely as a result of overestimating the absolute size of Soviet consumption as well as rates of Soviet economic growth. I also observed that CIA's results on the size of the USSR economy are notably incompatible with its results on the structure of the economy in AFC prices.

Chapter 2

Soviet National Accounting Methods and Concepts

In this chapter I analyze major features of NEB, input-output and financial methods of compiling Soviet national accounts. My objective is to outline how these methods differ from each other and to explore whether published data compiled with these methods can be analyzed as an integrated system of national accounts.

The NEB Method

Basic Concepts

The NEB method is used to compile key Soviet national accounting tables on production and national income flows as well as on capital and labor resources. According to Soviet planning manuals, the NEB method has these distinguishing features:[1]

- national income--the aggregate of primary incomes in the economy--is generated by material production sectors

- all other sectors--referred to as non-material sectors--are financed by distributing primary incomes earned by material production sectors

- any enterprise that maintains an independent banking account is the basic production unit

- industrial production performed under the aegis of non-industrial enterprises is aggregated with that of the industrial sector, and

- the same as the above applies to subsidiary agricultural and construction activities performed in the industrial and other sectors

The central feature of the NEB method that sets it apart from the commonly used GNP method is the special treatment of material production sectors. Soviet government officials believe that, as an aggregate measure of economic activities, GNP provides an exaggerated impression of national economic power which supposedly depends on the size and quality of produced material wealth. They also believe that the GNP format simplifies the production and income distribution process which must be analyzes as the annual economic cycle consisting of four stages.

These stages are: 1) production and distribution of new material wealth; 2) interindustry flows, i.e. the exchange of intermediate products between material production sectors; 3) primary and secondary distributions of national income earned by material production sectors; and 4) the end use of national income--purchases of material wealth for current consumption, net investment and defense purposes.

Aggregate Measures of Produced Material Wealth

In accordance with their Marxist views, Soviet government officials prefer to operate with their own aggregate measures of national economic activities. Instead of GNP, they estimate gross social product (GSP) as the annual output of material production sectors. Compared to GNP, GSP includes interindustry flows but excludes the output of those sectors that do not directly participate in the production and delivery of new material wealth. Instead of net national product (NNP), they estimate net material product (NMP) at both production and end use stages of the economic cycle.

At the production stage, NMP is called produced national income (PNI) and at the end use stage--utilized national income (UNI). GSP exceeds

---

[1] Planning manuals are various publications of the USSR Gosplan and other planning agencies containing instructions on how to collect and compile data for NEB tables. None of these publications has been translated into English. One manual which is most commonly used is USSR Gosplan, <u>Metodicheskie Ukazaniya k Razrabotke Gosudarstvennykh Planov Razvitiya Narodnogo Khozyaistva SSSR</u>. This manual had 1969, 1974 and 1980 known editions.

PNI by the cost of intermediate and capital goods which are used as inputs to produce NMP. This cost is referred to as "c". The cost of these capital goods equals capital depreciation plus the unamortized value of liquidated capital.

As the total new value generated in the economy during a given year, NMP is divided into two parts: incomes of material production labor and revenues of enterprises employing this labor. These two NMP components are referred to as "v" and "m"). Following Marx, GSP is estimated as c + v + m. At the production stage, NMP is estimated as v + m, and at the end use stage--as the sum of private and public consumption, net fixed investment, and additions to working capital (inventories, reserves, unfinished construction, and defense uses).

PNI differs from UNI not only numerically but also structurally. PNI exceeds UNI by the sum of losses occured outside the production process and foreign currency earnings converted into domestic prices. At the end use stage, this sum is subtracted from NMP because it represents produced material wealth unavailable for end use. Whereas the PNI structure is determined by the wages/revenue ratio, the UNI structure is determined by patterns of consumption, investment and defense expenditure.

No division of national income into "v" and "m" and into PNI and UNI is usually performed in GNP accounts. NNP exceeds national income by the difference between indirect business taxes and subsidies. This differences constitutes the non-factor charge against NNP. Though Soviet officials saw no need in the past for estimating national income at factor cost, they recently began to publish estimates for NMP without domestic taxes, subsidies and foreign trade tariffs. In effect, these estimates lead to NMP at factor cost.

An attempt to estimate the agricultural NMP at factor cost was made by academic economists in the early 1980s and later by Goskomstat officials. They estimated that the ratio between established and factor cost prices declined in 1980-1983 from 50 to 25 percent due to price reforms that caused agricultural subsidies to double in size.[2] The size of

new value generated by the agricultural labor is supposedly underestimated in published national accounts because taxes and excessive profits contained in retail prices on goods made from agricultural raw materials outweigh agricultural subsidies.

In 1988, Soviet government officials announced that they will begin estimating GNP because it is the best aggregate measure of the overall economic development and because it a recognized standard against which international economic comparisons are made. The detailed discussion of published Soviet GNP estimates will be presented in the next chapter. What must be emphasized at this point is that in order to independently derive Soviet GNP using the officials statistics it is first necessary to understand the obscure accounting procedure by which all Soviet economic activities are divided into material and non-material.

Material and Non-Material Production Activities

Marx and other nineteenth century economists defined a material production activity as the transformation of natural resources by social labor for the purpose of satisfying the ever-growing need for material goods. Labor is social when it is engaged in regular work which is deemed gainful for the society at large as opposed to a limited group of individuals. For example, making movies for distribution is performed by social labor, while making movies for private use is not.

Material production entails the creation of new material products, improvements on the already created products, and the delivery of new products to their users. The creation of new products is performed in industrial, agricultural, construction and other small material production sectors. Industry is in turn divided into mining and manufacturing groups as well as into Group "A" and "B" which represent producer and consumer goods. Improvements on existing products are performed by capital and current repair services. Delivery of material goods is performed in transportation and communications (T&C) and trade and distribution (T&D) services.

In most cases the division between material and non-material services is straightforward. As a rule, services that do not increase the value of material products are treated as non-material: household

---

[2] USSR Goscomstat. Narodnoe Khozyaistvo SSSR v 1980 Godu. Moscow: Finansy i Statistika, 1981, p. 379; and the 1983 edition, p. 407. This major statistical manual is refered to by Western economists as Narkhoz.

services, science and state administration. As opposed to industrial services that enhance the value of consumer items, household services serve personal needs outside the production process. Major non-material services are performed in the following sectors:

- T&C serving household and service sectors
- TV and radio broadcasting
- housing and municipal services
- education, culture and arts
- health and recreational facilities
- insurance and banking, and
- personal care or "everyday" services (public baths, barber shops, legal aid, rental offices and religious centers).

Utilities which provide power, gas and water to households are treated as industrial services because they deliver material goods.

Household services are performed primarily by specialized organizations. Around 15 percent of these services is financed by state enterprises and collective farms from profits and production funds. Planning manuals stipulate that all production units are required to maintain separate accounts for keeping track of their non-material activities. These accounts make it possible to exclude the entire output of non-material services from GSP and PNI as well as to trace all sources of financing these services. Financing of non-material services which are small in size and which maintain no separate accounts is an exception to the rule. In comparison, all production activities performed under the aegis of service organizations are supposedly taken into account.

As opposed to household services, science and state administration are sectors of the Soviet economy where the distinction between material and non-material activities is blurred and where planners must introduce their own conventions to supplement the Marxist theory. Thus, gas and oil exploration and design works financed as capital investment are treated as material production, while all other such works are scientific services.

Research institutes and design bureaus that maintain independent banking accounts are all engaged in non-material production activities. As a rule, research activities performed under the administrative control of material production enterprises are included in their output. The same

principle is applied to administrative services. Until 1986 computer centers were registered according to the material production status of their customers.[3] Starting during that year all computer services were registered as serving material production sectors. A similar change took place with respect to the road service sector in 1985.

Defense Production

The above discussion of how material and non-material production activities are separated in Soviet national accounts was based on a careful reading of published segments of planning manuals that cover civilian sectors. Unfortunately, planning manuals are virtually silent about the treatment of defense sectors in Soviet national accounts. According to conventional wisdom, the division into material and non-material production has been applied similarly in civilian and defense sectors.

Production, repair and delivery of armaments is presumably treated as material production, while military personnel and employees of military science and administration sectors are perceived as performing non-material activities. Conventional wisdom has been accepted as a given, even though the issue as to whether defense production can be equated with material production has never been settled among Soviet economists.[4]

In Soviet literature, the distinction between material production and defense labor is discussed in terms of three departments of social production, where Departments I, II and III represent the output of producer, consumer and military goods.[5] Military goods are divided into two groups: goods with a dual function that can be used by both military and civilian sectors, and goods that are produced only for purposes of conducting war. The first group consists of producer goods purchased by defense industries and consumer goods purchased by the armed forces and science organizations performing military R&D services. The second group consists of armaments and other military hardware that have no civilian use.

[3] See Metodicheskie Ukazaniya... for 1974, p. 703.
[4] See A. Zalkind, ed., Dva Podrazdeleniya Obshchestvennogo Produkta. Moscow: Statistika, 1976, pp. 23-24.
[5] See A.I. Pozharov, Ekonomicheskie Osnovy Oboronnogo Mogushchestva Sotsialisticheskogo Gosudarstva. Moscow: Voenizdat, 1981, pp. 130-132.

It is clear that goods that have a dual function are hidden in the official statistics with producer and consumer goods purchased by civilian sectors. What is unclear is how Soviet government officials decided to hide defense goods that have no civilian use, given the fact that all planning manuals claim that GSP consists only of Departments I and II.[6]

In discussing how defense production activities ought to be covered in national accounts, Soviet economists have not advanced their public debates beyond providing general references to Marx's economic and philosophical writings. It appears that these debates never will be settled because Marx did not specify whether material production encompasses defense production.

Those Soviet economists who cite Marxist economic theories consider defense production as a special type of material production. Others emphasize the fact that Marxist economic theories constitute an integral part of his philosophical system whereby the aim of material production is not to accumulate material wealth in a mindless way but to advance human progress. Since the aim of defense production is to create means of destruction impeding human progress, it follows that the output of Department III is not part of produced material wealth.[7]

However, it is doubtful that Soviet national accounts are compiled by officials who are adept in Marxist philosophical nuances. Furthermore, even advocates of the philosophical approach contradict their arguments by admitting that for "reasons of convenience" Department III is hidden within Department I.[8] Undoubtedly, their expression "reasons of convenience" is a euphemism for concealing the output of defense industries. As I argued elsewhere, it would be a mistake to accept unconditionally the official expediency theory.[9]

There is an urgent need to develop an independent systematic approach that avoids quotes from Marx's works as well as hidden innuendos of Soviet economists.

It is thus necessary to test two equally valid hypotheses. The first hypothesis is that published totals for industrial production and UNI include the entire cost of produced weapons intended for domestic use.[10] The second hypothesis is that published totals for industrial production and UNI exclude a large part of this cost. An attempt will be made in the next chapter to test each hypothesis by examining the official statistics on wages and employment.

Comparison of the NEB Method with Input-Output and Financial Methods

Major Differences

In addition to the main NEB method, Soviet government officials also use input-outpu (i-o) and financial methods of compiling national economic data. As was noted above, the NEB method is designed to analyze activities of enterprises that maintain independent banking accounts. In comparison, the method used to compile detailed i-o tables is designed to analyze all material production activies, including various subsidiary activities performed by industrial enterprises.

For example, according to the NEB method, the production of metals at machine-building (MB) enterprises is registered together with the MB output. According to the i-o method, the same production is registered together with the output of the metallurgical sector. Sectors that are grouped within the framework of commodity production flows are often referred to as "net sectors" of material production.

The NEB method is designed to trace the flow of purchased goods and services measured in prices that are actually used in economic transactions. In contrast, most published i-o data on interindustry flows are measured in prices that 1) exclude subsidies on goods purchased from and sold to the

[6] Metodicheskie Ukazaniya...,1974, pp. 621-623.

[7] See M. Bor, Effektivnost' Obshchestvennogo Proizvodstva i Problemy Optimal'nogo Planirovaniya. Moscow: Mysl', 1972, pp. 70-71. Also refer to his earlier work Voprosy Metodologii Planovogo Balansa Narodnogo Khozyaistva. Moscow: Izdatel'stvo AN SSSR, 1960.

[8] Bor, 1972, op. cit., p. 72, and Zalkind, op. cit., pp. 24-25.

[9] See Dmitri Steinberg, "Estimating Total Soviet Military Expenditures: An Alternative Approach Based on Reconstructed Soviet National Accounts."
Edited by C.G. Jacobsen. The Soviet Defense Enigma. SIPRI. Oxford: Oxford University Press, 1987, p. 29.

[10] Becker, 1972, op. cit. pp. 90-91.

agricultural sector and 2) include T&C and T&D costs which are counted twice.

Compared to annual tables compiled with the NEB method, i-o tables are compiled every five years primarily for the purpose of determining key long-term changes in the structure of production and distribution of output between and within enterprises in a minute detail. The advantage of using detailed i-o tables is that they are convenient for analyzing interindustry flows.

Annual NEB tables compiled with the NEB method are based on information that is not detailed enough for performing such an analysis of commodity flows. In 1975, planners began to compile the aggregate version of the detailed i-o table using the NEB method. Their primary purpose was to establish a more precise linkage between forecasts based on the above analysis and the annual NEB data. Over the years Goskomstat officials have released only one aggregated i-o table for 1988. Some data were also published from detailed 1959, 1966, 1972 and 1987 i-o tables.

In the 1970s Soviet academic economists have offered sophisticated mathematical models based on i-o techniques. It was expected that these models would replace planning models based on the NEB method. However, the i-o method was unacceptable to central planners as the main accounting method for three reasons. First, the detailed i-o tables require a quite detailed information which cannot be collected on the annual basis. Second, employing the i-o method made it difficult for planners to focus on activities of enterprises as basic production units in the economy. Third, employing this method made it impossible for planners to integrate production and financial plans because of the incompatibility of data on commodity and financial flows.

Indeed, in order to make these data compatible planners would have to convert the i-o data into the NEB-based production data and the financial data into the NEB-based national income data. This cumbersome conversion procedure would destroy the whole purpose of abandoning the NEB method in the first place.

The incompatibility of the i-o and financial data stems from the fact that financial sectors are aggregated not on the basis of their production specialization but on the basis of their institutional affiliation. For example, in NEB tables compiled with the NEB method, the output of machinery enterprises is aggregated according to a particular production specialization principle and regardless of whether these enterprises affiliate with different ministries. In comparison, financial tables contain the output of MB enterprises subordinated to particular ministries which are responsible for production and financial plans of these enterprises.

Financial tables have two sections containing information on revenues and outlays and on annual changes in savings and reserves. The structure of financial tables corresponds to the division of financial plans prepared for:

- ministries, committees and other agencies
- collective farms
- cooperative and public organizations
- the state budget
- banking and insurance systems, and
- the household sector.

Due to the noted peculiarities of the financial method, it cannot be used for collecting data on production and service activities of particular ministries. Nor can it be used to collect data on civilian ministries' involvement in the defense effort. As a result, published financial statistics cannot be used to determine the proportion of income received from production, service and defense sectors. This explains why Soviet officials are willing to reveal more data compiled with the financial method than with the NEB method.

As opposed to NEB and i-o methods, the financial method is limited to compiling data on financial flows between sectors. Thus, incomes that do not involve the exchange of money are excluded from financial flows. At the same time, financial flows cover various distribution processes between sectors that do not generate any income. Other differences that are discussed below concern the estimation of particular components of national economic aggregates.

Foreign Trade

While all data on foreign trade activities are compiled in both foreign and domestic prices, NEB tables are designed to store foreign trade data only in domestic prices. Most published data on foreign trade, however, are reported in foreign prices. In

NEB tables, the primary income from foreign trade is estimated as the difference between imports and exports in domestic prices plus foreign currency earnings converted into domestic prices with the combined export conversion coeficient (the ratio between total exports in domestic and foreign prices). Imports increase the value of resources available in the economy during a given year, while exports are treated as goods which are not available for end use. The combined import conversion coefficient was used in 1972 and 1975 when imports exceeded exports in foreign trade prices.

In published i-o tables, PNI is estimated without revenues from foreign trade, while the final demand section contains two columns for exports and imports, where data on imports are entered with a minus sign. Published i-o tables also exclude foreign currency earnings. In some versions of the i-o table which are designed to balance total resources of commodities with their total uses imports in domestic prices are added to primary incomes of material production sectors, while exports in domestic prices are added to end uses without imports. These versions of the table cover foreign trade similarly to tables on the the supply and uses of goods compiled with the NEB method.

In financial tables, foreign trade is registered in two ways: as other net financial revenues of the state sector and as other revenues and outlays of the state budget. Total revenues are estimated as the sum of state revenues from: domestic sales of imports, foreign sales of exports, and revenues from foreign monetary transactions (sales of gold, tourism, etc.).

Total budgetary outlays on foreign trade equal the sum of import and export subsidies paid both to domestic manufacturers and foreign client states.

Wages and Profits

In NEB tables on national income and i-o flows, total primary income is divided into "v" and "m". As reported in planning manuals, "v" is the aggregate measure of:

- wages of workers and service persons
- other monetary earnings (one-time bonuses and business travel expenses)
- wages of collective farmers, and
- net income from private agricultural,

construction and other material production activities.

Until recently data on "v" were usually published every five years in connection with detailed i-o tables. Starting this year data on "v" are published in Goskomstat press releases on the Soviet economy.

Total "m" is estimated as the sum of:

- net profits (excluding bonus wages)
- social security deductions
- turnover and other taxes less subsidies
- other revenues from domestic production, and
- net revenues from foreign trade.

It must be noted that NEB tables on the supply and uses of goods are not designed to measure "v" and "m". In these tables, producer prices on goods are divided into production outlays and net profit that contain bonus wages. Published tables on the structure of industrial production outlays, which are extracted from these NEB tables, contain data on:

- production materials, fuels and power
- capital depreciation
- wages (excluding bonus wages) and social security deductions, and
- overhead outlays.

The latter encompass additional material, capital and labor costs connected with running the enterprise as well as payments to various services offered to employees which are excluded from net profit but which are still treated as part of "m". Published NEB data on the structure of production outlays differs from the i-o data on production outlays, which are aggregated for "net" production sectors.

Even original financial tables have a limited usefulness in estimating many components of "v" and "m" by sector. For example, all income in-kind is excluded from financial accounts. In addition, published excerpts from original financial tables exclude data on "m" registered as production outlays. In addition, published financial data on wages, net profit and capital of T&C services are never disaggregated into material and non-material sectors.

This poses the major obstacle to estimating total production factors and value added components of material production sectors. Difficulties also arise in using financial data on the turnover tax paid by various industrial sectors. These data were recently published in the Goskomstat's previously classified handbook Finansy SSSR. As was noted above, the NEB coverage of industrial sectors is incompatible with financial statistics.

Capital Depreciation

In NEB tables, capital depreciation is divided into material and non-material services. It is estimated for financially independent enterprises as deductions on capital repair and replacement plus other sources of financing capital replacement plus the unamortized value of liquidated capital. Budget-supported and private sectors are exempt from making obligatory banking deposits on depreciation accounts. Soviet officials make their own expert calculations of capital depreciation for these sectors using existing replacement and repair rates for particular types of capital stock.

Depreciation of capital operated by material production sectors is excluded from NMP. Capital depreciation in non-material sectors is grouped with private and public consumption and other uses. Depreciation of the housing stock is treated as part of private consumption, depreciation of capital operated by services--as part of public consumption, and depreciation of capital operated by the unidentified defense sectors--as other uses of UNI.

Published financial statistics on capital depreciation exclude data on a) the unamortized value of liquidated capital, b) performed capital repair works, c) collective farms, d) budgetary and private sectors. While the new publication Finansy SSSR contains data on (a) and (b), there is still no published data on depreciation deductions made by T&C sectors serving material production and non-material sectors. Nor was any depreciation data published on.

Capital Outlays

In original NEB tables, gross fixed investment (GFI) in material production and sectors is divided into capital depreciation, uninstalled capital, losses of capital and net fixed investment. In published excerpts from NEB tables, data on capital outlays on T&C services are usually aggregated with

material production sectors.

In financial tables, the GFI is defined in terms of sources of financing that consist of:

- capital investment
- capital repair
- purchases of machinery by existing budgetary organizations, and
- additions of productive livestock.

Capital investment is divided into installed and uninstalled capital, where the latter is estimated as the sum of investment writeoffs and additions of unfinished construction financed from capital investment funds. Financial tables also contain data on sources of financing various capital outlays. In national income tables, investment writeoffs are defined as losses in construction (the value of abandoned construction sites) which are excluded from PNI.

Data on net fixed investment are published in the official statistical tables on the UNI. Data on additions of unfinished construction had been published until 1984. Data on the total GFI, capital investment in current prices and budgetary purchases of machinery have never been published. Instead, the official statistics contain information on capital investment in 1969 and 1984 comparable prices and on purchases of machinery by household services with All-Republic budgetary funds.

Consumption

In published NEB tables, consumption has two categories 'private' and 'public', where 'private' pertains to households, the armed forces and the police. Private consumption includes:

- purchases of goods for current consumption at retail trade stores and at ex-village markets
- purchases of utilities (power, gas and water)
- food packages received as wages
- consumption-in-kind, and
- depreciation of residential housing.

Consumption-in-kind is measured as the value of consumed goods that are not exchanged for money. Public consumption consists of current purchases of goods by organizations serving households and by science and state administration sectors. Such purchases in effect constitute current material expenditures of non-material services.

All published data on retail sales are extracted from financial tables where goods are aggregated on the basis of the retail trade nomenclature. For example, sporting goods are produced in machine-building, wood-making and n.e.c. industry sectors. In addition, published data on retail sales are compiled in such a way that totals for particular categories of goods include sales to institutions, and sales of second-hand and producer goods and services. Both of these sales are excluded from current consumption.

As was noted above, all consumption activities not involving monetary transactions are excluded from financial tables. This explains why data on consumption-in-kind are never published in the official statistics. Published financial tables are also difficult to use because data on the structure of prices on purchased services have never been published.

Working Capital

In NEB tables, working capital is defined as uses of goods that are excluded from consumption and from additions to fixed capital. Working capital consists of:

- unfinished production in industrial sectors and in agriculture
- unfinished construction works
- uninstalled machinery
- inventories of finished producer goods
- inventories of finished consumer goods, and
- reserves of civilian and defense goods stored for emergency and strategic needs.

Due to its general definition, working capital is the most convenient category of material wealth where planners can hide defense goods. This fact explains why no information exists on the contents and uses of reserves, nor on their price structure. Goods held as reserves are produced primarily by chemical, machinery and agricultural sectors. Purchases of reserves are often referred to as 'other expenditures.'

Published excerpts from the original NEB tables contain data on total additions of working capital and other expenditure which are never defined in

planning manuals. Published data on components of working capital are usually extracted from financial tables of state-cooperative sectors. Information on the division of inventories of materials and reserves into sectors of production appear to be excluded not only from published but also from original financial tables.

Consolidated Soviet National Accounts

Types of Accounts

The detailed information on the Soviet economy is aggregated into consolidated production and financial accounts. Consolidated production accounts combine the detailed data on the supply and uses of goods with data on national income flows and with data on labor and capital resources. Consolidated financial accounts combine the detailed data on revenues, outlays and reserves (savings) of state enterprises, households, the state budget, and banking and insurance systems. At issue is whether it is possible to integrate consolidated production and financial accounts.

Consolidated accounts that are compiled with the NEB method are referred to as the "unified balance of the Soviet economy" (UB). No excerpts from the UB table have ever been published and this probably explains why no attempt has been made to reconstruct this table. Instead, both Westen and Soviet economists have focused their attention on consolidated material production accounts compiled using the format of the detailed i-o table. However, data from this table cannot be reconciled with the published financial data.

In order to utilize the annual financial data it is necessary to integrate consolidated production and financial accounts within the framework of the UB table compiled with the NEB method. As discussed by Soviet authors, UB table appears to have the same general format as the i-o table: Quadrant I contains data on interindustry flows, Quadrant II--on end uses of goods, Quadrant III--on value added in material production sectors, and Quadrant IV--on value added in non-material production sectors. However, as opposed to the i-o table, the UB table has four complete quadrants, including the south-eastern quadrant which is poorly represented in the i-o table.

I would like to propose a hypothesis that Soviet

officials are unwilling to reveal much information about the fourth quadrant in the hope of diverting attention from sensitive data contained in this quadrant. There are reasons to believe that NEB and i-o tables are intentionally designed to exclude data on foreign trade, services, state budgetary, banking and defense activities and that these data are consolidated in separate NEB tables which are never discussed in published literature. The fact that most Soviet economic secrets pertain to the above activities seems to support this hypothesis.

Data on these activities are apparently removed from NEB and i-o tables to narrow the circle of officials who are privy to Soviet economic secrets. In this way, only those officials who compile consolidated accounts can analyze in detail the full scope and contents of military buildup, the civilian economic support for the military, and the real role played by foreign trade in the economy.

Description of the UB Table

As in the i-o table, the UB data are stored on the intersection of rows and columns. The UB data are aggregated along rows and columns of the table (see Chart below) in four directions which are referred to here as vectors. In correspondance with four quadrants, these vectors are designated as 1.2, 1.3, 2.3 and 3.4. Data on labor and capital are entered at the intersection of rows under Quadrants III and IV with columns originating in Quadrants I and II.

Data aggregated along vector 1.3 represent components of prices on goods that are produced in material production sectors. The summary row in Quadrant I stands for total outlays on material inputs that are included in the total price of goods. The summary row in Quadrant III contains data on capital depreciation and primary incomes without net taxes (indirect taxes less subsidies) generated in material production sectors. The summary row in Quadrants I and III contains data on the gross value of output (GVO) of material production sectors in producers' prices.

Total supply of goods produced in these sectors exceed these sectors' GVO by the sum of net taxes, imports, and T&C and T&D charges. In published tables on produced national income all subsidies are subtracted from the industrial PNI. In comparison, in the UB table, subsidies reduce the

value of supplies and hence the purchasers' price of agricultural products.

Uses of goods are aggregated along vector 1.2. The summary column in Quadrant I contains data on total interindustry uses of goods and services supplied by material production sectors. The column containing data on agricultural and construction losses is located between Quadrants I and II because these losses reduce the value of goods available for end use. The summary column for Quadrant II contains data on the value of goods purchased for civilian consumption, investment, export, defense and reserve purposes.

The summary column for vector 1.2 is estimated as the sum total of supplied goods. Since the supply and uses are balanced, a summary column for vector 1.2 must be identical to the summary row for vector 1.3. Additions to foreign currency reserves--referred to here as net exports--must be added to supply and uses of goods to preserve the inner consistency of the UB table with GSP and PNI. The sum of total supplies of goods and additions to foreign currency reserves earned through foreign trade exceeds GSP by the value of exports.

Data aggregated along vector 2.4 represent the price structure for civilian and defense services. The aggregation along this vector is performed for the same price components--capital depreciation, wages and revenues--as along vector 1.3. Prices on services performed by budget-supported sectors exclude profit. The summary row for vector 2.4 contains data on the total volume of performed non-material and defense services.

Finally, data on the total value added in the Soviet economy are aggregated along vector 3.4 in the same way as in GNP accounts. The summary column for this vector contains data on the total for depreciation, wages and other earnings received by employees and self-employed, social security deductions, net profits, revenues from foreign trade and financial transactions, and other revenues of material and non-material sectors.

Gross revenues remaining at the disposal of enterprises--profits less taxes and interest payments--are allocated to finance bonus wages, new fixed investment, working capital (financial assets), research works, administrative and socio-cultural services. Published data on the distribution of profits pertain to enterprises that actually earn

UNIFIED BALANCE OF THE USSR ECONOMY

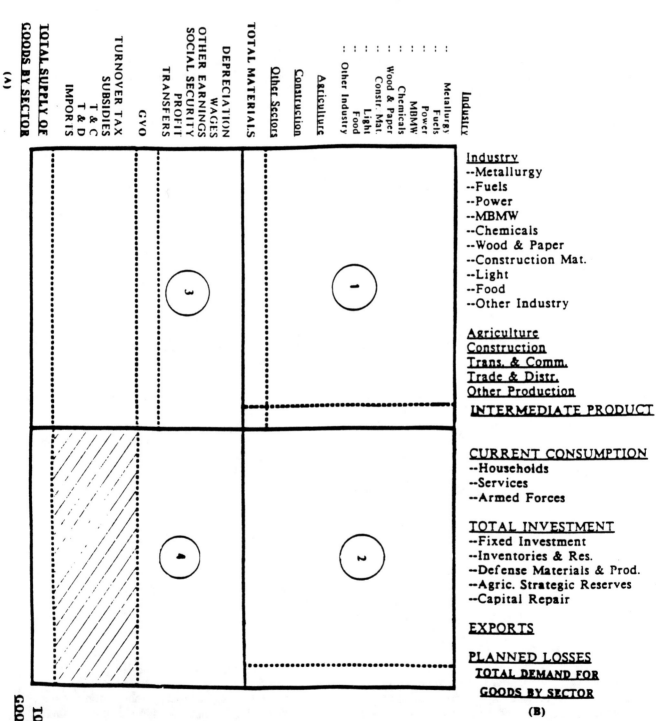

Industry
--Metallurgy
--Fuels
--Power
--MBMW
--Chemicals
--Wood & Paper
--Construction Mat.
--Light
--Food
--Other Industry

Agriculture
Construction
Trans. & Comm.
Trade & Distr.
Other Production
INTERMEDIATE PRODUCT

CURRENT CONSUMPTION
--Households
--Services
--Armed Forces

TOTAL INVESTMENT
--Fixed Investment
--Inventories & Res.
--Defense Materials & Prod.
--Agric. Strategic Reserves
--Capital Repair

EXPORTS

PLANNED LOSSES
TOTAL DEMAND FOR
GOODS BY SECTOR

(B)

Left side labels (rotated):

Industry:
: Metallurgy
: Fuels
: Power
: MBMW
: Chemicals
: Wood & Paper
: Constr. Mat.
: Light
: Food
: Other Industry

Agriculture
Construction
Other Sectors
TOTAL MATERIALS

DEPRECIATION
WAGES
OTHER EARNINGS
SOCIAL SECURITY
PROFIT
TRANSFERS
GVO

TURNOVER TAX
SUBSIDIES
T & C
T & D
IMPORTS
TOTAL SUPPLY OF GOODS BY SECTOR

(A)

TOTAL SUPPLY OF GOODS AND SERVICES

48

profit. Around one fifth of Soviet enterprises incur losses which are covered with emergency funds formed by ministries by means of taxing successful enterprises. Some production losses are also financed from the state budgetary.

The Expanded UB Table

The UB table can be described as an integrated system of national accounts designed to facilitate the analysis of inputs and outputs in conjunction with employed factors of production for the economy divided into material and non-material sectors. Such a comprehensive representation of linkages between economic sectors, however, still cannot be considered complete. In particular, it excludes data on a large portion of financial transactions as well as uses of non-material services outside the production sphere and defense outputs.

In an attempt to systematize all production and financial linkages between economic sectors, two groups of Soviet academic economists both working under the aegis of the USSR Academy of Sciences proposed in the mid-1970s two rival models based on the expanded version of the UB table. The first group was formed under direction of V. Belkin in the Economic Institute, while the second group was formed under direction of B. Isayev in the Central Institute of Mathematical Economics (TSEMI).[11]

In addition to the four quadrants, Belkin's group introduced three more sections of the table, each containing data on financial outlays, revenues and balances of payments. The version of the UB table proposed by Isayev's group appears similar to an expanded version of the national income table divided into production, distribution and end use stages of the annual economic cycle. This version is not analyzed here because it centers on the distribution stage rather than on production and end use stages covered in the regular UB table.

In the version of the UB table proposed by Belkin's group, rows under Quadrant 4 contain data on sources of financing investment and service activities. Added columns outside quadrant 2

---

[11] V. Belkin and A. Geronimus, eds., Model' "Dokhod-Tovary" i Balans Narodnogo Khozyaistva SSSR. Moscow: Nauka, 1978; and B. Isayev and A. Terushkin, eds., Svodnyi Material'no-Finansovyi Balans, Moscow: Nauka, 1978.

contain summary information on types of revenues received by household, state budgetary and banking sectors. Balances of payments are estimated as differences between revenues and outlays.

Balances of payments in the household sector represent additions to monetary savings held in banks and cash savings. Balances of payment in budgetary and banking sectors represent reserves or deficits of monetary funds. As in market economies, deficits are covered with the emission of extra monetary notes. Soviet state budgetary authorities follow practices of market economies and pay interest on government's borrowing from the economy.

Both of the above hypothetical models of the expanded UB table were designed to represent consolidated material-financial flows excluding defense sectors. Unfortunately, neither group was willing or able to demonstrate how their models could be applied to the analysis of Soviet economic trends using the real NEB data as opposed to hypothetical numbers. For this reason, both models still await rigorous testing.

For practical purposes, the expanded UB table will undoubtedly remain a distant theoretical possibility, especially considering that various Soviet government bureaucracies prefer to compile individual consolidated accounts for sectors of the economy that are under their direct supervision. As was noted above, these accounts contain data on foreign trade activities, household budgets, services, defense activities, the state budget and credit flows.

Consolidated Foreign Trade Accounts

There are several types of economic transactions with foreign countries: exports and imports of goods and material services (ship repair) and various financial transactions (sales of gold, banking and insurance operations, passenger transportation and tourism). Soviet foreign trade officials publish a detailed set of statistics on exports and imports of goods in foreign trade prices which are converted from dollars into foreign trade prices using the official exchange rate.

For purposes of estimating Soviet GNP, net exports of financial capital must be added to net exports of goods. Consolidated foreign accounts probably have three sections containing data on foreign trade prices, conversion coefficients for each individual group of goods and services, and financial transactions in foreign trade prices. Before compiling NEB and i-o tables, Goskomstat officials must adjust the foreign trade statistics to conform with the NEB and i-o classification of material production sectors.

Consolidated Household Accounts

Consolidated household accounts are based on combining national income and financial tables containing data on incomes, expenditures and savings of employees of public sectors and members of their families. Total household income consists of:

- wages and other earnings (bonus wages and compensations for business trips) in state-cooperative and collective sectors
- gross income from private production and service activities, and
- transfer payments received from public sectors.

Total outlays and savings consist of:

- purchases of goods and services
- net transfers to the insurance sector
- membership dues paid to public organizations
- income taxes and other transfers to the budget
- consumption and investment-in-kind
- interest on savings accounts
- additions to savings accounts, and
- additions to cash savings.

The value of socio-cultural services financed by the public sector is added to both household incomes and outlays.

Consolidated household accounts, which are never published even in an abridged form, should not be confused with published family budgets of workers and service persons and collective farmers. These budgets are based on information collected from interviews with 62 thousand families. As known, this information is much less precise than that collected for compiling NEB tables. However, all NEB data on household budgets can be derived using the official statistical data, with the exception of wages in defense sectors and annual additions to

unorganized savings.

Non-Material Services

NEB tables containing data on consolidated accounts of non-material sectors have two sections. The first section includes financial data on all sources of financing various services, while the second section includes data on the price structure for these services.

There are five sources of financing:

- revenues of production enterprises
- revenues of non-material services
- household incomes
- allocations from the state budget, and
- bank credits.

Prices on services have the following structure:

- purchases of materials
- purchases of food and uniforms
- purchases of non-capital durable goods
- capital depreciation
- wages and other earnings
- social security deductions
- profits and other revenues
- purchases of non-material services, and
- transfer payments to households (stipends, pensions and allowances).

The gross and net output of services excludes transfer payments.

Defense and Police Activities

As opposed to tables on non-material services, tables containing consolidated accounts of defense and police sectors probably have one rather than two sections. The reason for this is that defense activities have only a single source of financing--the All-Union budget. Militarized guards which safeguard production enterprises appear to be registered as employees of these enterprises. At the same time, military R&D works are registered as non-material services.

All published NEB tables exclude data on the military and police administrations (operational and maintanance expenditure or O&M). It appears that wages of the military and the police are treated as other (non-wage) income of households and thus are excluded from total wages paid in the Soviet economy. As a result, the control total for wages that is entered in the original UB table probably excludes wages of the military and the police.

A close analysis of the published data on consumption of national income leads to two conclusions: 1) current material purchases of the military and the police are excluded from public consumption but are included in the published total for private consumption of national income; and 2) current material purchases and capital depreciation of military R&D are treated as part of the total science sector.[12]

As was analyzed above, there are two possible ways of accounting for production and repair of weapons. The i-o method provides adequate tools for analyzing defense production as a type of material production. In this case, total defense production output can be identified as other secret expenditures of national income. Some economists also assert that Soviet officials may hide some of this output together with purchases of civilian goods for fixed investment and consumption. However, there is no evidence to support this assertion.

As will be argued in the next chapter, the reconstruction of the UB table's fourth quadrant leads to the conclusion that a sizable part of defense production appears to be excluded from other expenditures of national income. This indicates that some defense production activities are outside material production. What remains unclear is whether Quadrant IV contains value added of the entire secret sector that assembles and repairs finished military goods or part of this value added in the amount of hidden defense subsidies.

The State Budget

It is commonly assumed that purchases of armaments are financed together with budgetary allocations for industry and that the officially declared USSR defense budget covers operations and maintenance (O&M) expenditures of the armed forces. However, a close analysis of budgetary allocations leads to two conclusions: 1) the published total for budgetary outlays is too

---

[12] See D. Steinberg, op. cit., pp. 34-40.

small to cover both civilian and military programs, and 2) the small and stable size of the officially declared defense budget contradicts mounting evidence about rising O&M costs.

These conclusions lead to the hypothesis first suggested by Igor Birman that there must exist hidden extra-budgetary funds for financing a large part of the defense program. This hypothesis supports the possibility that material production excludes some defense production, providing that budgetary and national income accounts are compatible.

To test this hypothesis it is necessary to convert official budgetary statistics into the NEB-type format where state budgetary flows are estimated for production, service, defense and household sectors. Such a division must be performed for each type of budgetary revenue and outlay. In theory, the budgetary deficit equals the difference between real budgetary revenues received from public and household sectors and real outlays on these sectors.

The estimation of real budgetary revenues is hindered by the fact that other revenues are never disaggregated into sectors of origin. Any attempt to disaggregate other revenues in effect entails the integration of published data on the distribution of national income and itemized budgetary revenues.

Major published revenue items are:

- turnover and other taxes collected from public sectors
- payments from profit
- social security deductions
- unused All-Republic budgetary funds, and
- revenues from foreign trade and other foreign transactions
- income and other taxes collected from households.

Excluded from the above list are the following major items:

- social security payments of collectives
- social security payments of defense sectors
- various price surcharges
- unused funds of production enterprises, and
- unused All-Union budgetary funds.

The estimation of real budgetary outlays is hindered by the fact that the published data on total budgetary financing of public sectors are never disaggregated into allocations of resources by cost item. In the official statistics such a disaggregation is performed only for All-Republic budgets. A similar independent disaggregation can be performed for the All-Union budget using the complete set of data obtained from reconstructed NEB tables.

The list of budgetary cost items includes:

- wages and social security deductions
- capital investment and repair
- machinery purchases by existing budget-supported services
- agricultural, foreign trade and other subsidies
- industrial materials
- food, uniforms, and other consumer items
- research and administrative expenditures,
- defaulted bank loans made to collectives
- working assets and financial reserves, and
- financial transfers to households.

Whether procurement of weapons is included in the above list depends on the way production of weapons is treated in Soviet national income accounts. If planners exclude weapons from end uses of national income, it is likely that weapons are represented in the above list with wages, industrial materials and capital expenditures of industries manufacturing weapons. It is also likely that all these cost components are aggregated into one item called "subsidies to other sectors."

The Banking System

Consolidated accounts of the banking system are often referred to as credit flows. Revenues of the banking system include interest payments made by public and household sectors, the unused state budgetary funds as well as additions to household savings accounts. Outlays of the banking system include new credits which are issued to finance capital investment, additions to inventories and other working capital, and purchases of consumer goods by households. Outlays also include interest payments made on household savings accounts and financing of the state budgetary deficit.

As is evident from the official statistical publications, the sum of credits issued to the public sector far exceeds the ability of this sector to pay back all outstanding credits in the foreseeable future. Moreover, such sectors as agriculture and coal mining were for a long time unable to make interest payments.

Sizable annual additions to household savings accounts, which are regularly published, help reduce the gap between outlays and revenues of the banking system. Every year, however, the gap continues to rise. In estimating the total Soviet government borrowing from the economy a certain portion of this gap in the amount of delinquent loans must be added to budgetary deficits.

Conclusion

The objective of the above discussion was to analyze Soviet national accounting methods and practices and to examine whether it is possible to integrated published Soviet data compiled with these methods. I demonstrated that consolidated production accounts presented as the four-quadrant UB table provide a useful framework for integrating the input-output and financial data on material production, services and defense sectors. I also discussed how various consolidated financial accounts must be adjusted to make them fully compatible with consolidated production accounts.

•   •

Chapter 3

Integrating Soviet National Accounts

CHAPTER 3

INTEGRATING SOVIET NATIONAL ACCOUNTS

In the preceding chapter, I outlined a new integrated approach to analyzing published Soviet production and financial statistics. In this chapter, I apply this approach to compile an independent set of Soviet national product accounts for 1965, 1970-1990. All results presented in this study for 1989-1990 are preliminary and will be revised as soon as the Goskomstat releases final reports for these years.

In accord with Soviet national accounting methods, I compile two separate accounts for material production and non-material service sectors. I integrate these accounts for 1988 within the framework of the full 4-quadrant input-output (i-o) table which I referred to in the previous chapter as the Unified Balance (UB) table of the USSR economy.

My estimates for 1987-1989 were based to a large extent on the new production and financial data which were published for the first time in 1990. In comparison, my estimates for other years were based on reconciling the regularly published statistics with the new data. The purpose of this reconciliation is two-fold: 1) to determine how published statistics relate to the original NEB data and 2) to test the inner consistency of the Goskomstat's estimates of gross and net output and end uses.

After presenting the 1988 UB table, I reconcile the two sets of published data on the industrial production and on various components of value added and end uses. In Soviet national accounts, value added consists of five major components: depreciation, regular wages and other labor income, gross profits, taxes less subsidies, and net revenues from foreign trade. End uses include consumption of goods and services, fixed investment, inventories, losses and other uses. The results derived in the process of reconciling the two sets of data are enclosed in the detailed tables in the end of this chapter.

In an attempt to integrate Soviet production and financial statistics I used the following eight sets of new data:

- the ex-post 1988 i-o table in producers' prices which is reproduced below in its original format[1]

- excerpts from the ex-post 1988 i-o table in purchasers' prices[2]

- various excerpts from the original NEB data base, including data on the structure of intermediate purchases by sector, wages and other labor income, profit, subsidies and other nonlabor income, consumption in purchasers' prices and other national income flows, the growth of household income by type, and stocks of undepreciated and depreciated fixed capital--all for 1985-1989 and other selected years[3]

- the financial data on profits, depreciation, capital repair works and working assets for 1987-1988 and other selected years[4]

- the retail trade data on household purchases of second hand goods and institutional purchases[5]

- data on civilian and military labor resources for 1989, including 4.3 million workers and service persons who are employed in

---

[1] The i-o table in producers' prices was distributed by Goskomstat officials but it was never actually published.

[2] Data on intermediate purchases in purchasers' prices were published in Vestnik Statistiki, N9, 1990.

[3] These data were published in Goskomstat SSSR, Osnovnye Pokazateli Balansa Narodnogo Khozyaistva SSSR i Soyuznykh Respublik, Moscow: IIT, 1990; and in L. Mikhailov, "Natsional'nyi Dokhod: Proizvedennyi i Ispol'zovannyi," Ekonomika i Zhizn', N40, 1990, p. 16.

[4] See Goskomstat SSSR, Finansy SSSR. Statisticheskiy Sbornik (1987-1988). Moscow, 1990.

[5] Goskomstat SSSR, Torgovlya SSSR. Moscow: Finansy i Statistika, 1989.

previously unreported state sectors[6]

- data on total wages and other monetary income of households for 1987-1989,[7] and

- budgetary data on foreign trade, investment, subsidies, police, research and the official military expenditures for 1988-1990.[8]

The 1988 UB Table: A Summary

I integrated the new data listed above within the framework of the ex-post 1988 Soviet UB table in purchasers' prices. As was discussed in the preceding chapter, the UB is a Soviet-type i-o table: the coverage of supply and demand flows in quadrants I and III is limited to material production sectors. Hence, production of household and government services is listed in quadrants II and IV.

Quadrants I and III contain data on intermediate purchases made by material production sectors and on value added in these sectors. Quadrant II contains data on end uses of the material production output, including purchases made by non-material service, civilian government and defense sectors. Data on value added in these sectors are contained in Quadrant IV. Data on the size of employed resorces--average annual employment and stocks of fixed capital and inventories--are enclosed in the bottom of the table.

Seven sets of estimation results contained in the UB table are absent from all published Soviet statistics. These results were derived as follows:

1. *GNP by sector of origin (row 42).* The GNP data were derived in two ways: a) as the difference between gross value of output--which is referred to as the GVO (row 35)--and the sum of intermediate purchases of goods (row 19) and services (rows 24 and 31), or b) as the difference between value added components (rows 20 through 27) and the sum of rows 24 and 31 and column 20. Value added was estimated for all listed economic sectors, including secret sectors that are hidden in the column 35 called "other uses and errors".

2. *Unfinished construction and reserves (column 33).* In the original input-output tables prepared by the Goskomstat, these end uses are aggregated with additions of inventories (column 34). As listed in the Narkhoz and Finansy SSSR--the two main Goskomstat's annual statistical abstracts--additions of inventories equaled 9.3 b.r.[9] Starting in 1986 these additions began to exclude unfinished construction works. It was thus possible to estimate additions to reserves of machinery products as a residual. The uncertainty remains as to the exact nature of this residual. One possibility is that it pertains to military purchases of dual technologies and spacecraft.

3. *Other uses (column 35).* In the original input-output tables prepared by the Goskomstat, this column is not represented in Quadrant IV. During my meetings with Goskomstat officials I learned that simply they have had no information on secret NS sectors ever since they began compiling i-o tables in the 1950s. This information has been traditionally withheld from them for the apparent purpose of concealing the total volume of secret NS activities. Each figure for other sectors' value added was derived as the difference between the published control totals for the entire USSR economy (column 37) and the sum of columns 19 and 21 which contain published totals for material production and non-material service sectors of the economy.

4. *Distribution of gross profits (rows 29 through 33).* Data on financing of fixed investment as well as budgetary and credit payments were derived by combining published financial statistics. Data on total purchases of business services were derived as the difference between the total volume of services and their financing by households and the state budget (see the discussion of Table A10 below). These data were distributed among sectors in proportion to profits and wages.

5. *Disaggregated data on services (columns 24 through 29).* In the original input-output tables, the column for total services is not disaggregated into individual sectors. Data on these sectors were obtained from i-o tables as well as from published national income and financial tables, including Narkhoz tables on national income, household services, employment, average wages, net profits and capital depreciation.

---

[6] Goskomstat SSSR, "Sotsial'no-Ekonomicheskoe Razvitie SSSR v 1989 godu," Statisticheskiy Press-Bullten', N4, GKS SSSR, Moscow, 1990, p.2.
[7] Pravitel'stvennyi Vestnik, N12, 1990.
[8] Ekonomika i Zhizn', N15, 1990, p. 7.

[9] Narkhoz for 1988, p. 623.

(million rubles)

Page 1

	1 Electric. Power	2 Oil & Gas	3 Coal	4 Other Fuels	5 Ferrous Metals	6 Non Fer. Metals	7 Chemical	8 MBMW	9 Wood & Paper	10 Contr. Mat.
1 Electrical power	278.4	1910.7	567.8	25.7	1860.7	1585.2	3866.7	4213.2	871.6	1533.0
2 Oil and Gas	6659.9	14067.1	31.8	25.9	737.0	439.7	1593.9	1499.9	665.5	1306.6
3 Coal	2818.5	0.1	5988.6	2.5	3144.3	87.1	40.4	211.2	101.9	240.6
4 Other Fuel Ind	228.8	0.6	1.2	123.4	0.4	0.8	12.3	23.7	11.8	29.0
5 Ferrous Metals	80.3	69.8	139.9	12.6	17779.8	587.0	1144.4	17562.3	449.0	2114.1
6 Nonferrous Metals	152.3	13.2	1.9	0.8	1893.0	12798.2	1022.9	11340.9	51.0	134.7
7 Chemical Industry	101.1	501.3	225.0	21.6	496.4	483.2	17070.3	7857.5	1391.1	930.5
8 MBMW	1397.4	374.0	797.3	32.0	1501.8	763.8	1732.0	82236.6	1510.1	1128.8
9 Wood & Paper	44.6	31.3	397.8	6.5	182.0	141.8	1182.1	2793.4	11783.3	739.3
10 Const Mat & Glass	41.4	19.2	61.4	2.5	72.4	89.6	276.0	1090.7	385.4	6639.8
11 Light Industry	50.7	44.1	84.5	5.4	208.5	99.4	1842.9	2513.2	1101.4	372.3
12 Food Industry	33.3	38.1	9.7	0.9	50.8	37.6	943.0	393.7	83.3	84.5
13 Industry NEC	351.7	53.2	99.7	1.2	75.4	43.6	590.3	805.3	143.7	179.1
14 INDUSTRY TOTAL	12238.4	17122.7	8406.6	261.0	28002.5	17157.0	31317.2	132541.6	18549.1	15432.3
15 Construction	0.0	0.0	0.0	0.0	0.0	0.0	0.0	0.0	0.0	0.0
16 Agriculture	0.9		1.0	0.2	1.5	1.6	55.1	93.1	874.2	2.4
17 Transport & Com.	282.2	558.6	994.1	6.2	160.5	256.9	392.4	1655.3	474.9	1315.2
18 Other Branches	19.4	48.0	37.3	1.0	451.2	125.7	153.8	231.4	167.1	167.1
19 TOTAL MATERIALS	12540.9	17729.3	9439.0	268.4	28615.7	17541.2	31918.5	134607.3	20129.6	16917.0
20 Transport Markup	3363.4	3248.8	405.3	8.3	3322.6	870.8	2630.7	5183.7	2062.5	2408.1
21 Trade & Distr	569.9	83.5	40.5	3.4	325.3	130.3	379.3	1868.3	301.4	291.0
22 Turnover Tax	517.6	242.9	28.7	0.7	141.4	454.3	562.9	1153.5	406.3	238.0
23 Imports	189.2	199.1	201.5	4.8	573.2	1208.7	2789.3	5489.3	605.1	331.4
24 TOTAL MATERIALS	17181.0	21503.6	10115.0	285.6	32978.2	20205.3	38280.7	148302.1	23504.9	20185.5
25 Depreciation	6970.8	7633.5	2730.5	164.6	5035.3	2919.9	7151.7	20792.7	3768.2	4354.5
26 TOTAL COST	24151.8	29137.1	12845.5	450.2	38013.5	23125.2	45432.4	169094.8	27273.1	24540.0
27 Wages	2056.2	1014.7	4960.6	176.0	3840.6	2280.8	4612.9	42254.4	7186.6	6499.5
28 Other Payments	105.2	91.9	220.0	4.7	145.1	130.7	151.6	1699.1	283.8	189.7
29 Kolkhoz Pay	4.2	0.0	0.0	0.0	0.0	0.0	1.7	612.7	242.2	81.2
30 State Soc. Sec.	322.4	165.3	515.7	22.2	433.2	263.2	689.6	6115.1	620.2	785.1
31 Kolkhoz Soc. Sec.	0.5						0.2	77.5	30.4	10.2
32 State-Coop Profit	6680.3	10499.8	2893.7	45.5	9057.0	6521.9	11276.6	47310.5	6324.2	4446.1
33 Bonuses	339.5	250.6	349.2	21.1	462.1	284.2	654.9	4861.8	508.9	476.9
34 Turnover Tax										
35 Other Net Income	157.6	3970.1	174.2	40.8	181.6	347.5	257.9	2659.0	70.4	260.2
36 Kolkhoz Profit	3.1			-0.1			5.4	443.7	261.1	61.8
37 Private Income										
38 Undeprec. Assets	-172.4	-184.0	-67.5	-4.1	-126.6	-72.7	-176.8	-515.5	-93.2	-107.7
39 NATIONAL INCOME	9157.1	15557.8	8696.7	285.0	13530.9	9471.4	16819.1	100656.5	14925.7	12226.1
40 TOTAL OUTPUT	33308.9	44694.9	21542.2	735.2	51544.4	32596.6	62251.5	269751.3	42198.8	36766.1
41 Fixed Capital	126057	86896	29288	1565	62709	32454	78963	242147	36821	50911
42 Labor, 1,000s	835.2	320.6	1221.4	75.8	1302.7	672.2	1908.2	17420.4	2772.8	2511

ORIGINAL 1988 SOVIET EX-POST INPUT-OUTPUT TABLE IN PRODUCERS' PRICES
(million rubles)

	11 Light Ind	12 Food Ind	13 Industry NEC	14 INDUSTRY TOTAL	15 Construc	16 Agric	17 T & C	18 T & D	19 Other Branches	20 INTER-IND USE
1 Electrical power	898.8	986.6	589.2	19187.6	1636.0	1246.6	2270.1	1146.9	45.9	25533.1
2 Oil and Gas	147.6	713.1	277.6	28165.6	1508.6	1839.3	1671.3	258.6	25.7	33469.1
3 Coal	45.7	129.7	37.3	12847.9	69.2	200.3	115.3	135.9	3.6	13372.2
4 Other Fuel Ind	2.0	1.8	2.1	437.9	51.3	119.2	2.4	5.4		616.2
5 Ferrous Metals	97.8	307.0	178.1	40522.1	6165.7	151.8	209.0	31.1	2.3	47082.0
6 Nonferrous Metals	39.8	174.2	935.5	28558.4	215.1	9.1	30.4	7.7	3.8	28824.5
7 Chemical Industry	4481.3	722.5	1009.9	35291.7	2114.9	6561.7	1057.5	176.5	67.0	45269.3
8 MBMW	764.1	1184.6	724.8	94147.3	12475.2	9175.8	2527.5	769.4	84.6	119179.8
9 Wood & Paper	463.9	1386.4	801.1	19953.5	5697.8	891.6	297.1	435.0	511.1	27786.1
10 Const Mat & Glass	64.2	453.5	109.4	9305.5	24774.7	755.1	405.7	297.3	6.4	35544.5
11 Light Industry	60229.9	1112.3	1581.2	69245.8	855.9	818.9	387.5	510.4	214.6	72033.1
12 Food Industry	615.0	43730.5	2171.1	48191.5	104.3	5656.0	17.5	623.5	5.0	54597.8
13 Industry NEC	168.6	989.7	3246.5	6748.0	287.5	14590.7	337.1	328.2	820.5	23112.0
14 INDUSTRY TOTAL	68018.7	51891.9	11663.8	412602.8	55956.2	42016.1	9328.4	4725.7	1790.5	526419.7
15 Construction	0.0	0.0	0.0	0.0	0.0	0.0	0.0	0.0		
16 Agriculture	12597.5	49472.6	7125.7	70225.8	68.2	44981.5	9.0	702.1		115986.6
17 Transport & Com.	520.0	1604.5	156.2	8437.0	2826.5	745.5	138.8	469.0	57.0	12673.8
18 Other Branches	133.5	178.5	171.6	2035.8	367.9	239.9	167.3	496.7	224.7	3532.3
19 TOTAL MATERIALS	81269.7	103147.5	19117.3	493301.4	59218.8	87983.0	9643.5	6393.5	2072.2	658612.4
20 Transport Markup	1039.5	2440.1	853.9	27837.7	8250.9	1801.4	1550.5	264.5	76.7	39781.7
21 Trade & Distr	949.3	3600.9	1230.1	9773.2	1146.3	1197.2	493.0	50.9	27.2	12687.8
22 Turnover Tax	7450.6	2595.7	241.4	14034.0	1301.6	1159.9	3702.8	29.0	47.9	20275.2
23 Imports	5247.0	7915.4	1310.0	26064.0	3016.8	4392.9	269.4	162.7	93.9	33999.7
24 TOTAL MATERIALS	95956.1	119699.6	22812.7	571010.3	72934.4	96534.4	15659.2	6900.6	2317.9	765356.8
25 Depreciation	2639.6	5602.0	2143.7	71907.0	11924.0	20293.0	19046.6	5023.4	620.2	128814.2
26 TOTAL COST	98595.7	125301.6	24956.4	642917.3	84858.4	116627.4	34705.8	11924.0	2938.1	894171.0
27 Wages	10035.3	7123.0	3308.4	95349.0	44088.0	25769.0	16473.0	19541.0	1592	202812
28 Other Payments	202.5	266.4	91.3	3582.0	1678.0	509.0	328.0	217.0	41	6355
29 Kolkhoz Pay	76.6	392.4	313.0	1724.0	1782.0	22875.0	47.0	0.0	132	26560
30 State Soc. Sec.	1272.5	879.0	282.5	1236.0	4187.0	1427.0	1495.0	1335.0	87	20897
31 Kolkhoz Soc. Sec.	10.0	72.7	39.5	241.0	18.0	2914.0	3.0		15	3191
32 State-Coop Profit	15817.9	12189.3	2702.2	135765.0	22946.0	25387.0	17125.0	15145.0	2462	218830
33 Bonuses	1041.6	616.8	369.4	10237.0	3562.0	2348.0	1078.0	247.0		17472
34 Turnover Tax								30.3		30.3
35 Other Net Income	91.5	1387.2	19.5	9617.5	3183.1	1217.4	3575.5	3872.2	184.6	21650.3
36 Kolkhoz Profit	146.5	54.5	469.0	1445.0	-13.0	18643.0			19	20094
37 Private Income					3194.0	46543.0			1647	51384
38 Undeprec. Assets	-66.0	-138.5	-53.0	-1778.0	-475.0	-1304.4	-301.9	-126.4	-12.2	-3997.9
39 NATIONAL INCOME	27586.8	22226.0	7172.4	258311.5	80588.1	143980.0	38744.6	40014.1	6167.4	567805.7
40 TOTAL OUTPUT	126182.5	147527.6	32128.8	901228.8	165446.	260807.4	73450.4	51938.1	9105.5	1461976.
41 Fixed Capital	33028	58529	44536	883904	92002	362984	312156	91185	5959	1747190
42 Labor, 1,000s	4941	3235.4	1684.5	38901.2	15438.8	25987.7	5332.4	10735.4	812	97207.5

ORIGINAL 1988 SOVIET EX-POST INPUT-OUTPUT TABLE IN PRODUCERS' PRICES
(million rubles)

	21 Private Consump	22 Public Consump	23 TOTAL CONSUMP	24 Net Fixed Capital	25 Working Capital	26 Capital Repair& Replacmnt	27 Other	28 Replace-ment (Losses)	29 Exports	30 Budget Subsidy	31 TOTAL Final Demand
1 Electrical power	3454.9	3568.3	7023.2						752.6		7775.8
2 Oil and Gas	1177.7	1260.4	2438.1		345		357.7		8085		11225.8
3 Coal	256.6	1452.6	1709.2		48		18		1105.8	5289	8170
4 Other Fuel Ind	48.8	62	110.8		0.4		4.4		3.4		119
5 Ferrous Metals	82.7	1131	1213.7		-164.9		8.6		3405		4462.4
6 Nonferrous Metals		1078.4	1078.4		674				2019.7		3772.1
7 Chemical Industry	5087.3	3164.7	8252		1117.6		2919.1	19.3	3737.5	936.7	16982.2
8 MBMW	16625.8	8958.3	25584.1	14219.8	7755.7	65225.1	24729.9	137.9	12436.3	482.7	150571.5
9 Wood & Paper	8329.8	1345.3	9675.1	645.5	-19	986.6	1.9	61.3	3061.3		14412.7
10 Const Mat & Glass	1118.5	740.9	1859.4		-876.1		4.4	20.1	213.8		1221.6
11 Light Industry	42977.2	2016.7	44993.9	199.1	1755.4	508.3	576.2	1109.5	3872.2	1134.8	54149.4
12 Food Industry	83228.2	6077.8	89306		-2497.6		61		2217.2	3843.2	92929.8
13 Industry NEC	6039.4	1156.5	7195.9	31.8	988.9	88.2		80.8	631.2		9016.8
14 INDUSTRY TOTAL	168426.	32012.9	200439.	15096.2	9127.4	66808.2	28681.2	1428.9	41541	11686.4	374809.1
15 Construction				66242	15151.5	77121.8		6931.2			165446.5
16 Agriculture	55931.9	1095.1	57027	3036.8	3863.9	709.8		1857.8	247.9	78077.6	144820.8
17 Transport & Com.		4877.8	4877.8						131.9		5009.7
18 Other Branches	3087.8	962.9	4047.7	204.7	231.3				1089.5		5573.2
19 TOTAL MATERIALS	227443.	38948.7	266392.	84579.7	28374.1	144639.	28681.2	10217.9	43010.3	89764	695659.3
20 Transport Markup	7358.3	2206.1	9564.4	568.1	221.9	1707.7	687	45.4	3190.7		15985.2
21 Trade & Distr	35704.5	2347.7	38052.2	198.7	110.9	804.7	72.8	11	0		39250.3
22 Turnover Tax	73717.5	5996.5	79714	0	4.5	0	970.2	0	0		80688.7
23 Imports	35669.1	701.5	36370.6	4149.5	2.4	19098	3011.5	0	0		62632
24 TOTAL MATERIALS	379893	50200.5	430093.	89496	28713.8	166250.	33422.7	10274.3	46201	89764	894215.5
25 Depreciation	13139	22503.5	35642.5				1793.5				37436
26 TOTAL COST	393032	72704	465736	89496	28713.8	166250.	35216.2	10274.3	46201		841887.5

	21	22	23
27 Wages		75941	75941
28 Other Payments		2340	2340
29 Kolkhoz Pay		442	442
30 State Soc. Sec.		6072	6072
31 Kolkhoz Soc. Sec.		45	45
32 State-Coop Profit		4834	4834
33 Bonuses		4640	4640
34 Turnover Tax			
35 Other Net Income			
36 Kolkhoz Profit			
37 Private Income			
38 Undeprec. Assets		-102	-102
39 NATIONAL INCOME			

40 TOTAL OUTPUT

	21	22	23
41 Fixed Capital	506693	436683	943376
42 Labor, 1,000s			

49

CHART 3   RECONSTRUCTED 1988 UNIFIED BALANCE TABLE OF THE USSR ECONOMY IN PURCHASERS' PRICES
(billion rubles; million persons)

	1 POWER	2 P&G (FUELS)	3 COAL*	4 FERR (METALS)	5 N-FER	6 CHEM	7 MBMW	8 WOOD PAPER	9 CONS MAT.	10 N.E.C. IND.	11 LIGHT	12 FOOD	13 TOTAL IND.	14 AGRI. FOREST	15 CONSTR	16 T&C	17 T&D	18 OTHER PROD	19 INTER USE
1 POWER	0.3	1.9	0.6	1.9	1.6	3.9	4.2	0.9	1.5	0.6	0.9	1.0	19.2	1.2	1.6	2.3	1.1	0.0	25.5
2 PETRO-GAS	10.3	17.7	0.1	1.9	0.8	3.4	2.7	1.0	2.4	0.4	0.3	1.4	42.3	3.5	2.4	6.8	0.3	0.0	55.5
3 COAL*	3.9	0.0	6.5	4.0	0.1	0.1	0.3	0.1	0.3	0.1	0.1	0.2	15.6	0.4	0.1	0.2	0.2	0.0	16.5
4 FERRROUS M.	0.1	0.1	0.2	19.4	0.7	1.3	20.5	0.5	2.5	0.2	0.1	0.4	45.9	0.2	8.6	0.2	0.0	0.0	55.0
5 N-FERROUS M.	0.2	0.0	0.0	2.0	14.8	1.1	12.1	0.1	0.1	1.0	0.0	0.2	31.5	0.0	0.2	0.0	0.0	0.0	31.7
6 CHEMICALS	0.1	0.6	0.3	0.6	0.6	20.7	9.9	1.7	1.1	1.3	5.7	0.9	43.5	8.2	2.8	1.4	0.2	0.1	56.3
7 MBMW	1.5	0.4	0.9	1.9	0.9	2.1	88.3	1.6	1.3	0.8	0.8	1.3	101.7	9.3	13.8	2.6	0.7	0.4	128.5
8 WOOD & PAPER	0.0	0.0	0.5	0.2	0.2	1.5	3.2	13.9	0.8	1.0	0.5	1.8	23.6	1.0	6.5	0.3	0.6	0.7	32.8
9 CONST MAT*	0.0	0.0	0.1	0.1	0.1	0.3	1.3	0.4	8.0	0.1	0.1	0.5	11.2	0.9	32.2	0.6	0.4	0.0	45.2
10 N.E.C. IND.	0.4	0.1	0.1	0.1	0.0	0.3	0.8	0.2	0.1	3.5	0.2	1.0	6.8	15.3	0.4	0.4	0.5	0.5	23.9
11 LIGHT	0.1	0.0	0.1	0.2	0.1	1.9	2.6	1.4	0.4	1.6	72.6	1.1	82.1	0.8	0.9	0.4	0.5	0.2	84.9
12 FOOD	0.0	0.0	0.0	0.1	0.0	1.1	0.4	0.1	0.1	2.4	0.6	51.7	56.5	6.4	0.1	0.0	0.6	0.0	63.7
13 TOTAL INDUSTRY	16.9	20.9	9.4	32.3	19.8	37.7	146.2	21.9	18.7	12.9	82.0	61.4	479.8	47.4	69.6	15.3	5.2	2.1	619.5
14 AGRICULTURE	0.0	0.0	0.0	0.0	0.0	0.1	0.1	0.9	0.0	9.5	13.4	56.5	80.5	48.1	0.1	0.0	0.7	0.0	129.4
15 CONSTRUCTION	-	-	-	-	-	-	-	-	-	-	-	-	-	-	-	-	-	-	0.0
16 T&C	0.3	0.6	1.0	0.2	0.3	0.4	1.7	0.5	1.3	0.2	0.5	1.6	8.5	0.7	2.8	0.1	0.5	0.1	12.7
17 T&D	-	-	-	-	-	-	-	-	-	-	-	-	-	-	-	-	-	-	0.0
18 OTHER SECTORS	0.0	0.1	0.0	0.5	0.1	0.2	0.3	0.3	0.2	0.2	0.1	0.2	2.1	0.3	0.4	0.2	0.5	0.2	3.8
19 TOTAL OUTLAYS	17.2	21.5	10.4	33.0	20.2	38.3	148.3	23.5	20.2	22.8	96.0	119.8	571.0	96.5	72.9	15.6	6.9	2.4	765.4
20 DEPRECIATION	7.0	7.6	2.9	5.0	2.9	7.2	20.8	3.8	4.4	2.1	2.6	5.6	72.0	20.3	11.9	19.0	5.0	0.6	128.9
21 REPLACEMENT	5.0	5.4	1.7	3.2	1.7	5.0	14.5	2.5	2.9	1.3	1.7	3.8	48.8	9.1	6.8	12.7	3.7	0.4	81.6
22 CAP REPAIR	2.0	2.2	1.2	1.8	1.2	2.2	6.3	1.3	1.5	0.8	0.9	1.8	23.2	11.2	5.1	6.3	1.3	0.2	47.3
23 WAGES, etc.	2.1	1.1	4.9	4.0	2.3	4.4	42.3	7.1	6.5	3.5	9.9	7.5	95.6	48.4	45.9	16.5	19.5	1.6	227.5
24 BUSINESS WAGES	0.1	0.1	0.3	0.2	0.1	0.4	2.2	0.4	0.3	0.2	0.4	0.3	5.0	0.7	1.7	0.4	0.3	0.1	8.2
25 PRIVATE INCOME	-	-	-	-	-	-	-	-	-	-	-	-	-	46.5	3.2	-	-	1.6	51.3
26 SOC SECURITY	0.3	0.2	0.5	0.4	0.3	0.7	6.2	0.7	0.8	0.3	1.3	0.9	12.6	4.3	4.2	1.5	1.3	0.1	24.0
27 PROFIT, etc.	6.8	14.4	3.3	9.0	6.9	11.4	50.5	6.8	4.7	3.3	16.0	13.5	146.8	45.4	26.2	20.7	19.1	2.7	260.8
28 BONUS WAGES	0.3	0.3	0.4	0.5	0.3	0.7	4.9	0.5	0.5	0.4	1.0	0.6	10.3	2.3	3.6	1.1	0.3	0.0	17.5
29 FIX INVEST	1.0	1.9	0.4	0.9	0.6	1.6	5.1	0.6	0.8	0.2	0.9	1.6	15.6	7.4	3.1	6.6	2.1	0.1	34.9
30 WORK CAPITAL	0.7	1.3	0.3	0.7	0.6	1.0	9.5	1.0	0.8	0.1	1.7	1.3	19.0	16.7	5.2	0.9	9.9	0.3	51.9
31 BUS. SERVICES	1.0	1.5	0.5	0.9	0.8	1.7	10.0	1.4	0.9	0.4	1.7	1.2	22.0	8.3	5.0	2.0	1.2	0.1	38.6
32 BUDGET	3.4	9.2	1.5	5.6	4.3	5.9	19.0	3.0	1.4	2.1	10.2	8.4	74.0	8.1	7.7	9.8	4.8	2.1	106.2
33 CREDITS & INS	0.4	0.3	0.2	0.4	0.3	0.5	2.0	0.3	0.3	0.1	0.5	0.4	5.7	2.6	1.6	0.4	0.8	0.1	11.2
34 UNAMORTIZED W.	-0.2	-0.2	-0.1	-0.1	-0.1	-0.2	-0.5	-0.1	-0.1	-0.1	-0.1	-0.1	-1.9	-1.3	-0.5	-0.3	-0.1	0.0	-4.1
35 GVO w/o tax	33.3	44.7	22.3	51.5	32.6	62.3	269.7	42.2	36.8	32.1	126.1	147.5	901.1	260.8	165.5	73.5	51.9	9.1	1462.0
36 TURNOVER TAX	2.6	12.3	-	0.2	0.4	4.1	11.8	0.3	1.8	2.7	22.4	42.2	101.0	-	-	-	-	-	101.0
37 NET SUBSIDIES	-	-	-5.3	-	-	-0.9	-0.5	-	-	-	-1.1	-3.8	-11.7	-78.1	-	-	-	-	-89.8
38 IMPORTS	0.0	0.4	0.3	3.3	1.5	7.8	31.4	2.4	0.9	1.2	22.9	14.5	86.5	8.6	-	-	-	1.5	96.6
39 T & C	-	17.0	2.5	3.9	1.0	3.0	7.0	3.8	8.5	0.9	1.3	3.2	52.2	2.5	-	-	-	1.1	55.8
40 T & D	-	2.4	0.4	1.0	0.1	2.8	5.0	1.3	0.9	1.2	7.4	20.3	42.9	7.4	-	-	-	1.6	51.9
41 TOTAL SUPPLY	35.9	76.8	20.2	59.9	35.6	79.1	324.5	49.9	48.9	38.1	179.1	223.9	1172	201.3	165.5	17.7	-	13.2	1569.8
42 TOTAL GNP	17.5	33.8	5.7	17.4	11.8	24.9	120.0	17.1	17.1	11.3	48.1	64.5	390.6	74.0	85.4	55.2	43.5	6.5	653.6
43 LABOR	0.9	0.3	1.3	1.3	0.7	1.9	17.4	2.8	2.3	1.9	4.9	3.2	38.9	26.0	15.4	5.3	10.7	0.8	97.2
44 FIXED CAPITAL	126	86.9	30.9	62.7	32.5	79.0	242.1	36.8	50.9	44.5	33.0	58.5	883.8	363.0	92.0	312.2	90.2	6.0	1747.2
45 -/-Additions	6.8	10.5	3.4	3.6	2.0	5.2	17.7	2.7	2.9	1.7	2.2	4.0	62.7	37.6	8.0	14.6	3.8	0.4	127.1
46 -/-Undepr Net	6.2	8.7	2.5	2.6	1.4	3.7	12.9	1.2	1.3	0.9	1.5	2.5	45.3	22.9	4.6	11.5	3.1	0.3	87.7
47 -/-Scrap	0.6	1.8	0.9	1.0	0.6	1.5	4.8	1.5	1.6	0.8	0.7	1.5	17.4	14.7	3.4	3.1	0.7	0.1	39.4
48 -/-Net Fixed	1.8	5.1	1.7	0.4	0.3	0.2	3.2	0.2	0.0	0.4	0.5	0.2	13.9	28.5	1.2	1.9	0.1	0.0	45.5
49 INVENTORIES	4.6	2.1	1.6	10.7	5.1	6.7	82.3	5.5	3.9	3.0	21.8	19.7	167.0	113.1	24.7	5.2	140	2.6	452.6
50 -/-Additions													3.1	4.7	1.4	0.4	-2.9	0.1	6.8

RECONSTRUCTED 1988 UNIFIED BALANCE TABLE OF THE USSR ECONOMY IN PURCHASERS' PRICES
(billion rubles)

	20 PLAN LOSS	21 TOTAL CONSU	22 HOUS	23 SERVICES TOTAL	24 T&C	25 HC&E	26 ED&C	27 HEALT	28 R&D	29 ADM	30 NET FIXED	31 CAP REPL	32 CAP REPR	33 UNFIN CONST	34 INVEN &RESRV	35 OTHER &ERR	36 EXPORT	37 TOTAL USES	38
1 POWER	-	9.6	4.7	4.9	1.3	0.7	0.9	0.6	1.2	0.2	-	-	-	-	0.0	0.0	0.8	35.9	
2 PETRO-GAS	-	9.6	5.0	4.6	2.6	0.6	0.3	0.2	0.8	0.1	-	-	-	-	0.4	1.2	10.1	76.8	
3 COAL*	-	2.3	0.4	1.9	0.2	1.1	0.3	0.2	0.1	0.0	-	-	-	-	0.1	0.0	1.3	20.2	
4 FERRROUS M.	-	1.4	0.1	1.3	0.2	0.3	0.0	0.0	0.8	0.0	-	-	-	-	-0.2	0.0	3.6	59.9	
5 N-FERROUS M.	-	1.1	-	1.1	0.1	0.0	0.0	0.0	1.0	0.0	-	-	-	-	0.7	0.1	2.1	35.6	
6 CHEMICALS	0.0	14.3	9.7	4.6	0.7	0.4	0.4	1.6	1.4	0.1	-	-	-	-	1.2	3.3	4.0	79.1	
7 MBMW	0.1	41.2	31.1	10.1	0.8	0.9	0.5	0.2	7.6	0.1	19.1	63.7	23.3	0.7	7.1	28.2	12.6	324.5	
8 WOOD & PAPER	0.1	12.2	10.6	1.6	0.1	0.3	0.4	0.3	0.4	0.1	0.7	0.7	0.3	-	0.0	0.0	3.2	49.9	
9 CONST MAT*	0.0	4.5	3.4	1.1	0.2	0.5	0.0	0.0	0.4	0.0	-	-	-	-	-1.1	0.1	0.2	48.9	
10 N.E.C. IND.	0.1	12.5	10.6	1.9	0.2	0.6	0.4	0.2	0.4	0.1	0.0	0.1	-	-	1.0	-0.1	0.6	38.1	
11 LIGHT	1.1	86.1	83.6	2.5	0.1	0.3	1.0	0.9	0.1	0.1	0.2	0.5	-	-	1.8	0.6	3.9	179.1	
12 FOOD	-	160.4	153.1	7.3	0.0	0.1	5.3	1.9	0.0	0.0	-	-	-	-	-2.5	0.0	2.3	223.9	
13 TOTAL INDUSTRY	1.5	355.3	312.4	42.9	6.5	5.8	9.5	6.1	14.2	0.8	20.0	64.9	23.6	0.7	8.5	33.4	44.7	1172.0	
14 AGRICULTURE	1.9	62.0	60.8	1.2	0.0	0.0	0.9	0.3	0.0	0.0	3.1	0.7	-	-	3.9	0.0	0.3	201.3	
15 CONSTRUCTION	6.9	0.0	-	-	-	-	-	-	-	-	66.2	40.0	36.9	14.6	0.8	0.0	-	165.5	
16 T&C	-	4.9	-	4.9	0.0	0.3	0.0	0.0	0.0	4.6	-	-	-	-	-	0.0	0.1	17.7	
17 T&D	-	0.0	-	-	-	-	-	-	-	-	-	-	-	-	-	0.0	-	-	
18 OTHER SECTORS	-	7.9	6.7	1.2	0.1	0.1	0.4	0.3	0.2	0.1	0.2	-	-	-	0.3	0.0	1.1	13.2	
19 TOTAL OUTLAYS	10.3	430.1	379.9	50.2	6.6	6.2	10.8	6.7	14.4	5.5	89.5	105.8	60.5	15.3	13.5	33.4	46.2	1569.8	
20 DEPRECIATION	-	35.6	13.1	22.5	6.3	3.6	5.8	2.2	4.0	0.6	-	-	-	-	-	1.8	-	166.3	
21 REPLACEMENT	-	22.8	6.8	16.0	4.3	2.6	4.2	1.5	3.0	0.4	-	-	-	-	-	1.4	-	105.8	
22 CAP REPAIR	-	12.8	6.3	6.5	2.0	1.0	1.6	0.7	1.0	0.2	-	-	-	-	-	0.4	-	60.5	
23 WAGES, etc.	-	76.3		76.3	10.5	9.0	25.3	13.0	12.9	5.6	-	-	-	-	-	27.7	-	331.5	
24 BUSINESS WAGES		2.4		2.4	0.3	0.3	0.7	0.3	0.6	0.2	-	-	-	-	-	0.6	-	11.2	
25 PRIVATE INCOME	-	11.0		11.0	1.5	5.0	1.5	3.0	-	-	-	-	-	-	-	-	-	62.3	
26 SOC SECURITY	-	6.1		6.1	0.7	0.6	1.8	0.9	1.5	0.6	-	-	-	-	-	2.5	-	32.6	
27 PROFIT, etc.	-	18.6		18.6	7.5	4.1	-	-	-	7.0	-	-	-	-	-	1.7	-	281.1	
28 BONUS WAGES	-	4.6		4.6	1.1	0.7	0.6	0.2	1.6	0.4	-	-	-	-	-	1.0	-	23.1	
29 FIX INVEST	-	1.2		1.2	0.9	0.3	-	-	-	-	-	-	-	-	-	-	-	36.1	
30 WORK CAPITAL	-	-1.1		-1.1	0.7	0.3	-0.6	-0.2	-1.6	0.3	-	-	-	-	-	0.2	-	51.0	
31 BUS. SERVICES	-	0.9		0.9	0.6	0.3	-	-	-	-	-	-	-	-	-	-	-	39.5	
32 BUDGET	-	12.6		12.6	4.0	2.3	-	-	-	6.3	-	-	-	-	-	0.5	-	119.6	
33 CREDITS & INS	-	0.4		0.4	0.2	0.2	-	-	-	-	-	-	-	-	-	-	-	11.6	
34 UNAMORTIZED W.	-	-0.1		-0.1	-0.1	0.0	-	-	-	-	-	-	-	-	-	-	-	-4.2	
35 GVO w/o tax	-	189.8		189.8	33.3	28.8	46.5	26.3	35.0	19.9	-	-	-	-	-	67.7	-	1719.5	
36 TURNOVER TAX	-	0.0		-	-	-	-	-	-	-	-	-	-	-	-	-	-	101.0	
37 NET SUBSIDIES	-	-13.8		-13.8	-	-11.4	-0.6	-0.2	-1.6	-	-	-	-	-	-	-	-	-103.6	
38 IMPORTS	-	-		-	-	-	-	-	-	-	-	-	-	-	-	-	-	-	
39 T & C	-	-		-	-	-	-	-	-	-	-	-	-	-	-	-	-	-	
40 T & D	-	-		-	-	-	-	-	-	-	-	-	-	-	-	-	-	-	
41 TOTAL SUPPLY	-	189.8		189.8	33.3	28.8	46.5	26.3	35.0	19.9	-	-	-	-	-	67.7	1.4	1827.2	
42 TOTAL GNP	-	136.3		136.3	25.8	22.0	35.0	19.3	20.0	14.2	-	-	-	-	-	33.7	51.8	875.4	
43 LABOR	-	37.3		37.3	4.4	5.0	13.0	7.4	5.0	2.5	-	-	-	-	-	6.2	-	140.7	
44 FIXED CAPITAL	-	943	507	437	70.9	123	121	59.8	54.7	8.1	-	-	-	-	-	26.0	-	2716.6	
45 -/-Additions	-	77.5	31.7	32.5	7.7	8.8	8.6	2.5	4.4	0.5	-	-	-	-	-	4.0	-	208.6	
46 -/-Undepr Net	-	54.7	28.2	25.6	6.6	7.7	6.3	1.7	2.9	0.4	-	-	-	-	-	3.2	-	145.6	
47 -/-Scrap	-	10.5	3.5	6.9	1.1	1.1	2.3	0.8	1.5	0.1	-	-	-	-	-	0.8	-	50.7	
48 -/-Net Fixed	-	41.4	24.9	16.5	3.4	6.2	4.4	1.0	1.4	0.1	-	-	-	-	-	2.6	-	89.5	
49 INVENTORIES	-	20.8																473.4	
50 -/-Additions		2.5																9.3	

RECONSTRUCTED 1988 UNIFIED BALANCE TABLE OF THE USSR ECONOMY IN PURCHASERS' PRICES
(billion rubles)

	20 PLAN LOSS	21 TOTAL CONSU	22 HOUS	23 SERVICES TOTAL	24 T&C	25 HC&E	26 ED&C	27 HEALT	28 R&D	29 ADM	30 NET FIXED	31 CAP REPL	32 CAP REPR	33 UNFIN CONST	34 INVEN &RESRV	35 OTHER &ERR	36 EXPORT	37 TOTAL USES
51 GNP by End Use																		875.4
52 Consumption			350.7		22.9	26.1	37.8	26.5		8.5								472.5
53 Goods & PGW*			350.7			15.5												366.2
54 Paid Services+					15.2	10.6	5.0	5.7		8.5								45.0
55 Free Services					7.7		32.8	20.8										61.3
56 Investment											89.5	101.6	60.5	15.3	9.2			276.1
57 Government	6.9	10.8			2.2				30.5	3.0					4.3	67.7	1.4	126.8

* including repair of goods, PGW (power, gas & water)

6. Additions to fixed capital stock (row 45). These additions were derived for each sector using the Narkhoz data on capital investment and commissioned (installed) capital, productive livestock as well as budgetary data on purchases of machinery by government service sectors (primarily education and health). Control totals for material production and non-material service sectors were estimated as the sum of net fixed investment and capital replacement (the difference between total depreciation and capital repair).

7. GNP by end use (rows 51 through 57). Household consumption was derived as material consumption (column 22) less business travel expenditures plus household purchases of services plus government purchases of services and media. I extracted data on household purchases of services, except for so-called "administration services", from the Narkhoz and Finansy SSSR. In addition to insurance and banking services, households finance public organizations, such as the party, unions, etc., by paying membership dues. Purchases of transportation and housing-communal services were also reduced by the amount of business travel expenditures. Data on private services were taken from a reliable Soviet source.[10]

To avoid double counting of household purchases of such material production services as repairs of goods and utility services, I registered them separately. I extracted data on government purchases of household services without transfer payments--pensions, allowances, stipends, etc.

The total for R&D works was derived as the sum of state budgetary outlays on "science", space, oil and gas exploration, and agricultural programs.[11] The total for government purchases of communications services was derived as the difference between the total volume of these services and the sum of reported household purchases and state budgetary outlays on media.

The gross output and value added in other secret sectors (column 35) amounted to as much as 67.7

and 33.7 b.r. respectively. These sectors are almost entirely absent from Western studies of the USSR economy based on a standard GNP methodology. I make an attempt in the next chapter to determine the exact nature of activities performed in these sectors by compiling Soviet GNP accounts by end use.

Industrial Production

One of the key economic indicators represented in the UB table is GVO by sector. The Narkhoz regularly includes a table on the structure of the industrial GVO by sector in comparable 1967, 1975 and 1982 producers' prices. Using published data on total industrial GVO in current established prices, growth rates and official price indexes, I converted comparable prices into current prices for individual sectors. I present my estimation results for 1965, 1970-1990 in Tables A1 and A2 in the end of this chapter.

Table A1 contains data on total industrial GVO and nine industrial sectors in comparable and current prices. Official price indexes equal the ratio between current and comparable prices. As known, official price indexes are not designed to measure hidden price increases in the USSR economy. For this reason, these indexes cannot be used as price deflators in measuring real growth rates. However, these indexes appear quite useful for converting published data on GVO in comparable prices into current prices, data which are needed for the analysis of the annual allocation of resources in the USSR economy.

Indeed, the comparison of the i-o table with the results presented in Table A1 indicates that the Goskomstat's GVO data are internally consistent for all sectors, except for the machine-building and metal works (MBMW) sector. The MBMW GVO published in the Narkhoz is notably smaller than the MBMW GVO estimated using the i-o data.

In Table A2, I reconcile the two sets of published data by analyzing the industrial GVO residual which in the noted Narkhoz table supposedly contains the metallurgy and other sectors' GVOs. Data on these sectors are not published regularly even in comparable prices and hence must be obtained by alternative estimation methods.

I obtained data on the metallurgy sector by

---

[10] See T. Koryagina, Platnye Uslugi v SSSR, Moscow: Ekonomika, 1990.

[11] Narkhoz for 1988, pp. 241 and 627; Pravda, June 8, 1990, p. 3; and V.N. Semenov, Finansovo-Kreditnyi Mekhanizm v Razvitii Sel'skogo Khozyistva, Moscow: Finansy i Statistika, 1983, pp. 179-180.

combining the 1972 i-o data with the regularly published data on growth rates and price indexes. My estimates for other industrial sectors were based on aggregating data on these sectors' output of producer and consumer goods listed in various Narkhoz tables and other Soviet sources.

As estimation results presented in Table A2 indicate, the industrial GVO residual contains hidden machinery output which probably manufactures armaments for exports and some specialized defense electronic components. In the official statistics, data on the hidden machinery output is secretly aggregated with data on other sectors of industry. At the same time, it must be emphasized that the discovered machinery sector should not be confused with other secret machinery sectors that are excluded from all published statistics on the industrial production, national income and employment.

Further research must determine why hidden MBMW GVO doubled in size during the 1980s. Could there be unreported price reforms combined with large increases in defense production in 1982 and 1985-1988? Or has the Goskomstat itself derived hidden MBMW GVO as a residual, thus making unintentional estimation errors?

If the official statistics were internally consistent, then one could expect data on hidden machinery sectors to be excluded from all published MBMW data on the annual employment, wages and profits. In the 1988 i-o table, the total MBMW sector employed 17.4 million persons (m.p.), of which 0.3 m.p. worked on collective farms and the remaining 17.1 m.p.--in state sectors. In the Narkhoz table on employment by sector of industry, the total state MBMW sector employed 16.2 m.p. and other residual sectors (construction materials and other industries)--4.1 m.p.[12]

Part of the discrepancy between the two sets of the MBMW data on employment can be possibly explained by the fact that the Goskomstat employs different methods of aggregating production and financial statistics (see the discussion in the previous chapter). Specifically, it can be argued that 16.1 m.p. exclude employees of machinery repair enterprises administered by other industrial ministries, particularly those that are in charge of the metal and chemicals production. However, the comparison of

the two sets of data for all industrial sectors indicates that the noted methodological differences have a negligible effect on the estimation of the MBMW labor force.[13]

Employment in the hidden MBMW sector then amounted to around 0.9 (17.1-16.2) m.p. What work do they perform and where are they registered in the Narkhoz? Initially, I assumed that, similar to the hidden MBMW GVO, they are registered in the residual together with employees of construction materials and other industries. However, according to the 1988 i-o table, these industries employed 2.5 and 1.7 m.p. respectively. Hence, I concluded that 0.9 million employees of the hidden MBMW sector are excluded from the total industrial labor force published in the Narkhoz.

The question then arises as to whether the same conclusion can be reached with respect to the Narkhoz data on the entire labor force. A comparison of this data with the i-o data led me to another conclusion: employees of the hidden MBMW sector are registered as employees of other material production sectors. According to the Narkhoz, these sectors employed 1.74 m.p. in 1988 as opposed to 0.81 m.p. listed in the i-o table. It appears that the authors of the official statistics are themselves unaware who the mysterious 0.94 million machinery production employees are.

This observation is further confirmed by the comparison of the two sets of data on MBMW wages. According to the 1988 i-o table, total wages (including bonus wages) of the state MBMW sector amounted to 47.1 b.r. The average monthly salary of MBMW employees can be derived as:

229.5 = (47.1*1000:17.1:12).

Surprisingly, the Narkhoz lists another average monthly salary--241.3 rubles.[14] This figure was apparently derived using 16.2 m.p. as the official

---

12 Narkhoz for 1988, p. 366.

13 Total discrepancy between the two sets of data was as follows for individual industrial sectors: -0.1 m.p. for power and fuels, -0.1 m.p. for metals, 1.2 m.p. for MBMW, 0.1 m.p. for chemical and wood, 0.1 m.p. for light industry, 0.2 m.p. for food industry, and 0.1 m.p. for other industries. There were 0.6-0.7 million collective farmers engaged in industrial production whose existence explains some of the discrepancy.

14 ibid, p. 377.

statistical figure for the MBMW employment. The extent to which the official statistical data on the machinery production seems to lack inner consistency is indeed staggering.

Capital Depreciation

I reconcile the i-o and financial data on capital depreciation in Table A3. I present the data for material production, non-material service and other sectors in accord with the NEB format. I identify three sources of financing capital depreciation: 1) accounts of profit-seeking enterprises specifically created to service depreciation payments, 2) other sources (primarily the state budget), and 3) profits that cover capital losses. The exact nature of other non-budgetary sources is uncertain.

Until 1985 material production enterprises had enough funds on their accounts to cover all depreciation costs. Starting in that year, the transportation sector needed additional funds in the amount of 25 percent of the total to cover expenses connected with the road maintenance. All these expenses used to be covered by the state budget and other allocations on the so-called "state administration" sector. This suggests that depreciation in this sector declined by the same amount in 1985 as it increased in transportation sector. The increase of other funding in the industrial sector starting in 1987-1988 is connected with launching of agro-business enterprises.

Unlike material production sectors, most service sectors receive their capital formation funds from the state budget. Estimation results indicate, however, that depreciation accounts of service sectors grew faster than budgetary funds, especially in the 1980s. In 1988, there was a significant increase in the size of these accounts because of price reforms. As a result of these reforms, R&D and other budget organizations were required to finance their capital formation with their own contract funds.

There is still no information about capital formation in so-called "other non-material sectors" which are listed in i-o tables separately from state administration services. I propose a hypothesis that these sectors are engaged in secret production and repair work for the military. The fact that capital depreciation in these sectors is relatively small suggests two possibilities.

First, machinery and other producer durables purchased by these sectors are registered as working capital (current material purchases) as opposed to fixed capital. Second, most capital depreciation of defense production enterprises is hidden with that of the published MBMW sector. If my hypothesis is correct, then capital stock operated by the armed forces--barracks and other dwellings, facilities, etc.-- is hidden with the housing stock. In this case, total military expenditure must be increased by depreciation of capital operated by the armed forces.

Wages and Other Labor Income

To estimate GNP by sector of origin it is necessary to have data on total wages and other labor income. Until recently no such data has ever been published by the Goskomstat. This made it difficult to derive GNP of published sectors as well as of secret NS sectors that are excluded from all published statistics on average wages and annual employment. Difficulties existed because it has remained unclear how the Narkhoz employment data has been extracted from original NEB and i-o tables. In addition, it was extremely difficult to determine wages of NS sectors. The only procedure that was available entailed the reconstruction of the entire budget of Soviet households where the income residual supposedly represented unpublished wages of secret state sectors.

The new data published in 1990 now make it possible for the first time to estimate labor income of published sectors and to compare them with total labor income paid in the entire USSR economy for any year since 1965 (see Charts 4 and 5 below). Wages of other secret sectors then can be estimated as the residual labor income.

In Soviet national accounts, total labor income is estimated as the sum of a) regular wages paid to so called "workers and service persons" employed in state sectors and in trade cooperatives, b) regular wages paid to members of all other cooperatives owned by individuals, c) other earnings (compensations for business trips, one-time bonuses and other payments excluded from regular wages), d) labor income of collective farmers, and e) private income in agriculture, construction and other material production sectors.

Chart 4: New Soviet Data on Household Incomes
(billion rubles; non-monetary incomes excluded)

	1987	1988	1989
TOTAL INCOME	451.6	493.1	556.9
Total Wages	334.6	361.8	408.5
Regular Wages	299.6	322.1	351.6
Cooperatives	0.3	2.5	15.7
Other Wages	9.9	11.2	12.9
Collectives	24.8	26.0	28.3
Private Agriculture	16.7	19.1	22.7
Transfers	67.3	71.8	76.7
Other Income	33.0	40.4	49.0
From State	16.7	19.5	20.0
Other	16.3	20.9	29.0

Chart 5: New Soviet Data on Annual Growth Rates of Labor Income

1966	108.9	1974	106.8	1982	103.5
1967	108.9	1975	105.7	1983	103.3
1968	110.1	1976	105.1	1984	103.2
1969	107.0	1977	104.7	1985	103.6
1970	106.9	1978	104.9	1986	103.2

In 1988, total regular wages equaled 324.6 b.r. (see Chart 4 above), while regular wages paid by material production and non-material service sectors equaled 220.4 and 80.5 b.r. respectively (see Tables A4 and A5). Thus, I estimate regular wages of other sectors as:

325.6 - 220.4 - 80.5 = 24.7 b.r.

Following the same procedure, I estimate other earnings of other sectors as 2.5 b.r. Of this amount, compensations for business trips equaled around 0.5 b.r. Total wages of other sectors then equaled 26.7 b.r. As I discuss in the next chapter, these wages exceed the combined wages of the military and the police by around 14.5 b.r.

How reliable is the size of the derived wage gap? It is difficult to answer this question because there is still very little information on how the Goskomstat estimates total wages as well as wages of particular production and service sectors. As evident from Chart 5 above, total wages increased by 21.3 percent in 1987-1989, including by 8.2 percent in

1987-1988 and by 13.1 percent in 1988-1989. With the average annual rate of growth of 3.2 percent, it took seven years before 1988 to have a similar increase in wages.

This spectacular increase in wages in the 1987-1989 period is not, however, evident in Goskomstat's detailed reports on wage in production and service sectors which I present in Tables A4 and A5.

I estimated total wages of these sectors as the sum of regular wages plus the NEB adjustment plus other earnings. As a rule, I obtained data on regular wages of each published state-cooperative sector by multiplying the standard Narkhoz data on average salary and employment.[15] Data on agricultural collective farms are also published in the Narkhoz.[16] Data on total labor income by sector,

---

[15] Data on hired workers are presented in Narkhoz tables on employment and wages as the difference between data on total agriculture and state farms.
[16] Published NEB data on wages paid by collective

other earnings, other material production sectors, and industrial and construction wages paid by collective farms--were all extracted from the recently published i-o and NEB tables.[17]

I derived data on private agricultural incomes as the difference between total labor income and total wages in agriculture. I could not use the same procedure for construction because of uncertainties with the construction wage statistics. Instead, I estimated private income in construction as the difference between total private income and private incomes in agriculture and other sectors.

Estimates presented for some material production and non-material service sectors contain rows with NEB and i-o adjustments. The idea to use these adjustments arose after I compared estimates of wages derived using the Narkhoz data with the new wage data listed in published NEB and i-o tables for selected years. These adjustments in effect help reconcile discrepancies between the two sets of data.

The fact that these adjustments are required in the first place suggests that there is a significant lack of inner consistency in Goskomstat's estimates of total and average wages and employment. I already demonstrated this with respect to wages of machinery sectors. The same lack of consistency also can be observed when analyzing Goskomstat's estimates of employment in construction, transportation and trade and distribution (T&D) sectors and their average wages.

The Narkhoz employment data on transportation are never disaggregated into sectors serving various material production and households. As evident from explanatory notes enclosed in the Narkhoz table on employment, around 1.1 m.p. involved in loading and distribution were always registered as construction employees, even though they worked in transportation and distribution organizations. Until 1988 they were registered in the Narkhoz separately from construction employees. This explains why the reported totals for transportation and T&D sectors declined by 0.9 and 0.3 m.p. respectively in 1987-1988.

However, even after adjusting for loading works it is still difficult to explain the discrepancy between the i-o and Narkhoz data on state construction

---

[17] See fn. 6. Refer particularly to pp. 12-15.

wages. As evident from Table A4 (row 28), the discrepancy--observed for the first time in 1984--increased from 1.3 b.r. in 1985 to 3.7 b.r. in 1988 and to 6 b.r. in 1989. It is unclear who earns these wages and why their size has increased so dramatically in the last several years. It is also unclear why their increase has coincided with the decline of private construction incomes. The latter decreased by 50 percent in 1986-1987 from 5.4 to 2.8 b.r.

The difference between employment in T&D sectors reported in the i-o and Narkhoz tables amounted to 0.7 m.p. Of this amount, information services employed 0.4 m.p. and loading services--0.3 m.p. When I estimated regular wages of T&D sectors I did not even count wages of these 0.7 m.p. If I did, I would have arrived at a much greater discrepancy between the i-o and Narkhoz wage data. As in the case with industry, Goskomstat officials grossly overestimated the average monthly salary in T&D sectors.

For 1988, out of the total 9.5 million transportation sectors' employees, I could only account for 8.2 (4.9+2.9+0.3) million. No information exists on the remaining 1.3 m.p. It is difficult to determine the nature of their work because even i-o tables exclude information on service sectors' employees.

Reconciliation of the wage data for service sectors--see Table A5--also points to a number of inconsistencies in the Goskomstat's estimates. The comparison of the two sets of data suggests that it is necessary to make the following adjustments: wages should be decreased for communal (municipal) and state administration sectors and increased for education and science sectors. Overall, adjustments result in the increase of total wages of service sectors by 2 b.r. in 1988.

In Table 5 (rows 17 and 30) I list wages of other education and science sectors which I obtained by comparing the i-o and Narkhoz data. Were these wages earned by 0.8-0.9 m.p. in space centers? If this is indeed the case, then the remaining 0.4-0.5 m.p. were involved in transporting military hardware.

Estimation results presented in Tables A4 and A5 also make it possible to perform a consistency test for the entire state wage bill for "workers and service persons" determined by the Goskomstat. In 1988, according to the Narkhoz, 117.24 million "workers and service persons" supposedly earned 219.8 rubles a month.[18] Their total wage bill then should equal 309.2 b.r.

However, according to the original 1988 NEB and i-o tables, their total wage bill was 301.9 b.r.[19] I believe that the discrepancy in the amount of 7.3 b.r. results from the inability of Goskomstat's officials to measure correctly average monthly wages of employees who are registered in the Narkhoz. In particular, these officials appear to have compared two incomparable entities: the total wage bill and the Narkhoz data on employment which, as I demonstrated above, are incompatible with the NEB and i-o data.

In reality, there were around 123.5 million "workers and service persons". This includes 4.2 million state civilian employees and around 2 million military and police officials holding a military rank who are all missing from the Narkhoz. As was determined above, the total civilian wage bill amounted to 325.6 b.r. Indeed, the average monthly salary in the USSR economy was 219.8 rubles a month in 1988:

$$325.6*1000:123.5:12 = 219.8.$$

Why Goskomstat officials cannot make their estimates of labor income consistent remains unclear.

Interindustry Purchases

In the USSR economy, interindustry purchases comprise more than half of the GVO. I estimated total purchases for both material production and service sectors as the difference between the derived total for material expenditures and capital depreciation which I already discussed above. The total for material expenditures made by six major material production sectors can be easily obtained as the difference between the regularly published Narkhoz data on GVO and NMP. Similarly, the total for non-material service sectors can be obtained using the Narkhoz data on public consumption of national income.

---

[18] Narkhoz for 1988, pp. 34 and 81.

[19] See Tables A4 and A5. State-cooperative sectors paid 301 b.r., while hired workers received 0.9 b.r. from collective farms.

Major problems arise in obtaining data on material expenditures for particular industrial sectors for years for which the i-o data are unavailable. The standard Narkhoz table on the structure of production outlays are difficult to utilize because it is compiled using the financial method. In particular, it remains unclear what sectoral coefficients are applied by Goskomstat officials in disaggregating the item "other expenditures" into material, labor and capital inputs. Therefore, I decided to estimate total interindustry purchases by sector of industry as the difference between GVO and the sum of depreciation, wages and revenues.

I made adjustments in the Narkhoz data on wages and depreciation deductions similar to those that I described above in connection with the entire industry in Tables A3 and A4. I applied the same approach to converting the Narkhoz financial data into the NEB format.

I present my estimates in Tables A6 and A9. Table 6 consists of two parts. Part one contains estimates performed for the entire industry, agriculture, transportation and communications (T&C), trade and distribution (T&D), other material production sectors, and foreign trade. Part two contains estimates performed for ten major industrial sectors. In effect, Table 6 contains consolidated sector-of-origin accounts of material production sectors.

Revenues of Public Sectors

In addition to interindustry purchases, Table A6 contains estimates of total revenues which are referred to in Soviet national accounts as "m". I derived "m" as the difference between NMP and total labor income ("v"). Similar to labor income, Goskomstat officials make no attempt to reconcile NEB, i-o and official estimates of revenues. The Narkhoz and Finansy SSSR regularly contain data on: a) net profits (including bonus wages) of production and service sectors, b) turnover tax by sector of industry, and c) total social security payments to the state budget. As discussed in the previous chapter, all these data are collected following the financial method which differs significantly from NEB and i-o methods.

I used my estimates of total revenues derived in Table A6 as control totals for analyzing the structure of revenues in Table A7. I derived other revenues for industry, agriculture and construction as the difference between total revenues and the sum of net profits (without bonus wages), net taxes and social security deductions. Net taxes are presented in Table A8. I estimated social security deductions by multiplying wages and deduction rates.

I could not apply the same procedure for T&C and T&D sectors because published data on T&C profits are never disaggregated into sectors serving material and non-material activities and because data on T&D GVO are excluded from the official statistics. For these sectors, I extrapolated the i-o data with the margin of error around 0.5 b.r.

As presented in Table A8, net taxes equal turnover tax less net subsidies. Net taxes in industry equal the difference between two types of industrial GVO published in the official statistics.[20] Net subsidies equal the difference between subsidies and surcharges on agricultural and machinery products.

Data on turnover tax were openly published by sector of industry for the early 1970s and for 1987-1988.[21] Data were also published for selected industries throughout the observed period.[22] I determined changes in tax rates and then applied these rates to extrapolate published data for other years. There were significant increases in tax rates on oil products in 1978-1979 and on alcohol in 1982. The anti-alcohol campaign resulted in the loss of state budgetary revenues that averaged 10 b.r. in 1985-1987. As a result of 1982 price reforms, tax rates dropped in the oil industry. In addition, some taxes were cancelled on the oil industry and were introduced in the oil-processing chemical industry.

Until 1985 price reforms, subsidies were estimated for agricultural products sold to apparel and food industries and for machinery and chemical products sold to agriculture.[23] There were large

increases in subsidies in 1983 following agricultural price reforms.

In 1985-1986, producers' price increases in coal, apparel and food industries resulted in the introduction of subsidies on these industries' products, thus further aggravating the financial crisis. Between 1982-1989 subsidies tripled in size and approached 100 b.r. As result, net taxes dropped dramatically from 71.2 to 11.7 b.r.

Table A6 contains data on NMP at factor cost which the Goskomstat began to compile in 1988. I was able to duplicate Goskomstat's estimate using the following formula: NMP minus the sum of net taxes and foreign trade revenues. The ratio between NMP at factor cost and total NMP has grown from 75 to 90 percent during the 1980s largely as a result of increases in subsidies and the slowdown in foreign trade revenues.

In addition to data on revenues, Tables A6 and A7 contain estimates of total foreign trade revenues. All regularly published statistics on foreign trade are presented in foreign trade prices. This makes it difficult to analyze the role played by foreign trade in the allocation of resources in the USSR economy.

Total foreign trade revenues were derived in two ways: 1) as the difference between NMP and the sum of NMP at factors cost and net taxes, and 2) as residual revenues of T&D and other sectors. This total consists of two components: the difference between imports and exports as well as additions to foreign currency reserves estimated as net exports converted into established prices with the export conversion coefficient. Data derived in Table A7 indicate that foreign trade revenues increased rapidly till 1985 and stagnated since then. The detailed discussion of imports and exports will be presented later in the chapter.

Non-Material Services

As opposed to material sectors, no gross and products are estimated in the official statistics for non-material services. Therefore, I had to make my own independent estimates which I present as consolidated accounts on non-material services in Tables A9 and A10 below.

[20] Data on the industrial GVO estimated without net taxes were published in Goskomstat SSSR, Promyshlennost' SSSR. Moscow: Finansy i Statistika, 1988, p. 6.
[21] See Finansy SSSR for 1987-1988, op. cit., p. 5.
[22] For example, refer to Semenov, op. cit., p. 119.
[23] For the detailed and informative discussion of agricultural subsidies see ibid, pp. 159-177. Also refer to V. Semenov, Prodovol'stvennaya Programma i Finansy, Moscow: Finansy i Statistika, 1985, pp.
33-83.

Table A9 contains data on gross and net product of eight major service sectors: transportation, communications, housing-communal and everyday, education, culture and arts, health and social welfare, science, banking and insurance, and administration.

I determined gross product for each sector as the sum of current material expenditures, depreciation, wages, social security deductions and revenues. The correct procedure is to reduce revenues by state budgetary subsidies to service sectors that operate on a profit basis. Unfortunately, data on subsidies in service sectors are regularly published only for housing services. There is no regularly published information on the growth of subsidies in cultural, health and science services. Since I failed to account for subsidies in measuring the output of these services in Table A9, there is a discrepancy between my annual estimates and the i-o data.

Estimates in Table A9 are based on the following sources: depreciation and wages are from Tables A3, social security deductions were derived using data on wages and deduction rates, current material expenditures and revenues were derived by combining the Narkhoz, Finansy SSSR and i-o data.[24] The i-o data on current material expenditures by sector were also published for the early 1970s.[25]

I use estimates of gross products of service sectors derived in Table A9 as control totals in identifying sources of financing these sectors' activities in Table A10. Altogether I identify three major sources: the state budget, enterprises and households.

To simplify the estimation procedure I treat all financing of service activities from the social security budget as originating from one source--enterprises. As known, the state budget and households

subsidize the social security budget. State budgetary subsidies are allocated primarily to make pensions and other transfer payments, while household expenditures are accounted for separately in statistics on household budgets.

Published financial statistics contain information on a) household financing of various services, b) housing subsidies, and c) state budgetary financing of broadcasting (communications), culture, health, science, geology, and state administration.[26] Data on operational expenditures in agriculture, which are treated as other outlays on science, can be found in books and articles written by Semenov--a senior official in the USSR Finance Ministry.[27]

Excluded from published statistics are allocations made by enterprises from profit and other sources as well as budgetary financing of transportation, communications and space. I believe that outlays on space are also treated as other science expenditures in Soviet national accounts. There is simply no other place in NEB and i-o tables where current expenditures on space can be hidden. I derive all of the above missing statistics using the residual approach.

The derived residual for the science sectors consists of outlays on space and financing of research contracts by enterprises. Information on

---

[24] In the Narkhoz, data on current material expenditures in household service sectors and in science and state administration are regularly included together with depreciation as public consumption of national income. My procedure was based on extrapolating the i-o data using the annual NEB data published in the Narkhoz as a control total.

[25] V. Rutgeizer, Resursy Razvitiya Neproizvodstvennoi Sfery, Moscow: Mysl', 1975, pp. 157-158.

[26] Data on household purchases of services have been published so far for the early 1970s and the late 1980s. I extrapolated the published data for intervening years. I believe that the margin of error is insignificant because transportation services comprise around half of all household purchases of services. The state budgetary statistics on education and health are presented with and without capital investment in these services. The major difficulty with using budgetary statistics is that it remains unclear how capital investment financed by the state budget can be divided into depreciation and net fixed investment components. This division is necessary to perform because gross output of service sectors excludes net fixed investment. After comparing published budgetary and investment data I discovered that, given the existing Soviet national acounting practices, this division is impossible to perform with any precision. Therefore, I assumed that depreciation in service sectors is financed proportionately to their overall financing.

[27] See Semenov (1985), op. cit., p. 104.

space outlays exists only for 1989-1990.[28] I disaggregate the science residual for the preceding years arbitrarily on the basis of the ratio between outlays on space and research contracts observed for 1989-1990.

As will be discussed in the next chapter, the derived data on financing of non-material service activities are indispensable in measuring Soviet GNP by sector of origin and end use. To estimate GNP by sector of origin it is necessary to have data on business purchases of services, including those connected with business trips.

Enterprises purchase five types of business services: 1) various services connected with employees' business travel, 2) education services to improve professional qualifications of their employees, 3) research services, 4) banking, insurance and advertising services, and 5) administration services. In addition, enterprises subsidize housing, entertainment and health services offered to their employees.

In Table A10, business purchases of services are estimated for individual sectors and for the entire economy using a residual approach after accounting for purchases made by the state budget and households. I estimated business trip purchases as 30 and 10 percent of total household purchases of transportation and housing-communal services respectively.

Material Inputs by Sector

So far, the discussion has focused on sector-of-origin accounts which pertain to the structure of prices on domestically produced goods and services. In this section, the subject changes to sector-of-use accounts. There are two major uses: interindustry and end uses. In Soviet i-o tables, sector-of-use accounts pertain to purchases of material goods and services.

I present the detailed data on the structure of interindustry purchases (material inputs) made by individual sectors in Table B1. The data on material production sectors have been published for 1972, 1975, 1980 and 1985-1988.[29] The data on non-

material service sectors have been published for 1970 and 1988.[30] I extrapolated these data for other years using information on industrial producers' price indexes which I systematized in Table A1.

Sometime in the mid-1970s Goskomstat officials introduced a new format for disaggregating data on T&C surcharges contained in producers' prices on industrial goods. In the 1972 i-o table, most T&C surcharges were disaggregated into sectors supplying material inputs. As evident from the 1988 i-o table, around 24 percent of T&C services were registered as a distinct component of material expenditures in production and service sectors.

As was noted above, this new format was introduced in service sectors in 1985 when road maintenace services were registered for the first time as material production services. To preserve continuity in the presentation of data on material inputs, I applied the new format in Table B1 starting in 1985.

The detailed data on the structure of material inputs are aggregated in Table B2. The derived results are presented in established prices which distort the real role played by each sector in the USSR economy. Thus, agricultural material inputs are registered in prices which exclude subsidies. The latter increased from 15 to 82 b.r. (see Table A8). In established prices, the share of the agricultural sector in total interindustry purchases fell from 18.5 percent in 1970 to 16 percent in 1988. However, the same share increases from 22 to 23 percent when subsidies are taken into account.

After adjusting established prices for subsidies, one observes a remarkable stability in the structure of material inputs. A small increase in the share of fuels, chemicals, metals and MBMW sectors was a result of the decreasing role played by apparel and food sectors whose share fell from 23.5 to 17.5 percent. There was also a small decrease in the share of wood and construction materials. Overall, the results confirm the well known fact that in the allocation of resources Soviet policy makers have traditionally favored investment and defense sectors at the expense of consumer sectors.

---

[28] See fn. 11.

[29] Goskomstat, _Osnovnye Pokazateli_, op. cit., pp. 81-111. Data on 1972 were reproduced in D. Gallik,

et. al., "The Input-Output Structure of the Soviet Economy," op. cit.

[30] See fn. 25.

Household Consumption

In Soviet national accounts, household consumption is estimated as the sum of retail trade purchases of goods for current consumption, purchases of services and consumption-in-kind (goods that are consumed by producers of these goods). There are difficulties with estimating household consumption because production and financial statistics are never integrated in the official Goskomstat publications.

The Narkhoz regularly includes tables on retail trade purchases of particular goods and on purchases of particular services. No data on consumption-in-kind have been published by the Goskomstat in value terms. The retail trade turnover includes purchases of goods that are used for production purposes, purchases of services which are double counted in the table on services, institutional purchases, and purchases of second hand goods (primarily used automobiles). All these purchases must be taken into account in order to avoid overestimation of household consumption.

The new Goskomstat publication Torgovlya SSSR includes data on institutional purchases, including purchases of food items, and purchases of second hand goods and services included in the retail trade turnover. This publication significantly simplifies the estimation of retail trade purchases of goods for current consumption. However, several problems with integrating published production and financial statistics on retail trade and household consumption still remain.

1. Published table on the retail trade contains a large residual whose contents are inadequately explained in Goskomstat and other Soviet publications. Thus, it remains unclear how to determine the size of production and some consumer goods included in the residual. A note in Torgovlya SSSR lists the size of two residual purchases: medical supplies and fuels. I assumed that other chemicals and other industry (everyday services and crafts) products comprise most of the remaining purchases.

2. In published tables on purchases of services, utilities are never disaggregated by sector of origin, such as power, gas, water and housing-communal services. I determined purchases of power using the new data on combined household purchases of fuels and power.

3. There is no information on the structure of institutional purchases of non-food items since the late 1960s. This makes it difficult to derive the structure of household consumption by sector of origin with the desirable precision. I assumed that the structure of institutional purchases reported for the late 1960s has remained the same. I believe that the margin of error is below 0.2 b.r. because most institutional purchases consist of food items.

4. Such items as office supplies, sporting goods and haberdashery are all produced in three different sectors: wood-making, MBMW and apparel. At the same time, retail trade purchases of food items are never divided into processed food and raw agricultural products. I made the following assumptions: 80 percent of office supply products are manufactured in wood processing sector, 100 percent of sporting goods--in the MBMW sector, 80 percent of haberdashery--in the apparel sector; the MBMW sector also manufactured the remaining office supply and haberdashery products.

I integrated the retail trade data with the i-o data on household consumption of material goods and services in Tables B3 and B4. In table B3, I estimated total consumption (row 1) as the difference between the Narkhoz data on private consumption of national income and depreciation of residential housing (see Table A3). I estimated retail trade purchases (row 2) as the difference between total retail trade turnover and the sum of a) institutional purchases, b) household purchases of construction materials that are used for production purposes, and c) purchases of second hand goods. I estimated utilities (row 3) as household payments for power, heating gas and water.

Initially, I assumed that consumption-in-kind (row 4) comprised most of the residual consumption of material goods and services. Afterward, I performed a test which entailed making independent estimates of consumption-in-kind. These estimates were based on the new i-o data on household consumption published in the Goskomstat publication containing excerpts from original NEB tables.[31] This publication includes data on total household consumption of food, agricultural, power and fuels, MBMW, light industry (apparel, footware, etc.) products for 1975, 1980,

---

[31] Goskomstat, Osnovnye Pokazateli, op. cit., p. 128.

1985-1988. Thus, agricultural consumption-in-kind can be estimated for these years as the published control total for agriculture plus institutional purchases minus total retail trade purchases of unprocessed food products. As the available i-o data indicate, consumption-in-kind of other products increased slowly from 0.7 to 1.6 b.r. in 1972-1988.

Test results indicate that before 1976 there had existed other household consumption whose nature remains mysterious. Although its size had been relatively small--1.6 b.r. in 1970 and 2.1 b.r. in 1975, the fact that it had existed is quite surprising. Nothing has ever been mentioned about it in the Soviet literature. One possibility is that it represents wholesale purchases by the armed forces which since 1976 have been registered as other uses of national income.

Test results also indicate that in 1986-1988 there were large retail trade purchases of production goods other than construction materials. A close sector-of-origin analysis of the retail trade data leads to the conclusion that in 1988 households purchased processed food items for agricultural production purposes as well as other unidentified production goods in the amount of 4.1 and 2.6 b.r. respectively. These production goods were excluded from Goskomstat's estimates of household consumption.

I repeated the above test for individual industrial sectors using the 1988 i-o data to check the accuracy of results. The test proved remarkably successful for each sector as I found all Goskomstat's estimates of household consumption by sector of origin internally consistent. At issue is whether this consistency was achieved by Goskomstat officials artificially. Specifically, it remains unclear whether they bothered to find out the actual contents of residual retail trade purchases of other chemicals and fuels which are quite large in size. Thus, in 1988 these purchases amounted to 6.9 and 4 b.r. respectively. I found no evidence in the openly published Soviet literature that points to such large household purchases of other chemical and fuels products.

Common sense suggests that these combined purchases should not exceed 3 b.r. Did the military make secret retail trade purchases of chemicals and fuels in the amount of 8 b.r.? Whether this is indeed the case will be explored in the next chapter.

Fixed Investment

Fixed investment is the least analyzed segment of Soviet national accounts. The reason for this is that most published statistics on fixed investment are presented in comparable prices which differ from both established and real constant prices. On the one hand, data on fixed investment in established prices are needed for analyzing end uses of Soviet GNP and national income. On the other hand, it is necessary to have data in real constant prices in order to measure the growth of Soviet investment. In this respect, the official data in comparable prices have been often perceived as serving no useful analytical purpose.

I would like to demonstrate that this common perception is quite misleading because it prevents analysts from performing a close analysis of the allocation of resources and price trends in the Soviet investment sector. The usefulness of comparable prices becomes apparent when one attempts to integrate published production, financial and foreign trade statistics on fixed investment. No such integration has ever been attempted by both Soviet and Western economists. I believe that this fact provides an additonal explanation for the present lack of knowledge on Soviet fixed investment.

I attempt to integrate capital flows in the entire USSR economy and in production and service sectors in Table B5. The objective of estimates performed in this table is to determine comparable price indexes for new construction and new investment in machinery and equipment (M&E). Once these indexes are known, it becomes possible to convert the entire published statistics on capital flows from comparable to established prices. Estimates performed in this table are based on two methods for integrating capital flows that are described in Soviet planning manuals.[32] These two methods are as follows:

1. New construction plus M&E investment plus agricultural and other capital plus M&E purchases made by the existing education and health organizations financed by the state budget plus capital repair of buildings, installations and M&E.

2. Additions to unfinished construction plus losses of fixed capital (investment writeoffs) plus net

---

[32] See, for example, Metodicheskie Ukazaniya, 1974, op. cit., p. 617.

fixed investment plus depreciation.

Difficulties with using the first method is that published statistics exclude information on a) domestic and imported machinery in current established prices, b) capital repair works performed by construction and machinery production enterprises, c) All-Union budgetary purchases of M&E, and d) additions to agricultural capital in private farms.

I estimated total capital repair works as the sum of a) published depreciation deductions and All-Republic Budgetary allocations on capital repair of buildings and installations, and b) independently estimated other allocations from the All-Union Budget and household budgets.[33] The available i-o data indicate that the share of machinery repair works in total capital repair has decreased from 41 to 39 percent during the observed period. Using published data on the construction and machinery repair GVO I assumed that this share decreased from 40 to 38 percent in 1975-1984 but then increased from 38 to 39 percent in 1984-1988.

Except for the machinery price index, other mentioned difficulties with using the first method are quite minor; they can be overcome by making assumptions that lead to estimation results with an insignificant margin of error.[34]

Difficulties with using the second method are that there is no accurate data on additions to unfinished construction in current established prices

since 1980 and capital losses prior to 1980.[35] I made independent estimates of unfinished construction in 1981-1988 using the Narkhoz data on the ratio between unfinished construction and capital investment.[36] I also made independent estimates of capital losses for 1970-1980 by combining two approaches: a) extrapolating data published for 1981-1988 and b) continuing the trend in the share of these losses in total losses of national income (see Table B10).

After performing all of these estimates I was able to derive total capital outlays, new construction works and capital investment in M&E--all in current prices. I estimated total capital outlays using the second method; new construction--as the difference between construction GVO and the sum of a) capital repair of buildings and installations, and b) unfinished production in construction (see Table B8); and capital investment in M&E--as a residual.

Estimation results presented in Table B5 indicate that:

1) the combined capital investment index has fluctuated between 0.98 and 1.01 throughout the observed period;

2) the 1969 construction index has fluctuated between 0.99-1.01 until 1981; it increased to 1.03 in 1982 and fell to 1.01 in 1983;

3) the 1984 costruction price index has remained around 0.93-0.94;

4) the 1969 machinery price index fell from 0.99 in 1970-1972 to 0.96-0.94 in 1973-1975,

5) the 1973 machinery price index fluctuated between 0.97-1.02 in 1975-1983; and

6) the 1984 machinery price index has increased from around 1.06 in 1984 to 1.10 in 1988.

I successfully tested the above results for 1973 by

[33] See Narkhoz for 1987, p. 585, and Ministerstvo Finansov SSSR, Gosudarstvennyi Byudget SSSR, 1981-1985, p. 49. I assumed that All-Union budgetary allocations on capital repair comprise 10 percent of All-Republic budgetary allocations. To estimate capital repair works financed by households I applied the published repair rate (2 percent) to the average annual value of privately held housing stock. Estimation results indicate that the size of other allocations has grown slowly during the observed period from 2.2 b.r. in 1970 to around 3 b.r. in 1988.

[34] I assumed that All-Union budgetary purchases of M&E have increased from 0.3 to 0.5 b.r. during the observed period. I estimated additions to fixed capital in private agriculture by comparing the Narkhoz data on this capital in public and private farms reported in physical units.

[35] Data on unfinished construction presented in the Narkhoz for 1983, p. 367, in current rubles seem to lack inner consistency. For data on losses refer to Ekonomicheskaya Gazeta, N50, 1989, p. 11.

[36] Narkhoz for 1988, pp. 551 and 558. I made adjustments in estimates for some years in order to have a consistent machinery price index for each year of the observed period.

comparing the Narkhoz data published in editions before and after introduction of new investment prices.[37] The comparison indicates that 1969 official construction and machinery price indexes equaled 0.99 and 0.96 in 1973. In turn, 1973 construction and machinery price indexes equaled 1.19 and 1.06 in January 1984. Between January and July 1984, construction prices fell by 6 percent, while machinery prices remained the same. I thus estimated that in 1983-1984 the official construction price index increased by 11 percent. Since 1984 the official construction price index declined by 0.2 percent, while the official machinery price index increased by 7 percent largely as a result of price increases on imported machinery.

Table B5 also contains estimates for investment in domestic and imported M&E. The share of imported M&E has more than doubled from 13 percent in 1970 to 27.5 percent in 1988. My estimates of machinery imports in current and constant prices are based on a) the detailed data published by the USSR Ministry of Foreign Trade in foreign trade prices and b) published data on the price index for imported machinery.[38] As I discuss below, the coefficient for converting machinery imports from foreign trade to domestic prices has been around 1.00.

Estimates for domestic machinery were derived as the difference between total M&E investment and imported machinery. Estimation results indicate a remarkable stability in the domestic machinery price index. It declined by 5 percent in 1972-1973 and in 1975-1976 and reamined the same since then. It appears that Goskomstat officials make no attempt to estimate real price changes for producer durables.

The derived construction and machinery price indexes make it possible to convert the entire set of published statistics on capital investment in production and service sectors into current established prices. Published data on capital

investment in major sectors are systematized in Table B6. I converted these data into current prices and then estimated additions to fixed capital stock by sector in Table B7. These estimates can be combined with data on scrap to derive data on fixed capital stock in current prices for any year of the observed perod.

Inventories

Goskomstat collects information on stocks of inventories by sector of origin and end use and by type. The latter include production materials, unfinished production, finished goods, trade inventories, state reserves and other. Published statistics contain information on stocks of inventories by sector of use and by type.[39] These statistics are reproduced in Tables B8 and B9.

As evident from Table B8, since 1985 Goskomstat officials have introduced several changes in the presentation of data by sector of use. First, most unfinished production was removed from stocks of inventories held by construction organizations in 1986. This change was justified because unfinished production was double counted as part of unfinished construction. What is unclear is whether the remaining stocks of unfinished production are double counted with stocks of unfinished construction. Second, in 1986 trade inventories were divided into stocks held by trade and other organizations whose activities remain unknown. Third, in 1987 state agricultural reserves were removed from stocks of inventories held by supply organizations. Fourth, in 1988 agro-business enterprises were registered independently from state farms.

Until 1981 the Narkhoz included data on stocks of inventories by type. These data were again published in Finansy SSSR. I combined these two sets of data in Table B9. In Soviet national accounts, supplies held as reserves are registered as other commodities. In 1986-1987, their size fell from 56 to 15 b.r. Some supplies were registered in 1987 as the unfinished industrial production, while others--as reserves of finished goods in 1988.

[37] See Narkhoz for 1975, p. 503, Narkhoz for 1980, p. 334 and Narkhoz for 1985, p. 364.
[38] See Ministerstvo Vneshnei Torgovli, Vneshnyaya Torgovlya SSSR, 1922-1981, and other editions. For data on the price index refer to N. Glushkov and A. Dryabin, eds., Tsena V Khozyaistvennom Mekhanizme, Moscow: Nauka, 1983, p. 378; and V. Sel'tsovsky, "Sovershenstvovanie Analiza Effektivnosti Vneshnei Torgovli," Vestnik Statistiki, N6, 1983.

[39] Prior to 1986 published statistics on inventories were limited to state-cooperative sectors. Data on collective farms have been reported in the Narkhoz section on these farms' capital stock.

Goskomstat officials have never incorporated data on inventories by sector of origin within the format of published statistics. This peculiarity in Goskomstat's reporting practices has made it difficult for economists who do not have access to annual i-o tables to analyze end uses of goods in the USSR economy. However, as evident from those i-o tables that have become available in the West, other (residual) uses of goods consist primarily of petroleum, chemical and machinery products. Moreover, machinery products comprise as much as 85 percent of other uses. Therefore, providing the availability of data on total supplies, consumption and fixed investment by sector of origin, missing data on inventories can be derived using a residual approach. Though it must be noted that this approach is quite cumbersome as it requires the reconstruction of the i-o table for every given year.

Losses of National Income

In Soviet national accounts, losses are esstimated to determine the size of national income available for end use. As a national accounting concept, losses differ little from interindustry purchases and hence should be treated as charges against profit that reduce total value added in given sectors. However, the established Soviet practice is to distinguish between losses incurred in and outside the process of producing national income whereby the latter (sometime referred to as planned losses) are treated as a charge against the entire national income.

In published statistics, planned losses are hidden together with foreign trade earnings in the difference between produced and utilized national income. Until 1988 there were two types of planned losses: 1) perished livestock and agricultural raw materials purchased by the state and 2) abandoned construction projects, primarily in the oil and gas industry. In 1988 planned losses began to include losses incurred by trade organizations as a result of reevaluating inventories of overstocked consumer goods, primarily poor quality apparel and footware.

I present estimates of planned losses in Table B10. The detailed data on both production and planned losses were recently published for for 1981-1988.[40] The accuracy of these estimates before 1981 depend on the accuracy of my estimates of foreign trade earnings which are discussed below. I

---

[40] See fn. 35.

determined agricultural losses as the difference between total losses and reported losses in construction.

Other Uses of National Income

The exact nature of other uses of national income remains probably the most debated aspect of Soviet national accounts. The objective of the debate has been to establish beyond any doubt whether other uses contain the entire value of procured weapons. As this debate will be discussed in the next chapter, the objective of this section is to integrate other expenditures within the general framework of Soviet national accounts.

In published statistics, other uses are hidden as part of the investment residual together with additions to working capital--inventories, reserves and unfinished construction. I estimate other uses in Table B11 as the residual after accounting for additions to published working capital--inventories (see Table B8) and unfinished construction (see Table B5). Published data on inventories in construction include unfinished production works in construction which increased by 0.8 b.r. in 1988. Until 1985 these works were double counted in the official statistics as inventories and unfinished construction works. As was discussed above, in 1986 most of these works were removed from stocks inventories. What remains unclear is whether construction works which were registered as inventories after 1985 were still double counted as unfinished construction in published statistics on capital investment.

According to the 1988 i-o table, additions to working capital (28.7 b.r.) plus other expenditures (35.2 b.r.) equaled 63.9 b.r.--the same amount published in the Narkhoz table on end uses of national income. However, according to Narkhoz tables, additions to inventories (9.2 b.r.) plus unfinished construction (18.9 b.r.) equaled 28.1 b.r. Additions to reserves probably account for the discrepancy in the amount of 0.6 b.r.

As estimates presented in Table B11 indicate, other uses (row 6) have fluctuated during the observed period in a way which is uncharacteristic for the USSR economy. I believe that these fluctuations have been caused by the fact that other uses are estimated by the Goskomstat using a residual method. Specifically, the precision of the residual method depends on the accuracy of the Goskomstat's estimate of a) the total supply of goods manufactured in material production sectors, particularly in secretive defense industries, and b) the value of agricultiral commodities registered as inventories as opposed to state strategic reserves.

As I discussed above, Goskomstat officials have difficulties estimating the total output of defense industries. This apparently causes the Goskomstat to make estimation errors. The correct way of estimating other uses is based on increasing both produced and utilized national income by the amount of value added in defense industries which is excluded each year from published indicators for the industrial production.

Similarly, Goskomstat officials appear to have difficulties collecting information on state strategic agricultural reserves which in Soviet national accounts are registered separately from inventories. In the 1988 i-o table, all other agricultural uses are grouped in one column "inventories and reserves." Large annual fluctuations in inventories of agricultural commodities held by procurement and supply agencies suggests that there might be annual fluctuations in strategic agricultural reserves of a similarly large size.

The analysis of annual changes in other uses (see Table B11) suggests the existence of a cycle in the depletion and accumulation of strategic reserves. The depletion occured during bad harvest years (1972, 1975 and 1988-1989) and in the last year of the Five-Year Plan (1970, 1975, 1980 and 1985). It is also possible to detect five-year cycles: accumulation took place in 1966-1969, 1976-1979 and 1981-1984, while depletion took place in 1970-1974, 1986-1989. One the basis of these observations, I first made speculative estimates of annual changes in strategic reserves and then derived other defense uses as a residual.

Other defense uses are divided in Soviet national income accounts into current expenditures and depreciation of capital stock operated by the military. Although data on other depreciation can be obtained only from the latest i-o tables, its relatively small size can be extrapolated for the preceding years with an insignificant margin of error. Other current expenditures, which are determined as residual uses of national income, increased from 14.6 b.r. in 1970 to around 38-39 b.r. in 1987-1988. As can be deduced from preliminary

Goskomstat reports for 1989-1990, the size of other current expenditures decreased to around 35-36 b.r.

Foreign Trade

Before the dissolution of the CEMA trading bloc Soviet foreign trade organizations and planning agencies collected information on foreign trade in four sets of prices: world, CEMA contract, Soviet foreign trade and domestic prices. World prices were used to registered transactions in convertible currencies. Transactions with CEMA-member countries were registered in contract prices determined on a bilateral basis. Transactions between Soviet foreign trade organizations and enterprises within the USSR were registered in domestic prices. Foreign trade prices were created artificially to keep track of all transactions in so-called "gold rubles" which existed only as a medium of exchange and as a buffer between the domestic ruble and foreign currencies.

Similar to production and financial statistics, foreign trade statistics are never integrated in Soviet national accounts. As was discussed in the previous chapter, Soviet planning and statistical agencies estimate the value of exports and imports in foreign trade and domestic prices as well as revenues and outlays connected with foreign economic transactions.

The Foreign Trade Ministry collects information on exports and imports in foreign trade prices which is regularly published in Vneshnyaya Torgovlya SSSR (VT) and in Narkhoz. Goskomstat and Gosplan collect information in domestic prices which they use in estimating gross and net national product. The Finance Ministry and State Bank collect information on budgetary and credit operations connected with foreign financial transactions. All this information remains unintegrated for bureaucratic reasons: the aforementioned Soviet agencies simply refuse to share their data not only with the outside world but also with each other.

Until recently all information in domestic prices has been kept secret. During the last two years Soviet officials began to release data on a) the foreign trade balance in domestic prices counted as part of national income, b) the value of exports and imports included in i-o tables, and c) state budgetary revenues and outlays connected with foreign economic transactions.

It has become possible to estimate total revenues from foreign trade in domestic rubles with precision since the Goskomstat began to publish data on NMP at factor cost (NMP less the sum of net taxes and foreign trade revenues). Data on NMP at factor cost has been published for 1980 and 1985-1989.[41] Foreign trade revenues thus can be estimated as the difference between total NMP and NMP at factor cost plus net taxes.[42] I estimated foreign trade revenues for other years as a residual NMP sector after accounting for value added in trade, distribution and other sectors (see Table A7).

In 1988, the foreign trade balance equaled 51.8 b.r., while foreign trade revenues and outlays equaled 62.6 and 26.0 b.r. respectively.[43] The foreign trade balance consisted of the import-export balance (50.4 b.r.) and foreign currency earnings (1.4 b.r.). In theory, the import-export balance exceeds net revenues from foreign trade by the difference between expenditures and revenues incurred in connection with foreign credit and other monetary operations. Credits and aid to Soviet client states accounts for most of this difference--13.8 b.r. in 1988.

Two major difficulties arise in integrating data on exports and imports in foreign trade and domestic prices. First, all data in foreign trade prices are compiled following the nomenclature of the Foreign Trade Ministry which differs in several respects from the NEB and i-o nomenclature which is used by Goskomstat and Gosplan. Second, Soviet officials have not published conversion coefficients for particular exports and imports since the late 1960s. These coefficients are needed to convert the detailed VT data from foreign trade to domestic prices.

Reconciliation of the two nomenclatures requires the following adjustments in the VT and Narkhoz data:

1. Food products must be grouped into raw agricultural and processed food products.

2. Consumer goods must be grouped into

chemical, MBMW, wood and paper, apparel and n.e.c. industry products.

3. Other exports must be disaggregated into other machinery (armaments) and other material production (precious stones).

4. Other imports must be disaggregated into other chemical, machinery and other material production.

The last two adjustments can be performed only approximately because no data on other exports and imports have ever been published in foreign trade prices. I made these adjustments using the i-o data on exports and imports of other material products in domestic prices. I assumed that the export conversion coefficient for these products is the same as for civilian and defense machinery--it decreased from 0.9 to 0.56 in 1970-1990. I also assumed that the import conversion coefficient for these products is around 2.00.

The only attempt to obtain detailed statistics on foreign trade coefficients was made by Treml and Kostinsky as part of a large-scale effort to reconstruct the 1972 i-o table.[44] Their work is practically impossible to update before Goskomstat officials make available the results of its last large i-o study completed for 1987. Despite this fact, the availability of the aggregate data on foreign trade in domestic and foreign trade prices for 1972 and 1988 combined with published physical output and price indexes make it possible to derive conversion coefficients at least for major sectors of the USSR economy.

Estimates of Soviet export and import activities are presented in Tables C1 and C2. Conversion coefficients were derived in both tables as the ratio between the value of traded goods in domestic and foreign trade prices. Export of fuels in domestic prices was derived by combining the i-o data with the Narkhoz data on exports of fuels in physical quantities of output. Data on other export conversion coefficients were derived by comparing published world, CEMA and domestic price indexes. I applied these coefficients to determine the value of

[41] Narkhoz for 1989, p. 6.
[42] Net taxes equal the difference between industrial GVO reported with and without net taxes.
[43] Pravitel'stvennyi Vestnik, N18, 1989, p. 6.

[44] V. Treml and B. Kostinsky, "Domestic Value of Foreign Trade: Exports and Imports in the 1972 Input-Output Table," Foreign Economic Report N20, U.S. Department of Commerce, Bureau of the Census, October 1982.

exports for individual sectors and then total exports in domestic prices.

Afterward, I estimated total imports in domestic prices using the following formula:

$$Fd = (Md - Ed) + (Eg - Mg)*e,$$

where Fd is total foreign trade revenues included in national income, Md and Eg are total imports and exports in domestic prices, Eg and Mg are exports and imports in foreign trade prices (in published gold rubles), and e is the combined conversion coefficient. I used the export conversion coefficient for all years, except for 1972, 1975-1976 and 1989-1990. Because for these years imports exceeded exports in gold rubles, I solved the above formula with e representing the import conversion coeffiecient.

I disaggregated total imports in domestic prices into individual sectors in five stages. First, I observed that the import conversion coefficient has remained approximately the same for machinery (around 1.00), apparel (around 4-4.5) and other consumer goods since the early 1970s. Second, I used published data on fuels, textiles, food and agricultural products in physical quantities of output. Third, I compared world, contract and domestic price indexes for metals, wood and chemical products. Fourth, I made the initial estimate of total imports as the sum of component parts. Fifth, I determined the difference between the initial estimate and the control total derived with the above formula and then distributed it proportionately among chemical, food and agricultural sectors for which initial estimates were least reliable.

Estimation results indicate that both export and import conversion coefficients dropped significantly in 1972-1973, 1974-1977 and 1979-1980 as a result of large world price increases which were followed by increases in CEMA contract prices. The reverse trend can be observed in 1981-1982 and 1985-1988 when domestic prices increased faster than world and CEMA prices. The overall export and import conversion coefficients decreased from 1.4 to 0.7 and from 2.4 to 1.5 respectively during the observed period.

Export conversion coefficients changed the following way for major sectors: fuels--from 1.27 to 0.42, metals--from 1.3 to 0.9, machinery--from 0.92 to 0.56, and wood products--from 1.5 to 1.35. Import conversion coefficients changed the following way for major sectors: metals--from 1.0 to 0.92, chemicals--from 2.46 to 1.85, wood products--from 3.0 to 1.7, processed food--4.0 to 2.1, agricultural products--2.4 to 2.1, all non-food consumer items--from 4.9 to 3.1.

Conclusion

The objective of the above discussion was to integrate published production, national and financial statistics for the purpose of compiling an internally consistent set of Soviet national accounts. In the course of the discussion I compared published excerpts from original Soviet national accounting tables and detected a number of inconsistencies in the estimation of value added and major end uses. I tried to reconcile these inconsistencies and offered an explanation as to why they exist.

In general, it appears that for some unknown reason Goskomstat officials make no effort to reconcile statistics compiled with different national accounting methods. This results in their inability to account for the total value of defense goods produced in the country, thus undermining the reliability of their estimates of the total size and structure of the USSR economy.

The above discussion suggests that to improve its organizational image Goskomstat officials will have to make a concerted effort to integrate statistics on the defense sector within the overall framework of Soviet national accounts. This pertains to published statistics on employment, average wages, value added in production and service sectors, and end uses of national income.

Soviet Data Base

I. Allocation of Resources

Part A

Sector-of-Origin Accounts

TABLE A1: INDUSTRIAL GVO BY SECTOR IN ENTERPRISE PRICES
      (billion rubles)

		July 1967 Prices							January 1975 Prices					Page 1
		1965	1970	1971	1972	1973	1974	1975	1975	1976	1977	1978	1979	1980
1	TOTAL GVO	239.0	360.0	387.7	412.9	443.9	479.6	515.5	505.5	529.8	560.0	586.8	606.8	628.6
2	annual growth	66.4	100.0	107.7	106.5	107.5	108.0	107.5	100.0	104.8	105.7	104.8	103.4	103.6
3	current prices	229.4	374.3	395.7	420.0	447.3	479.6	511.2	511.2	527.9	553.7	577.7	595.1	616.3
4	price index	0.96	1.04	1.02	1.02	1.01	1.00	0.99	1.01	1.00	0.99	0.98	0.98	0.98
6	'75/'67, '82/'81								0.98					
7	POWER	6.8	10.4	11.2	12.1	12.9	13.6	14.4	15.3	16.4	17.0	17.7	18.3	18.9
8	% to total	2.8	2.9	2.9	2.9	2.9	2.8	2.8	3.0	3.1	3.0	3.0	3.0	3.0
9	price index	0.90	1.10	1.10	1.10	1.10	1.10	1.10	1.04	1.04	1.04	1.04	1.04	1.04
10	current prices	6.1	11.5	12.4	13.4	14.2	15.0	15.9	15.9	17.1	17.7	18.4	19.0	19.7
11	'75/'67, '82/'81								1.06					
12	FUELS	16.8	22.3	23.7	24.9	26.2	27.8	29.5	30.1	31.7	33.0	34.0	34.6	35.3
13	% to total	7.0	6.2	6.1	6.0	5.9	5.8	5.7	6.0	6.0	5.9	5.8	5.7	5.6
14	price index	0.75	1.02	1.00	1.00	1.00	1.00	1.00	0.98	0.98	0.98	0.98	0.98	0.99
15	current prices	12.6	22.8	23.7	24.9	26.2	27.8	29.5	29.5	31.0	32.4	33.4	33.9	35.0
16	'75/'67, '82/'81								1.02					
17	CHEMICALS	12.0	21.6	23.7	26.2	28.9	32.1	35.4	33.4	36.0	38.6	40.5	41.9	44.1
18	% to total	5.0	6.0	6.1	6.3	6.5	6.7	6.9	6.6	6.8	6.9	6.9	6.9	7.0
19	price index	1.00	1.04	1.03	1.02	1.00	0.98	0.96	1.02	1.01	1.00	1.00	1.00	1.00
20	current prices	12.0	22.5	24.4	26.6	28.8	31.5	33.9	33.9	36.2	38.6	40.5	41.9	44.1
21	'75/'67, '82/'81								0.94					
22	PUBLISHED MBMW	47.7	82.8	92.4	102.8	115.1	128.9	143.5	121.3	133.0	145.0	157.5	169.1	180.7
23	% to total	19.9	23.0	23.8	24.9	25.9	26.9	27.8	24.0	25.1	25.9	26.8	27.9	28.7
24	price index	1.00	1.07	1.02	1.00	0.92	0.90	0.89	1.05	0.98	0.96	0.94	0.93	0.93
25	current prices	47.7	88.6	94.2	102.8	106.3	116.5	127.0	127.0	130.7	139.0	148.7	157.6	167.5
26	'75/'67, '82/'81								0.84					
27	WOOD AND PAPER	14.1	18.7	19.8	20.7	21.8	22.7	23.7	23.7	24.3	24.6	25.2	25.1	25.8
28	% to total	5.9	5.2	5.1	5.0	4.9	4.7	4.6	4.7	4.6	4.4	4.3	4.1	4.1
29	price index	0.85	1.04	1.03	1.02	1.01	1.01	1.00	1.00	1.00	1.00	1.00	1.01	1.01
30	current prices	12.0	19.5	20.3	21.2	22.0	22.8	23.7	23.7	24.3	24.6	25.2	25.2	25.9
31	'75/'67, '82/'81								1.00					
32	CONSTR MATERIALS	9.9	14.8	15.9	16.8	18.2	19.7	21.1	20.7	21.6	21.9	22.3	22.5	22.9
33	% to total	4.1	4.1	4.1	4.1	4.1	4.1	4.1	4.1	4.1	3.9	3.8	3.7	3.6
34	price index	0.87	1.00	0.98	0.98	0.98	0.98	0.98	1.00	1.00	1.00	1.00	1.00	1.00
35	current prices	8.6	14.8	15.6	16.5	17.8	19.3	20.7	20.7	21.6	21.9	22.3	22.5	22.9
36	'75/'67, '82/'81								1.00					
37	LIGHT INDUSTRY	40.6	61.2	65.1	66.8	69.5	72.9	76.7	84.4	87.9	91.3	94.1	96.5	100.2
38	% to total	17.0	17.0	16.8	16.2	15.7	15.2	14.9	16.7	16.6	16.3	16.0	15.9	15.9
39	price index	1.00	1.02	1.02	1.02	1.10	1.10	1.10	1.00	1.00	1.00	1.00	1.00	1.00
40	current prices	40.6	62.7	66.4	68.1	76.4	80.2	84.4	84.4	87.9	91.3	94.1	96.5	100.2
41	'75/'67, '82/'81								1.10					
42	FOOD INDUSTRY	57.6	76.3	80.3	82.8	87.0	93.5	98.2	103.6	101.7	106.6	109.0	111.6	112.3
43	% to total	24.1	21.2	20.7	20.1	19.6	19.5	19.0	20.5	19.2	19.0	18.6	18.4	17.9
44	price index	1.00	1.04	1.04	1.05	1.05	1.05	1.06	1.00	1.01	1.01	1.01	1.01	1.01
45	current prices	57.6	79.0	83.1	86.9	91.4	98.2	103.6	103.6	102.5	107.5	109.9	112.5	113.9
46	Food Products	51.8	71.1	74.9	78.3	82.4	88.7	93.7	93.7	92.2	96.3	98.3	100.6	101.7
47	Flour & Cereals	5.8	7.9	8.2	8.6	9.0	9.5	9.9	9.9	10.3	11.2	11.6	11.9	12.2
48	'75/'67, '82/'81								1.06					
49	TOTAL RESIDUAL	33.6	51.8	55.8	59.7	64.3	68.3	73.0	73.0	77.1	81.8	86.4	87.3	88.4
50	% to total	14.0	14.4	14.4	14.5	14.5	14.2	14.2	14.4	14.6	14.6	14.7	14.4	14.1
51	price index	0.96	1.02	1.00	1.00	1.00	1.00	0.99	0.99	0.99	0.99	0.99	0.99	0.99
52	current prices	32.2	53.1	55.8	59.7	64.3	68.3	72.5	72.5	76.5	80.7	85.2	86.1	87.1
53	'75/'67, '82/'81								1.00					

74

TABLE A1: INDUSTRIAL GVO BY SECTOR IN ENTERPRISE PRICES
        (billion rubles)

		January 1982 prices					Adjusted 1982 Prices				Page 2
		1981	1982	1983	1984	1985	1986	1987	1988	1989	1990
1	TOTAL GVO	650.0	721.5	751.3	781.8	811.5	837.0	868.5	903.1	918.3	927.0
2	annual growth	103.4	102.9	104.2	104.1	103.8	104.0	103.8	103.9	101.7	101.0
3	current prices	635.3	721.5	751.3	779.0	803.8	836.1	866.9	901.2	918.3	927.0
4	price index	0.98	1.00	1.00	1.00	0.99	1.00	1.00	1.00	1.00	1.00
6	'75/'67, '82/'81		1.11								
7	POWER	19.8	28.0	28.9	30.2	31.2	31.6	32.7	33.3	33.8	34.3
8	% to total	3.0	3.9	3.8	3.9	3.8	3.8	3.8	3.7	3.7	3.7
9	price index	1.04	0.97	0.97	0.97	0.97	1.00	1.00	1.00	1.00	1.00
10	current prices	20.6	27.0	27.9	29.2	30.1	31.6	32.7	33.3	33.8	34.3
11	'75/'67, '82/'81		1.24								
12	FUELS	35.9	57.6	58.9	59.5	59.7	60.3	62.2	63.5	64.3	64.9
13	% to total	5.5	8.0	7.8	7.6	7.4	7.2	7.2	7.0	7.0	7.0
14	price index	0.99	1.00	1.00	1.00	1.00	1.06	1.06	1.06	1.06	1.06
15	current prices	35.6	57.6	58.9	59.5	59.7	63.9	65.9	67.0	67.9	68.5
16	'75/'67, '82/'81		1.62								
17	CHEMICALS	46.5	46.5	49.1	51.6	53.1	55.2	57.3	59.3	60.3	60.9
18	% to total	7.2	6.4	6.5	6.6	6.5	6.6	6.6	6.6	6.6	6.6
19	price index	1.00	1.05	1.05	1.05	1.05	1.05	1.05	1.05	1.05	1.05
20	current prices	46.5	48.8	51.6	54.2	55.7	58.0	60.2	62.3	63.3	63.9
21	'75/'67, '82/'81		1.00								
22	PUBLISHED MBMW	190.6	183.1	194.1	206.7	222.6	236.9	249.0	262.1	268.1	271.6
23	% to total	29.3	25.4	25.8	26.4	27.4	28.3	28.7	29.0	29.2	29.3
24	price index	0.92	1.00	1.00	0.99	0.97	0.96	0.96	0.96	0.96	0.96
25	current prices	175.9	182.4	193.5	205.0	215.9	226.4	238.0	250.5	258.5	261.8
26	'75/'67, '82/'81		0.99								
27	WOOD AND PAPER	26.5	32.6	33.9	34.9	36.1	37.7	39.1	40.5	41.1	41.7
28	% to total	4.1	4.5	4.5	4.5	4.5	4.5	4.5	4.5	4.5	4.5
29	price index	1.01	1.04	1.04	1.04	1.04	1.04	1.04	1.04	1.04	1.04
30	current prices	26.7	33.9	35.2	36.3	37.6	39.2	40.6	42.2	42.8	43.4
31	'75/'67, '82/'81		1.23								
32	CONSTR MATERIALS	23.2	27.7	28.9	29.8	30.8	31.7	32.6	34.3	34.4	35.0
33	% to total	3.6	3.8	3.9	3.8	3.8	3.8	3.8	3.8	3.8	3.8
34	price index	1.00	0.96	0.96	0.96	0.96	0.96	0.96	0.96	0.96	0.96
35	current prices	23.2	26.6	27.8	28.6	29.6	30.5	31.3	32.9	33.1	33.6
36	'75/'67, '82/'81		1.22								
37	LIGHT INDUSTRY	102.7	113.3	114.3	115.1	118.5	119.7	120.7	124.9	125.3	126.1
38	% to total	15.8	15.7	15.2	14.7	14.6	14.3	13.9	13.8	13.7	13.6
39	price index	1.00	1.00	1.00	1.00	1.00	1.01	1.01	1.01	1.01	1.01
40	current prices	102.7	113.3	114.3	115.1	118.5	120.9	121.9	126.1	126.6	127.3
41	'75/'67, '82/'81		1.11								
42	FOOD INDUSTRY	113.8	123.9	131.0	135.7	137.5	138.4	143.8	147.5	151.1	152.0
43	% to total	17.5	17.2	17.4	17.4	16.9	16.5	16.6	16.3	16.5	16.4
44	price index	1.02	1.00	1.00	1.00	1.00	1.00	1.00	1.00	1.00	1.00
45	current prices	116.0	123.9	131.0	135.7	137.5	138.4	143.8	147.5	151.1	152.0
46	Food Products	104.0	111.4	118.3	122.8	123.8	123.8	128.8	131.9	135.2	135.8
47	Flour & Cereals	12.0	12.5	12.7	12.9	13.7	14.6	15.0	15.6	15.9	16.2
48	'75/'67, '82/'81		1.05								
49	TOTAL RESIDUAL	91.0	108.8	112.3	118.4	122.0	125.6	131.1	137.6	139.8	140.4
50	% to total	14.0	15.1	14.9	15.1	15.0	15.0	15.1	15.2	15.2	15.2
51	price index	0.97	0.99	0.99	0.98	0.98	1.01	1.01	1.01	1.01	1.01
52	current prices	88.3	108.0	111.1	115.5	119.2	127.3	132.4	139.2	141.3	142.0
53	'75/'67, '82/'81		1.16								

75

TABLE A2: INDUSTRIAL GVO RESIDUAL
        (billion rubles)

		1965	1970	1971	1972	1973	1974	1975	1976	1977	1978	1979	1980
1	TOTAL RESIDUAL	32.2	53.1	55.8	59.7	64.3	68.3	72.5	76.5	80.7	85.2	86.1	87.1
2	METALLURGY	21.6	36.4	38.0	40.4	42.7	45.2	47.8	50.7	53.2	54.9	54.9	54.8
3	growth rate	0.65	1.00	1.04	1.11	1.17	1.24	1.31					
								1.00	1.03	1.08	1.11	1.11	1.11
4	N.E.C. INDUSTRY	7.0	12.0	12.9	13.7	15.6	16.6	17.9	18.9	20.3	21.8	22.5	23.7
5	Group "A"	4.4	7.6	8.3	8.5	9.9	10.7	11.4	12.2	13.0	14.1	14.5	15.0
6	Group "B"	2.6	4.4	4.6	5.2	5.7	6.0	6.5	6.7	7.4	7.7	8.0	8.7
7	HIDDEN MBMW	3.6	4.7	4.9	5.6	6.0	6.4	6.8	6.9	7.2	8.5	8.7	8.6

Supporting Data on GVO of Other (N.E.C.) Industrial Sectors:

		1965	1970	1971	1972	1973	1974	1975	1976	1977	1978	1979	1980
8	Agri. Supplies	2.0	4.0	4.3	4.5	5.5	6.1	6.4	6.9	7.4	8.0	8.2	8.4
9	Processed Feed	1.4	3.0	3.2	3.4	4.2	4.6	4.9	5.2	5.7	6.1	6.3	6.5
10	Microbiology	0.4	0.7	0.8	0.8	0.9	1.0	1.0	1.1	1.1	1.2	1.2	1.2
11	Other	0.1	0.3	0.3	0.3	0.4	0.5	0.5	0.5	0.6	0.6	0.6	0.7
12	Glass and China	1.0	1.6	1.6	1.8	1.9	2.1	2.3	2.4	2.6	2.8	2.9	2.9
13	Producer	0.7	1.1	1.1	1.2	1.3	1.4	1.5	1.6	1.7	1.8	1.9	1.9
14	Consumer	0.3	0.5	0.5	0.6	0.6	0.7	0.8	0.8	0.9	1.0	1.0	1.0
15	a) China	0.2	0.3	0.3	0.4	0.4	0.5	0.5	0.5	0.5	0.6	0.6	0.6
16	b) Chrystal et	0.1	0.1	0.1	0.2	0.2	0.3	0.3	0.3	0.4	0.4	0.4	0.4
17	Pharmaceuticals	0.5	0.9	1.0	1.1	1.1	1.2	1.3	1.4	1.4	1.6	1.7	1.8
18	Producer	0.0	0.1	0.1	0.1	0.1	0.1	0.2	0.2	0.2	0.3	0.3	0.3
19	Consumer	0.5	0.8	0.9	1.0	1.0	1.1	1.1	1.2	1.2	1.3	1.4	1.5
20	Printing & Copy	0.7	1.0	1.1	1.1	1.2	1.2	1.3	1.3	1.4	1.5	1.6	1.7
21	Producer	0.4	0.6	0.7	0.7	0.7	0.7	0.8	0.8	0.8	0.9	0.9	1.0
22	Consumer	0.3	0.4	0.4	0.4	0.5	0.5	0.5	0.5	0.6	0.6	0.7	0.7
23	Water Supply	1.2	1.7	1.8	1.8	2.0	2.1	2.2	2.3	2.5	2.7	2.8	2.9
24	Producer	1.1	1.5	1.6	1.6	1.8	1.9	2.0	2.1	2.3	2.5	2.6	2.7
25	Consumer	0.1	0.2	0.2	0.2	0.2	0.2	0.2	0.2	0.2	0.2	0.2	0.2
26	Personal Items	1.4	2.5	2.6	3.0	3.3	3.5	3.8	4.0	4.4	4.6	4.8	5.2
27	Toys	0.3	0.6	0.6	0.7	0.7	0.7	0.8	0.8	0.9	0.9	1.0	1.0
28	Musical Instr.	0.1	0.2	0.2	0.2	0.2	0.2	0.2	0.2	0.2	0.2	0.2	0.3
29	Personal Care	0.3	0.5	0.5	0.5	0.6	0.6	0.7	0.7	0.7	0.8	0.8	0.8
30	Jewelry	0.2	0.5	0.6	0.9	1.1	1.2	1.3	1.5	1.7	1.8	1.9	2.2
31	Crafts	0.5	0.7	0.7	0.7	0.7	0.8	0.8	0.8	0.9	0.9	0.9	0.9
32	Other Producer	0.2	0.3	0.4	0.4	0.5	0.5	0.5	0.6	0.6	0.6	0.6	0.7

TABLE A2: INDUSTRIAL GVO RESIDUAL
       (billion rubles)

	1981	1982	1983	1984	1985	1986	1987	1988	1989	1990
1 TOTAL RESIDUAL	88.3	108.0	111.1	115.5	119.2	127.3	132.4	139.2	141.3	142.0
2 METALLURGY	54.8	67.8	69.8	71.9	74.2	79.6	81.3	84.1	84.9	85.5
3 growth rate										
	1.11	1.12	1.15	1.18	1.21	1.25	1.27	1.30	1.31	1.32
4 N.E.C. INDUSTRY	24.7	27.9	28.8	30.1	31.1	32.5	33.8	35.9	36.8	37.6
5 Group "A"	15.7	17.8	18.4	19.3	20.1	20.9	21.9	23.6	24.0	24.6
6 Group "B"	9.0	10.1	10.4	10.8	11.0	11.6	11.9	12.3	12.8	13.0
7 HIDDEN MBMW	8.8	12.3	12.5	13.5	13.9	15.2	17.3	19.2	19.5	19.0

Supporting Data on GVO of Other (N.E.C.) Industrial Sectors:

	1981	1982	1983	1984	1985	1986	1987	1988	1989	1990
8 Agri. Supplies	8.8	10.0	10.2	10.7	11.1	11.4	11.9	12.9	13.1	13.6
9 Processed Feed	6.8	7.8	8.0	8.4	8.7	9.0	9.4	10.2	10.4	10.7
10 Microbiology	1.3	1.4	1.4	1.4	1.5	1.5	1.6	1.7	1.7	1.8
11 Other	0.7	0.8	0.8	0.8	0.9	0.9	0.9	1.0	1.0	1.1
12 Glass and China	3.1	3.3	3.5	3.6	3.6	3.8	3.9	4.0	4.2	4.2
13 Producer	2.0	2.2	2.3	2.4	2.5	2.6	2.7	2.7	2.8	2.8
14 Consumer	1.1	1.1	1.2	1.2	1.1	1.2	1.2	1.3	1.4	1.4
15 a) China	0.7	0.7	0.8	0.8	0.7	0.7	0.7	0.8	0.8	0.8
16 b) Chrystal et	0.4	0.4	0.4	0.4	0.4	0.5	0.5	0.5	0.6	0.6
17 Pharmaceuticals	1.8	2.0	2.0	2.2	2.3	2.3	2.4	2.6	2.7	2.8
18 Producer	0.3	0.4	0.4	0.5	0.5	0.5	0.5	0.6	0.6	0.6
19 Consumer	1.5	1.6	1.6	1.7	1.8	1.8	1.9	2.0	2.1	2.2
20 Printing & Copy	1.8	2.1	2.2	2.3	2.4	2.6	2.7	2.9	2.9	2.9
21 Producer	1.0	1.2	1.3	1.3	1.4	1.5	1.6	1.7	1.7	1.7
22 Consumer	0.8	0.9	0.9	1.0	1.0	1.1	1.1	1.2	1.2	1.2
23 Water Supply	3.1	3.5	3.7	3.8	4.0	4.2	4.4	4.8	4.9	5.0
24 Producer	2.9	3.2	3.4	3.5	3.7	3.9	4.1	4.5	4.6	4.7
25 Consumer	0.2	0.3	0.3	0.3	0.3	0.3	0.3	0.3	0.3	0.3
26 Personal Items	5.5	6.2	6.4	6.6	6.8	7.2	7.4	7.5	7.8	7.9
27 Toys	1.1	1.2	1.3	1.4	1.5	1.6	1.7	1.7	1.9	2.0
28 Musical Instr.	0.3	0.4	0.4	0.5	0.5	0.5	0.5	0.5	0.5	0.5
29 Personal Care	0.9	1.1	1.2	1.3	1.4	1.5	1.5	1.6	1.7	1.8
30 Jewelry	2.3	2.5	2.5	2.4	2.4	2.5	2.6	2.5	2.5	2.4
31 Crafts	0.9	1.0	1.0	1.0	1.0	1.1	1.1	1.2	1.2	1.2
32 Other Producer	0.7	0.8	0.8	0.9	0.9	1.0	1.1	1.2	1.2	1.2

TABLE A3: DEPRECIATION IN MATERIAL PRODUCTION, SERVICE AND OTHER SECTORS
(billion rubles)

	1965	1970	1971	1972	1973	1974	1975	1976	1977	1978	1979	1980
1 TOTAL	29.6	44.5	49.3	52.7	58.1	63.0	72.6	78.2	81.6	87.7	93.4	101.8
2 Dep. Accounts	20.4	31.5	34.7	38.1	42.0	46.1	53.6	58.1	62.8	67.6	72.3	77.7
3 NEB Adjustment	8.1	11.2	12.1	12.6	13.9	14.6	16.3	17.4	16.0	17.2	18.2	21.1
4 Capital Losses	1.1	1.8	2.5	2.0	2.2	2.3	2.7	2.7	2.8	2.9	2.9	3.0
5 MATERIAL PROD	20.3	31.2	34.8	37.2	41.2	45.0	52.8	56.9	58.9	63.5	68.0	75.0
6 Dep. Accounts	18.9	29.1	32.0	34.9	38.7	42.4	49.7	53.8	55.7	60.2	64.6	71.6
7 NEB Adjustment	0.3	0.4	0.4	0.4	0.4	0.4	0.5	0.5	0.5	0.5	0.5	0.5
8 Capital Losses	1.1	1.7	2.4	1.9	2.1	2.2	2.6	2.6	2.7	2.8	2.9	2.9
9 INDUSTRY	11.1	16.8	18.7	20.1	22.2	24.2	29.2	31.5	31.6	34.1	36.6	41.8
10 Dep. Accounts	10.2	15.5	17.1	18.7	20.6	22.5	27.3	29.6	29.6	32.1	34.5	39.7
11 NEB Adjustment	0.3	0.4	0.4	0.4	0.4	0.4	0.5	0.5	0.5	0.5	0.5	0.5
12 Capital Losses	0.6	0.9	1.2	1.0	1.2	1.3	1.4	1.4	1.5	1.5	1.6	1.6
13 AGRICULTURE	3.9	5.7	6.5	6.9	7.6	8.4	9.5	10.3	11.2	12.1	12.9	13.6
14 Dep. Accounts	3.6	5.3	5.9	6.4	7.1	7.9	8.9	9.7	10.6	11.4	12.2	12.9
15 State Farms	2.0	2.9	3.3	3.6	4.0	4.5	5.2	5.8	6.4	6.9	7.5	7.9
16 Collectives	1.6	2.4	2.6	2.8	3.1	3.4	3.7	3.9	4.2	4.5	4.7	5.0
17 Capital Losses	0.3	0.4	0.6	0.5	0.5	0.5	0.6	0.6	0.6	0.7	0.7	0.7
18 PROD T&C	2.6	4.0	4.4	4.7	5.2	5.7	6.4	6.9	7.3	7.9	8.4	9.1
19 Dep. Accounts	2.5	3.9	4.2	4.7	5.1	5.6	6.2	6.7	7.1	7.7	8.2	8.9
20 NEB Adjustment												
21 Capital Losses	0.1	0.1	0.2	0.1	0.1	0.1	0.2	0.2	0.2	0.2	0.2	0.2
22 CONSTRUCTION	1.6	2.9	3.3	3.5	3.9	4.2	4.6	4.9	5.3	5.7	6.1	6.4
23 Dep. Accounts	1.5	2.7	3.0	3.3	3.7	4.0	4.3	4.6	5.0	5.4	5.8	6.1
24 Construction	1.3	2.3	2.6	2.9	3.2	3.5	3.7	4.0	4.4	4.7	5.1	5.4
25 Oil & Gas Exp	0.2	0.4	0.4	0.4	0.5	0.5	0.6	0.6	0.6	0.7	0.7	0.7
26 Capital Losses	0.1	0.2	0.3	0.2	0.2	0.2	0.3	0.3	0.3	0.3	0.3	0.3
27 TRADE, etc.	1.0	1.7	1.8	1.8	2.1	2.3	2.8	3.0	3.2	3.4	3.7	3.8
28 Dep. Accounts	1.0	1.6	1.7	1.7	2.0	2.2	2.7	2.9	3.1	3.3	3.6	3.7
29 Capital Losses	0.0	0.1	0.1	0.1	0.1	0.1	0.1	0.1	0.1	0.1	0.1	0.1
30 OTHER PROD	0.1	0.1	0.1	0.1	0.2	0.2	0.3	0.3	0.3	0.3	0.3	0.3
31 SERVICES	8.6	12.5	13.5	14.7	15.9	17.0	18.7	20.2	21.5	23.1	24.2	25.6
32 Dep. Accounts	1.5	2.4	2.7	3.2	3.3	3.7	3.9	4.3	7.1	7.4	7.7	6.2
33 NEB Adjustment	7.1	10.1	10.9	11.4	12.6	13.3	14.8	15.9	14.4	15.6	16.5	19.5
34 Non-Mat. T&C	1.1	1.6	1.8	2.1	2.3	2.5	2.7	3.0	3.2	3.5	3.7	3.9
35 Dep. Accounts	1.1	1.5	1.7	1.9	2.0	2.2	2.5	2.7	2.9	3.1	3.4	3.6
36 NEB Adjustment	0.0	0.1	0.2	0.2	0.2	0.2	0.2	0.3	0.3	0.3	0.3	0.4
37 Communal, etc.	0.6	0.8	0.9	1.0	1.1	1.3	1.5	1.7	1.8	2.0	2.1	2.3
38 Education	1.0	1.6	1.7	1.8	2.0	2.2	2.6	2.9	3.1	3.3	3.5	3.7
39 Culture & Arts	0.0	0.1	0.1	0.1	0.1	0.1	0.2	0.2	0.2	0.3	0.3	0.4
40 Health	0.3	0.5	0.6	0.7	0.8	0.8	0.9	1.0	1.1	1.2	1.3	1.4
41 Science	0.4	0.7	0.8	0.9	1.0	1.1	1.3	1.5	1.7	1.9	2.0	2.1
42 Administration	0.5	0.8	0.9	1.0	1.2	1.3	1.5	1.7	1.9	2.1	2.3	2.5
43 Housing	4.7	6.4	6.7	7.1	7.4	7.7	8.0	8.2	8.5	8.8	9.0	9.3
44 OTHER SECTORS	0.6	0.8	0.9	0.9	1.0	1.0	1.1	1.1	1.2	1.2	1.2	1.3

TABLE A3: DEPRECIATION IN MATERIAL PRODUCTION, SERVICE AND OTHER SECTORS
      (billion rubles)   Page 2

		1981	1982	1983	1984	1985	1986	1987	1988	1989	1990
1	TOTAL	108.8	114.9	122.8	130.8	139.3	148.1	159.9	166.3	173.5	182.3
2	Dep. Accounts	83.5	89.5	96.1	102.7	109.9	117.1	125.7	133.6	140.4	148.5
3	NEB Adjustment	22.3	23.0	24.4	25.6	27.0	27.5	30.2	28.5	29.0	29.4
4	Capital Losses	3.0	2.4	2.3	2.5	2.4	3.5	4.0	4.1	4.1	4.4
5	MATERIAL PROD	80.6	85.1	91.4	97.7	107.9	115.0	124.5	128.8	134.3	141.6
6	Dep. Accounts	77.1	82.2	88.6	94.7	101.7	107.5	114.2	118.0	123.2	129.9
7	NEB Adjustment	0.6	0.6	0.6	0.6	3.9	4.1	6.5	6.8	7.1	7.4
8	Capital Losses	2.9	2.3	2.2	2.4	2.3	3.4	3.9	4.0	4.0	4.3
9	INDUSTRY	45.1	47.8	51.2	55.0	58.7	62.6	68.7	71.9	75.3	79.5
10	Dep. Accounts	42.9	46.2	49.7	53.3	57.1	60.2	63.9	66.9	70.0	73.7
11	NEB Adjustment	0.6	0.6	0.6	0.6	0.6	0.7	3.0	3.2	3.5	3.8
12	Capital Losses	1.6	1.0	0.9	1.1	1.0	1.7	1.8	1.8	1.8	2.0
13	AGRICULTURE	14.8	15.7	16.9	17.9	19.2	20.2	21.4	20.1	20.7	21.7
14	Dep. Accounts	14.1	15.0	16.2	17.2	18.5	19.2	20.2	18.8	19.4	20.3
15	State Farms	8.8	9.4	10.2	10.9	11.8	12.1	12.6	10.8	11.2	11.8
16	Collectives	5.3	5.6	6.0	6.3	6.7	7.1	7.6	8.0	8.2	8.5
17	Capital Losses	0.7	0.7	0.7	0.7	0.7	1.0	1.2	1.3	1.3	1.4
18	PROD T&C	9.7	10.2	10.9	11.7	15.9	16.9	17.8	19.0	19.8	20.8
19	Dep. Accounts	9.5	10.0	10.7	11.5	12.4	13.3	14.0	15.1	15.9	16.9
20	NEB Adjustment					3.3	3.4	3.5	3.6	3.6	3.6
21	Capital Losses	0.2	0.2	0.2	0.2	0.2	0.2	0.3	0.3	0.3	0.3
22	CONSTRUCTION	6.7	7.1	7.7	8.1	8.7	9.3	11.2	11.9	12.5	13.3
23	Dep. Accounts	6.4	6.8	7.4	7.8	8.4	8.9	10.7	11.4	12.0	12.8
24	Construction	5.7	6.1	6.6	7.0	7.5	8.0	9.7	10.4	11.0	11.7
25	Oil & Gas Exp	0.7	0.7	0.8	0.8	0.9	0.9	1.0	1.0	1.0	1.1
26	Capital Losses	0.3	0.3	0.3	0.3	0.3	0.4	0.5	0.5	0.5	0.5
27	TRADE, etc.	3.9	3.9	4.2	4.5	4.8	5.4	4.8	5.0	5.2	5.4
28	Dep. Accounts	3.8	3.8	4.1	4.4	4.7	5.3	4.7	4.9	5.1	5.3
29	Capital Losses	0.1	0.1	0.1	0.1	0.1	0.1	0.1	0.1	0.1	0.1
30	OTHER PROD	0.4	0.4	0.5	0.5	0.6	0.6	0.7	0.8	0.8	0.9
31	SERVICES	26.9	28.4	30.0	31.6	29.8	31.5	33.6	35.6	37.3	38.7
32	Dep. Accounts	6.4	7.3	7.5	8.0	8.2	9.6	11.5	15.6	17.2	18.6
33	NEB Adjustment	20.5	21.2	22.5	23.6	21.6	21.9	22.1	20.0	20.1	20.1
34	Non-Mat. T&C	4.2	4.6	4.9	5.2	5.4	5.5	6.0	6.3	6.6	6.7
35	Dep. Accounts	3.8	4.1	4.3	4.6	4.7	4.8	5.0	5.2	5.4	5.5
36	NEB Adjustment	0.4	0.5	0.6	0.6	0.7	0.7	1.0	1.1	1.2	1.2
37	Communal, etc.	2.4	2.6	2.7	2.9	3.1	3.2	3.4	3.6	3.8	3.9
38	Education	3.9	4.1	4.3	4.4	4.5	4.7	5.0	5.3	5.5	5.7
39	Culture & Arts	0.4	0.4	0.4	0.4	0.4	0.4	0.5	0.5	0.5	0.5
40	Health	1.5	1.6	1.7	1.8	1.9	2.0	2.1	2.2	2.3	2.4
41	Science	2.2	2.4	2.6	2.8	3.0	3.3	3.6	4.0	4.3	4.6
42	Administration	2.7	2.9	3.1	3.4	0.4	0.4	0.5	0.6	0.6	0.6
43	Housing	9.6	9.9	10.3	10.7	11.1	12.0	12.5	13.1	13.7	14.3
44	OTHER SECTORS	1.3	1.4	1.4	1.5	1.5	1.6	1.7	1.8	1.9	2.0

TABLE A4: WAGES AND OTHER LABOR INCOME OF MATERIAL PRODUCTION SECTORS

(billion rubles)

	1965	1970	1971	1972	1973	1974	1975	1976	1977	1978	1979	1980
1 TOTAL	100.7	141.6	149.4	155.4	164.1	173.9	181.4	192.7	200.5	209.1	216.4	225.0
2 Wages	75.1	110.9	117.1	123.8	130.1	138.5	146.4	155.4	162.3	169.3	175.7	183.4
3 State-Coop	63.9	96.1	101.9	108.2	113.4	121.3	129.5	137.4	143.6	150.2	156.4	163.9
4 NEB Adjustment	-0.6	-0.9	-0.8	-0.8	-0.8	-0.8	-0.8	-0.7	-0.7	-0.7	-0.7	-0.8
5 Collectives	11.8	15.7	16.0	16.4	17.5	18.1	17.7	18.6	19.4	19.9	20.0	20.3
6 Other Earnings	1.9	2.9	3.1	3.5	3.7	3.9	4.3	4.6	4.9	5.0	5.3	5.5
7 Private Income	23.7	27.9	29.2	28.1	30.3	31.4	30.7	32.7	33.4	34.7	35.4	36.1
8 INDUSTRY	35.5	52.0	54.8	57.5	60.6	64.9	69.2	74.1	76.9	80.2	82.9	86.2
9 Wages	34.9	51.0	53.6	56.1	59.0	63.2	67.3	71.9	74.6	77.7	80.3	83.5
10 State-Coop	34.3	50.5	53.0	55.4	58.2	62.3	66.3	70.8	73.4	76.4	79.0	82.1
11 NEB Adjustment	-0.6	-0.8	-0.7	-0.7	-0.7	-0.7	-0.7	-0.6	-0.6	-0.6	-0.6	-0.6
12 Collectives	0.6	0.5	0.6	0.7	0.8	0.9	1.0	1.1	1.2	1.3	1.3	1.4
13 Other Earnings	1.2	1.8	1.9	2.1	2.3	2.4	2.6	2.8	2.9	3.1	3.2	3.3
14 AGRICULTURE	38.0	48.9	50.9	51.1	54.6	57.1	56.7	60.7	62.7	65.0	66.2	68.0
15 Wages	17.8	25.0	25.9	27.1	28.6	30.1	30.5	32.5	34.0	35.2	35.9	37.0
16 State Farms	7.3	10.4	11.2	12.1	12.8	14.0	14.9	16.2	17.1	18.0	18.6	19.6
17 Collective	10.0	13.8	13.9	14.1	14.8	15.1	14.5	15.1	15.6	15.8	15.9	15.9
18 Hired Workers	0.5	0.8	0.8	0.9	1.0	1.0	1.1	1.2	1.3	1.4	1.4	1.5
19 Other Earnings	0.1	0.2	0.2	0.3	0.3	0.3	0.3	0.4	0.4	0.4	0.4	0.5
20 Private Income	20.1	23.7	24.8	23.6	25.7	26.8	25.9	27.8	28.3	29.4	30.0	30.5
21 T & C	5.5	7.9	8.6	9.2	9.6	10.5	11.3	12.1	12.6	13.0	13.9	14.8
22 Wages	5.4	7.8	8.4	9.0	9.4	10.3	11.0	11.8	12.3	12.7	13.6	14.5
23 Other Earnings	0.1	0.2	0.2	0.2	0.2	0.2	0.3	0.3	0.3	0.3	0.3	0.3
24 CONSTRUCTION	14.9	22.5	24.3	25.8	27.0	28.5	30.5	31.5	33.1	34.3	36.0	37.2
25 Wages	11.5	18.3	19.9	21.3	22.4	23.8	25.6	26.5	27.8	28.9	30.3	31.4
26 State-Coop.	9.9	16.3	17.7	19.0	19.8	21.0	22.5	23.2	24.3	25.2	26.4	27.3
27 Transport	0.9	1.4	1.5	1.6	1.7	1.8	2.0	2.1	2.2	2.3	2.5	2.7
28 NEB Adjustment												
29 Collectives	0.7	0.6	0.7	0.7	0.9	1.0	1.1	1.2	1.3	1.4	1.4	1.5
30 Other Earnings	0.4	0.6	0.6	0.7	0.7	0.8	0.9	0.9	1.0	1.1	1.2	1.2
31 Private Income	3.0	3.6	3.8	3.8	3.9	3.9	4.0	4.1	4.2	4.4	4.5	4.6
32 TRADE & DISTRIB.	5.6	8.7	9.2	10.0	10.4	10.9	11.7	12.2	13.0	14.1	14.8	16.1
33 Wages	5.5	8.7	9.1	9.9	10.3	10.8	11.6	12.1	12.9	14.0	14.7	16.1
34 NEB Adjustment	0.0	-0.1	-0.1	-0.1	-0.1	-0.1	-0.1	-0.1	-0.1	-0.1	-0.1	-0.2
35 Other Earnings	0.1	0.1	0.2	0.2	0.2	0.2	0.2	0.2	0.2	0.2	0.2	0.2
36 OTHER SECTORS	1.3	1.6	1.6	1.8	1.9	1.9	2.1	2.2	2.3	2.5	2.5	2.7
37 Wages	0.7	1.0	1.0	1.1	1.2	1.2	1.3	1.3	1.4	1.5	1.6	1.6
38 State-Coop	0.7	1.0	1.0	1.1	1.1	1.2	1.3	1.3	1.4	1.5	1.6	1.6
39 Collectives	0.0	0.0	0.0	0.0	0.0	0.0	0.0	0.0	0.0	0.0	0.0	0.0
40 Other Earnings	0.0	0.0	0.0	0.0	0.0	0.0	0.0	0.0	0.0	0.0	0.0	0.0
41 Private Income	0.6	0.6	0.6	0.7	0.7	0.7	0.8	0.8	0.9	0.9	0.9	1.0

Supporting Data on Employment in T&C Sectors Serving Material Production:

	1965	1970	1971	1972	1973	1974	1975	1976	1977	1978	1979	1980
42 TOTAL T & C	5.0	5.5	5.8	5.9	6.0	6.1	6.3	6.4	6.5	6.6	6.8	6.9
43 Transportation	4.1	4.4	4.5	4.6	4.6	4.7	4.8	4.9	5.0	5.1	5.2	5.3
44 Communications	0.2	0.3	0.4	0.4	0.5	0.5	0.5	0.5	0.5	0.5	0.5	0.5
45 Construction	0.7	0.8	0.9	0.9	0.9	0.9	1.0	1.0	1.0	1.0	1.1	1.1

TABLE A4: WAGES AND OTHER LABOR INCOME OF MATERIAL PRODUCTION SECTORS
(billion rubles)   Page 2

		1981	1982	1983	1984	1985	1986	1987	1988	1989	1990
1	TOTAL	231.6	242.4	252.4	261.7	271.4	282.2	289.8	304.6	334.3	354.6
2	Wages	188.6	196.9	203.9	211.4	218.2	228.0	235.9	246.9	275.9	293.7
3	State-Coop	169.0	176.5	181.6	187.3	193.6	201.0	206.5	219.4	241.5	256.4
4	NEB Adjustment	-1.0	-1.2	-1.4	-0.5	-0.9	0.0	2.3	1.0	6.0	7.3
5	Collectives	20.6	21.6	23.7	24.6	25.4	27.0	27.0	26.6	28.3	30.0
6	Other Earnings	5.7	6.0	6.3	6.5	6.9	6.8	6.1	6.3	6.6	6.8
7	Private Income	37.3	39.4	42.2	43.8	46.4	47.4	47.7	51.4	51.9	54.1
8	INDUSTRY	88.7	92.5	94.7	97.4	100.1	102.6	105.6	110.9	121.3	128.7
9	Wages	86.0	89.8	92.0	94.8	97.8	100.6	103.3	109.8	117.6	124.9
10	State-Coop	84.6	88.4	90.5	93.3	96.3	99.0	101.6	108.1	115.8	123.0
11	NEB Adjustment	-0.8	-1.0	-1.2	-1.5	-2.0	-2.2	-1.2	-2.5		
12	Collectives	1.4	1.4	1.5	1.5	1.5	1.6	1.6	1.7	1.8	1.9
13	Other Earnings	3.5	3.7	3.9	4.1	4.3	4.2	3.5	3.6	3.7	3.8
14	AGRICULTURE	69.8	73.5	79.5	83.0	86.2	90.3	94.1	97.4	102.9	107.0
15	Wages	37.7	39.7	43.1	45.2	46.7	49.2	50.1	50.4	55.9	58.5
16	State Farms	20.1	21.2	22.6	23.8	24.6	25.6	26.6	27.5	31.3	32.5
17	Collective	16.0	16.8	18.6	19.4	20.0	21.4	21.3	22.0	23.6	25.0
18	Hired Workers	1.6	1.7	1.9	2.0	2.1	2.2	2.2	0.9	1.0	1.0
19	Other Earnings	0.5	0.5	0.5	0.5	0.5	0.5	0.5	0.5	0.5	0.5
20	Private Income	31.6	33.3	35.9	37.3	39.0	40.6	43.5	46.5	46.5	48.0
21	T & C	15.3	16.0	16.4	16.8	17.1	17.8	16.9	17.7	18.6	19.5
22	Wages	15.0	15.7	16.1	16.5	16.8	17.5	16.5	17.4	18.3	19.2
23	Other Earnings	0.3	0.3	0.3	0.3	0.3	0.3	0.3	0.3	0.3	0.3
24	CONSTRUCTION	38.4	40.6	41.6	43.8	45.8	49.4	50.6	54.4	63.4	69.5
25	Wages	32.7	34.3	35.2	37.2	39.0	42.5	46.3	49.5	57.9	63.3
26	State-Coop.	28.3	29.5	30.3	31.1	32.7	34.9	37.2	44.0	50.0	54.0
27	Transport	2.8	3.2	3.3	3.3	3.4	3.5	3.6			
28	NEB Adjustment				1.2	1.3	2.4	3.7	3.7	6.0	7.3
29	Collectives	1.5	1.6	1.6	1.6	1.6	1.7	1.8	1.8	1.9	2.0
30	Other Earnings	1.1	1.3	1.3	1.4	1.5	1.5	1.5	1.7	1.8	1.9
31	Private Income	4.6	5.0	5.1	5.2	5.3	5.4	2.8	3.2	3.7	4.3
32	TRADE & DISTRIB.	16.5	16.8	17.0	17.5	18.0	18.5	19.0	20.0	23.7	25.2
33	Wages	16.5	16.8	17.0	17.5	18.0	18.5	19.0	20.0	23.5	25.0
34	NEB Adjustment	-0.2	-0.2	-0.2	-0.2	-0.2	-0.2	-0.2	-0.2		
35	Other Earnings	0.2	0.2	0.2	0.2	0.2	0.2	0.2	0.2	0.2	0.2
36	OTHER SECTORS	2.9	2.9	3.1	3.2	3.4	3.5	3.5	4.2	4.5	4.7
37	Wages	1.8	1.8	1.9	1.9	2.0	2.1	2.1	2.5	2.7	2.8
38	State-Coop	1.7	1.7	1.8	1.8	1.9	2.0	2.0	2.4	2.6	2.7
39	Collectives	0.1	0.1	0.1	0.1	0.1	0.1	0.1	0.1	0.1	0.1
40	Other Earnings	0.0	0.0 `	0.0	0.0	0.0	0.0	0.0	0.0	0.1	0.1
41	Private Income	1.1	1.1	1.2	1.3	1.4	1.4	1.4	1.6	1.7	1.8

Supporting Data on Employment in T&C Sectors Serving Material Production:

		1981	1982	1983	1984	1985	1986	1987	1988	1989	1990
42	TOTAL T & C	6.9	7.1	7.1	7.2	7.2	7.1	6.5	5.3	5.2	5.1
43	Transportation	5.3	5.4	5.4	5.4	5.4	5.4	4.9	4.9	4.8	4.7
44	Communications	0.5	0.5	0.5	0.5	0.5	0.5	0.4	0.4	0.4	0.4
45	Construction	1.1	1.2	1.2	1.2	1.2	1.2	1.1			

81

TABLE A5: WAGES AND OTHER LABOR INCOME OF SERVICE SECTORS
        (billion rubles)

		1965	1970	1971	1972	1973	1974	1975	1976	1977	1978	1979	1980
1	TOTAL	25.0	35.9	37.9	40.3	43.3	46.0	47.9	49.7	52.0	55.2	57.8	60.9
2	State Wages	24.1	34.5	36.4	38.7	41.5	44.2	46.1	47.8	50.1	53.2	55.8	58.9
3	Collectives	0.3	0.3	0.3	0.3	0.3	0.3	0.3	0.3	0.3	0.3	0.3	0.3
4	Other Earnings	0.6	1.1	1.2	1.3	1.4	1.4	1.5	1.6	1.6	1.7	1.7	1.7
5	TRANSPORTATION	2.9	4.3	4.6	4.9	5.1	5.5	5.9	6.2	6.4	6.6	6.8	7.0
6	Wages	2.8	4.2	4.5	4.7	5.0	5.4	5.7	6.0	6.2	6.4	6.6	6.8
7	Other Earnings	0.1	0.1	0.1	0.1	0.1	0.1	0.2	0.2	0.2	0.2	0.2	0.2
8	COMMUNICATIONS	0.7	1.3	1.3	1.3	1.5	1.6	1.7	1.9	2.0	2.1	2.1	2.2
9	Wages	0.7	1.2	1.2	1.2	1.4	1.5	1.6	1.8	1.9	2.0	2.0	2.1
10	Other Earnings	0.0	0.1	0.1	0.1	0.1	0.1	0.1	0.1	0.1	0.1	0.1	0.1
11	HOUS-COM., etc.	2.1	3.6	3.8	4.1	4.4	4.8	5.1	5.4	5.8	6.3	6.8	7.1
12	Wages	2.1	3.5	3.7	4.1	4.3	4.7	5.0	5.3	5.7	6.2	6.7	7.1
13	I-O Adjustment				-0.1	-0.1	-0.1	-0.1	-0.1	-0.1	-0.1	-0.1	-0.2
14	Other Earnings	0.0	0.1	0.1	0.1	0.2	0.2	0.2	0.2	0.2	0.2	0.2	0.2
15	EDUCATION	7.7	10.0	10.5	11.2	12.2	12.7	13.0	13.4	13.8	14.6	15.0	15.8
16	State Wages	7.2	9.4	9.8	10.5	11.5	12.0	12.3	12.7	13.0	13.8	14.2	15.0
17	Other State	0.2	0.3	0.3	0.3	0.4	0.4	0.4	0.4	0.4	0.4	0.5	0.5
18	Collectives	0.2	0.2	0.2	0.2	0.2	0.2	0.2	0.2	0.2	0.2	0.2	0.2
19	Other Earnings	0.0	0.1	0.1	0.1	0.1	0.1	0.1	0.1	0.1	0.1	0.1	0.1
20	CULTURE & ARTS	0.9	1.4	1.5	1.6	1.6	1.7	1.8	2.0	2.1	2.2	2.3	2.5
21	Culture	0.5	0.8	0.9	1.0	1.0	1.1	1.2	1.3	1.4	1.5	1.6	1.7
22	Arts	0.4	0.5	0.5	0.5	0.5	0.5	0.5	0.6	0.6	0.6	0.6	0.7
23	Other Earnings	0.0	0.0	0.1	0.1	0.1	0.1	0.1	0.1	0.1	0.1	0.1	0.1
24	HEALTH, etc.	4.2	5.8	6.1	6.4	6.8	7.0	7.3	7.6	8.0	8.6	9.1	9.6
25	State	4.1	5.6	5.9	6.2	6.6	6.8	7.1	7.4	7.8	8.4	8.9	9.4
26	Collectives	0.1	0.1	0.1	0.1	0.1	0.1	0.1	0.1	0.1	0.1	0.1	0.1
27	Other Earnings	0.0	0.1	0.1	0.1	0.1	0.1	0.1	0.1	0.1	0.1	0.1	0.1
28	SCIENCE	4.6	6.6	7.1	7.7	8.2	8.9	9.3	9.4	9.7	10.3	11.0	11.6
29	State Wages	3.6	5.3	5.7	6.1	6.6	7.2	7.5	7.5	7.8	8.3	8.9	9.4
30	Other State	0.6	0.8	0.9	1.0	1.0	1.1	1.2	1.2	1.2	1.3	1.4	1.5
31	Other Earnings	0.4	0.5	0.5	0.6	0.6	0.6	0.6	0.7	0.7	0.7	0.7	0.7
32	ADMINISTRATION	2.0	2.9	3.1	3.2	3.5	3.7	3.8	3.8	4.2	4.6	4.7	5.1
33	State Administ.	1.9	2.6	2.8	3.0	3.2	3.4	3.5	3.5	3.8	4.1	4.3	4.7
34	I-O Adjustment	-0.3	-0.4	-0.4	-0.5	-0.5	-0.6	-0.6	-0.6	-0.7	-0.8	-0.9	-1.0
35	Credit & Insur.	0.3	0.6	0.6	0.6	0.7	0.8	0.8	0.8	1.0	1.1	1.1	1.2
36	Other Earnings	0.1	0.1	0.1	0.1	0.1	0.1	0.1	0.1	0.1	0.2	0.2	0.2

Supporting Data on Employment in Transportation and Communications Sectors:

		1965	1970	1971	1972	1973	1974	1975	1976	1977	1978	1979	1980
37	NON-MATERIAL T&C	3.3	3.9	3.9	3.9	4.0	4.1	4.2	4.2	4.3	4.3	4.4	4.5
38	Transportation	2.2	2.5	2.5	2.6	2.6	2.6	2.7	2.7	2.7	2.8	2.8	2.8
39	Communication	0.8	1.0	1.0	1.0	1.1	1.1	1.1	1.1	1.2	1.2	1.2	1.2
40	Space	0.3	0.4	0.4	0.4	0.4	0.4	0.4	0.4	0.4	0.4	0.5	0.5

TABLE A5: WAGES AND OTHER LABOR INCOME OF SERVICE SECTORS
     (billion rubles)

		1981	1982	1983	1984	1985	1986	1987	1988	1989	1990
1	TOTAL	63.1	65.3	66.7	68.6	70.8	73.4	77.9	83.3	90.1	96.7
2	State Wages	61.0	63.0	64.4	66.2	68.4	70.9	75.3	80.5	87.3	93.9
3	Collectives	0.3	0.3	0.3	0.3	0.3	0.4	0.4	0.4	0.4	0.4
4	Other Earnings	1.8	2.0	2.0	2.1	2.1	2.1	2.2	2.4	2.4	2.4
5	TRANSPORTATION	7.4	7.6	7.7	7.8	7.9	8.2	8.5	9.0	9.6	10.1
6	Wages	7.1	7.3	7.4	7.5	7.6	7.9	8.1	8.6	9.2	9.7
7	Other Earnings	0.3	0.3	0.3	0.3	0.3	0.3	0.4	0.4	0.4	0.4
8	COMMUNICATIONS	2.3	2.3	2.4	2.4	2.5	2.5	2.7	2.9	3.1	3.4
9	Wages	2.2	2.2	2.3	2.3	2.4	2.4	2.6	2.8	3.0	3.3
10	Other Earnings	0.1	0.1	0.1	0.1	0.1	0.1	0.1	0.1	0.1	0.1
11	HOUS-COM., etc	7.5	7.8	8.0	8.3	8.6	8.9	9.4	10.0	11.0	12.0
12	Wages	7.5	7.8	8.0	8.2	8.6	8.9	9.4	10.0	11.0	12.0
13	I-O Adjustment	-0.2	-0.2	-0.2	-0.2	-0.3	-0.3	-0.3	-0.3	-0.3	-0.3
14	Other Earnings	0.2	0.2	0.2	0.3	0.3	0.3	0.3	0.3	0.3	0.3
15	EDUCATION	16.1	16.6	17.1	17.5	18.7	20.0	21.8	23.2	24.3	25.4
16	State Wages	15.3	15.7	16.1	16.6	17.7	18.9	20.7	22.0	23.1	24.2
17	Other State	0.5	0.5	0.5	0.5	0.5	0.6	0.6	0.7	0.7	0.7
18	Collectives	0.2	0.2	0.2	0.2	0.3	0.3	0.3	0.3	0.3	0.3
19	Other Earnings	0.1	0.2	0.2	0.2	0.2	0.2	0.2	0.2	0.2	0.2
20	CULTURE & ARTS	2.6	2.7	2.7	2.8	2.9	2.9	3.2	3.4	3.8	4.2
21	Culture	1.8	1.9	1.9	2.0	2.0	2.0	2.2	2.3	2.5	2.7
22	Arts	0.7	0.7	0.7	0.7	0.8	0.8	0.9	1.0	1.2	1.4
23	Other Earnings	0.1	0.1	0.1	0.1	0.1	0.1	0.1	0.1	0.1	0.1
24	HEALTH, etc.	9.9	10.2	10.5	10.8	10.9	11.4	12.2	13.5	14.9	15.4
25	State	9.7	10.0	10.3	10.6	10.8	11.2	12.0	13.3	14.7	15.2
26	Collectives	0.1	0.1	0.1	0.1	0.0	0.1	0.1	0.1	0.1	0.1
27	Other Earnings	0.1	0.1	0.1	0.1	0.1	0.1	0.1	0.1	0.1	0.1
28	SCIENCE	12.0	12.6	12.8	13.2	13.5	13.7	14.1	15.1	16.8	18.3
29	State Wages	9.8	10.2	10.4	10.7	11.0	11.2	11.5	12.4	14.0	15.5
30	Other State	1.5	1.6	1.6	1.7	1.7	1.7	1.8	1.8	1.9	1.9
31	Other Earnings	0.7	0.8	0.8	0.8	0.8	0.8	0.8	0.9	0.9	0.9
32	ADMINISTRATION	5.3	5.5	5.6	5.8	5.8	5.8	6.0	6.2	6.6	7.9
33	State Administ	4.9	5.0	5.1	5.2	5.3	5.1	4.5	4.5	4.8	6.0
34	I-O Adjustment	-1.1	-1.1	-1.1	-1.1	-1.2	-1.1	-0.4	-0.4	-0.4	-0.4
35	Credit & Insur	1.3	1.4	1.4	1.5	1.5	1.6	1.7	1.8	1.9	2.0
36	Other Earnings	0.2	0.2	0.2	0.2	0.2	0.2	0.2	0.3	0.3	0.3

Supporting Data on Trasportation and Communication Sectors:

		1981	1982	1983	1984	1985	1986	1987	1988	1989	1990
37	Labor	4.2	4.3	4.3	4.3	4.4	4.4	4.4	4.4	4.4	4.5
38	Transportation	2.9	2.9	2.9	2.9	2.9	2.9	2.9	2.9	2.9	2.9
39	Communication	1.2	1.2	1.2	1.2	1.2	1.2	1.2	1.2	1.2	1.2

TABLE A6: GROSS AND NET MATERIAL PRODUCT BY SECTOR

(billion rubles)   Page 1

	1965	1970	1971	1972	1973	1974	1975	1976	1977	1978	1979	1980
1 TOTAL GSP	420.0	643.5	685.3	717.4	770.9	816.7	862.6	903.9	949.6	995.7	1032	1079
2 Total "c"	226.5	353.6	380.3	403.8	433.1	462.7	499.3	518.2	544.0	569.4	591.7	616.2
3 Materials	206.2	322.4	345.5	366.7	391.9	417.7	446.5	461.3	485.1	505.9	523.7	541.3
4 Depreciation	20.3	31.2	34.8	37.2	41.2	45.0	52.8	56.9	58.9	63.5	68.0	75.0
5 TOTAL NMP	193.5	289.9	305.0	313.6	337.8	354.0	363.3	385.7	405.6	426.3	440.7	462.3
6 Total "v"	100.7	141.6	149.4	155.4	164.1	173.9	181.4	192.7	200.5	209.1	216.4	225.0
7 Total "m"	92.8	148.3	155.6	158.2	173.7	180.1	181.9	193.0	205.1	217.2	224.3	237.3
8 NMP factor cost	150.1	243.8	254.0	262.2	280.0	292.2	297.4	311.5	327.0	338.7	344.7	349.5
9 INDUSTRY	266.0	409.0	434.3	458.4	490.1	525.6	558.3	578.4	605.5	636.3	657.1	685.5
10 Total "c"	165.9	260.7	277.4	294.8	316.8	339.3	367.1	378.7	398.5	416.7	430.6	447.4
11 Materials	154.8	243.9	258.7	274.7	294.6	315.1	337.9	347.2	366.9	382.6	394.0	405.6
12 Depreciation	11.1	16.8	18.7	20.1	22.2	24.2	29.2	31.5	31.6	34.1	36.6	41.8
13 NMP	100.1	148.3	156.9	163.6	173.3	186.3	191.2	199.7	207.0	219.6	226.5	238.1
14 Total "v"	35.5	52.0	54.8	57.5	60.6	64.9	69.2	74.1	76.9	80.2	82.9	86.2
15 Total "m"	64.6	96.3	102.1	106.1	112.7	121.4	122.0	125.6	130.1	139.4	143.6	151.9
16 AGRICULTURE	71.0	103.8	108.1	108.8	121.9	122.1	122.3	132.4	141.7	147.0	151.9	152.6
17 Total "c"	27.4	40.7	45.2	49.2	53.5	56.5	60.8	66.2	70.1	73.4	78.7	83.6
18 Materials	23.5	35.0	38.7	42.3	45.9	48.1	51.3	55.9	58.9	61.3	65.8	70.0
19 Depreciation	3.9	5.7	6.5	6.9	7.6	8.4	9.5	10.3	11.2	12.1	12.9	13.6
20 NMP	43.6	63.1	62.9	59.6	68.4	65.6	61.5	66.2	71.6	73.6	73.2	69.0
21 Total "v"	38.0	48.9	50.9	51.0	54.6	57.1	56.7	60.7	62.7	65.0	66.2	68.0
22 Total "m"	5.6	14.2	12.0	8.6	13.8	8.5	4.8	5.5	8.9	8.6	7.0	1.0
23 T and C	18.0	25.7	27.7	29.5	31.7	34.1	36.7	38.6	41.1	43.7	45.2	47.6
24 Total "c"	6.5	9.4	10.2	11.0	11.9	12.7	13.7	14.7	16.0	17.8	19.0	20.6
25 Materials	3.9	5.4	5.8	6.3	6.7	7.0	7.4	7.8	8.7	9.9	10.6	11.5
26 Depreciation	2.6	4.0	4.4	4.7	5.2	5.7	6.3	6.9	7.3	7.9	8.4	9.1
27 NMP	11.5	16.3	17.5	18.5	19.8	21.4	23.0	23.9	25.1	25.9	26.2	27.0
28 Total "v"	5.5	7.9	8.6	9.2	9.6	10.5	11.3	12.1	12.6	13.0	13.9	14.8
29 Total "m"	6.0	8.4	8.9	9.3	10.2	10.9	11.7	11.8	12.5	12.9	12.3	12.2
30 CONSTRUCTION	40.0	67.6	74.7	77.4	80.9	86.4	91.7	94.2	96.2	99.6	101.1	103.4
31 Total "c"	22.1	37.6	41.7	42.7	44.7	47.5	50.4	50.8	51.6	53.2	54.4	55.8
32 Materials	20.5	34.7	38.4	39.2	40.9	43.3	45.9	45.9	46.3	47.5	48.3	49.4
33 Depreciation	1.6	2.9	3.3	3.5	3.8	4.2	4.5	4.9	5.3	5.7	6.1	6.4
34 NMP	17.9	30.0	33.0	34.7	36.2	38.9	41.3	43.4	44.6	46.4	46.7	47.6
35 Total "v"	14.9	22.5	24.3	25.8	27.0	28.5	30.5	31.5	33.1	34.3	36.0	37.2
36 Total "m"	3.0	7.5	8.7	8.9	9.2	10.4	10.8	11.9	11.5	12.1	10.7	10.4
37 TRADE, etc.	25.0	37.4	40.5	43.3	46.3	48.5	53.6	60.3	65.1	69.1	77.1	89.4
38 Total "c"	4.6	5.3	5.8	6.1	6.2	6.7	7.3	7.8	7.8	8.3	8.5	8.8
39 Materials	3.5	3.5	3.9	4.2	3.9	4.2	4.2	4.5	4.3	4.6	4.5	4.7
40 Depreciation	1.1	1.8	1.9	1.9	2.3	2.5	3.1	3.3	3.5	3.7	4.0	4.1
41 NMP	20.4	32.2	34.7	37.2	40.1	41.8	46.3	52.5	57.3	60.8	68.6	80.6
42 Total "v"	6.8	10.3	10.8	11.8	12.3	12.8	13.8	14.4	15.3	16.6	17.3	18.8
43 T and D	5.6	8.7	9.2	10.0	10.4	10.9	11.7	12.2	13.0	14.1	14.8	16.1
44 Other Prod	1.3	1.6	1.6	1.8	1.9	1.9	2.1	2.2	2.3	2.5	2.5	2.7
45 Total "m"	13.6	21.9	23.9	25.4	27.8	29.0	32.5	38.1	42.0	44.2	51.3	61.8
46 T and D	6.6	9.9	10.9	11.7	12.0	12.2	12.7	13.4	14.0	14.1	16.0	16.9
47 Other Prod	0.2	0.5	0.5	0.7	0.9	1.0	1.0	1.1_	1.2	1.2	1.3	1.4
48 Foreign Trade	6.8	11.4	12.4	13.0	15.0	15.8	18.8	23.7	26.8	29.0	34.0	43.5

84

TABLE A6: GROSS AND NET MATERIAL PRODUCT BY SECTOR
        (billion rubles)   Page 2

		1981	1982	1983	1984	1985	1986	1987	1988	1989	1990
1	TOTAL GSP	1123	1236	1293	1346	1384	1426	1465	1525	1569	1606
2	Total "c"	636.1	712.6	744.6	776.2	805.1	838.4	864.9	894.2	912.2	929.5
3	Materials	555.5	627.5	653.2	678.5	697.2	723.4	740.4	765.3	777.9	787.9
4	Depreciation	80.6	85.1	91.4	97.7	107.9	115.0	124.5	128.8	134.3	141.6
5	TOTAL NMP	486.7	523.4	548.1	569.6	578.5	587.4	599.6	630.8	656.8	676.0
6	Total "v"	231.6	242.4	252.4	261.7	271.4	282.2	289.8	304.6	334.3	354.6
7	Total "m"	255.1	281.0	295.7	307.9	307.1	305.2	309.8	326.2	322.5	321.4
8	NMP factor cost	363.1	399.1	442.6	462.8	475.0	502.8	519.7	567.9	589.2	611.2
9	INDUSTRY	709.0	792.7	800.0	826.7	844.6	862.5	892.3	912.4	930.0	940.0
10	Total "c"	461.0	525.9	545.9	564.5	581.5	604.5	623.7	642.9	652.9	657.5
11	Materials	415.9	478.1	494.7	509.5	522.8	541.9	555.0	571.0	577.6	578.0
12	Depreciation	45.1	47.8	51.2	55.0	58.7	62.6	68.7	71.9	75.3	79.5
13	NMP	248.0	266.8	254.1	262.2	263.1	258.0	268.6	269.5	277.1	282.5
14	Total "v"	88.7	92.5	94.7	97.4	100.1	102.6	105.6	110.9	121.3	128.7
15	Total "m"	159.3	174.3	159.4	164.8	163.0	155.4	163.0	158.6	155.8	153.8
16	AGRICULTURE	160.0	170.3	207.9	217.0	219.5	232.6	234.9	259.7	270.1	283.0
17	Total "c"	86.9	90.0	97.8	104.3	106.7	111.4	112.3	116.4	120.1	128.0
18	Materials	72.1	74.3	80.9	86.4	87.5	91.2	90.9	96.2	99.4	106.3
19	Depreciation	14.8	15.7	16.9	17.9	19.2	20.2	21.4	20.1	20.7	21.7
20	NMP	73.1	80.3	110.1	112.7	112.8	121.2	122.6	143.3	150.0	155.0
21	Total "v"	69.8	73.5	79.5	83.0	86.2	90.3	94.1	97.4	102.9	107.0
22	Total "m"	3.3	6.8	30.6	29.7	26.6	30.9	28.5	45.9	47.1	48.0
23	T and C	49.8	55.2	58.0	59.5	66.0	68.8	70.3	73.5	76.0	80.0
24	Total "c"	21.7	23.6	24.7	25.5	31.0	32.3	33.7	34.8	36.5	39.0
25	Materials	12.0	13.4	13.8	13.8	15.1	15.4	15.9	15.8	16.7	18.2
26	Depreciation	9.7	10.2	10.9	11.7	15.9	16.9	17.8	19.0	19.8	20.8
27	NMP	28.1	31.6	33.3	34.0	35.0	36.5	36.6	38.7	39.5	41.0
28	Total "v"	15.3	16.0	16.4	16.8	17.1	17.8	16.9	17.7	18.6	19.5
29	Total "m"	12.8	15.6	16.9	17.2	17.9	18.7	19.7	21.0	20.9	21.5
30	CONSTRUCTION	106.4	115.1	119.3	132.3	136.3	147.9	155.9	165.5	170.0	179.0
31	Total "c"	57.4	63.2	66.1	71.6	74.0	77.6	81.2	84.9	87.0	89.0
32	Materials	50.7	56.1	58.4	63.5	65.3	68.3	70.0	73.0	74.5	75.7
33	Depreciation	6.7	7.1	7.7	8.1	8.7	9.3	11.2	11.9	12.5	13.3
34	NMP	49.0	51.9	53.2	60.7	62.3	70.3	74.7	80.6	83.0	90.0
35	Total "v"	38.4	40.6	41.6	43.8	45.8	49.4	50.6	54.4	63.4	69.5
36	Total "m"	10.6	11.3	11.6	16.9	16.5	20.9	24.1	26.2	19.6	20.5
37	TRADE, etc.	97.6	102.7	107.5	110.3	117.2	114.0	111.1	113.9	122.9	123.5
38	Total "c"	9.1	9.9	10.1	10.3	11.9	12.6	14.0	15.2	15.7	16.0
39	Materials	4.8	5.5	5.4	5.3	6.5	6.6	8.5	9.4	9.7	9.7
40	Depreciation	4.3	4.4	4.7	5.0	5.4	6.0	5.5	5.8	6.0	6.3
41	NMP	88.5	92.8	97.4	100.0	105.3	101.4	97.1	98.7	107.2	107.5
42	Total "v"	19.4	19.7	20.1	20.7	21.4	22.0	22.6	24.2	28.2	29.9
43	T and D	16.5	16.8	17.0	17.5	18.0	18.5	19.0	20.0	23.7	25.2
44	Other Prod	2.9	2.9	3.1	3.2	3.4	3.5	3.5	4.2	4.5	4.7
45	Total "m"	69.1	73.1	77.3	79.3	83.9	79.4	74.5	74.6	79.0	77.6
46	T and D	17.4	18.1	18.6	18.1	19.0	19.0	22.8	19.8	20.0	20.9
47	Other Prod	1.6	1.8	1.8	2.1	2.1	2.2	2.3	3.0	3.3	3.3
48	Foreign Trade	50.0	53.1	56.8	59.1	62.8	58.2	49.5	51.8	55.7	53.5

TABLE A6: GROSS AND NET MATERIAL PRODUCT
    (billion rubles)   Page 3

		1965	1970	1971	1972	1973	1974	1975	1976	1977	1978	1979	1980
1	POWER	6.1	11.5	12.4	13.4	14.2	15.0	15.9	17.1	17.7	18.4	19.0	19.7
2	Materials	3.3	5.2	5.6	6.2	6.7	7.0	7.4	7.9	8.1	8.4	8.7	8.9
3	Depreciation	1.2	1.7	1.9	2.0	2.2	2.4	3.0	3.1	3.3	3.5	3.7	3.9
4	Wages	0.7	1.1	1.1	1.2	1.3	1.4	1.4	1.5	1.6	1.6	1.7	1.8
5	Revenues	0.9	3.5	3.8	4.0	4.0	4.2	4.1	4.6	4.7	4.9	4.9	5.1
6	FUELS	12.6	22.8	23.7	24.9	26.2	27.8	29.5	31.0	32.4	33.4	33.9	35.0
7	Materials	8.3	10.9	11.1	11.9	12.6	13.7	15.1	15.8	16.6	17.4	18.0	18.7
8	Depreciation	1.6	2.4	2.6	2.7	2.9	3.1	3.6	3.9	4.3	4.7	5.0	5.6
9	Wages	3.4	3.8	3.9	3.9	4.0	4.2	4.3	4.7	4.9	5.1	5.3	5.5
10	Revenues	-0.6	5.7	6.1	6.5	6.7	6.9	6.5	6.6	6.6	6.2	5.6	5.1
11	METALLURGY	21.6	36.4	38.0	40.4	42.7	45.2	47.8	50.7	53.2	54.9	54.9	54.8
12	Materials	14.3	23.3	24.5	26.5	28.0	29.4	31.5	33.2	35.5	36.8	37.2	37.0
13	Depreciation	1.7	2.2	2.4	2.5	2.8	3.1	3.7	4.1	4.4	4.6	4.9	5.2
14	Wages	3.3	4.1	4.3	4.5	4.7	4.9	5.2	5.3	5.5	5.6	5.8	6.0
15	Revenues	2.3	6.8	6.8	6.9	7.3	7.8	7.4	8.1	7.8	7.9	7.0	6.6
16	CHEMICALS	12.0	22.5	24.4	26.6	28.8	31.5	33.9	36.2	38.6	40.5	41.9	44.1
17	Materials	7.6	14.2	15.3	16.6	17.6	19.0	20.1	21.5	22.9	24.0	25.0	26.3
18	Depreciation	0.7	1.4	1.6	1.7	1.9	2.1	2.8	3.1	3.4	3.7	4.2	4.7
19	Wages	1.7	2.7	2.9	3.0	3.1	3.3	3.6	3.8	4.0	4.2	4.3	4.4
20	Revenues	2.0	4.2	4.6	5.3	6.2	7.1	7.4	7.8	8.3	8.6	8.4	8.7
21	TOTAL MBMW	51.3	93.3	99.1	108.4	112.2	123.0	133.9	137.6	146.2	157.2	166.2	176.1
22	Materials	24.0	50.6	54.0	58.7	61.5	67.1	74.5	76.0	83.8	90.2	93.4	97.4
23	Depreciation	2.5	4.1	4.6	5.0	5.6	6.1	7.2	7.7	5.8	6.7	7.2	10.2
24	Wages	12.9	19.6	21.4	22.8	24.3	26.3	28.5	31.0	32.2	34.3	35.5	37.2
25	Revenues	11.9	19.0	19.1	21.9	20.7	23.5	23.7	22.9	24.3	26.0	30.1	31.3
26	WOOD & PAPER	12.0	19.5	20.3	21.2	22.0	22.8	23.7	24.3	24.6	25.2	25.2	25.9
27	Materials	7.0	10.4	10.9	11.5	11.9	12.6	13.1	13.7	13.8	14.3	14.3	14.7
28	Depreciation	0.9	1.2	1.4	1.6	1.7	1.8	2.1	2.2	2.3	2.4	2.6	2.7
29	Wages	3.4	4.9	5.0	5.2	5.5	5.7	6.0	6.2	6.3	6.4	6.5	6.6
30	Revenues	0.7	3.0	3.0	2.9	2.9	2.7	2.5	2.2	2.2	2.1	1.8	1.9
31	CONSTR. MATER.	8.6	14.8	15.6	16.5	17.8	19.3	20.7	21.6	21.9	22.3	22.5	22.9
32	Materials	5.2	8.4	8.9	9.4	10.3	11.1	12.1	12.7	13.0	13.4	13.8	14.0
33	Depreciation	0.8	1.2	1.3	1.4	1.5	1.7	2.0	2.1	2.3	2.4	2.5	2.7
34	Wages	2.1	3.4	3.5	3.7	3.9	4.2	4.4	4.6	4.7	4.8	4.9	5.0
35	Revenues	0.5	1.8	1.9	2.0	2.1	2.3	2.2	2.2	1.9	1.7	1.3	1.2
36	LIGHT INDUSTRY	40.6	62.7	66.4	68.1	76.4	80.2	84.4	87.9	91.3	94.1	96.5	100.2
37	Materials	32.3	48.6	51.9	53.5	61.2	64.2	67.7	70.4	73.3	75.5	77.5	80.3
38	Depreciation	0.4	0.7	0.7	0.8	0.9	0.9	1.1	1.2	1.3	1.4	1.5	1.6
39	Wages	4.1	6.4	6.5	6.7	6.9	7.4	7.9	8.4	8.7	9.0	9.3	9.7
40	Revenues	3.8	7.0	7.2	7.1	7.4	7.7	7.7	7.9	8.0	8.2	8.2	8.6
41	FOOD INDUSTRY	57.6	79.0	83.1	86.9	91.4	98.2	103.6	102.5	107.5	109.9	112.5	113.9
42	Materials	48.3	65.3	69.1	72.5	75.4	81.0	85.3	84.3	87.4	89.1	92.0	93.4
43	Depreciation	1.0	1.4	1.5	1.6	1.8	2.0	2.5	2.8	3.1	3.2	3.4	3.5
44	Wages	2.8	4.3	4.4	4.6	4.7	5.1	5.5	5.9	6.0	6.1	6.3	6.5
45	Revenues	5.5	8.0	8.1	8.2	9.5	10.1	10.3	9.5	11.0	11.5	10.8	10.5
46	N.E.C.INDUSTRY	7.0	12.0	12.9	13.7	15.6	16.6	17.9	18.9	20.3	21.8	22.5	23.7
47	Materials	4.5	6.9	7.4	7.9	9.3	9.9	11.0	11.7	12.5	13.5	14.1	14.9
48	Depreciation	0.3	0.6	0.7	0.8	0.9	1.0	1.3	1.3	1.4	1.5	1.6	1.7
49	Wages	1.1	1.8	1.9	2.0	2.2	2.4	2.5	2.7	3.0	3.1	3.3	3.5
50	Revenues	1.1	2.7	2.9	3.0	3.2	3.3	3.2	3.2	3.4	3.7	3.5	3.6

TABLE A6: GROSS AND NET MATERIAL PRODUCT
        (billion rubles)   Page 4

		1981	1982	1983	1984	1985	1986	1987	1988	1989	1990
1	POWER	20.6	27.0	27.9	29.2	30.1	31.6	32.7	33.3	33.8	34.3
2	Materials	9.2	13.4	13.6	14.0	14.2	14.6	14.9	17.2	17.4	17.6
3	Depreciation	4.1	4.5	4.8	5.2	5.5	6.1	6.7	7.0	7.3	7.6
4	Wages	1.8	1.9	2.0	2.1	2.2	2.3	2.4	2.5	2.6	2.8
5	Revenues	5.5	7.2	7.5	7.9	8.2	8.6	8.7	6.6	6.5	6.3
6	FUELS	35.6	57.6	58.9	59.5	59.7	63.9	65.9	67.0	67.9	68.5
7	Materials	19.4	27.7	28.0	28.2	28.5	28.8	29.3	31.9	32.1	32.3
8	Depreciation	6.0	6.4	7.2	7.8	8.4	9.3	10.0	10.5	11.0	11.5
9	Wages	5.8	6.3	6.4	6.5	6.7	6.8	7.0	7.1	7.4	7.7
10	Revenues	4.4	17.2	17.3	17.0	16.1	19.0	19.6	17.5	17.4	17.0
11	METALLURGY	54.8	67.8	69.8	71.9	74.2	79.6	81.3	84.1	84.9	85.5
12	Materials	37.3	44.1	44.7	45.8	47.5	52.2	52.7	53.2	53.5	53.0
13	Depreciation	5.4	5.6	5.9	6.2	6.5	7.0	7.6	8.0	8.4	8.9
14	Wages	6.1	6.3	6.5	6.7	6.8	6.9	7.0	7.2	7.5	7.8
15	Revenues	6.0	11.8	12.7	13.2	13.4	13.5	14.0	15.7	15.5	15.8
16	CHEMICALS	46.5	48.8	51.6	54.2	55.7	58.0	60.2	62.3	63.3	63.9
17	Materials	27.6	32.0	33.7	35.2	35.9	36.6	37.8	38.3	38.8	38.0
18	Depreciation	5.2	5.5	6.0	6.3	6.5	6.8	7.0	7.2	7.5	7.7
19	Wages	4.6	4.7	4.8	4.9	5.1	5.3	5.5	5.7	6.0	6.2
20	Revenues	9.1	6.6	7.1	7.8	8.2	9.3	9.9	11.1	11.0	12.0
21	TOTAL MBMW	184.6	194.7	206.0	218.5	229.8	241.6	255.3	269.7	278.0	280.8
22	Materials	100.5	116.6	121.0	126.8	131.9	137.6	142.5	148.3	152.2	151.4
23	Depreciation	11.3	12.1	12.6	13.7	15.3	16.0	19.5	20.8	21.7	23.3
24	Wages	38.3	40.0	41.1	42.4	43.5	44.5	46.0	49.7	57.8	62.6
25	Revenues	34.5	26.0	31.3	35.7	39.1	43.5	47.3	51.0	46.3	43.5
26	WOOD & PAPER	26.7	33.9	35.2	36.3	37.6	39.2	40.6	42.2	42.8	43.4
27	Materials	15.1	18.9	19.7	20.4	21.2	22.2	23.2	23.5	23.7	24.2
28	Depreciation	2.9	3.0	3.2	3.4	3.5	3.6	3.7	3.8	3.9	4.1
29	Wages	6.8	7.1	7.2	7.3	7.5	7.7	7.8	8.0	8.3	8.6
30	Revenues	1.9	4.9	5.1	5.2	5.4	5.7	5.9	6.9	6.9	6.5
31	CONSTR. MATER.	23.2	26.6	27.8	28.6	29.6	30.5	31.3	32.9	33.1	33.6
32	Materials	14.1	15.8	16.4	16.7	17.3	17.7	18.0	18.6	18.8	19.0
33	Depreciation	2.8	3.0	3.3	3.5	3.6	3.8	3.9	4.0	4.2	4.4
34	Wages	5.2	5.4	5.5	5.7	5.9	6.1	6.3	6.5	6.7	7.0
35	Revenues	1.1	2.4	2.6	2.7	2.8	2.9	3.1	3.8	3.4	3.2
36	LIGHT INDUSTRY	102.7	113.3	114.3	115.1	118.5	119.7	120.7	124.9	125.3	126.1
37	Materials	81.9	90.2	90.9	91.4	93.2	95.0	95.4	96.0	96.1	96.3
38	Depreciation	1.7	1.8	1.9	2.1	2.2	2.4	2.5	2.6	2.8	3.0
39	Wages	9.8	10.0	10.1	10.3	10.6	10.8	11.1	11.3	11.6	12.0
40	Revenues	9.2	11.3	11.4	11.3	12.5	11.5	11.7	15.0	14.8	14.8
41	FOOD INDUSTRY	116.0	123.9	131.0	135.7	137.5	138.4	143.8	147.5	151.1	152.0
42	Materials	94.8	101.1	107.4	110.8	112.3	115.5	118.3	119.8	120.0	120.9
43	Depreciation	3.8	4.0	4.3	4.7	5.0	5.3	5.4	5.6	5.9	6.3
44	Wages	6.6	6.9	7.1	7.3	7.5	7.6	7.8	7.9	8.1	8.4
45	Revenues	10.8	11.9	12.2	12.9	12.7	10.0	12.3	14.2	17.1	16.4
46	N.E.C.INDUSTRY	24.7	27.9	28.8	30.1	31.1	32.5	33.8	35.9	36.8	37.6
47	Materials	16.0	18.3	19.3	20.2	20.8	21.7	22.9	24.2	25.0	25.3
48	Depreciation	1.8	1.9	2.0	2.1	2.2	2.3	2.4	2.5	2.6	2.7
49	Wages	3.6	3.9	4.0	4.2	4.4	4.6	4.7	5.0	5.3	5.6
50	Revenues	3.3	3.8	3.5	3.6	3.7	3.9	3.8	4.2	3.9	4.0

TABLE A7: PRODUCTION REVENUES
(billion rubles)

	1965	1970	1971	1972	1973	1974	1975	1976	1977	1978	1979	1980
1 TOTAL	92.8	148.3	155.6	158.2	173.7	180.1	181.9	193.0	205.1	217.2	224.3	237.3
2 Net Profit	41.3	88.2	89.3	90.9	96.0	99.2	98.0	99.5	103.7	107.0	105.5	101.7
3 State-Coop.	34.6	84.8	87.1	90.4	94.7	99.4	100.8	102.6	106.6	111.2	111.8	112.5
4 Bonus Wages	0.0	5.2	5.9	6.3	7.0	7.7	8.4	8.9	9.3	9.5	9.9	10.3
5 Collectives	6.7	8.6	8.1	6.8	8.3	7.5	5.6	5.8	6.4	5.3	3.6	-0.5
6 Net Taxes	36.6	34.7	38.6	38.4	42.8	46.0	47.1	50.5	51.8	58.6	62.0	69.2
7 Foreign Trade	6.8	11.4	12.4	13.0	15.0	15.8	18.8	23.7	26.8	29.0	34.0	43.5
8 Social Sec.	4.8	7.6	8.3	8.8	9.0	9.7	10.2	10.7	11.1	11.9	12.2	12.5
9 State	4.2	6.4	6.8	7.2	7.4	8.0	8.4	9.0	9.4	9.9	10.3	10.7
10 Collectives	0.6	1.2	1.5	1.6	1.6	1.7	1.8	1.7	1.7	2.0	1.9	1.8
11 Other Revenues	3.4	6.5	7.0	7.2	10.9	9.4	7.8	8.7	11.7	10.8	10.7	10.4
12 INDUSTRY	64.6	96.3	102.1	106.1	112.7	121.4	122.0	125.6	130.1	139.4	143.6	151.9
13 Net Profit	24.1	56.0	55.8	58.9	59.2	63.3	64.6	63.2	65.7	69.2	68.6	71.5
14 Published	22.5	56.0	56.2	59.4	60.0	64.2	65.9	64.8	67.2	70.7	70.3	73.3
15 Bonus Wages	0.0	3.7	4.1	4.4	4.8	5.2	5.7	5.9	6.1	6.2	6.4	6.7
16 Other	1.1	2.8	2.8	3.0	3.0	3.2	3.3	3.2	3.4	3.5	3.5	3.7
17 Public Orgs.	0.3	0.5	0.5	0.5	0.6	0.6	0.6	0.6	0.7	0.7	0.7	0.7
18 Collectives	0.2	0.4	0.4	0.4	0.4	0.5	0.5	0.5	0.5	0.5	0.5	0.5
19 Net Taxes	36.6	34.7	38.6	38.4	42.8	46.0	47.1	50.5	51.8	58.6	62.0	69.2
20 Social Sec.	2.5	3.8	4.0	4.2	4.4	4.7	5.0	5.3	5.5	5.8	5.9	6.2
21 State	2.5	3.7	3.9	4.1	4.3	4.6	4.9	5.2	5.4	5.7	5.8	6.1
22 Collectives	0.0	0.1	0.1	0.1	0.1	0.1	0.1	0.1	0.1	0.1	0.1	0.1
23 Other Revenues	1.4	1.8	3.7	4.6	6.3	7.4	5.3	6.6	7.1	5.8	7.1	5.0
24 AGRICULTURE	5.6	14.2	12.0	8.6	13.8	8.5	4.8	5.5	8.9	8.6	7.0	1.0
25 Net Profit	5.9	12.9	11.9	8.5	11.6	9.0	4.2	6.6	6.8	6.3	4.5	-2.3
26 State Farms	-0.6	4.9	4.6	2.5	4.1	2.5	-0.5	1.9	1.4	2.1	2.1	-0.5
27 Bonus Wages	0.0	0.4	0.6	0.6	0.7	0.8	0.8	1.0	1.1	1.2	1.3	1.3
28 Collectives	6.5	8.2	7.7	6.4	7.9	7.0	5.1	5.3	5.9	4.8	3.1	-1.0
29 Other	0.0	0.2	0.2	0.2	0.3	0.3	0.4	0.4	0.6	0.6	0.6	0.5
30 Social Sec.	1.0	1.6	1.9	2.1	2.1	2.3	2.4	2.3	2.4	2.7	2.7	2.6
31 State	0.4	0.5	0.5	0.6	0.6	0.7	0.7	0.7	0.8	0.8	0.9	0.9
32 Collectives	0.6	1.1	1.4	1.5	1.5	1.6	1.7	1.6	1.6	1.9	1.8	1.7
33 Other Revenues	-1.3	-0.3	-1.8	-2.0	0.1	-2.8	-1.7	-3.4	-0.2	-0.4	-0.3	0.8
34 T AND C	6.0	8.4	8.9	9.3	10.2	10.9	11.7	11.8	12.5	12.9	12.3	12.2
35 Net Profit	5.4	7.5	7.9	8.3	9.0	9.7	10.5	10.4	11.1	11.5	11.0	10.7
36 Total Profit	5.4	7.9	8.4	8.8	9.6	10.3	11.2	11.1	11.9	12.3	11.8	11.6
37 Bonus Wages	0.0	0.4	0.5	0.5	0.6	0.6	0.7	0.7	0.8	0.8	0.8	0.9
38 Social Sec.	0.4	0.5	0.6	0.6	0.6	0.6	0.6	0.7	0.7	0.8	0.8	0.8
39 Other Revenues	0.2	0.4	0.4	0.5	0.6	0.6	0.6	0.7	0.7	0.6	0.6	0.7
40 CONSTRUCTION	3.0	7.5	8.7	8.9	9.2	10.4	10.8	11.9	11.5	12.1	10.7	10.4
41 Net Profit	1.8	4.5	5.5	6.0	6.5	7.1	8.0	8.0	8.2	7.8	7.3	6.8
42 Published	1.6	4.7	5.6	6.2	6.8	7.6	8.6	8.5	8.7	8.2	7.8	7.3
43 Bonus Wages	0.0	0.6	0.6	0.7	0.8	1.0	1.1	1.1	1.1	1.1	1.2	1.2
44 Oil-Gas Expl.	0.2	0.4	0.5	0.5	0.5	0.5	0.5	0.6	0.6	0.7	0.7	0.7
45 Social Sec.	0.7	1.2	1.3	1.4	1.4	1.5	1.6	1.7	1.8	1.9	2.0	2.1
46 Other Revenues	0.5	1.8	1.9	1.5	1.3	1.8	1.2	2.2	1.5	2.4	1.4	1.5

TABLE A7: PRODUCTION REVENUES
   (billion rubles)

	1981	1982	1983	1984	1985	1986	1987	1988	1989	1990
1 TOTAL	255.1	281.0	295.7	307.9	307.1	305.2	309.8	326.2	322.5	321.4
2 Net Profit	111.9	129.6	148.0	150.7	158.2	178.8	186.4	212.2	204.9	202.9
3 State-Coop.	116.1	132.3	151.4	156.3	164.6	186.9	197.6	223.8	217.1	217.4
4 Bonus Wages	10.9	11.3	11.5	12.4	14.7	15.6	16.8	17.4	18.6	19.8
5 Collectives	1.5	2.0	11.5	12.8	11.7	13.2	13.8	19.0	19.5	19.6
6 Net Taxes	73.7	71.2	48.7	47.7	40.8	26.4	30.5	11.2	11.9	11.3
7 Foreign Trade	50.0	53.1	56.8	59.1	62.8	56.9	50.6	51.8	55.3	53.5
8 Social Sec.	13.0	18.7	19.4	20.2	20.8	21.5	22.3	23.5	25.1	26.4
9 State	11.2	16.5	16.9	17.6	18.1	18.7	19.3	20.4	21.7	22.8
10 Collectives	1.8	2.2	2.5	2.6	2.7	2.8	3.0	3.1	3.4	3.6
11 Other Revenues	6.4	8.4	22.8	30.2	24.5	21.6	20.1	27.5	25.3	27.4
12 INDUSTRY	159.3	174.3	159.4	164.8	163.0	155.4	163.0	158.6	155.8	153.8
13 Net Profit	73.2	86.5	91.7	94.9	99.0	110.7	114.5	127.5	124.7	122.1
14 Published	75.0	87.8	92.7	96.3	100.6	112.3	116.6	128.9	127.0	125.0
15 Bonus Wages	7.0	7.2	7.3	8.0	8.5	9.2	10.0	10.2	11.2	11.8
16 Other	3.8	4.4	4.6	4.8	5.0	5.6	5.8	6.4	6.4	6.3
17 Public Orgs.	0.8	0.8	0.9	0.9	0.9	0.9	0.9	1.0	1.0	1.0
18 Collectives	0.6	0.7	0.8	0.9	1.0	1.1	1.2	1.4	1.5	1.6
19 Net Taxes	73.7	71.2	48.7	47.7	40.8	26.4	30.5	11.2	11.7	11.0
20 Social Sec.	6.5	10.2	10.4	10.7	11.0	11.3	11.6	12.2	13.4	14.2
21 State	6.3	10.0	10.2	10.5	10.8	11.1	11.4	12.0	13.1	13.9
22 Collectives	0.2	0.2	0.2	0.2	0.2	0.2	0.2	0.2	0.3	0.3
23 Other Revenues	6.0	6.4	8.5	11.5	12.1	7.0	6.4	7.8	6.1	6.6
24 AGRICULTURE	3.3	6.8	30.6	29.7	26.6	30.9	28.5	45.9	47.1	48.0
25 Net Profit	0.3	1.2	21.4	19.8	20.2	24.0	27.4	41.8	43.0	43.0
26 State Farms	0.3	0.7	11.6	8.9	10.6	13.0	16.0	19.3	20.0	20.0
27 Bonus Wages	1.5	1.6	1.7	1.8	1.9	2.0	2.1	2.2	2.3	2.5
28 Collectives	0.9	1.3	10.7	11.9	10.7	12.1	12.6	17.6	18.0	18.0
29 Other	0.6	0.8	0.8	0.8	0.8	0.9	0.9	7.1	7.3	7.5
30 Social Sec.	2.6	3.0	3.3	3.5	3.6	3.8	4.0	4.2	4.4	4.7
31 State	1.0	1.0	1.0	1.1	1.1	1.2	1.2	1.3	1.3	1.4
32 Collectives	1.6	2.0	2.3	2.4	2.5	2.6	2.8	2.9	3.1	3.3
33 Other Revenues	0.4	2.6	5.9	6.4	2.8	3.1	-2.9	-0.1	-0.3	0.3
34 T AND C	12.8	15.6	16.9	17.2	17.9	18.7	19.7	21.0	20.9	21.5
35 Net Profit	11.2	13.6	14.8	15.1	15.7	16.4	17.4	18.6	18.5	19.1
36 Total Profit	12.1	14.6	15.8	16.1	16.7	17.4	18.5	19.7	19.6	20.3
37 Bonus Wages	0.9	1.0	1.0	1.0	1.0	1.0	1.1	1.1	1.1	1.2
38 Social Sec.	0.9	1.3	1.3	1.3	1.4	1.4	1.4	1.5	1.5	1.5
39 Other Revenues	0.7	0.7	0.8	0.8	0.8	0.9	0.9	0.9	0.9	0.9
40 CONSTRUCTION	10.6	11.3	11.6	16.9	16.5	20.9	24.1	26.2	19.6	20.5
41 Net Profit	6.6	6.9	7.3	10.0	9.8	15.6	16.2	20.2	13.2	14.0
42 Published	7.1	7.3	7.8	10.3	11.7	17.6	18.3	22.7	15.7	16.8
43 Bonus Wages	1.3	1.3	1.3	1.4	3.1	3.2	3.4	3.6	3.7	4.0
44 Oil-Gas Expl.	0.8	0.9	0.8	1.1	1.2	1.2	1.3	1.1	1.2	1.2
45 Social Sec.	2.2	3.1	3.3	3.5	3.6	3.8	4.0	4.2	4.4	4.6
46 Other Revenues	1.8	1.3	1.0	3.4	3.1	1.5	3.9	1.8	2.0	2.0

TABLE A7: PRODUCTION REVENUES
      (billion rubles)

	1965	1970	1971	1972	1973	1974	1975	1976	1977	1978	1979	1980
47 T AND D	6.6	9.9	10.9	11.7	12.0	12.2	12.7	13.4	14.0	14.1	16.0	16.9
48 Net Profit	3.8	6.9	7.8	8.7	9.1	9.4	10.0	10.4	11.0	11.2	13.1	13.9
49 Total Profit	3.8	7.0	7.9	8.8	9.2	9.5	10.1	10.6	11.2	11.4	13.3	14.1
50 a) Trade	1.4	2.9	3.0	3.6	3.7	4.0	4.4	4.7	4.9	5.2	6.7	7.5
51 b) Supply	0.9	1.7	2.2	2.5	2.7	2.7	3.0	3.2	3.3	3.3	3.7	3.5
52 c) Agri Proc	0.7	1.2	1.3	1.2	1.3	1.3	1.2	1.2	1.4	1.3	1.4	1.4
53 d) Coop	0.8	1.2	1.4	1.5	1.5	1.5	1.5	1.5	1.6	1.6	1.5	1.7
54 Bonus Wages	0.0	0.1	0.1	0.1	0.1	0.1	0.1	0.2	0.2	0.2	0.2	0.2
55 Social Sec.	0.2	0.4	0.4	0.5	0.5	0.5	0.5	0.5	0.6	0.6	0.7	0.7
56 Other Revenues	2.6	2.7	2.7	2.6	2.4	2.3	2.2	2.4	2.4	2.2	2.2	2.3
57 OTHER PROD.	0.2	0.5	0.5	0.7	0.9	1.0	1.0	1.1	1.2	1.2	1.3	1.4
58 Total Profit	0.2	0.4	0.4	0.6	0.7	0.8	0.8	0.9	1.0	1.0	1.1	1.2
59 Forestry	0.0	0.0	0.0	0.1	0.1	0.1	0.1	0.1	0.1	0.1	0.1	0.1
60 Other Prod	0.2	0.4	0.4	0.5	0.6	0.7	0.7	0.8	0.9	0.9	1.0	1.1
61 Social Sec.	0.0	0.1	0.1	0.1	0.1	0.1	0.1	0.1	0.1	0.1	0.1	0.1
62 Other Revenues	0.0	0.0	0.0	0.0	0.1	0.1	0.1	0.1	0.1	0.1	0.1	0.1
63 FOREIGN TRADE	6.8	11.4	12.4	13.0	15.0	15.8	18.8	23.7	26.8	29.0	34.0	43.5
64 Import-Export	6.6	10.1	10.7	14.3	14.6	13.6	23.6	24.9	24.2	28.1	30.8	40.4
65 Imports	14.8	26.7	28.0	31.8	34.2	36.7	46.5	49.5	51.2	56.3	60.5	71.4
66 Exports	8.3	16.6	17.3	17.5	19.6	23.1	22.9	24.6	27.0	28.2	29.7	31.0
67 Foreign Curr.	0.2	1.3	1.7	-1.3	0.4	2.2	-4.8	-1.2	2.6	0.9	3.2	3.1

TABLE A7: PRODUCTION REVENUES
        (billion rubles)

		1981	1982	1983	1984	1985	1986	1987	1988	1989	1990
47	T AND D	17.4	18.1	18.6	18.1	19.0	19.0	22.8	19.8	20.0	20.9
48	Net Profit	14.3	13.4	14.8	15.3	15.3	16.1	17.2	14.9	15.8	16.3
49	Total Profit	14.5	13.6	15.0	15.5	15.5	16.3	17.4	15.1	16.1	16.6
50	a) Trade	8.0	7.5	7.8	8.0	8.0	8.1	9.9	8.0	8.5	9.0
51	b) Supply	3.3	3.0	4.2	4.3	4.2	4.5	3.1	2.8	3.0	2.8
52	c) Agri Proc	1.4	1.2	1.1	1.3	1.3	1.1	1.4	1.1	1.2	1.2
53	d) Coop	1.8	1.9	1.9	1.9	2.0	2.6	3.0	3.2	3.4	3.6
54	Bonus Wages	0.2	0.2	0.2	0.2	0.2	0.2	0.2	0.2	0.3	0.3
55	Social Sec.	0.7	1.0	1.0	1.1	1.1	1.2	1.2	1.3	1.3	1.3
56	Other Revenues	2.4	3.7	2.8	1.8	2.6	1.7	4.4	3.6	2.9	3.3
57	OTHER PROD.	1.6	1.8	1.8	2.1	2.1	2.2	2.3	3.0	3.3	3.3
58	Total Profit	1.3	1.5	1.5	1.8	1.8	1.9	2.0	2.7	3.0	3.0
59	Forestry	0.1	0.1	0.1	0.2	0.2	0.2	0.2	0.2	0.2	0.3
60	Other Prod	1.2	1.4	1.4	1.6	1.6	1.7	1.8	2.5	2.8	2.7
61	Social Sec.	0.1	0.1	0.1	0.1	0.1	0.1	0.1	0.1	0.1	0.1
62	Other Revenues	0.2	0.2	0.2	0.2	0.2	0.2	0.2	0.2	0.2	0.2
63	FOREIGN TRADE	50.0	53.1	56.8	59.1	62.8	58.2	49.5	51.8	55.7	53.5
64	Import-Export	47.4	48.9	51.8	53.7	60.7	54.2	44.5	50.4	60.7	58.5
65	Imports	80.8	88.3	92.2	97.0	102.8	98.5	88.9	96.6	107.8	106.5
66	Exports	33.4	39.4	40.4	43.3	42.1	44.3	44.4	46.2	47.1	48.0
67	Foreign Curr.	2.6	4.2	5.0	5.4	2.1	4.0	5.0	1.4	-5.0	-5.0

TABLE A8: TAXES AND SUBSIDIES BY SECTOR
(billion rubles)

	1965	1970	1971	1972	1973	1974	1975	1976	1977	1978	1979	1980
1 TURNOVER TAX	38.7	49.4	54.5	55.6	59.1	63.5	66.6	70.7	74.6	84.1	88.3	94.1
2 TOTAL INDUSTRY	38.2	49.0	54.3	55.4	58.9	63.3	66.4	70.6	74.5	84.0	88.2	94.0
3 Heavy Industry	7.6	9.9	11.5	13.5	15.4	17.0	18.7	21.2	22.1	28.2	31.1	33.7
4 Power	0.3	0.5	0.5	0.6	0.7	0.7	0.8	0.9	0.9	1.3	1.4	1.6
5 Oil & Gas	4.1	4.4	5.1	5.9	6.2	6.7	7.3	8.2	8.6	13.3	15.4	16.6
6 Ferrous Metals	0.0	0.1	0.1	0.1	0.1	0.1	0.2	0.2	0.2	0.2	0.2	0.2
7 Non-Ferrous M.					0.1	0.1	0.1	0.2	0.2	0.3	0.3	0.4
8 MBMW	2.2	3.1	3.7	4.4	5.6	6.5	7.3	8.2	8.6	9.2	9.7	10.4
9 Chemicals	0.3	0.5	0.6	0.7	0.7	0.8	0.8	0.9	0.9	1.0	1.1	1.2
10 Wood & Paper	0.1	0.2	0.2	0.3	0.3	0.3	0.3	0.4	0.4	0.4	0.4	0.4
11 Const. Mater.	0.1	0.2	0.2	0.2	0.2	0.2	0.2	0.3	0.3	0.3	0.3	0.3
12 N.E.C.	0.5	0.9	1.1	1.3	1.5	1.6	1.7	1.9	2.0	2.2	2.3	2.6
13 Apparel, etc.	11.3	13.8	15.0	15.7	15.5	16.0	16.7	17.5	18.6	19.0	20.1	21.9
14 Food Industry	19.3	25.3	27.8	26.2	28.0	30.3	31.0	31.9	33.8	36.8	37.0	38.4
15 TRADE, etc.	0.5	0.4	0.2	0.2	0.2	0.2	0.2	0.1	0.1	0.1	0.1	0.1
16 BUDGET PAYMENT												
17 SUBSIDIES	4.0	15.5	17.8	19.3	18.6	20.9	21.8	22.5	25.1	27.8	29.3	28.2
18 Industry	0.1	0.2	0.2	0.2	0.2	0.2	0.2	0.2	0.2	0.2	0.2	0.2
19 Coal												
20 Light												
21 Fish	0.1	0.2	0.2	0.2	0.2	0.2	0.2	0.2	0.2	0.2	0.2	0.2
22 Sugar												
23 Agriculture	3.9	15.3	17.6	19.1	18.4	20.7	21.6	22.3	24.9	27.6	29.1	28.0
24 Industry Prod	0.4	1.6	1.5	1.9	2.1	2.5	2.0	2.0	2.1	3.4	3.5	3.8
25 MBMW	0.2	0.5	0.5	0.6	0.7	0.9	0.8	0.9	0.9	0.9	0.9	1.0
26 Chemicals	0.2	0.4	0.4	0.6	0.7	0.8	1.0	1.0	1.1	1.1	1.2	1.3
27 Fuels	0.0	0.0	0.0	0.0	0.0	0.0	0.0	0.0	0.0	1.2	1.3	1.4
28 Process Feed	0.0	0.7	0.6	0.7	0.7	0.8	0.2	0.1	0.1	0.1	0.1	0.1
29 Agri. Prod	3.5	13.7	16.1	17.2	16.3	18.2	19.6	20.3	22.8	24.2	25.6	24.2
30 Meats	2.8	8.8	10.9	11.7	11.5	13.5	14.2	13.0	14.6	15.8	15.3	14.0
31 Dairy	0.0	2.1	2.6	2.7	3.2	3.4	4.0	5.3	6.0	5.9	7.7	7.5
32 Crops	0.3	0.8	0.7	0.6	0.7	0.6	0.6	0.7	0.7	0.8	0.8	0.8
33 Raw Fiber	0.4	1.7	1.6	1.9	0.6	0.5	0.5	1.0	1.2	1.4	1.5	1.7
34 Other	0.0	0.3	0.3	0.3	0.3	0.2	0.3	0.3	0.3	0.3	0.3	0.2

Net Taxes, Net Subsidies and Surcharges:

	1965	1970	1971	1972	1973	1974	1975	1976	1977	1978	1979	1980
35 Industry GVO1	266.0	409.0	434.3	458.4	490.1	525.6	558.3	578.4	605.5	636.3	657.1	685.5
36 Industry GVO2	229.4	374.3	395.7	420.0	447.3	479.6	511.2	527.9	553.7	577.7	595.1	616.3
37 Net Taxes	36.6	34.7	38.6	38.4	42.8	46.0	47.1	50.5	51.8	58.6	62.0	69.2
38 Net Subsidies	1.6	14.3	15.7	17.0	16.1	17.3	19.3	20.1	22.7	25.4	26.2	24.8
39 Surcharges	2.4	1.2	2.1	2.3	2.5	3.6	2.5	2.4	2.4	2.4	3.1	3.4
40 Agriculture	1.6	0.7	1.6	1.8	1.9	3.0	1.9	1.8	1.8	1.8	2.5	2.8
41 MBMW	0.8	0.5	0.5	0.5	0.6	0.6	0.6	0.6	0.6	0.6	0.6	0.6

Net Taxes by Sector:

	1965	1970	1971	1972	1973	1974	1975	1976	1977	1978	1979	1980
42 Fuels	4.1	4.4	5.1	5.9	6.2	6.7	7.3	8.2	8.6	12.1	14.1	15.2
43 MBMW	2.8	3.1	3.7	4.3	5.5	6.2	7.1	7.9	8.3	8.9	9.4	10.0
44 Chemicals	0.1	0.1	0.2	0.1	0.0	0.0	-0.2	-0.1	-0.2	-0.1	-0.1	-0.1
44 Apparel	11.3	13.8	15.0	15.7	15.5	16.0	16.7	17.5	18.6	19.0	20.1	21.9
45 Processed Food	19.2	25.1	27.6	26.0	27.8	30.1	30.8	31.7	33.6	36.6	36.8	38.2
46 N.E.C. Ind.	0.5	0.2	0.5	0.6	0.8	0.8	1.5	1.8	1.9	2.1	2.2	2.5
47 Agriculture	-1.9	-13.0	-14.5	-15.4	-14.4	-15.2	-17.7	-18.5	-21.0	-22.4	-23.1	-21.4

TABLE A8: TAXES AND SUBSIDIES BY SECTOR
(billion rubles)

		1981	1982	1983	1984	1985	1986	1987	1988	1989	1990
1	TURNOVER TAX	100.4	100.6	102.9	102.7	97.7	91.5	94.4	101.0	110.6	115.0
2	TOTAL INDUSTRY	100.3	100.5	102.8	102.6	97.6	91.9	94.8	101.4	111.0	115.4
3	Heavy Industry	35.1	31.8	32.6	33.3	33.7	34.5	34.8	35.5	36.8	38.1
4	Power	1.8	1.9	2.0	2.1	2.2	2.4	2.5	2.6	2.7	2.8
5	Oil & Gas	16.9	11.6	11.7	11.8	12.0	12.1	12.2	12.3	12.4	12.6
6	Ferrous Metals	0.2	0.2	0.2	0.2	0.2	0.2	0.2	0.2	0.2	0.2
7	Non-Ferrous M.	0.4	0.4	0.4	0.4	0.4	0.4	0.4	0.5	0.5	0.5
8	MBMW	10.9	10.9	11.1	11.2	11.4	11.5	11.6	11.8	12.4	13.1
9	Chemicals	1.4	3.4	3.6	3.7	3.8	4.0	4.1	4.2	4.4	4.5
10	Wood & Paper	0.4	0.4	0.4	0.4	0.4	0.4	0.4	0.4	0.4	0.4
11	Const. Mater.	0.3	0.3	0.3	0.4	0.4	0.5	0.5	0.5	0.6	0.6
12	N.E.C.	2.8	2.7	2.9	3.1	2.9	2.9	2.9	3.0	3.2	3.4
13	Apparel, etc.	22.6	20.7	20.8	20.8	20.9	21.0	21.1	22.8	24.5	25.5
14	Food Industry	42.6	48.0	49.4	48.5	43.0	36.4	38.9	43.1	49.7	51.8
15	TRADE, etc.	0.1	0.1	0.1	0.1	0.1	0.1	0.1	0.1	0.1	0.1
16	BUDGET PAYMENT						-0.5	-0.5	-0.5	-0.5	-0.5
17	SUBSIDIES	30.2	32.9	56.8	57.8	59.8	69.1	73.1	92.0	99.1	104.1
18	Industry	0.2	0.2	0.2	0.2	3.3	8.7	9.6	10.2	10.9	11.7
19	Coal						4.8	5.0	5.3	5.8	6.2
20	Light						0.9	1.0	1.1	1.2	1.4
21	Fish	0.2	0.2	0.2	0.2	2.3	1.8	2.5	2.6	2.7	2.8
22	Sugar						1.0	1.2	1.1	1.2	1.2	1.3
23	Agriculture	30.0	32.7	56.6	57.6	56.5	60.4	63.5	81.8	88.2	92.4
24	Industry Prod	3.8	4.5	4.2	4.8	5.1	5.5	5.7	1.5	1.5	1.6
25	MBMW	1.0	1.3	1.9	2.4	2.6	2.8	2.9	0.5	0.5	0.5
26	Chemicals	1.3	1.6	2.3	2.4	2.5	2.7	2.8	1.0	1.0	1.1
27	Fuels	1.4	1.5								
28	Process Feed	0.1	0.1								
29	Agri. Prod	26.2	28.2	52.4	52.8	51.4	54.9	57.8	80.3	86.7	90.8
30	Meats	14.5	15.3	21.4	24.5	25.6	27.8	26.0	39.4	43.5	48.0
31	Dairy	8.1	9.0	13.8	15.2	16.9	17.6	21.5	28.6	30.5	32.5
32	Crops	0.9	2.0	3.7	4.3	5.1	5.6	6.4	7.1	7.3	4.6
33	Raw Fiber	2.0	1.1	3.2	3.1	3.0	3.1	3.2	4.1	4.3	4.6
34	Other	0.7	0.8	10.3	5.7	0.8	0.8	0.7	1.1	1.1	1.1

Net Taxes, Net Subsidies and Surcharges:

		1981	1982	1983	1984	1985	1986	1987	1988	1989	1990
35	Industry GVO1	709.0	792.7	800.0	826.7	844.6	862.5	892.3	912.4	930.0	940.0
36	Industry GVO2	635.3	721.5	751.3	779.0	803.8	836.1	866.9	901.2	918.3	929.0
37	Net Taxes	73.7	71.2	48.7	47.7	40.8	26.4	25.4	11.2	11.7	11.0
38	Net Subsidies	26.6	29.3	54.1	54.9	56.8	65.0	68.9	89.8	98.8	103.9
39	Surcharges	3.6	3.6	2.7	2.9	3.0	4.1	4.2	2.2	0.3	0.2
40	Agriculture	2.9	2.9	2.0	2.2	3.0	4.1	4.2	2.2	0.3	0.2
41	MBMW	0.7	0.7	0.7	0.7						

Net Taxes by Sector:

		1981	1982	1983	1984	1985	1986	1987	1988	1989	1990
42	Fuels	15.5	10.1	11.7	11.8	12.0	7.3	7.2	7.0	6.6	6.4
43	MBMW	10.6	10.3	9.9	9.5	8.8	8.7	8.7	11.3	11.9	12.6
44	Chemicals	0.1	1.8	1.3	1.3	1.3	1.3	1.3	3.2	3.4	3.4
44	Apparel	22.6	20.7	20.8	20.8	20.9	19.7	19.7	21.2	22.8	23.6
45	Processed Food	42.4	47.8	49.2	48.3	39.7	33.4	35.3	39.3	45.8	47.7
46	N.E.C. Ind.	2.7	2.6	2.9	3.1	2.9	2.9	2.9	3.0	3.2	3.4
47	Agriculture	-23.3	-25.3	-50.4	-50.6	-48.4	-50.8	-53.6	-78.1	-86.4	-90.6

93

TABLE A9: GROSS AND NET PRODUCT OF NON-MATERIAL SERVICE SECTORS
        (billion rubles)

	1965	1970	1971	1972	1973	1974	1975	1976	1977	1978	1979	1980
1 GROSS PRODUCT	45.2	66.7	71.4	76.7	83.3	89.1	94.8	100.0	105.2	112.4	117.8	124.6
2 Materials	11.4	17.2	18.4	19.8	21.5	22.9	24.5	25.7	26.9	28.5	29.8	31.5
3 Depreciation	4.0	6.2	6.8	7.5	8.4	9.3	10.7	12.0	13.0	14.3	15.2	16.3
4 NET PRODUCT	29.8	43.3	46.2	49.4	53.4	56.9	59.6	62.3	65.3	69.6	72.8	76.7
5 Wages	25.0	35.9	37.9	40.3	43.3	46.0	47.9	49.7	52.0	55.2	57.8	60.9
6 Social Security	1.4	2.0	2.1	2.2	2.4	2.5	2.6	2.8	2.9	3.0	3.2	3.4
7 Revenues	3.4	5.5	6.2	6.9	7.7	8.4	9.0	9.8	10.5	11.4	11.8	12.5
8 TRANSPORTATION	5.8	8.6	9.2	9.8	10.8	11.5	12.5	13.4	14.0	14.7	15.7	16.4
9 Materials	1.2	1.8	1.9	2.0	2.3	2.4	2.7	2.9	3.0	3.2	3.5	3.7
10 Depreciation	1.1	1.5	1.7	1.9	2.0	2.2	2.5	2.7	2.9	3.1	3.4	3.6
11 NET PRODUCT	3.5	5.3	5.6	6.0	6.4	6.9	7.3	7.8	8.1	8.4	8.8	9.1
12 Wages	2.9	4.3	4.6	4.9	5.1	5.5	5.9	6.2	6.4	6.6	6.8	7.0
13 Social Security	0.2	0.3	0.3	0.3	0.3	0.3	0.4	0.4	0.4	0.4	0.4	0.4
14 Revenues	0.4	0.7	0.7	0.8	1.0	1.0	1.0	1.2	1.3	1.4	1.6	1.7
15 COMMUNICATIONS	1.7	2.6	2.7	3.0	3.1	3.5	3.8	4.2	4.3	4.7	4.8	5.2
16 Materials	0.3	0.4	0.4	0.5	0.5	0.5	0.6	0.6	0.6	0.7	0.7	0.7
17 Depreciation	0.1	0.2	0.2	0.2	0.2	0.2	0.2	0.3	0.3	0.3	0.3	0.4
18 NET PRODUCT	1.3	2.0	2.2	2.3	2.4	2.8	3.0	3.3	3.4	3.7	3.8	4.1
19 Wages	0.7	1.3	1.3	1.3	1.5	1.6	1.7	1.9	2.0	2.1	2.1	2.2
20 Social Security	0.0	0.1	0.1	0.1	0.1	0.1	0.1	0.1	0.1	0.1	0.1	0.1
21 Revenues	0.5	0.7	0.8	0.9	0.9	1.1	1.2	1.3	1.3	1.5	1.6	1.8
22 HOUSING-COMMUNAL	5.2	8.0	8.4	9.0	9.6	10.5	11.5	12.3	13.1	14.2	15.1	15.9
23 Materials	1.5	2.2	2.3	2.4	2.6	2.8	3.1	3.4	3.6	3.8	4.0	4.1
24 Depreciation	0.6	0.8	0.9	1.0	1.1	1.3	1.5	1.7	1.8	2.0	2.1	2.3
25 NET PRODUCT	3.1	5.0	5.2	5.6	5.9	6.4	6.9	7.2	7.7	8.4	9.0	9.5
26 Wages	2.1	3.6	3.8	4.1	4.4	4.8	5.1	5.4	5.8	6.3	6.8	7.1
27 Social Security	0.1	0.2	0.2	0.2	0.2	0.2	0.3	0.3	0.3	0.3	0.3	0.4
28 Revenues	0.9	1.2	1.2	1.3	1.3	1.4	1.5	1.5	1.6	1.8	1.9	2.0
29 EDUCATION	11.7	16.2	17.0	18.2	19.8	20.9	21.8	23.0	23.9	25.4	26.3	27.7
30 Materials	2.6	4.0	4.3	4.6	4.9	5.2	5.5	5.9	6.3	6.7	7.0	7.3
31 Depreciation	1.0	1.6	1.7	1.8	2.0	2.2	2.6	2.9	3.1	3.3	3.5	3.7
32 NET PRODUCT	8.1	10.6	11.0	11.8	12.9	13.4	13.8	14.2	14.5	15.4	15.8	16.7
33 Wages	7.7	10.0	10.5	11.2	12.2	12.7	13.0	13.4	13.8	14.6	15.0	15.8
34 Social Security	0.4	0.6	0.6	0.6	0.7	0.7	0.7	0.7	0.8	0.8	0.8	0.9
35 CULTURE & ARTS	1.1	1.8	2.1	2.2	2.2	2.4	2.7	3.0	3.1	3.4	3.5	3.8
36 Materials	0.2	0.3	0.4	0.4	0.4	0.5	0.6	0.7	0.7	0.8	0.8	0.8
37 Depreciation	0.0	0.1	0.1	0.1	0.1	0.1	0.2	0.2	0.2	0.3	0.3	0.4
38 NET PRODUCT	0.9	1.4	1.6	1.7	1.7	1.8	1.9	2.1	2.2	2.3	2.4	2.6
39 Wages	0.9	1.4	1.5	1.6	1.6	1.7	1.8	2.0	2.1	2.2	2.3	2.5
40 Social Security	0.0	0.1	0.1	0.1	0.1	0.1	0.1	0.1	0.1	0.1	0.1	0.1

TABLE A9: GROSS AND NET PRODUCT OF NON-MATERIAL SERVICE SECTORS
        (billion rubles)

		1981	1982	1983	1984	1985	1986	1987	1988	1989	1990
1	GROSS PRODUCT	129.9	138.4	143.4	148.2	153.8	159.6	169.4	178.9	191.6	202.5
2	Materials	32.8	35.0	36.8	38.1	42.8	44.5	47.3	50.2	53.6	56.3
3	Depreciation	17.3	18.5	19.8	20.9	18.7	19.5	21.1	22.5	23.6	24.4
4	NET PRODUCT	79.8	84.9	86.8	89.2	92.3	95.6	101.0	106.2	114.4	121.8
5	Wages	63.1	65.3	66.7	68.6	70.8	73.4	77.9	83.3	90.1	96.7
6	Social Security	3.5	4.8	4.9	5.0	5.2	5.3	5.7	6.1	6.5	7.0
7	Revenues	13.2	14.8	15.2	15.6	16.4	16.8	17.4	16.8	17.7	18.1
8	TRANSPORTATION	17.5	18.7	19.3	19.8	20.4	21.1	21.9	23.5	24.3	25.4
9	Materials	4.0	4.6	4.7	4.8	5.0	5.2	5.3	5.5	5.7	5.9
10	Depreciation	3.8	4.1	4.4	4.6	4.7	4.8	5.1	5.4	5.7	5.9
11	NET PRODUCT	9.6	10.1	10.2	10.4	10.7	11.1	11.5	12.6	12.9	13.6
12	Wages	7.4	7.6	7.7	7.8	7.9	8.2	8.5	9.0	9.6	10.1
13	Social Security	0.4	0.5	0.5	0.5	0.5	0.5	0.6	0.6	0.6	0.7
14	Revenues	1.8	2.0	2.0	2.1	2.3	2.4	2.4	3.0	2.7	2.8
15	COMMUNICATIONS	5.4	6.1	6.4	6.6	6.9	7.3	7.9	8.4	8.8	9.5
16	Materials	0.7	0.9	0.9	0.9	0.9	1.0	1.0	1.1	1.1	1.2
17	Depreciation	0.4	0.5	0.6	0.6	0.7	0.7	0.8	0.8	0.9	0.9
18	NET PRODUCT	4.3	4.7	4.9	5.1	5.3	5.6	6.1	6.5	6.8	7.4
19	Wages	2.3	2.3	2.4	2.4	2.5	2.5	2.7	2.9	3.1	3.4
20	Social Security	0.1	0.1	0.1	0.1	0.1	0.2	0.2	0.2	0.2	0.2
21	Revenues	1.9	2.2	2.4	2.5	2.7	2.9	3.2	3.4	3.5	3.8
22	HOUSING-COMMUNAL	16.7	17.7	18.3	19.1	20.0	21.1	22.2	23.8	25.5	27.4
23	Materials	4.3	4.6	4.8	5.0	5.2	5.6	5.9	6.2	6.4	6.7
24	Depreciation	2.4	2.6	2.7	2.9	3.1	3.2	3.4	3.6	3.8	3.9
25	NET PRODUCT	10.0	10.5	10.8	11.2	11.7	12.3	12.9	14.0	15.3	16.8
26	Wages	7.5	7.8	8.0	8.3	8.6	8.9	9.4	10.0	11.0	12.0
27	Social Security	0.4	0.4	0.4	0.5	0.5	0.5	0.5	0.6	0.6	0.7
28	Revenues	2.1	2.3	2.4	2.4	2.6	2.9	3.0	3.4	3.7	4.1
29	EDUCATION	28.4	29.7	30.6	31.5	33.0	34.9	37.7	39.8	41.6	43.5
30	Materials	7.5	7.8	8.1	8.3	8.5	8.8	9.3	9.7	10.1	10.7
31	Depreciation	3.9	4.1	4.3	4.4	4.5	4.7	5.0	5.3	5.5	5.7
32	NET PRODUCT	17.0	17.8	18.2	18.8	20.0	21.4	23.4	24.8	26.0	27.2
33	Wages	16.1	16.6	17.1	17.5	18.7	20.0	21.8	23.2	24.3	25.4
34	Social Security	0.9	1.2	1.2	1.2	1.3	1.4	1.5	1.6	1.7	1.8
35	CULTURE & ARTS	3.9	4.1	4.2	4.3	4.4	4.4	4.9	5.2	5.6	6.2
36	Materials	0.8	0.8	0.9	0.9	0.9	0.9	1.0	1.1	1.1	1.2
37	Depreciation	0.4	0.4	0.4	0.4	0.4	0.4	0.5	0.5	0.5	0.5
38	NET PRODUCT	2.7	2.9	2.9	3.0	3.1	3.1	3.4	3.6	4.0	4.5
39	Wages	2.6	2.7	2.7	2.8	2.9	2.9	3.2	3.4	3.8	4.2
40	Social Security	0.1	0.2	0.2	0.2	0.2	0.2	0.2	0.2	0.2	0.3

TABLE A9: GROSS AND NET PRODUCT OF NON-MATERIAL SERVICE SECTORS
(billion rubles)

	1965	1970	1971	1972	1973	1974	1975	1976	1977	1978	1979	1980
41 HEALTH	7.0	9.8	10.4	11.0	11.7	12.1	12.8	13.4	14.1	15.0	16.0	16.7
42 Materials	2.3	3.2	3.4	3.6	3.8	4.0	4.2	4.4	4.6	4.8	5.1	5.2
43 Depreciation	0.3	0.5	0.6	0.7	0.8	0.8	0.9	1.0	1.1	1.2	1.3	1.4
44 NET PRODUCT	4.4	6.1	6.4	6.7	7.1	7.3	7.7	8.0	8.4	9.0	9.6	10.1
45 Wages	4.2	5.8	6.1	6.4	6.8	7.0	7.3	7.6	8.0	8.6	9.1	9.6
46 Social Security	0.2	0.3	0.3	0.3	0.4	0.4	0.4	0.4	0.4	0.5	0.5	0.5
47 SCIENCE	8.0	12.4	13.4	14.6	15.9	17.3	18.2	18.5	19.5	20.6	21.5	23.1
48 Materials	2.8	4.6	5.0	5.5	6.1	6.7	6.9	6.9	7.2	7.5	7.7	8.6
49 Depreciation	0.4	0.7	0.8	0.9	1.0	1.1	1.3	1.5	1.7	1.9	2.0	2.1
50 NET PRODUCT	4.8	7.1	7.6	8.2	8.8	9.5	10.0	10.1	10.6	11.2	11.8	12.4
51 Wages	4.6	6.6	7.1	7.7	8.2	8.9	9.3	9.4	9.7	10.3	11.0	11.6
52 Social Security	0.3	0.4	0.4	0.4	0.5	0.5	0.5	0.5	0.5	0.6	0.6	0.6
53 Revenues	0.0	0.1	0.1	0.1	0.1	0.1	0.2	0.2	0.3	0.3	0.2	0.2
54 ADMINISTRATION	4.7	7.4	8.3	9.0	10.1	10.8	11.5	12.3	13.2	14.4	14.8	15.8
55 Materials	0.5	0.7	0.7	0.8	0.8	0.8	0.9	0.9	0.9	1.0	1.0	1.1
56 Depreciation	0.5	0.8	0.9	1.0	1.2	1.3	1.5	1.7	1.9	2.1	2.3	2.5
57 NET PRODUCT	3.7	5.9	6.7	7.2	8.1	8.7	9.1	9.7	10.4	11.3	11.5	12.2
58 Wages	2.0	2.9	3.1	3.2	3.5	3.7	3.8	3.9	4.2	4.6	4.7	5.1
59 Social Security	0.1	0.2	0.2	0.2	0.2	0.2	0.2	0.2	0.2	0.3	0.3	0.3
60 Revenues	1.6	2.8	3.4	3.8	4.4	4.8	5.1	5.6	6.0	6.4	6.5	6.8

TABLE A9: GROSS AND NET PRODUCT OF NON-MATERIAL SERVICE SECTORS
          (billion rubles)

		1981	1982	1983	1984	1985	1986	1987	1988	1989	1990
41	HEALTH	17.2	18.1	18.7	19.3	19.6	20.4	21.5	23.3	25.2	26.2
42	Materials	5.3	5.6	5.8	5.9	6.0	6.2	6.4	6.7	7.0	7.3
43	Depreciation	1.5	1.6	1.7	1.8	1.9	2.0	2.1	2.2	2.3	2.4
44	NET PRODUCT	10.4	10.9	11.2	11.6	11.7	12.2	13.0	14.4	15.9	16.5
45	Wages	9.9	10.2	10.5	10.8	10.9	11.4	12.2	13.5	14.9	15.4
46	Social Security	0.5	0.7	0.7	0.8	0.8	0.8	0.9	0.9	1.0	1.1
47	SCIENCE	24.2	26.1	27.4	28.6	29.8	30.7	32.7	35.3	39.8	42.8
48	Materials	9.1	9.5	10.4	11.0	11.6	11.9	13.1	14.3	16.2	17.1
49	Depreciation	2.2	2.4	2.6	2.8	3.0	3.3	3.7	4.0	4.3	4.6
50	NET PRODUCT	12.9	14.2	14.4	14.8	15.2	15.5	15.9	17.0	19.3	21.1
51	Wages	12.0	12.6	12.8	13.2	13.5	13.7	14.1	15.1	16.8	18.3
52	Social Security	0.7	1.3	1.3	1.3	1.4	1.4	1.4	1.5	1.7	1.8
53	Revenues	0.2	0.3	0.3	0.3	0.3	0.4	0.4	0.4	0.8	1.0
54	ADMINISTRATION	16.6	18.0	18.4	19.1	19.8	19.7	20.6	19.5	20.7	21.7
55	Materials	1.1	1.2	1.2	1.2	4.7	4.9	5.3	5.6	6.0	6.2
56	Depreciation	2.7	2.9	3.1	3.4	0.4	0.4	0.5	0.6	0.6	0.6
57	NET PRODUCT	12.8	13.9	14.1	14.5	14.7	14.4	14.8	13.3	14.1	14.9
58	Wages	5.3	5.5	5.6	5.8	5.8	5.8	6.0	6.2	6.6	7.9
59	Social Security	0.3	0.4	0.4	0.4	0.4	0.4	0.4	0.5	0.5	0.6
60	Revenues	7.2	8.0	8.1	8.3	8.5	8.2	8.4	6.6	7.0	6.4

TABLE A10: SOURCES OF FINANCING NON-MATERIAL SERVICES
(billion rubles)

		1965	1970	1971	1972	1973	1974	1975	1976	1977	1978	1979	1980
1	TOTAL	45.2	66.7	71.4	76.7	83.3	89.1	94.8	100.0	105.2	112.4	117.8	124.6
2	State Budget	28.9	40.8	42.9	45.5	48.6	51.1	53.5	55.1	57.7	60.9	63.2	67.0
3	Free Services+	19.1	26.6	27.7	29.3	31.7	33.2	34.8	36.2	37.8	40.0	41.4	43.5
4	Science & Adm.*	9.8	14.2	15.2	16.2	16.9	17.9	18.7	18.9	19.9	20.9	21.8	23.5
5	Enterprises	6.1	10.0	11.4	12.5	14.7	16.6	18.2	20.2	22.0	24.4	26.7	28.3
6	Business Exp.	3.6	6.2	7.0	7.8	9.6	10.9	11.9	13.3	14.7	16.6	18.3	19.3
7	Subsidies	2.5	3.8	4.4	4.7	5.1	5.6	6.3	6.9	7.3	7.9	8.4	8.9
8	Households	10.3	16.0	17.2	18.7	20.0	21.4	23.1	24.7	25.6	27.1	27.9	29.3
9	Business Trips	1.6	2.4	2.6	2.8	2.9	3.1	3.3	3.6	3.7	4.0	4.2	4.4
10	Consumption	8.6	13.5	14.6	15.9	17.1	18.3	19.7	21.1	21.8	23.1	23.7	24.9
11	TRANSPORTATION	5.8	8.6	9.2	9.8	10.8	11.5	12.5	13.4	14.0	14.7	15.7	16.4
12	State Budget	1.2	1.9	2.0	2.1	2.6	2.7	2.9	3.1	3.2	3.3	3.5	3.6
13	Paid Services	4.6	6.7	7.2	7.7	8.2	8.8	9.6	10.3	10.8	11.4	12.2	12.8
14	Households	3.9	5.7	6.0	6.5	6.9	7.4	8.1	8.7	9.1	9.6	10.3	10.7
15	Enterprises	0.7	1.0	1.1	1.2	1.3	1.4	1.5	1.7	1.7	1.8	2.0	2.0
16	COMMUNICATIONS	1.6	2.5	2.7	3.0	3.2	3.5	3.8	4.1	4.3	4.7	4.8	5.2
17	State Budget	0.8	1.3	1.4	1.5	1.7	1.8	2.0	2.1	2.3	2.5	2.6	2.8
18	Media	0.3	0.5	0.6	0.7	0.8	0.9	1.0	1.1	1.2	1.3	1.4	1.5
19	Other	0.5	0.8	0.8	0.8	0.9	0.9	1.0	1.0	1.1	1.2	1.2	1.3
20	Households	0.8	1.2	1.4	1.5	1.5	1.7	1.8	2.0	2.0	2.2	2.2	2.4
21	HOUS-COMMUNAL	5.2	8.0	8.4	9.0	9.6	10.5	11.5	12.3	13.1	14.2	15.1	15.9
22	Hous. Subsidies	2.3	3.4	3.6	3.8	4.2	4.6	4.9	5.2	5.6	6.0	6.5	6.9
23	Enterprises	0.5	0.6	0.7	0.8	0.8	1.2	1.6	1.8	1.9	2.3	2.5	2.7
24	Households	2.4	4.0	4.1	4.4	4.6	4.7	5.0	5.3	5.6	5.9	6.1	6.3
25	Hous-Com.	2.0	3.1	3.2	3.4	3.5	3.6	3.8	4.0	4.2	4.4	4.5	4.6
26	Everyday	0.4	0.9	0.9	1.0	1.1	1.1	1.2	1.3	1.4	1.5	1.6	1.7
27	EDUCATION	11.7	16.2	17.0	18.2	19.8	20.9	21.8	23.0	24.0	25.4	26.3	27.7
28	State Budget	10.8	14.6	15.2	16.2	17.4	18.2	19.0	19.7	20.4	21.3	21.9	22.7
29	Enterprises	0.6	0.9	1.1	1.2	1.6	1.8	1.8	2.2	2.5	2.9	3.2	3.7
30	Households	0.3	0.7	0.7	0.8	0.8	0.9	1.0	1.1	1.1	1.2	1.2	1.3
31	CULTURE & ARTS	1.1	1.8	2.1	2.2	2.2	2.4	2.7	3.0	3.1	3.4	3.5	3.8
32	State Budget	0.4	0.8	0.9	0.9	0.9	1.0	1.1	1.1	1.2	1.4	1.4	1.6
33	Enterprises	0.1	0.2	0.3	0.3	0.3	0.4	0.6	0.8	0.8	0.9	0.9	1.0
34	Households	0.6	0.8	0.9	1.0	1.0	1.0	1.0	1.1	1.1	1.1	1.2	1.2
35	HEALTH, etc.	7.0	9.8	10.4	11.0	11.7	12.1	12.8	13.4	14.1	15.0	15.9	16.7
36	State Budget	6.4	8.8	9.0	9.4	10.0	10.4	10.8	11.2	11.8	12.7	13.2	14.1
37	Enterprises	0.1	0.2	0.5	0.6	0.6	0.6	0.8	0.9	0.9	0.9	1.0	1.0
38	Households	0.5	0.8	0.9	1.0	1.1	1.1	1.2	1.3	1.3	1.4	1.6	1.6

+including free transportation services
*including other communications

TABLE A10: SOURCES OF FINANCING NON-MATERIAL SERVICES
        (billion rubles)

		1981	1982	1983	1984	1985	1986	1987	1988	1989	1990
1	TOTAL	129.9	138.4	143.4	148.2	153.8	159.6	169.4	178.9	191.6	202.5
2	State Budget	69.8	74.5	76.8	79.5	82.8	86.1	90.1	97.0	103.1	109.7
3	Free Services+	45.0	47.6	48.5	50.1	52.2	54.7	56.5	61.3	64.0	67.0
4	Science & Adm.*	24.8	26.9	28.3	29.4	30.5	31.4	33.6	35.7	39.1	42.7
5	Enterprises	29.6	31.5	32.9	33.7	34.7	35.4	39.0	39.4	44.1	45.7
6	Business Exp.	20.2	21.6	22.6	22.8	23.3	23.5	26.3	26.8	30.3	30.6
7	Subsidies	9.4	9.9	10.4	10.9	11.3	11.9	12.6	12.6	13.8	15.1
8	Households	30.5	32.3	33.6	35.0	36.4	38.0	40.4	42.5	44.4	47.1
9	Business Trips	4.6	4.9	5.0	5.2	5.4	5.6	5.9	6.2	6.4	6.8
10	Consumption	25.9	27.4	28.6	29.9	31.0	32.4	34.5	36.3	37.9	40.3
11	TRANSPORTATION	17.5	18.7	19.3	19.8	20.4	21.1	21.9	23.5	24.3	25.4
12	State Budget	3.8	4.1	4.2	4.3	4.4	4.5	4.6	5.1	5.3	5.4
13	Paid Services	13.7	14.6	15.1	15.5	16.0	16.6	17.3	18.4	19.0	20.0
14	Households	11.5	12.2	12.6	13.0	13.5	14.0	14.6	15.4	16.0	16.8
15	Enterprises	2.2	2.4	2.5	2.5	2.6	2.6	2.7	3.0	3.0	3.2
16	COMMUNICATIONS	5.4	6.0	6.3	6.6	6.9	7.3	7.9	8.4	8.8	9.5
17	State Budget	3.0	3.3	3.5	3.7	3.9	4.2	4.5	4.7	4.8	5.0
18	Media	1.6	1.7	1.8	1.9	2.0	2.2	2.4	2.5	2.6	2.7
19	Other	1.4	1.6	1.7	1.8	1.9	2.0	2.1	2.2	2.2	2.3
20	Households	2.4	2.7	2.8	2.9	3.0	3.1	3.4	3.7	4.0	4.5
21	HOUS-COMMUNAL	16.7	17.7	18.3	19.1	20.0	21.1	22.2	23.8	25.5	27.4
22	Hous. Subsidies	7.4	7.9	8.3	8.8	9.3	9.8	10.4	11.4	12.5	13.7
23	Enterprises	2.9	3.1	3.1	3.1	3.1	3.3	3.4	3.6	3.7	3.9
24	Households	6.4	6.7	6.9	7.1	7.6	8.0	8.4	8.8	9.3	9.8
25	Hous-Com.	4.7	4.9	5.0	5.1	5.4	5.7	6.0	6.3	6.7	7.0
26	Everyday	1.7	1.8	1.9	2.0	2.2	2.3	2.4	2.5	2.6	2.8
27	EDUCATION	28.4	29.7	30.6	31.5	33.0	34.9	37.7	39.8	41.6	43.5
28	State Budget	23.4	24.7	25.0	25.9	27.4	29.2	29.7	31.0	32.0	33.8
29	Enterprises	3.7	3.7	4.3	4.3	4.3	4.4	6.5	7.2	7.9	8.0
30	Households	1.3	1.3	1.3	1.3	1.3	1.3	1.5	1.6	1.7	1.7
31	CULTURE & ARTS	3.9	4.1	4.2	4.3	4.4	4.4	4.9	5.2	5.6	6.2
32	State Budget	1.7	1.8	1.8	1.8	1.9	1.9	2.0	2.1	2.2	2.4
33	Enterprises	0.9	1.0	1.0	1.0	1.0	1.0	1.1	1.2	1.3	1.4
34	Households	1.3	1.3	1.4	1.5	1.5	1.5	1.8	1.9	2.1	2.4
35	HEALTH, etc.	17.2	18.1	18.7	19.3	19.6	20.4	21.5	23.3	25.2	26.2
36	State Budget	14.5	15.3	15.7	16.2	16.6	16.9	17.9	20.6	21.9	22.7
37	Enterprises	1.1	1.1	1.1	1.1	1.1	1.1	1.1	0.0	0.0	0.0
38	Households	1.6	1.7	1.9	2.0	2.0	2.3	2.5	2.7	3.3	3.5

+including free transportation services
*including other communications

TABLE A10: SOURCES OF FINANCING NON-MATERIAL SERVICES
       (billion rubles)

	1965	1970	1971	1972	1973	1974	1975	1976	1977	1978	1979	1980
39 SCIENCE	8.0	12.4	13.4	14.6	15.9	17.3	18.2	18.5	19.5	20.6	21.5	23.1
40 State Budget	8.0	11.7	12.6	13.6	14.1	15.1	15.7	15.9	16.6	17.4	18.2	19.7
41 Science	4.1	6.4	6.9	7.3	7.5	7.9	7.9	8.0	8.3	8.9	9.4	10.1
42 Geology	1.2	1.6	1.8	2.0	2.1	2.2	2.4	2.6	2.7	2.7	2.8	3.0
43 Agri. Services	1.1	1.7	1.8	1.9	2.0	2.2	2.4	2.4	2.5	2.5	2.7	3.2
44 Space	1.6	2.0	2.1	2.4	2.5	2.8	3.0	2.9	3.1	3.3	3.3	3.4
45 Enterprises	0.0	0.7	0.8	1.0	1.8	2.2	2.5	2.6	2.9	3.2	3.3	3.4
46 BANKING & INSUR.	1.9	3.4	4.0	4.4	5.1	5.6	5.9	6.4	7.0	7.5	7.6	8.0
47 Enterprises	1.6	2.6	3.0	3.2	3.4	3.6	3.6	4.1	4.6	5.0	5.6	5.8
48 Households	0.3	0.8	1.0	1.2	1.7	2.0	2.3	2.3	2.4	2.5	2.0	2.2
49 Banking	0.1	0.2	0.2	0.2	0.2	0.1	0.2	0.2	0.2	0.2	0.2	0.2
50 Insurance	0.2	0.6	0.8	1.0	1.5	1.9	2.1	2.1	2.2	2.3	1.8	2.0
51 ADMINISTRATION	2.8	4.0	4.3	4.6	5.0	5.2	5.6	5.9	6.2	6.9	7.2	7.8
52 State Budget	1.3	1.7	1.8	1.8	1.9	1.9	2.0	2.0	2.2	2.3	2.4	2.5
53 Enterprises	0.0	0.3	0.3	0.4	0.7	0.7	0.9	1.0	1.0	1.4	1.5	1.7
54 Households	1.5	2.0	2.2	2.3	2.4	2.6	2.7	2.9	3.0	3.2	3.3	3.6
55 Dues	1.5	2.0	2.1	2.2	2.3	2.5	2.6	2.8	2.9	3.1	3.2	3.4
56 Other	0.0	0.0	0.1	0.1	0.1	0.1	0.1	0.1	0.1	0.1	0.1	0.2

TABLE A10: SOURCES OF FINANCING NON-MATERIAL SERVICES
          (billion rubles)

		1981	1982	1983	1984	1985	1986	1987	1988	1989	1990
39	SCIENCE	24.2	26.1	27.4	28.6	29.8	30.7	32.7	35.3	39.8	42.8
40	State Budget	20.8	22.5	23.7	24.7	25.6	26.4	28.4	30.5	33.7	35.2
41	Science	10.9	11.7	12.7	13.2	13.4	14.2	15.3	16.7	18.2	19.1
42	Geology	3.3	3.5	3.5	3.6	3.9	3.9	4.1	4.3	4.6	4.4
43	Agri. Services	3.1	3.5	3.6	3.6	3.7	3.6	3.8	3.8	4.0	3.7
44	Space	3.5	3.8	3.9	4.3	4.6	4.7	5.2	5.7	6.9	8.0
45	Enterprises	3.4	3.6	3.7	3.9	4.2	4.3	4.3	4.8	6.1	7.6
46	BANKING & INSUR.	8.5	9.4	9.5	9.8	10.0	9.8	10.1	8.5	8.9	8.4
47	Enterprises	6.3	6.9	6.8	6.7	6.7	6.4	6.5	4.8	6.0	5.3
48	Households	2.2	2.5	2.7	3.1	3.3	3.4	3.6	3.7	2.9	3.1
49	Banking	0.2	0.2	0.2	0.3	0.3	0.3	0.4	0.4	0.5	0.5
50	Insurance	2.0	2.3	2.5	2.8	3.0	3.1	3.2	3.3	2.4	2.5
51	ADMINISTRATION	8.1	8.6	8.9	9.3	9.8	9.9	10.5	11.0	11.8	13.3
52	State Budget	2.6	2.8	2.9	2.9	3.0	3.0	3.0	3.0	3.2	5.2
53	Enterprises	1.7	1.9	2.0	2.3	2.5	2.5	2.9	3.3	3.5	2.8
54	Households	3.8	3.9	4.0	4.1	4.3	4.4	4.6	4.7	5.1	5.3
55	Dues	3.6	3.7	3.8	3.9	4.1	4.2	4.3	4.4	4.8	5.0
56	Other	0.2	0.2	0.2	0.2	0.2	0.2	0.3	0.3	0.3	0.3

Part B

Intermediate and End-Use Accounts

TABLE B1: MATERIAL INPUTS BY SECTOR
     (billion rubles)

		1965	1970	1971	1972	1973	1974	1975	1976	1977	1978	1979	1980
1	POWER	3.3	5.2	5.6	6.2	6.7	7.0	7.4	7.9	8.1	8.4	8.7	8.9
2	Power	0.1	0.1	0.2	0.2	0.2	0.2	0.2	0.2	0.2	0.2	0.2	0.2
3	Fuels	2.9	4.6	5.0	5.3	5.8	6.0	6.3	6.5	6.7	6.9	7.2	7.3
4	Metallurgy	0.0	0.0	0.0	0.0	0.1	0.1	0.1	0.2	0.2	0.2	0.2	0.2
5	Chemicals	0.0	0.1	0.1	0.1	0.1	0.1	0.1	0.1	0.1	0.1	0.1	0.1
6	MBMW	0.2	0.3	0.3	0.4	0.4	0.5	0.6	0.7	0.7	0.8	0.8	0.9
7	N.E.C. Industry	0.1	0.1	0.1	0.1	0.1	0.1	0.1	0.1	0.1	0.1	0.1	0.1
8	Light Industry	0.0	0.0	0.0	0.0	0.0	0.0	0.0	0.1	0.1	0.1	0.1	0.1
9	T&C												
10	FUELS	8.3	10.9	11.1	11.9	12.6	13.7	15.1	15.8	16.6	17.4	18.0	18.7
11	Power	0.8	1.0	1.0	1.0	1.1	1.2	1.3	1.4	1.4	1.5	1.6	1.6
12	Fuels	6.6	8.1	8.3	9.0	9.5	10.4	11.4	12.0	12.7	13.3	13.6	14.2
13	Metallurgy	0.1	0.2	0.2	0.2	0.2	0.2	0.2	0.2	0.2	0.2	0.2	0.2
14	Chemicals	0.2	0.5	0.5	0.5	0.5	0.5	0.6	0.6	0.6	0.6	0.7	0.7
15	MBMW	0.3	0.5	0.5	0.5	0.5	0.6	0.7	0.7	0.8	0.9	1.0	1.1
16	Wood & Paper	0.2	0.4	0.4	0.4	0.5	0.5	0.5	0.5	0.5	0.5	0.5	0.5
17	Constr. Mat.	0.0	0.0	0.0	0.1	0.1	0.1	0.1	0.1	0.1	0.1	0.1	0.1
18	N.E.C. Industry	0.1	0.1	0.1	0.1	0.1	0.1	0.1	0.1	0.1	0.1	0.1	0.1
19	Light Industry	0.0	0.1	0.1	0.1	0.1	0.1	0.1	0.1	0.1	0.1	0.1	0.1
20	Other Prod.	0.0	0.0	0.0	0.0	0.1	0.1	0.1	0.1	0.1	0.1	0.1	0.1
21	T&C												
22	METALLURGY	14.3	23.3	24.5	26.5	28.0	29.4	31.5	33.2	35.5	36.8	37.2	37.0
23	Power	0.6	1.4	1.5	1.6	1.7	1.8	1.9	2.0	2.1	2.2	2.3	2.4
24	Fuels	2.0	3.9	4.0	4.2	4.3	4.4	4.5	4.6	4.7	4.8	4.8	4.8
25	Metallurgy	9.8	15.1	15.9	17.3	18.5	19.7	21.1	22.4	24.3	25.4	25.6	25.3
26	Chemicals	0.3	0.5	0.5	0.6	0.6	0.7	0.7	0.7	0.8	0.8	0.8	0.9
27	MBMW	0.6	1.2	1.3	1.3	1.4	1.4	1.6	1.6	1.7	1.8	1.9	1.9
28	Wood & Paper	0.1	0.2	0.2	0.2	0.2	0.2	0.3	0.3	0.3	0.3	0.3	0.3
29	Constr. Mat.	0.0	0.1	0.1	0.1	0.1	0.1	0.1	0.2	0.2	0.2	0.2	0.2
30	N.E.C. Industry	0.1	0.1	0.1	0.1	0.1	0.1	0.1	0.1	0.1	0.1	0.1	0.1
31	Light Industry	0.1	0.1	0.2	0.2	0.2	0.2	0.3	0.3	0.3	0.3	0.3	0.3
32	Food Industry	0.0	0.0	0.1	0.1	0.1	0.1	0.1	0.1	0.1	0.1	0.1	0.1
33	Other Prod.	0.7	0.8	0.8	0.8	0.8	0.8	0.8	0.8	0.8	0.8	0.8	0.7
34	T&C												
35	CHEMICALS	7.6	14.2	15.3	16.6	17.6	19.0	20.1	21.5	22.9	24.0	25.2	26.3
36	Power	0.6	1.3	1.4	1.5	1.6	1.7	1.8	1.9	2.0	2.1	2.2	2.3
37	Fuels	0.4	0.8	0.8	0.9	1.0	1.1	1.1	1.2	1.3	1.5	1.6	1.8
38	Metallurgy	0.5	0.8	0.9	1.0	1.1	1.2	1.3	1.4	1.4	1.4	1.4	1.5
39	Chemicals	3.5	7.0	7.6	8.2	8.7	9.6	10.2	11.5	12.3	12.8	13.7	14.4
40	MBMW	0.7	1.4	1.4	1.6	1.6	1.7	1.8	1.8	1.8	1.9	1.9	2.0
41	Wood & Paper	0.4	0.7	0.8	0.9	1.0	1.0	1.1	1.1	1.1	1.2	1.2	1.2
42	Constr. Mat.	0.1	0.1	0.1	0.1	0.2	0.2	0.2	0.3	0.3	0.3	0.3	0.3
43	N.E.C. Industry	0.1	0.2	0.2	0.2	0.2	0.2	0.2	0.2	0.2	0.2	0.2	0.2
44	Light Industry	0.6	0.9	1.0	1.0	1.1	1.2	1.3	1.3	1.4	1.5	1.6	1.6
45	Food Industry	0.7	0.9	1.0	1.0	1.0	1.0	1.0	1.0	1.0	1.0	1.0	1.0
46	Agriculture												
47	Other Prod.	0.0	0.1	0.1	0.1	0.1	0.1	0.1	0.1	0.1	0.1	0.1	0.1
48	T&C												

TABLE B1: MATERIAL INPUTS BY SECTOR
     (billion rubles)

		1981	1982	1983	1984	1985	1986	1987	1988	1989	1990
1	POWER	9.2	13.4	13.6	14.0	14.2	14.6	14.9	17.2	17.4	17.6
2	Power	0.2	0.3	0.3	0.3	0.3	0.3	0.3	0.3	0.3	0.3
3	Fuels	7.5	11.3	11.5	11.7	11.6	11.8	12.0	14.2	14.4	14.6
4	Metallurgy	0.2	0.2	0.2	0.2	0.2	0.3	0.3	0.3	0.3	0.3
5	Chemicals	0.1	0.1	0.1	0.1	0.1	0.1	0.1	0.1	0.1	0.1
6	MBMW	1.0	1.0	1.1	1.2	1.2	1.3	1.4	1.5	1.5	1.5
7	N.E.C. Industry	0.1	0.4	0.3	0.4	0.4	0.4	0.4	0.4	0.4	0.4
8	Light Industry	0.1	0.1	0.1	0.1	0.1	0.1	0.1	0.1	0.1	0.1
9	T&C					0.3	0.3	0.3	0.3	0.3	0.3
10	FUELS	19.4	27.7	28.0	28.2	28.5	28.8	29.3	31.9	32.1	32.3
11	Power	1.7	2.3	2.4	2.4	2.4	2.5	2.5	2.5	2.5	2.5
12	Fuels	14.7	22.2	22.3	22.5	21.3	21.4	21.9	24.3	24.5	24.7
13	Metallurgy	0.2	0.3	0.3	0.3	0.3	0.3	0.3	0.3	0.3	0.3
14	Chemicals	0.8	0.8	0.8	0.8	0.8	0.8	0.8	0.9	0.9	0.9
15	MBMW	1.1	1.1	1.2	1.2	1.3	1.3	1.3	1.3	1.3	1.3
16	Wood & Paper	0.5	0.5	0.5	0.5	0.5	0.5	0.5	0.5	0.5	0.5
17	Constr. Mat.	0.1	0.1	0.1	0.1	0.1	0.1	0.1	0.1	0.1	0.1
18	N.E.C. Industry	0.1	0.2	0.2	0.2	0.2	0.2	0.2	0.2	0.2	0.2
19	Light Industry	0.1	0.1	0.1	0.1	0.1	0.1	0.1	0.1	0.1	0.1
20	Other Prod.	0.1	0.1	0.1	0.1	0.1	0.1	0.1	0.1	0.1	0.1
21	T&C					1.4	1.5	1.5	1.6	1.6	1.6
22	METALLURGY	37.3	44.1	44.7	45.8	47.5	52.2	52.7	53.2	53.5	53.0
23	Power	2.4	2.9	2.9	3.1	3.3	3.5	3.5	3.6	3.6	3.6
24	Fuels	4.8	5.6	5.7	5.8	6.0	6.3	6.6	6.8	6.8	6.7
25	Metallurgy	25.6	30.8	31.2	31.8	32.5	36.4	36.5	36.6	36.8	36.4
26	Chemicals	0.9	1.0	1.0	1.1	1.1	1.2	1.2	1.2	1.2	1.2
27	MBMW	1.9	2.0	2.1	2.2	2.4	2.6	2.7	2.8	2.9	2.9
28	Wood & Paper	0.3	0.4	0.4	0.4	0.4	0.4	0.4	0.4	0.4	0.4
29	Constr. Mat.	0.2	0.2	0.2	0.2	0.2	0.2	0.2	0.2	0.2	0.2
30	N.E.C. Industry	0.1	0.1	0.1	0.1	0.1	0.1	0.1	0.1	0.1	0.1
31	Light Industry	0.3	0.3	0.3	0.3	0.3	0.3	0.3	0.3	0.3	0.3
32	Food Industry	0.1	0.1	0.1	0.1	0.1	0.1	0.1	0.1	0.1	0.1
33	Other Prod.	0.7	0.7	0.7	0.7	0.6	0.6	0.6	0.6	0.6	0.6
34	T&C					0.5	0.5	0.5	0.5	0.5	0.5
35	CHEMICALS	27.6	32.0	33.7	35.2	35.9	36.6	37.8	38.3	38.8	38.0
36	Power	2.4	3.3	3.4	3.5	3.6	3.7	3.8	3.9	4.0	4.0
37	Fuels	1.9	2.7	2.9	3.1	2.9	3.1	3.4	3.5	3.6	3.6
38	Metallurgy	1.5	1.9	2.0	2.1	2.2	2.4	2.5	2.5	2.5	2.5
39	Chemicals	15.4	17.0	18.2	19.1	19.4	19.6	20.2	20.5	20.8	20.0
40	MBMW	2.0	2.0	2.0	2.1	2.1	2.1	2.1	2.1	2.1	2.1
41	Wood & Paper	1.2	1.4	1.4	1.5	1.5	1.5	1.5	1.5	1.5	1.5
42	Constr. Mat.	0.3	0.3	0.3	0.3	0.3	0.3	0.3	0.3	0.3	0.3
43	N.E.C. Industry	0.2	0.3	0.3	0.3	0.3	0.3	0.3	0.3	0.3	0.3
44	Light Industry	1.6	1.8	1.8	1.8	1.8	1.8	1.9	1.9	1.9	1.9
45	Food Industry	1.0	1.1	1.1	1.1	1.1	1.1	1.1	1.1	1.1	1.1
46	Agriculture			0.1	0.1	0.1	0.1	0.1	0.1	0.1	0.1
47	Other Prod.	0.1	0.2	0.2	0.2	0.2	0.2	0.2	0.2	0.2	0.2
48	T&C					0.4	0.4	0.4	0.4	0.4	0.4

TABLE B1: MATERIAL INPUTS BY SECTOR
(billion rubles)

		1965	1970	1971	1972	1973	1974	1975	1976	1977	1978	1979	1980
49	TOTAL MBMW	24.0	50.6	54.0	58.7	61.5	67.1	74.5	76.0	83.8	90.2	93.4	97.4
50	Power	0.7	1.5	1.6	1.8	1.9	2.1	2.2	2.2	2.3	2.4	2.5	2.6
51	Fuels	0.5	1.2	1.3	1.3	1.4	1.5	1.6	1.6	1.7	1.8	1.9	2.1
52	Metallurgy	7.6	13.2	14.0	15.0	15.5	17.6	18.8	19.2	20.6	22.1	22.8	23.5
53	Chemicals	1.8	3.7	4.0	4.3	4.9	5.4	6.0	6.2	6.6	7.0	7.3	7.7
54	MBMW	11.3	27.1	29.0	32.1	33.2	35.5	40.6	41.5	47.1	51.0	53.0	55.4
55	Wood & Paper	0.7	1.4	1.5	1.7	1.7	1.8	2.0	2.0	2.0	2.1	2.1	2.2
56	Constr. Mat.	0.2	0.4	0.5	0.5	0.5	0.6	0.6	0.6	0.7	0.8	0.8	0.8
57	N.E.C. Industry	0.3	0.5	0.5	0.5	0.6	0.6	0.6	0.6	0.6	0.7	0.7	0.7
58	Light Industry	0.5	1.1	1.2	1.2	1.4	1.5	1.6	1.6	1.7	1.8	1.8	1.9
59	Food Industry	0.2	0.2	0.2	0.2	0.2	0.3	0.3	0.3	0.3	0.3	0.3	0.3
60	Other Prod.	0.2	0.2	0.2	0.2	0.2	0.2	0.2	0.2	0.2	0.2	0.2	0.2
61	T&C												
62	WOOD AND PAPER	7.0	10.4	10.9	11.5	11.9	12.6	13.1	13.7	13.8	14.3	14.4	14.7
63	Power	0.2	0.4	0.4	0.4	0.4	0.5	0.5	0.5	0.5	0.6	0.6	0.6
64	Fuels	0.4	0.7	0.7	0.7	0.8	0.8	0.8	0.9	0.9	0.9	0.9	0.9
65	Metallurgy	0.2	0.2	0.2	0.3	0.3	0.4	0.4	0.4	0.4	0.4	0.4	0.4
66	Chemicals	0.3	0.5	0.6	0.7	0.7	0.8	0.9	1.0	1.0	1.1	1.1	1.2
67	MBMW	0.4	0.5	0.5	0.6	0.6	0.7	0.8	0.8	0.8	0.9	0.9	1.0
68	Wood & Paper	4.5	6.7	7.1	7.3	7.5	7.7	8.1	8.2	8.3	8.5	8.5	8.6
69	Constr. Mat.	0.1	0.1	0.1	0.1	0.1	0.1	0.1	0.2	0.2	0.2	0.3	0.3
70	N.E.C. Industry	0.1	0.1	0.1	0.2	0.2	0.2	0.2	0.2	0.2	0.2	0.2	0.2
71	Light Industry	0.4	0.6	0.6	0.6	0.6	0.6	0.6	0.7	0.7	0.7	0.7	0.7
72	Food Industry	0.0	0.1	0.1	0.1	0.1	0.1	0.1	0.1	0.1	0.1	0.1	0.1
73	Agriculture	0.0	0.1	0.1	0.1	0.1	0.2	0.2	0.3	0.3	0.4	0.4	0.4
74	Other Prod.	0.4	0.5	0.5	0.5	0.5	0.5	0.5	0.5	0.4	0.4	0.3	0.3
75	T&C												
76	CONSTR. MATER.	5.6	9.1	9.7	10.2	11.2	12.1	13.1	13.8	14.1	14.6	15.0	15.2
77	Power	0.3	0.7	0.7	0.6	0.7	0.7	0.8	0.8	0.8	0.9	1.0	1.0
78	Fuels	0.7	1.5	1.4	1.6	1.5	1.5	1.6	1.7	1.8	2.0	2.1	2.1
79	Metallurgy	0.9	1.3	1.4	1.4	1.7	1.5	1.7	2.1	2.1	1.8	1.8	1.8
80	Chemicals	0.2	0.4	0.4	0.5	0.6	0.6	0.7	0.7	0.7	0.8	0.8	0.7
81	MBMW	0.5	0.7	0.7	0.9	1.0	0.8	0.9	0.9	0.9	0.9	0.9	1.0
82	Wood & Paper	0.2	0.4	0.4	0.4	0.4	0.4	0.5	0.6	0.6	0.6	0.6	0.6
83	Constr. Mat.	2.6	3.8	4.2	4.3	4.8	6.0	6.4	6.5	6.6	7.0	7.2	7.4
84	N.E.C. Industry	0.1	0.1	0.1	0.1	0.1	0.1	0.1	0.1	0.1	0.1	0.1	0.1
85	Light Industry	0.1	0.2	0.2	0.2	0.2	0.2	0.2	0.3	0.3	0.3	0.3	0.3
86	Food Industry	0.1	0.1	0.1	0.1	0.1	0.1	0.1	0.1	0.1	0.1	0.1	0.2
87	Other Prod.	0.0	0.0	0.1	0.1	0.1	0.1	0.1	0.1	0.1	0.1	0.1	0.1
88	T&C												

TABLE B1: MATERIAL INPUTS BY SECTOR
       (billion rubles)

	1981	1982	1983	1984	1985	1986	1987	1988	1989	1990
49 TOTAL MBMW	100.5	116.6	121.0	126.8	131.9	137.6	141.3	148.3	151.0	150.0
50 Power	2.7	3.4	3.5	3.6	3.7	3.9	4.0	4.2	4.3	4.4
51 Fuels	2.3	2.9	3.0	3.2	3.4	2.6	2.8	3.0	3.1	3.0
52 Metallurgy	24.4	29.5	30.0	30.4	30.8	31.6	32.0	32.6	33.1	33.6
53 Chemicals	7.9	8.2	8.5	8.7	8.9	9.2	9.5	9.9	10.0	10.0
54 MBMW	56.7	65.4	68.7	73.2	75.9	80.7	83.2	88.3	91.9	90.5
55 Wood & Paper	2.3	2.6	2.6	2.7	2.8	2.9	3.0	3.2	3.2	3.1
56 Constr. Mat.	0.9	1.0	1.0	1.1	1.1	1.2	1.2	1.3	1.3	1.3
57 N.E.C. Industry	0.7	0.7	0.7	0.8	0.8	0.8	0.8	0.8	0.8	0.8
58 Light Industry	2.0	2.2	2.3	2.4	2.4	2.5	2.5	2.6	2.6	2.6
59 Food Industry	0.3	0.4	0.4	0.4	0.4	0.4	0.4	0.4	0.4	0.4
60 Other Prod.	0.3	0.3	0.3	0.3	0.3	0.3	0.3	0.3	0.3	0.3
61 T&C					1.4	1.5	1.6	1.7	1.7	1.7
62 WOOD AND PAPER	15.1	18.9	19.7	20.4	21.2	22.2	23.2	23.5	23.7	24.2
63 Power	0.6	0.8	0.8	0.8	0.9	0.9	0.9	0.9	0.9	0.9
64 Fuels	1.0	1.2	1.2	1.3	1.0	1.1	1.1	1.1	1.1	1.1
65 Metallurgy	0.4	0.5	0.5	0.5	0.5	0.5	0.6	0.6	0.6	0.6
66 Chemicals	1.3	1.4	1.4	1.5	1.6	1.7	1.7	1.7	1.8	1.8
67 MBMW	1.1	1.3	1.4	1.4	1.5	1.6	1.6	1.6	1.6	1.6
68 Wood & Paper	8.6	11.3	11.9	12.3	12.5	13.1	13.8	13.9	13.9	14.3
69 Constr. Mat.	0.3	0.4	0.4	0.4	0.4	0.4	0.4	0.4	0.4	0.4
70 N.E.C. Industry	0.2	0.2	0.2	0.2	0.2	0.2	0.2	0.2	0.2	0.2
71 Light Industry	0.7	0.9	0.9	1.0	1.1	1.2	1.3	1.4	1.4	1.4
72 Food Industry	0.1	0.1	0.1	0.1	0.1	0.1	0.1	0.1	0.1	0.1
73 Agriculture	0.5	0.5	0.6	0.6	0.7	0.7	0.8	0.9	0.9	1.0
74 Other Prod.	0.3	0.3	0.3	0.3	0.3	0.3	0.3	0.3	0.3	0.3
75 T&C					0.5	0.5	0.5	0.5	0.5	0.5
76 CONSTR. MATER.	15.3	17.2	17.8	18.1	18.8	19.2	19.5	20.2	20.3	20.4
77 Power	1.0	1.2	1.2	1.2	1.3	1.3	1.4	1.5	1.5	1.5
78 Fuels	2.1	2.5	2.6	2.7	2.4	2.5	2.6	2.7	2.7	2.7
79 Metallurgy	1.8	2.2	2.2	2.3	2.3	2.5	2.5	2.6	2.6	2.6
80 Chemicals	0.7	0.7	0.8	0.9	1.0	1.0	1.0	1.1	1.1	1.1
81 MBMW	1.0	1.1	1.1	1.1	1.2	1.2	1.2	1.3	1.3	1.3
82 Wood & Paper	0.6	0.6	0.6	0.6	0.7	0.7	0.7	0.8	0.8	0.8
83 Constr. Mat.	7.5	8.1	8.5	8.5	8.0	8.0	8.1	8.1	8.2	8.3
84 N.E.C. Industry	0.1	0.1	0.1	0.1	0.1	0.1	0.1	0.1	0.1	0.1
85 Light Industry	0.3	0.4	0.4	0.4	0.4	0.4	0.4	0.4	0.4	0.4
86 Food Industry	0.1	0.1	0.1	0.1	0.1	0.1	0.1	0.1	0.1	0.1
87 Other Prod.	0.1	0.2	0.2	0.2	0.2	0.2	0.2	0.2	0.2	0.2
88 T&C					1.1	1.2	1.2	1.3	1.3	1.3

TABLE B1: MATERIAL INPUTS BY SECTOR
(billion rubles)

		1965	1970	1971	1972	1973	1974	1975	1976	1977	1978	1979	1980
89	N.E.C. INDUSTRY	4.1	6.2	6.6	7.1	8.4	8.9	9.8	10.5	11.4	12.3	12.8	13.6
90	Power	0.1	0.2	0.2	0.3	0.3	0.3	0.3	0.3	0.3	0.4	0.4	0.5
91	Fuels	0.1	0.2	0.2	0.2	0.3	0.3	0.3	0.3	0.4	0.4	0.4	0.4
92	Metallurgy	0.3	0.4	0.4	0.4	0.4	0.5	0.5	0.6	0.6	0.7	0.7	0.7
93	Chemicals	0.4	0.6	0.7	0.6	0.8	0.8	0.9	0.9	1.0	1.0	1.0	1.0
94	MBMW	0.1	0.2	0.2	0.3	0.3	0.3	0.4	0.4	0.5	0.6	0.6	0.7
95	Wood & Paper	0.2	0.5	0.5	0.5	0.5	0.5	0.5	0.6	0.6	0.7	0.7	0.7
96	Constr. Mat.	0.0	0.1	0.1	0.1	0.1	0.1	0.1	0.1	0.1	0.1	0.1	0.1
97	N.E.C. Industry	0.4	0.5	0.5	0.6	0.8	0.9	1.0	1.1	1.2	1.3	1.4	1.5
98	Light Industry	0.4	0.4	0.5	0.5	0.6	0.7	0.8	0.9	1.0	1.1	1.1	1.2
99	Food Industry	0.9	1.0	1.0	1.0	1.1	1.2	1.3	1.5	1.7	1.8	1.9	1.9
100	Agriculture	1.1	2.1	2.3	2.4	3.2	3.3	3.5	3.7	3.8	4.2	4.4	4.7
101	Other Prod.	0.0	0.0	0.1	0.1	0.1	0.1	0.1	0.1	0.1	0.1	0.1	0.1
102	T&C												
103	LIGHT INDUSTRY	32.3	48.6	51.9	53.5	61.2	64.2	67.7	70.4	73.3	75.5	77.8	80.3
104	Power	0.2	0.4	0.4	0.5	0.5	0.6	0.6	0.6	0.6	0.6	0.6	0.6
105	Fuels	0.1	0.2	0.2	0.2	0.2	0.2	0.2	0.3	0.3	0.3	0.3	0.3
106	Metallurgy	0.0	0.1	0.1	0.1	0.1	0.1	0.1	0.1	0.1	0.1	0.1	0.1
107	Chemicals	1.3	2.4	2.7	2.9	3.2	3.5	3.8	4.0	4.2	4.6	4.8	5.0
108	MBMW	0.2	0.3	0.3	0.4	0.4	0.4	0.5	0.5	0.5	0.6	0.6	0.6
109	Wood & Paper	0.2	0.3	0.3	0.4	0.4	0.4	0.4	0.4	0.4	0.4	0.4	0.4
110	Constr. Mater.	0.0	0.0	0.0	0.0	0.1	0.1	0.1	0.1	0.1	0.1	0.1	0.1
111	N.E.C. Industry	0.0	0.0	0.0	0.0	0.0	0.1	0.1	0.1	0.1	0.1	0.1	0.1
112	Light Industry	25.0	38.1	40.6	41.5	47.8	49.9	52.7	54.7	57.0	58.3	60.0	62.1
113	Food Industry	0.3	0.4	0.4	0.5	0.5	0.5	0.5	0.5	0.5	0.5	0.5	0.5
114	Agriculture	5.0	6.4	6.9	7.1	8.1	8.5	8.8	9.2	9.6	10.0	10.4	10.6
115	Other Prod.												
116	T&C												
117	FOOD INDUSTRY	48.3	65.3	69.1	72.5	75.4	81.0	85.3	84.3	87.4	89.1	92.0	93.4
118	Power	0.2	0.4	0.4	0.5	0.5	0.5	0.5	0.5	0.5	0.5	0.6	0.6
119	Fuels	0.4	0.8	0.9	0.9	0.9	1.0	1.0	1.0	1.1	1.1	1.1	1.1
120	Metallurgy	0.2	0.3	0.3	0.3	0.3	0.3	0.3	0.4	0.4	0.4	0.4	0.4
121	Chemicals	0.1	0.3	0.3	0.4	0.4	0.5	0.5	0.6	0.6	0.7	0.7	0.7
122	MBMW	0.4	0.7	0.7	0.8	0.8	0.9	0.9	0.9	0.9	1.0	1.0	1.1
123	Wood & Paper	0.4	1.0	1.1	1.1	1.1	1.2	1.3	1.3	1.3	1.4	1.4	1.4
124	Constr. Mater.	0.1	0.1	0.1	0.2	0.2	0.2	0.2	0.2	0.2	0.2	0.2	0.3
125	N.E.C. Industry	0.4	0.5	0.5	0.5	0.6	0.6	0.7	0.7	0.7	0.7	0.7	0.8
126	Light Ind.	0.4	0.7	0.7	0.8	0.8	0.9	0.9	0.9	1.0	1.0	1.0	1.0
127	Food Industry	22.1	27.2	28.9	30.0	31.1	33.6	35.4	34.7	36.3	37.2	39.7	40.8
128	Agriculture	23.5	33.2	35.1	36.9	38.6	41.2	43.5	43.0	44.3	44.7	45.0	45.0
129	Other Prod.	0.1	0.1	0.1	0.1	0.1	0.1	0.1	0.1	0.1	0.2	0.2	0.2
130	T&C												

TABLE B1: MATERIAL INPUTS BY SECTOR
(billion rubles)

		1981	1982	1983	1984	1985	1986	1987	1988	1989	1990
89	N.E.C.INDUSTRY	14.8	17.0	17.8	18.7	19.3	20.2	21.4	22.8	23.0	23.4
90	Power	0.5	0.6	0.6	0.6	0.6	0.6	0.6	0.6	0.7	0.7
91	Fuels	0.5	0.6	0.6	0.6	0.6	0.5	0.5	0.5	0.5	0.5
92	Metallurgy	0.8	0.9	0.9	1.0	1.0	1.1	1.1	1.2	1.2	1.2
93	Chemicals	1.0	1.1	1.2	1.2	1.2	1.2	1.2	1.3	1.3	1.3
94	MBMW	0.7	0.7	0.7	0.8	0.8	0.8	0.8	0.8	0.8	0.8
95	Wood & Paper	0.8	0.9	0.9	0.9	0.9	0.9	0.9	1.0	1.0	1.0
96	Constr. Mat.	0.1	0.1	0.1	0.1	0.1	0.1	0.1	0.1	0.1	0.1
97	N.E.C. Industry	1.7	2.1	2.3	2.6	2.7	2.9	3.1	3.5	3.5	3.6
98	Light Industry	1.3	1.4	1.4	1.5	1.5	1.6	1.6	1.6	1.6	1.6
99	Food Industry	1.9	2.1	2.1	2.2	2.2	2.3	2.4	2.4	2.4	2.4
100	Agriculture	5.4	6.4	6.8	7.0	7.3	8.0	8.8	9.5	9.5	9.7
101	Other Prod.	0.1	0.2	0.2	0.2	0.2	0.2	0.2	0.2	0.2	0.2
102	T&C					0.2	0.2	0.2	0.2	0.2	0.2
103	LIGHT INDUSTRY	81.9	90.2	90.9	91.4	93.2	95.0	95.4	96.0	96.1	96.3
104	Power	0.6	0.8	0.8	0.8	0.8	0.9	0.9	0.9	0.9	0.9
105	Fuels	0.3	0.4	0.4	0.4	0.4	0.4	0.4	0.5	0.5	0.5
106	Metallurgy	0.1	0.2	0.2	0.2	0.2	0.2	0.2	0.2	0.2	0.2
107	Chemicals	5.2	5.3	5.4	5.4	5.5	5.6	5.6	5.7	5.7	5.7
108	MBMW	0.6	0.7	0.7	0.7	0.8	0.8	0.8	0.8	0.8	0.8
109	Wood & Paper	0.4	0.5	0.5	0.5	0.5	0.5	0.5	0.5	0.5	0.5
110	Constr. Mater.	0.1	0.1	0.1	0.1	0.1	0.1	0.1	0.1	0.1	0.1
111	N.E.C. Industry	0.1	0.2	0.2	0.2	0.2	0.2	0.2	0.2	0.2	0.2
112	Light Industry	63.2	70.2	70.1	70.4	71.2	72.4	72.6	72.6	72.8	72.9
113	Food Industry	0.5	0.6	0.6	0.6	0.6	0.6	0.6	0.6	0.6	0.6
114	Agriculture	10.9	11.4	12.1	12.3	12.5	12.9	13.1	13.4	13.4	13.5
115	Other Prod.					0.1	0.1	0.1	0.1	0.1	0.1
116	T&C					0.5	0.5	0.5	0.5	0.5	0.5
117	FOOD INDUSTRY	94.8	101.1	107.4	110.8	112.3	115.5	118.3	119.8	120.0	120.9
118	Power	0.6	0.8	0.8	0.9	0.9	0.9	0.9	1.0	1.0	1.1
119	Fuels	1.1	1.4	1.4	1.5	1.5	1.5	1.6	1.6	1.7	1.7
120	Metallurgy	0.4	0.6	0.6	0.6	0.6	0.6	0.6	0.6	0.6	0.6
121	Chemicals	0.7	0.8	0.8	0.8	0.8	0.9	0.9	0.9	0.9	1.0
122	MBMW	1.1	1.2	1.2	1.3	1.3	1.3	1.3	1.3	1.3	1.3
123	Wood & Paper	1.4	1.7	1.7	1.8	1.8	1.8	1.8	1.8	1.8	1.8
124	Constr. Mater.	0.3	0.4	0.4	0.4	0.4	0.4	0.5	0.5	0.5	0.5
125	N.E.C. Industry	0.8	0.9	0.9	0.9	0.9	0.9	1.0	1.0	1.0	1.0
126	Light Ind.	1.0	1.1	1.1	1.1	1.1	1.1	1.1	1.1	1.1	1.1
127	Food Industry	41.3	45.5	46.8	48.3	48.6	49.0	50.0	51.7	52.0	53.4
128	Agriculture	45.9	46.6	51.5	53.0	52.8	55.5	56.9	56.5	56.3	55.6
129	Other Prod.	0.2	0.2	0.2	0.2	0.2	0.2	0.2	0.2	0.2	0.2
130	T&C					1.4	1.4	1.5	1.6	1.6	1.6

TABLE B1: MATERIAL INPUTS BY SECTOR
(billion rubles)

		1965	1970	1971	1972	1973	1974	1975	1976	1977	1978	1979	1980
131	AGRICULTURE	23.5	35.0	38.7	42.3	45.9	48.1	51.3	55.9	58.9	61.3	65.8	70.0
132	Power	0.2	0.3	0.3	0.3	0.4	0.4	0.4	0.5	0.5	0.5	0.6	0.6
133	Fuels	1.4	2.1	2.3	2.5	2.6	2.7	2.8	3.0	3.2	3.3	3.4	3.5
134	Metallurgy	0.1	0.1	0.1	0.1	0.1	0.1	0.1	0.1	0.1	0.2	0.2	0.2
135	Chemicals	1.2	2.0	2.2	2.5	2.8	3.1	3.3	3.6	3.9	4.3	4.6	5.0
136	MBMW	1.7	3.2	3.4	3.9	4.2	4.5	5.0	5.2	5.6	6.1	6.4	6.9
137	Wood & Paper	0.2	0.5	0.5	0.6	0.6	0.7	0.7	0.7	0.8	0.8	0.8	0.8
138	Constr. Mater.	0.1	0.1	0.1	0.1	0.2	0.2	0.2	0.2	0.3	0.3	0.3	0.4
139	N.E.C. Industry	1.4	2.8	3.2	3.4	4.0	4.4	4.9	5.4	6.0	6.6	7.2	7.9
140	Light Ind.	0.3	0.3	0.4	0.4	0.4	0.4	0.4	0.5	0.5	0.5	0.5	0.5
141	Food Industry	1.2	3.5	3.9	4.3	4.5	4.6	4.7	4.7	4.9	5.0	5.2	5.4
142	Agriculture	15.6	20.0	22.2	24.1	26.0	26.9	28.7	31.9	32.7	33.4	36.3	38.5
143	Other Prod.	0.1	0.1	0.1	0.1	0.1	0.2	0.2	0.2	0.3	0.3	0.3	0.3
144	T&C												
145	T & C	3.9	5.4	5.8	6.3	6.7	7.0	7.4	7.8	8.7	9.9	10.6	11.5
146	Power	0.4	0.7	0.8	0.9	1.0	1.1	1.1	1.2	1.3	1.4	1.6	1.7
147	Fuels	1.9	2.6	2.7	3.0	3.2	3.3	3.4	3.6	4.0	4.8	5.0	5.7
148	Metallurgy	0.1	0.2	0.2	0.2	0.2	0.2	0.2	0.2	0.2	0.2	0.2	0.2
149	Chemicals	0.3	0.5	0.6	0.7	0.7	0.7	0.8	0.8	0.9	0.9	1.0	1.0
150	MBMW	0.8	0.9	0.9	1.0	1.0	1.1	1.2	1.2	1.4	1.6	1.7	1.8
151	Wood & Paper	0.2	0.2	0.2	0.2	0.2	0.2	0.2	0.2	0.2	0.2	0.2	0.2
152	Constr. Mat.	0.0	0.1	0.1	0.1	0.1	0.1	0.2	0.2	0.2	0.3	0.3	0.3
153	N.E.C. Industry	0.1	0.1	0.1	0.1	0.1	0.1	0.1	0.1	0.2	0.2	0.2	0.2
154	Light Industry	0.2	0.2	0.2	0.2	0.2	0.2	0.2	0.2	0.2	0.2	0.2	0.2
155	Other Prod.	0.0	0.0	0.0	0.0	0.0	0.0	0.0	0.1	0.1	0.1	0.2	0.2
156	T&C												
157	CONSTRUCTION	20.5	34.7	38.4	39.2	40.9	43.3	45.9	45.9	46.3	47.5	48.3	49.4
158	Power	0.2	0.4	0.5	0.5	0.6	0.6	0.7	0.7	0.7	0.8	0.9	0.9
159	Fuels	0.5	1.3	1.7	1.7	1.8	1.9	2.1	2.1	2.2	2.3	2.4	2.5
160	Metallurgy	2.4	3.7	4.1	4.2	4.3	4.5	4.6	4.6	4.6	4.7	4.8	5.0
161	Chemicals	0.5	1.0	1.3	1.4	1.5	1.6	1.7	1.7	1.7	1.8	1.8	1.9
162	MBMW	4.3	7.3	7.7	7.8	7.8	8.0	8.3	8.3	8.4	8.6	8.7	9.0
163	Wood & Paper	2.7	4.3	4.6	4.7	4.8	4.9	4.9	4.9	4.9	5.0	5.0	5.1
164	Constr. Mater.	9.2	15.7	17.3	17.7	18.9	20.4	22.1	22.1	22.3	22.7	23.1	23.4
165	N.E.C. Ind.	0.1	0.2	0.2	0.2	0.2	0.2	0.3	0.3	0.3	0.3	0.3	0.3
166	Light Ind.	0.5	0.6	0.6	0.6	0.6	0.6	0.7	0.7	0.7	0.7	0.7	0.7
167	Food Industry	0.0	0.0	0.1	0.1	0.1	0.1	0.1	0.1	0.1	0.1	0.1	0.1
168	Agriculture	0.0	0.0	0.0	0.0	0.1	0.1	0.1	0.1	0.1	0.1	0.1	0.1
169	Other Prod.	0.1	0.2	0.2	0.2	0.2	0.3	0.3	0.3	0.3	0.4	0.4	0.4
170	T&C												

TABLE B1: MATERIAL INPUTS BY SECTOR
           (billion rubles)

		1981	1982	1983	1984	1985	1986	1987	1988	1989	1990
131	AGRICULTURE	72.1	74.3	80.9	86.4	87.5	91.2	92.2	96.2	99.4	106.3
132	Power	0.7	0.9	1.0	1.1	1.1	1.2	1.2	1.2	1.3	1.4
133	Fuels	3.6	4.0	4.1	4.2	3.6	3.7	3.8	3.9	4.1	4.2
134	Metallurgy	0.2	0.2	0.2	0.2	0.2	0.2	0.2	0.3	0.3	0.3
135	Chemicals	5.5	5.9	6.4	6.7	6.9	7.4	7.6	8.1	8.5	9.0
136	MBMW	7.3	7.6	8.0	8.4	8.5	8.8	8.9	9.2	9.5	9.8
137	Wood & Paper	0.8	0.9	0.9	0.9	0.9	1.0	1.0	1.0	1.0	1.0
138	Constr. Mater.	0.4	0.5	0.6	0.7	0.7	0.8	0.8	0.9	0.9	0.9
139	N.E.C. Industry	8.5	9.4	9.8	10.4	11.0	12.4	13.8	15.3	16.0	16.8
140	Light Ind.	0.5	0.6	0.6	0.7	0.7	0.8	0.8	0.8	0.9	0.9
141	Food Industry	5.4	5.6	5.7	5.8	5.8	6.0	6.1	6.4	6.6	6.7
142	Agriculture	38.9	38.4	43.3	47.0	47.1	47.9	47.0	48.0	49.3	54.3
143	Other Prod.	0.3	0.3	0.3	0.3	0.3	0.3	0.3	0.3	0.3	0.3
144	T&C					0.7	0.7	0.7	0.7	0.7	0.7
145	T & C	12.0	13.4	13.8	13.8	15.1	15.4	15.9	15.7	16.7	18.2
146	Power	1.7	2.0	2.1	2.1	2.2	2.2	2.3	2.3	2.4	2.5
147	Fuels	6.0	6.7	6.9	6.9	7.3	7.4	7.4	7.2	7.9	8.8
148	Metallurgy	0.2	0.2	0.2	0.2	0.2	0.2	0.2	0.2	0.2	0.2
149	Chemicals	1.1	1.1	1.2	1.2	1.3	1.3	1.4	1.4	1.4	1.5
150	MBMW	1.9	2.0	2.0	2.0	2.3	2.4	2.6	2.6	2.8	3.0
151	Wood & Paper	0.2	0.2	0.2	0.2	0.3	0.3	0.3	0.3	0.3	0.3
152	Constr. Mat.	0.3	0.4	0.4	0.4	0.5	0.5	0.6	0.6	0.6	0.7
153	N.E.C. Industry	0.2	0.3	0.3	0.3	0.4	0.4	0.4	0.4	0.4	0.4
154	Light Industry	0.2	0.3	0.3	0.3	0.3	0.4	0.4	0.4	0.4	0.4
155	Other Prod.	0.2	0.2	0.2	0.2	0.2	0.2	0.2	0.2	0.2	0.2
156	T&C					0.1	0.1	0.1	0.1	0.1	0.2
157	CONSTRUCTION	50.7	56.1	58.4	63.5	65.3	68.3	70.0	72.9	74.5	75.7
158	Power	0.9	1.1	1.2	1.3	1.4	1.4	1.5	1.6	1.6	1.7
159	Fuels	2.6	3.4	3.6	3.7	2.2	2.3	2.4	2.5	2.7	2.9
160	Metallurgy	5.2	6.4	6.5	7.1	7.5	8.1	8.5	8.8	9.0	9.2
161	Chemicals	2.0	2.1	2.2	2.3	2.4	2.5	2.6	2.8	2.9	2.9
162	MBMW	9.3	9.6	10.2	11.0	11.5	12.6	13.0	13.8	14.1	14.2
163	Wood & Paper	5.2	5.5	5.6	5.7	5.8	6.1	6.3	6.5	6.6	6.7
164	Constr. Mater.	23.9	26.3	27.5	30.7	30.2	30.8	31.1	32.2	32.8	33.2
165	N.E.C. Ind.	0.3	0.3	0.3	0.3	0.4	0.4	0.4	0.4	0.4	0.4
166	Light Ind.	0.7	0.7	0.7	0.8	0.8	0.9	0.9	0.9	0.9	0.9
167	Food Industry	0.1	0.1	0.1	0.1	0.1	0.1	0.1	0.1	0.1	0.1
168	Agriculture	0.1	0.1	0.1	0.1	0.1	0.1	0.1	0.1	0.1	0.1
169	Other Prod.	0.4	0.4	0.4	0.4	0.4	0.4	0.4	0.4	0.4	0.4
170	T&C					2.5	2.6	2.7	2.8	2.9	3.0

TABLE B1: MATERIAL INPUTS BY SECTOR
(billion rubles)

		1965	1970	1971	1972	1973	1974	1975	1976	1977	1978	1979	1980
171	T&D & OTHER PROD	3.5	3.5	3.9	4.2	3.9	4.2	4.2	4.5	4.3	4.6	4.5	4.7
172	Power	0.1	0.2	0.3	0.3	0.3	0.4	0.4	0.4	0.4	0.5	0.5	0.5
173	Fuels	0.1	0.2	0.3	0.3	0.3	0.4	0.4	0.5	0.4	0.5	0.5	0.5
174	Chemicals	0.1	0.1	0.2	0.2	0.2	0.3	0.3	0.3	0.3	0.3	0.3	0.3
175	MBMW	0.3	0.4	0.5	0.6	0.5	0.5	0.5	0.5	0.5	0.5	0.5	0.6
176	Wood & Paper	0.6	0.5	0.5	0.5	0.4	0.5	0.5	0.6	0.5	0.5	0.5	0.6
177	Constr. Mater.	0.2	0.2	0.2	0.2	0.2	0.2	0.2	0.2	0.2	0.2	0.2	0.2
178	N.E.C. Industry	0.4	0.4	0.4	0.4	0.4	0.4	0.4	0.4	0.4	0.4	0.4	0.4
179	Light Industry	0.4	0.4	0.4	0.4	0.4	0.4	0.4	0.4	0.4	0.4	0.4	0.4
180	Food Industry	0.9	0.7	0.7	0.7	0.6	0.6	0.6	0.6	0.6	0.7	0.6	0.6
181	Agriculture	0.2	0.2	0.3	0.3	0.3	0.3	0.3	0.3	0.3	0.3	0.3	0.3
182	Other Sectors	0.2	0.2	0.2	0.3	0.2	0.2	0.2	0.3	0.3	0.3	0.3	0.4
182	T&C												
183	HOUSEHOLD SERV.	8.2	12.0	12.7	13.4	14.5	15.4	16.7	17.9	18.9	20.0	21.1	21.8
184	Power	0.4	0.6	0.7	0.8	1.0	1.1	1.2	1.4	1.5	1.7	1.8	1.9
185	Fuels	0.9	1.2	1.4	1.6	2.1	2.4	2.6	2.8	3.1	3.4	3.7	3.9
186	Metallurgy	0.2	0.2	0.2	0.2	0.2	0.2	0.3	0.3	0.3	0.3	0.3	0.3
187	Chemicals	0.3	0.6	0.7	0.8	0.9	1.1	1.3	1.5	1.7	1.9	2.1	2.2
188	MBMW	0.2	0.6	0.6	0.7	0.7	0.8	0.9	0.9	1.0	1.2	1.3	1.4
189	Wood and Paper	0.3	0.4	0.4	0.5	0.5	0.5	0.6	0.7	0.7	0.7	0.8	0.8
190	Constr. Mater.	0.2	0.2	0.2	0.2	0.2	0.2	0.3	0.3	0.3	0.3	0.4	0.4
191	N.E.C. Ind.	0.7	1.0	1.0	1.0	1.0	1.0	1.1	1.1	1.1	1.2	1.2	1.2
192	Light Industry	1.2	1.7	1.8	1.8	1.9	1.9	1.9	2.0	2.0	2.1	2.1	2.1
193	Food Industry	3.1	4.4	4.6	4.8	5.0	5.2	5.4	5.7	5.9	6.0	6.1	6.2
194	Agriculture	0.6	0.8	0.8	0.8	0.8	0.8	0.8	0.9	0.9	0.9	1.0	1.0
195	Other Prod.	0.1	0.2	0.2	0.2	0.2	0.2	0.3	0.3	0.3	0.3	0.3	0.4
196	T&C												
197	SCIENCE & ADMIN.	3.3	5.3	5.7	6.3	6.9	7.5	7.8	7.8	8.1	8.5	8.7	9.7
198	Power	0.2	0.3	0.4	0.4	0.5	0.5	0.5	0.5	0.5	0.6	0.6	0.6
199	Fuels	0.2	0.3	0.3	0.4	0.4	0.4	0.4	0.4	0.5	0.5	0.5	0.6
200	Metallurgy	0.4	0.6	0.7	0.8	0.8	0.9	1.0	1.0	1.1	1.2	1.2	1.2
201	Chemicals	0.5	0.6	0.7	0.7	0.8	0.8	0.9	0.9	0.9	0.9	0.9	1.0
202	MBMW	1.3	2.3	2.3	2.7	3.0	3.5	3.6	3.6	3.6	3.7	3.9	4.6
203	Wood and Paper	0.2	0.3	0.4	0.4	0.4	0.4	0.4	0.4	0.4	0.4	0.4	0.4
204	Constr. Mater.	0.2	0.3	0.3	0.3	0.3	0.3	0.3	0.3	0.4	0.4	0.4	0.4
205	N.E.C. Ind.	0.2	0.3	0.3	0.3	0.3	0.3	0.3	0.3	0.3	0.4	0.4	0.4
206	Light Industry	0.1	0.2	0.2	0.2	0.2	0.2	0.2	0.2	0.2	0.2	0.2	0.3
207	Other Prod.	0.1	0.1	0.1	0.1	0.2	0.2	0.2	0.2	0.2	0.2	0.2	0.2
208	T&C												

TABLE B1: MATERIAL INPUTS BY SECTOR
      (billion rubles)

		1981	1982	1983	1984	1985	1986	1987	1988	1989	1990
171	T & D and OTHER	4.8	5.5	5.4	5.3	6.5	6.6	8.5	9.5	9.7	9.8
172	Power	0.5	0.6	0.6	0.6	0.7	0.7	1.0	1.1	1.1	1.1
173	Fuels	0.5	0.6	0.6	0.5	0.6	0.6	0.6	0.6	0.6	0.6
174	Chemicals	0.3	0.3	0.3	0.3	0.3	0.3	0.3	0.3	0.3	0.3
175	MBMW	0.6	0.6	0.6	0.6	0.9	0.9	1.1	1.2	1.3	1.3
176	Wood & Paper	0.6	0.7	0.7	0.7	0.9	0.9	1.2	1.4	1.4	1.4
177	Constr. Mater.	0.2	0.3	0.3	0.3	0.3	0.4	0.4	0.4	0.4	0.4
178	N.E.C. Industry	0.4	0.5	0.5	0.5	0.6	0.6	0.8	1.0	1.0	1.0
179	Light Industry	0.4	0.5	0.5	0.5	0.6	0.6	0.7	0.7	0.8	0.8
180	Food Industry	0.6	0.6	0.6	0.6	0.6	0.6	0.6	0.6	0.6	0.6
181	Agriculture	0.3	0.4	0.3	0.3	0.4	0.4	0.6	0.7	0.7	0.7
182	Other Sectors	0.4	0.4	0.4	0.4	0.4	0.4	0.6	0.9	0.9	1.0
182	T&C					0.2	0.2	0.5	C.6	0.6	0.6
183	HOUSEHOLD SERV.	22.6	24.3	25.3	25.8	26.5	27.7	28.8	30.3	31.4	33.0
184	Power	2.0	2.7	2.8	2.9	3.0	3.2	3.3	3.5	3.6	3.8
185	Fuels	4.1	4.6	4.8	4.9	4.5	4.7	5.1	5.4	5.8	6.2
186	Metallurgy	0.3	0.3	0.4	0.4	0.4	0.5	0.5	0.6	0.6	0.6
187	Chemicals	2.3	2.4	2.5	2.6	2.6	2.8	2.9	3.1	3.2	3.3
188	MBMW	1.6	1.7	1.8	1.9	2.0	2.1	2.2	2.3	2.4	2.5
189	Wood and Paper	0.8	0.9	0.9	0.9	1.0	1.0	1.0	1.1	1.1	1.2
190	Constr. Mater.	0.5	0.5	0.5	0.5	0.6	0.7	0.7	0.8	0.8	0.8
191	N.E.C. Ind.	1.2	1.2	1.3	1.3	1.4	1.4	1.5	1.5	1.6	1.7
192	Light Industry	2.1	2.2	2.2	2.2	2.2	2.3	2.3	2.3	2.4	2.4
193	Food Industry	6.3	6.4	6.5	6.6	6.8	6.9	7.1	7.3	7.5	7.9
194	Agriculture	1.0	1.0	1.1	1.1	1.1	1.1	1.1	1.2	1.2	1.3
195	Other Prod.	0.4	0.4	0.5	0.5	0.6	0.7	0.8	0.9	0.9	1.0
196	T&C					0.3	0.3	0.3	0.3	0.3	0.3
197	SCIENCE & ADMIN.	10.2	10.7	11.6	12.2	16.3	16.8	18.5	19.9	22.2	23.3
198	Power	0.6	0.8	0.9	1.0	1.1	1.1	1.3	1.4	1.6	1.7
199	Fuels	0.6	0.7	0.8	0.9	0.9	0.9	1.0	1.0	1.1	1.1
200	Metallurgy	1.2	1.3	1.4	1.5	1.5	1.6	1.7	1.8	1.9	1.9
201	Chemicals	1.0	1.0	1.2	1.3	1.3	1.3	1.4	1.5	1.7	1.8
202	MBMW	5.2	5.3	5.6	5.8	6.0	6.2	6.9	7.8	9.1	9.9
203	Wood and Paper	0.4	0.4	0.4	0.4	0.4	0.4	0.5	0.5	0.6	0.6
204	Constr. Mater.	0.4	0.4	0.4	0.4	0.4	0.4	0.4	0.4	0.4	0.4
205	N.E.C. Ind.	0.4	0.4	0.4	0.4	0.4	0.4	0.4	0.4	0.4	0.4
206	Light Industry	0.2	0.2	0.2	0.2	0.2	0.2	0.2	0.2	0.2	0.2
207	Other Prod.	0.2	0.2	0.3	0.3	0.3	0.3	0.3	0.3	0.3	0.3
208	T&C					3.8	4.0	4.4	4.6	4.9	5.0

TABLE B2: STRUCTURE OF MATERIAL INPUTS IN PRODUCTION AND SERVICE SECTORS
       (billion rubles)

		1965	1970	1971	1972	1973	1974	1975	1976	1977	1978	1979	1980
1	TOTAL	217.7	339.7	363.8	386.4	413.3	440.6	471.0	487.0	512.1	534.5	553.5	572.8
2	Percentage	100	100	100	100	100	100	100	100	100	100	100	100
3	Power	5.2	10.0	10.8	11.6	12.6	13.6	14.3	15.1	15.7	16.8	18.0	18.6
4	(3/1)*100	2.40	2.93	2.97	3.00	3.05	3.09	3.04	3.09	3.07	3.14	3.24	3.24
5	Fuels	19.0	29.7	31.6	34.0	36.0	38.2	40.5	42.6	45.0	47.8	49.4	51.8
6	(5/1)*100	8.72	8.73	8.67	8.79	8.72	8.68	8.60	8.74	8.79	8.95	8.93	9.04
7	Metallurgy	22.8	36.4	38.6	41.4	43.9	47.5	50.7	53.1	56.6	59.2	60.3	61.0
8	(7/1)*100	10.46	10.70	10.62	10.72	10.61	10.77	10.77	10.91	11.06	11.08	10.89	10.65
9	Chemicals	11.0	20.7	23.0	25.1	27.5	30.3	32.7	34.9	37.3	39.6	41.7	43.8
10	(9/1)*100	5.06	6.10	6.33	6.49	6.64	6.87	6.95	7.17	7.29	7.40	7.54	7.64
11	MBMW	23.2	47.6	50.3	55.6	57.3	61.1	68.2	69.5	76.2	82.0	85.0	89.9
12	(11/1)*100	10.66	14.00	13.83	14.40	13.87	13.87	14.48	14.26	14.88	15.35	15.36	15.69
13	Wood & Paper	11.0	17.8	18.9	20.0	20.3	20.9	22.0	22.4	22.6	23.2	23.4	23.8
14	(13/1)*100	5.07	5.24	5.19	5.18	4.91	4.74	4.68	4.59	4.42	4.35	4.23	4.16
15	Constr. Material	13.1	21.2	23.3	24.0	26.0	28.9	31.3	31.6	32.3	33.2	34.0	34.7
16	(15/1)*100	6.01	6.24	6.40	6.22	6.30	6.55	6.64	6.50	6.30	6.22	6.14	6.06
17	N.E.C. Industry	4.5	6.9	7.4	7.7	8.7	9.4	10.2	10.8	11.7	12.7	13.4	14.3
18	(17/1)*100	2.08	2.04	2.03	2.00	2.11	2.13	2.16	2.22	2.28	2.38	2.42	2.49
19	Light Industry	30.2	45.5	48.6	49.7	56.5	59.1	62.3	64.9	67.6	69.3	71.1	73.5
20	(19/1)*100	13.89	13.41	13.36	12.85	13.67	13.40	13.22	13.32	13.21	12.97	12.84	12.83
21	Food Industry	29.6	38.5	41.0	42.8	44.5	47.3	49.6	49.4	51.7	52.9	55.7	57.2
22	(21/1)*100	13.58	11.34	11.26	11.08	10.75	10.75	10.54	10.15	10.09	9.91	10.06	9.98
23	Agriculture	46.0	62.9	67.7	71.8	77.1	81.3	85.9	89.4	91.9	93.9	98.0	100.6
24	(23/1)*100	21.13	18.51	18.62	18.58	18.66	18.45	18.24	18.35	17.95	17.58	17.70	17.56
25	Other Prod.	2.0	2.5	2.6	2.8	2.9	3.0	3.1	3.4	3.4	3.6	3.6	3.7
26	(25/1)*100	0.93	0.74	0.71	0.72	0.70	0.69	0.67	0.69	0.66	0.68	0.65	0.65
27	T&C												
28	(28/1)*100												

TABLE B2: STRUCTURE OF MATERIAL INPUTS IN PRODUCTION AND SERVICE SECTORS
        (billion rubles)

		1981	1982	1983	1984	1985	1986	1987	1988	1989	1990
1	TOTAL	588.3	662.4	690.1	716.6	740.0	767.9	787.6	815.5	831.5	844.2
2	Percentage	100	100	100	100	100	100	100	100	100	100
3	Power	19.1	24.4	25.3	26.2	27.3	28.3	29.4	30.4	31.3	32.1
4	(3/1)*100	3.25	3.68	3.66	3.65	3.68	3.68	3.73	3.73	3.76	3.80
5	Fuels	53.5	70.7	72.4	73.9	70.2	70.8	73.2	78.8	81.1	82.9
6	(5/1)*100	9.10	10.68	10.49	10.32	9.49	9.22	9.29	9.66	9.76	9.82
7	Metallurgy	62.5	75.5	76.8	78.8	80.4	86.5	87.7	89.2	90.2	90.6
8	(7/1)*100	10.62	11.40	11.14	10.99	10.87	11.26	11.14	10.94	10.85	10.73
9	Chemicals	46.2	49.2	52.0	54.0	55.2	56.9	58.4	60.5	61.8	61.9
10	(9/1)*100	7.86	7.43	7.53	7.54	7.46	7.41	7.41	7.42	7.43	7.33
11	MBMW	93.0	103.2	108.3	114.9	119.6	126.6	131.0	138.7	144.6	144.7
12	(11/1)*100	15.82	15.58	15.69	16.03	16.16	16.49	16.64	17.01	17.39	17.14
13	Wood & Paper	24.1	28.5	29.2	30.0	30.9	32.0	33.4	34.4	34.6	35.1
14	(13/1)*100	4.10	4.30	4.23	4.19	4.17	4.16	4.24	4.22	4.17	4.16
15	Constr. Materia	35.5	39.1	40.9	44.2	43.4	44.4	45.1	46.3	47.1	47.7
16	(15/1)*100	6.04	5.90	5.92	6.16	5.87	5.78	5.72	5.68	5.66	5.65
17	N.E.C. Industry	15.1	17.3	17.9	19.0	20.1	21.7	23.7	25.8	26.6	27.6
18	(17/1)*100	2.56	2.61	2.59	2.65	2.71	2.82	3.01	3.16	3.20	3.27
19	Light Industry	74.7	83.0	83.0	83.8	84.8	86.7	87.2	87.4	87.9	88.0
20	(19/1)*100	12.70	12.53	12.02	11.70	11.46	11.29	11.07	10.72	10.57	10.42
21	Food Industry	57.7	62.7	64.2	66.0	66.5	67.3	68.7	70.9	71.6	73.5
22	(21/1)*100	9.81	9.47	9.30	9.21	8.99	8.76	8.72	8.70	8.61	8.71
23	Agriculture	103.0	104.8	115.9	121.5	122.0	126.7	128.5	130.4	131.5	136.3
24	(23/1)*100	17.51	15.82	16.80	16.95	16.49	16.50	16.32	15.98	15.81	16.15
25	Other Prod.	3.8	4.1	4.3	4.3	4.3	4.4	4.7	5.1	5.1	5.3
26	(25/1)*100	0.64	0.62	0.62	0.60	0.58	0.57	0.60	0.62	0.61	0.63
27	T&C					15.3	15.7	16.7	17.6	18.0	18.6
28	(28/1)*100					2.07	2.05	2.12	2.16	2.16	2.21

115

TABLE B2: STRUCTURE OF MATERIAL INPUTS IN PRODUCTION AND SERVICE SECTORS
         (billion rubles)

		1965	1970	1971	1972	1973	1974	1975	1976	1977	1978	1979	1980
29	MAT. PRODUCTION	206.2	322.4	345.5	366.7	391.9	417.7	446.5	461.3	485.1	505.9	523.7	541.3
30	Power	4.6	9.0	9.7	10.3	11.1	12.0	12.6	13.1	13.7	14.5	15.6	16.1
31	Fuels	18.0	28.1	29.8	32.0	33.6	35.4	37.5	39.4	41.3	43.9	45.2	47.2
32	Metallurgy	22.2	35.6	37.7	40.4	42.9	46.3	49.4	51.8	55.2	57.7	58.8	59.5
33	Chemicals	10.2	19.5	21.6	23.6	25.8	28.4	30.5	32.5	34.7	36.8	38.7	40.6
34	MBMW	21.8	44.7	47.4	52.2	53.6	56.8	63.7	65.0	71.6	77.1	79.8	83.9
35	Wood	10.5	17.1	18.1	19.1	19.4	20.0	21.0	21.3	21.5	22.1	22.2	22.6
36	Con. MAt.	12.7	20.7	22.8	23.5	25.5	28.4	30.7	31.0	31.6	32.5	33.2	33.9
37	NEC	3.6	5.6	6.1	6.4	7.4	8.1	8.8	9.4	10.3	11.1	11.8	12.7
38	Light	28.9	43.6	46.6	47.7	54.4	57.0	60.2	62.7	65.4	67.0	68.8	71.1
39	Food	26.4	34.1	36.4	38.0	39.5	42.1	44.2	43.7	45.8	46.9	49.6	51.0
40	Agri	45.4	62.1	66.9	71.0	76.3	80.5	85.1	88.5	91.0	93.0	97.0	99.6
41	Other Prod.	1.8	2.2	2.3	2.5	2.5	2.6	2.6	2.9	2.9	3.1	3.1	3.1
42	T&C	0.0	0.0	0.0	0.0	0.0	0.0	0.0	0.0	0.0	0.0	0.0	0.0
43	SERVICES	11.5	17.3	18.4	19.7	21.4	22.9	24.5	25.7	27.0	28.5	29.8	31.5
44	Power	0.6	1.0	1.1	1.3	1.5	1.6	1.7	1.9	2.0	2.3	2.4	2.5
45	Fuels	1.0	1.5	1.7	2.0	2.5	2.8	3.0	3.2	3.6	3.9	4.2	4.5
46	Metallurgy	0.6	0.8	0.9	1.0	1.0	1.1	1.3	1.3	1.4	1.5	1.5	1.5
47	Chemicals	0.8	1.2	1.4	1.5	1.7	1.9	2.2	2.4	2.6	2.8	3.0	3.2
48	MBMW	1.5	2.8	2.9	3.4	3.7	4.3	4.5	4.5	4.6	4.9	5.2	6.0
49	Wood and Paper	0.5	0.7	0.8	0.9	0.9	0.9	1.0	1.1	1.1	1.1	1.2	1.2
50	Constr. Mater.	0.4	0.5	0.5	0.5	0.5	0.5	0.6	0.6	0.7	0.7	0.8	0.8
51	N.E.C. Ind.	0.9	1.3	1.3	1.3	1.3	1.3	1.4	1.4	1.4	1.6	1.6	1.6
52	Light Industry	1.3	1.9	2.0	2.0	2.1	2.1	2.1	2.2	2.2	2.3	2.3	2.4
53	Food Industry	3.1	4.4	4.6	4.8	5.0	5.2	5.4	5.7	5.9	6.0	6.1	6.2
54	Agriculture	0.6	0.8	0.8	0.8	0.8	0.8	0.8	0.9	0.9	0.9	1.0	1.0
55	Other Prod.	0.2	0.3	0.3	0.3	0.4	0.4	0.5	0.5	0.5	0.5	0.5	0.6
56	T&C												

TABLE B2: STRUCTURE OF MATERIAL INPUTS IN PRODUCTION AND SERVICE SECTORS
(billion rubles)

		1981	1982	1983	1984	1985	1986	1987	1988	1989	1990
29	MAT. PRODUCTION	555.5	627.5	653.2	678.5	697.2	723.4	740.4	765.4	777.9	787.9
30	Power	16.5	20.9	21.6	22.3	23.2	24.0	24.8	25.5	26.1	26.6
31	Fuels	48.9	65.5	66.8	68.1	64.8	65.2	67.1	72.4	74.2	75.6
32	Metallurgy	61.0	73.9	75.0	76.9	78.5	84.4	85.5	86.8	87.7	88.1
33	Chemicals	42.9	45.8	48.3	50.1	51.3	52.8	54.1	55.9	56.9	56.8
34	MBMW	86.2	96.2	100.9	107.1	111.6	118.3	121.9	128.6	133.1	132.3
35	Wood	22.9	27.2	27.9	28.7	29.5	30.6	31.9	32.8	32.9	33.3
36	Con. MAt.	34.6	38.2	40.0	43.3	42.4	43.3	44.0	45.1	45.9	46.5
37	NEC	13.5	15.7	16.2	17.3	18.3	19.9	21.8	23.9	24.6	25.5
38	Light	72.4	80.6	80.6	81.4	82.4	84.2	84.7	84.9	85.3	85.4
39	Food	51.4	56.3	57.7	59.4	59.7	60.4	61.6	63.6	64.1	65.6
40	Agri	102.0	103.8	114.8	120.4	120.9	125.6	127.4	129.2	130.3	135.0
41	Other Prod.	3.2	3.5	3.5	3.5	3.4	3.4	3.6	3.9	3.9	4.0
42	T&C					11.2	11.4	12.0	12.7	12.9	13.2
43	SERVICES	32.8	35.0	36.9	38.1	42.8	44.5	47.3	50.2	53.6	56.3
44	Power	2.6	3.5	3.7	3.9	4.1	4.3	4.6	4.9	5.2	5.5
45	Fuels	4.7	5.3	5.6	5.8	5.4	5.6	6.1	6.4	6.9	7.3
46	Metallurgy	1.5	1.6	1.8	1.9	1.9	2.1	2.2	2.4	2.5	2.5
47	Chemicals	3.3	3.4	3.7	3.9	3.9	4.1	4.3	4.6	4.9	5.1
48	MBMW	6.8	7.0	7.4	7.7	8.0	8.3	9.1	10.1	11.5	12.4
49	Wood and Paper	1.2	1.3	1.3	1.3	1.4	1.4	1.5	1.6	1.7	1.8
50	Constr. Mater.	0.9	0.9	0.9	0.9	1.0	1.1	1.1	1.2	1.2	1.2
51	N.E.C. Ind.	1.6	1.6	1.7	1.7	1.8	1.8	1.9	1.9	2.0	2.1
52	Light Industry	2.3	2.4	2.4	2.4	2.4	2.5	2.5	2.5	2.6	2.6
53	Food Industry	6.3	6.4	6.5	6.6	6.8	6.9	7.1	7.3	7.5	7.9
54	Agriculture	1.0	1.0	1.1	1.1	1.1	1.1	1.1	1.2	1.2	1.3
55	Other Prod.	0.6	0.6	0.8	0.8	0.9	1.0	1.1	1.2	1.2	1.3
56	T&C					4.1	4.3	4.7	4.9	5.1	5.4

TABLE B3: HOUSEHOLD CONSUMPTION OF MATERIAL GOODS AND SERVICES
        (billion rubles)

	1965	1970	1971	1972	1973	1974	1975	1976	1977	1978	1979	1980
1 TOTAL (2+3+4+5)	120.2	171.5	181.1	191.0	199.7	210.4	223.2	233.7	244.0	256.2	269.4	288.2
2 Retail	101.6	150.0	159.2	169.7	177.9	187.9	201.4	211.1	221.3	232.3	245.0	261.7
3 Utilities	1.1	1.6	1.8	1.8	1.9	2.0	2.2	2.5	2.6	2.7	2.8	3.0
4 Consump. In-kind	16.2	18.3	18.4	17.8	18.0	18.5	17.5	20.0	20.1	21.2	21.6	23.5
5 Other	1.2	1.6	1.7	1.8	1.9	2.0	2.1	0.0	0.0	0.0	0.0	0.0
6 POWER	0.9	1.2	1.3	1.3	1.4	1.5	1.7	1.8	1.9	2.0	2.1	2.2
7 FUELS (8+9)	0.7	1.3	1.5	1.6	1.8	1.9	1.9	2.1	2.4	2.6	2.7	2.8
8 Retail	0.6	1.1	1.2	1.3	1.5	1.6	1.6	1.7	2.0	2.2	2.3	2.4
9 Heating Gas	0.1	0.2	0.3	0.3	0.3	0.3	0.3	0.4	0.4	0.4	0.4	0.4
10 FERROUS METALS												
11 CHEMICALS	2.7	3.8	4.0	4.2	4.3	4.4	4.4	4.6	4.8	4.9	5.1	5.2
12 MBMW (13-14-15)	5.1	8.7	9.4	11.0	11.9	13.4	15.0	15.8	17.1	18.5	19.6	21.3
13 Retail	5.4	9.3	10.3	12.3	13.6	15.4	17.2	18.1	19.5	20.9	22.2	24.1
14 Second Hand	0.1	0.3	0.6	1.0	1.4	1.7	1.8	1.9	2.0	2.1	2.2	2.3
15 Orgs	0.3	0.3	0.3	0.3	0.3	0.4	0.4	0.4	0.4	0.4	0.4	0.4
16 WOOD&PAPER (17-18)	2.5	3.9	4.2	4.5	4.8	5.0	5.2	5.5	5.7	6.0	6.5	7.2
17 Retail	2.6	4.1	4.4	4.7	5.0	5.3	5.6	5.9	6.1	6.4	6.9	7.6
18 Orgs	0.1	0.2	0.2	0.2	0.2	0.3	0.4	0.4	0.4	0.4	0.4	0.4
19 CONST. MAT.(20+21)	0.5	0.7	0.9	1.1	1.1	1.3	1.5	1.6	1.8	2.1	2.1	2.3
20 Glass & China	0.4	0.6	0.7	0.9	0.9	1.0	1.2	1.3	1.5	1.7	1.8	1.9
21 Constr. Materials	0.1	0.1	0.2	0.2	0.2	0.2	0.3	0.3	0.3	0.4	0.4	0.4
22 N.E.C. (23-24+25)	1.6	2.7	3.0	3.3	3.5	3.8	4.4	5.1	5.4	6.5	7.5	8.9
23 Retail	1.6	2.6	3.0	3.3	3.5	3.9	4.5	5.2	5.5	6.6	7.6	8.9
24 Orgs	0.1	0.1	0.2	0.2	0.2	0.3	0.4	0.4	0.4	0.4	0.4	0.4
25 Water Supply	0.1	0.2	0.2	0.2	0.2	0.2	0.3	0.3	0.3	0.3	0.3	0.4
26 LIGHT (27-28-29)	25.0	39.0	41.2	43.4	45.1	47.4	51.2	53.7	56.8	58.8	62.4	67.4
27 Retail	25.5	39.7	42.0	44.2	46.1	48.4	52.3	54.9	58.1	60.1	63.8	68.9
28 Second Hand	0.2	0.3	0.4	0.4	0.5	0.5	0.6	0.6	0.7	0.7	0.7	0.7
29 Orgs	0.3	0.4	0.4	0.4	0.5	0.5	0.5	0.6	0.6	0.6	0.7	0.8
30 FOOD (31-32-33)	55.4	78.0	83.2	87.7	91.8	96.3	101.9	106.1	109.3	114.1	118.5	124.8
31 Retail	58.9	83.0	88.4	93.1	97.5	102.2	108.1	112.6	116.1	121.0	125.6	132.0
32 Producer												
33 Orgs	3.5	5.0	5.2	5.4	5.7	5.9	6.2	6.5	6.8	6.9	7.1	7.2
34 AGRI (35-36+37)	23.2	28.7	28.9	29.1	30.0	31.2	31.6	34.8	35.9	37.8	39.7	42.5
35 Retail	8.1	11.5	11.7	12.7	13.4	14.1	15.6	16.3	17.4	18.2	19.7	20.8
36 Orgs	0.5	0.5	0.6	0.7	0.7	0.7	0.7	0.7	0.7	0.7	0.7	0.7
37 In-kind	15.6	17.7	17.8	17.1	17.3	17.8	16.7	19.2	19.2	20.3	20.7	22.4
38 OTHER(39-40-41+42)	1.6	2.2	2.3	2.5	2.7	2.8	3.0	3.1	3.4	3.6	3.8	4.1
39 Printed Matter	1.1	1.7	1.8	2.0	2.1	2.2	2.3	2.4	2.6	2.8	3.0	3.2
40 Orgs	0.1	0.1	0.1	0.1	0.1	0.1	0.1	0.1	0.1	0.1	0.1	0.1
41 Second Hand												
42 In-kind	0.6	0.6	0.6	0.7	0.7	0.7	0.8	0.8	0.9	0.9	0.9	1.0

TABLE B3: HOUSEHOLD CONSUMPTION OF MATERIAL GOODS AND SERVICES
(billion rubles)

	1981	1982	1983	1984	1985	1986	1987	1988	1989	1990
1 TOTAL (2+3+4+5)	305.0	314.9	325.9	337.4	345.8	351.6	361.0	379.9	410.6	429.0
2 Retail	277.7	286.8	295.6	304.9	311.8	317.5	324.4	340.2	370.1	387.3
3 Utilities	3.2	3.8	4.0	4.3	4.7	5.2	5.7	5.9	6.2	6.4
4 Consump. In-kind	24.1	24.3	26.3	28.2	29.3	28.9	30.9	33.8	34.3	35.3
5 Other	0.0	0.0	0.0	0.0	0.0	0.0	0.0	0.0	0.0	0.0
6 POWER	2.3	2.8	3.0	3.3	3.6	4.1	4.5	4.6	4.9	5.1
7 FUELS (8+9)	3.0	3.2	3.4	3.5	3.8	4.3	4.6	5.4	5.6	5.6
8 Retail	2.5	2.6	2.8	2.9	3.1	3.6	3.9	4.6	4.8	4.8
9 Heating Gas	0.5	0.6	0.6	0.6	0.7	0.7	0.7	0.8	0.8	0.8
10 FERROUS METALS	0.1	0.1	0.1	0.1	0.1	0.1	0.1	0.1	0.1	0.1
11 CHEMICALS	5.4	5.5	5.8	6.2	7.2	8.5	9.0	9.7	10.1	10.3
12 MBMW (13-14-15)	23.7	23.9	25.7	26.2	27.5	29.1	30.0	31.1	34.4	36.7
13 Retail	26.7	27.3	29.4	30.3	32.2	34.6	36.3	38.8	42.6	45.2
14 Second Hand	2.6	2.9	3.2	3.5	3.8	4.7	5.5	6.8	7.2	7.5
15 Orgs	0.4	0.5	0.5	0.6	0.8	0.8	0.9	0.9	1.0	1.0
16 WOOD&PAPER (17-18)	7.6	7.8	8.1	8.4	8.8	9.4	9.8	10.6	11.1	11.4
17 Retail	8.0	8.2	8.5	8.9	9.4	10.0	10.4	11.2	11.8	12.1
18 Orgs	0.4	0.4	0.4	0.5	0.6	0.6	0.6	0.6	0.7	0.7
19 CONST. MAT.(20+21)	2.6	2.8	2.7	2.7	2.8	2.9	3.1	3.4	3.5	3.5
20 Glass & China	2.1	2.3	2.2	2.2	2.3	2.3	2.5	2.6	2.7	2.7
21 Constr. Materials	0.5	0.5	0.5	0.5	0.5	0.6	0.6	0.8	0.8	0.8
22 N.E.C. (23-24+25)	9.3	8.2	8.1	8.1	8.4	9.0	9.7	10.6	11.1	11.5
23 Retail	9.3	8.2	8.1	8.1	8.4	9.0	9.6	10.6	11.1	11.5
24 Orgs	0.4	0.4	0.4	0.4	0.4	0.4	0.4	0.4	0.5	0.5
25 Water Supply	0.4	0.4	0.4	0.4	0.4	0.4	0.5	0.5	0.5	0.5
26 LIGHT (27-28-29)	73.0	74.1	75.1	77.6	80.9	83.6	81.3	83.6	89.3	92.0
27 Retail	74.5	75.6	76.7	79.2	82.5	85.3	83.0	86.1	92.2	95.1
28 Second Hand	0.7	0.7	0.7	0.7	0.7	0.7	0.8	1.3	1.4	1.5
29 Orgs	0.8	0.8	0.9	0.9	0.9	1.0	1.0	1.2	1.5	1.6
30 FOOD (31-32-33)	129.5	136.0	140.4	145.1	144.6	140.8	145.3	153.1	172.0	182.6
31 Retail	136.8	143.4	148.0	152.9	152.6	149.4	155.8	166.3	181.3	192.4
32 Producer							0.4	2.1	4.1	
33 Orgs	7.3	7.4	7.6	7.8	8.0	8.2	8.4	9.0	9.3	9.8
34 AGRI (35-36+37)	44.9	46.5	49.2	51.5	53.0	54.6	57.5	60.8	62.0	63.4
35 Retail	22.6	24.0	24.8	25.3	25.8	27.8	28.7	29.2	30.2	30.7
36 Orgs	0.7	0.7	0.7	0.7	0.7	0.7	0.7	0.7	0.8	0.8
37 In-kind	23.0	23.2	25.1	26.9	27.9	27.5	29.5	32.3	32.6	33.5
38 OTHER(39-40-41+42)	4.2	4.6	5.0	5.3	5.7	6.0	6.3	6.7	7.4	7.8
39 Printed Matter	3.3	3.7	4.1	4.3	4.6	4.9	5.2	5.6	6.1	6.4
40 Orgs	0.1	0.1	0.2	0.2	0.2	0.2	0.2	0.3	0.3	0.3
41 Second Hand	0.1	0.1	0.1	0.1	0.1	0.1	0.1	0.1	0.1	0.1
42 In-kind	1.1	1.1	1.2	1.3	1.4	1.4	1.4	1.6	1.7	1.8

TABLE B4: RETAIL TRADE
     (billion rubles)

	1965	1970	1971	1972	1973	1974	1975	1976	1977	1978	1979	1980
1 TOTAL RETAIL	108.4	159.4	169.7	181.0	190.3	201.2	215.6	225.9	236.4	247.8	260.9	278.0
2 State-Coop	104.8	155.2	165.6	176.4	185.7	196.7	210.4	220.1	230.6	241.3	254.0	270.5
3 Ex-Village	3.6	4.2	4.1	4.6	4.6	4.5	5.2	5.8	5.8	6.5	6.9	7.5
4 FUELS	0.9	1.5	1.6	1.8	2.0	2.2	2.3	2.4	2.7	2.9	3.0	3.1
5 Private Use	0.6	1.1	1.2	1.3	1.5	1.6	1.6	1.7	2.0	2.2	2.3	2.4
6 Public Use	0.3	0.4	0.5	0.5	0.5	0.6	0.7	0.7	0.7	0.7	0.7	0.7
7 CHEMICALS	2.7	3.8	4.0	4.2	4.3	4.4	4.4	4.6	4.8	4.9	5.1	5.2
8 Rubber Footwear	0.7	0.9	0.9	0.9	0.9	0.9	0.9	1.0	1.0	1.1	1.1	1.1
9 Detergents	0.1	0.4	0.5	0.6	0.6	0.7	0.7	0.7	0.8	0.8	0.9	0.9
10 Other Chemicals	1.9	2.5	2.6	2.7	2.8	2.8	2.8	2.9	3.0	3.0	3.1	3.2
11 MBMW CONSUMER	5.4	9.3	10.3	12.3	13.6	15.4	17.2	18.1	19.5	20.9	22.2	24.1
12 Automobiles	0.2	0.5	1.2	2.3	3.4	4.7	5.7	6.0	6.5	7.3	7.8	8.8
13 Bikes &Electronics	3.8	6.7	6.9	7.5	7.6	8.1	8.6	9.1	9.8	10.2	10.8	11.4
14 Metal Works	0.8	1.0	1.1	1.2	1.3	1.3	1.4	1.5	1.6	1.7	1.8	1.9
15 Sporting Goods	0.3	0.5	0.5	0.6	0.6	0.6	0.7	0.7	0.7	0.8	0.8	0.9
16 Supplies, etc.	0.3	0.6	0.6	0.7	0.7	0.7	0.8	0.8	0.9	0.9	1.0	1.1
17 WOOD & PAPER	2.6	4.1	4.4	4.7	5.0	5.3	5.6	5.9	6.1	6.4	6.9	7.6
18 Furniture	2.0	2.9	3.2	3.4	3.6	3.9	4.2	4.4	4.6	4.9	5.3	6.0
19 Logging	0.4	0.5	0.5	0.5	0.5	0.5	0.5	0.5	0.5	0.5	0.5	0.5
20 Office Supplies	0.2	0.7	0.7	0.8	0.9	0.9	0.9	1.0	1.0	1.0	1.1	1.1
21 CONSTR. MATERIALS	1.5	2.2	2.5	2.8	2.8	3.0	3.3	3.4	3.6	3.9	4.0	4.2
22 Glass & China	0.4	0.6	0.7	0.9	0.9	1.0	1.2	1.3	1.5	1.7	1.8	1.9
23 Constr. Materials	1.1	1.6	1.8	1.9	1.9	2.0	2.1	2.1	2.1	2.2	2.2	2.3
24 N.E.C. INDUSTRY	1.6	2.6	3.0	3.3	3.5	3.9	4.5	5.2	5.5	6.6	7.6	8.9
25 Music & Toys	0.6	0.9	1.0	1.0	1.1	1.2	1.3	1.4	1.5	1.6	1.6	1.7
26 Medical Supplies	0.5	0.7	0.8	0.8	0.8	0.9	1.0	1.1	1.2	1.3	1.3	1.4
27 Jewelry	0.2	0.5	0.7	1.0	1.1	1.3	1.6	2.0	2.0	2.8	3.7	4.6
28 Everyday Services	0.2	0.3	0.3	0.3	0.3	0.3	0.4	0.4	0.5	0.5	0.6	0.6
29 Crafts, etc.	0.1	0.2	0.2	0.2	0.2	0.2	0.2	0.3	0.3	0.4	0.5	0.6
30 LIGHT INDUSTRY	25.5	39.7	42.0	44.2	46.1	48.4	52.3	54.9	58.1	60.1	63.8	68.9
31 Textiles & Apparel	23.1	35.7	37.5	39.3	40.9	42.9	46.4	48.6	50.7	52.4	55.0	58.2
32 Rugs	0.3	0.6	0.7	0.8	0.9	1.0	1.1	1.2	1.8	2.0	2.7	3.9
33 Haberdashery, etc.	2.1	3.4	3.8	4.1	4.3	4.5	4.8	5.1	5.6	5.7	6.1	6.8
34 FOOD INDUSTRY	58.9	83.0	88.4	93.1	97.5	102.2	108.1	112.6	116.1	121.0	125.6	132.0
35 Food Products	55.9	78.7	83.7	88.1	92.2	96.7	102.1	106.4	109.5	114.0	118.0	123.8
36 Tobacco	1.9	2.8	3.0	3.1	3.3	3.4	3.7	3.8	4.0	4.1	4.3	4.6
37 Cosmetics & Soap	1.0	1.4	1.5	1.7	1.8	1.9	2.1	2.2	2.4	2.7	3.1	3.4
38 Ex-Village	0.1	0.1	0.2	0.2	0.2	0.2	0.2	0.2	0.2	0.2	0.2	0.2
39 Cooperatives												
40 AGRICULTURE	8.1	11.5	11.7	12.7	13.4	14.1	15.6	16.3	17.4	18.2	19.7	20.8
41 State-Coop	4.1	6.8	7.1	7.7	8.2	9.0	9.8	9.8	10.9	10.9	12.0	12.5
42 Coop. & Dining	0.5	0.7	0.7	0.7	0.8	0.8	0.8	0.9	0.9	1.0	1.0	1.0
43 Ex-Village	3.5	4.0	3.9	4.3	4.4	4.3	5.0	5.6	5.6	6.3	6.7	7.3

TABLE B4: RETAIL TRADE
(billion rubles)

	1981	1982	1983	1984	1985	1986	1987	1988	1989	1990
1 TOTAL RETAIL	294.5	304.3	314.5	324.8	333.0	341.2	350.8	375.7	413.2	431.5
2 State-Coop	286.0	295.7	305.8	316.1	324.2	332.1	341.5	366.4	403.5	421.5
3 Ex-Village	8.5	8.6	8.7	8.7	8.8	9.1	9.3	9.3	9.7	10.0
4 FUELS	3.2	3.4	3.6	3.7	4.0	4.5	4.8	5.5	5.7	5.7
5 Private Use	2.5	2.6	2.8	2.9	3.1	3.6	3.9	4.6	4.8	4.8
6 Public Use	0.7	0.8	0.8	0.8	0.9	0.9	0.9	0.9	0.9	0.9
7 CHEMICALS	5.4	5.5	5.8	6.2	7.2	8.5	9.0	9.7	10.1	10.3
8 Rubber Footwear	1.1	1.2	1.1	1.2	1.5	1.4	1.5	1.5	1.6	1.6
9 Detergents	1.0	0.9	1.0	1.0	1.0	1.1	1.2	1.3	1.5	1.6
10 Other Chemicals	3.3	3.4	3.7	4.0	4.7	6.0	6.3	6.9	7.0	7.1
11 MBMW CONSUMER	26.7	27.3	29.4	30.3	32.2	34.6	36.3	38.8	42.6	45.2
12 Automobiles	10.7	10.7	11.7	11.8	12.3	13.0	13.6	14.5	15.6	16.7
13 Bikes &Electronics	12.0	12.4	13.3	14.0	15.1	16.6	17.2	18.4	20.6	21.7
14 Metal Works	2.0	2.1	2.2	2.3	2.3	2.4	2.6	2.7	2.8	2.8
15 Sporting Goods	0.9	0.9	1.0	1.0	1.1	1.2	1.3	1.4	1.6	1.7
16 Supplies, etc.	1.1	1.2	1.2	1.2	1.3	1.4	1.5	1.8	2.0	2.3
17 WOOD & PAPER	8.0	8.2	8.5	8.9	9.4	10.0	10.4	11.2	11.8	12.1
18 Furniture	6.4	6.5	6.8	7.1	7.6	8.1	8.4	9.1	9.6	9.8
19 Logging	0.5	0.5	0.5	0.5	0.5	0.5	0.5	0.6	0.6	0.6
20 Office Supplies	1.1	1.2	1.2	1.3	1.3	1.4	1.5	1.5	1.6	1.7
21 CONSTR. MATERIALS	4.6	4.8	5.4	5.7	6.3	7.1	8.1	9.3	12.0	12.3
22 Glass & China	2.1	2.3	2.2	2.2	2.3	2.3	2.5	2.6	2.7	2.7
23 Constr. Materials	2.5	2.6	3.2	3.5	4.0	4.8	5.7	6.7	9.3	9.6
24 N.E.C. INDUSTRY	9.3	8.2	8.1	8.1	8.4	9.0	9.6	10.6	11.1	11.5
25 Music & Toys	1.8	1.9	2.0	2.1	2.2	2.4	2.5	2.6	2.8	2.9
26 Medical Supplies	1.4	1.5	1.6	1.7	1.8	2.0	2.1	2.2	2.3	2.4
27 Jewelry	4.8	3.5	3.2	3.0	3.0	3.1	3.4	4.2	4.3	4.5
28 Everyday Services	0.7	0.7	0.7	0.7	0.7	0.8	0.8	0.8	0.9	0.9
29 Crafts, etc.	0.6	0.6	0.6	0.6	0.7	0.7	0.8	0.8	0.8	0.8
30 LIGHT INDUSTRY	74.5	75.6	76.7	79.2	82.5	85.3	83.0	86.9	92.2	95.1
31 Textiles & Apparel	62.2	62.9	64.1	66.8	70.0	72.0	69.8	73.2	78.2	80.6
32 Rugs	4.8	5.4	5.2	4.8	4.6	4.9	4.6	4.8	4.9	5.0
33 Haberdashery, etc.	7.5	7.3	7.4	7.6	7.9	8.5	8.6	8.9	9.1	9.5
34 FOOD INDUSTRY	136.8	143.4	148.0	152.9	152.6	149.4	155.8	166.3	181.3	192.4
35 Food Products	127.8	133.3	137.6	142.3	141.4	137.6	143.6	152.6	165.5	175.4
36 Tobacco	5.3	6.3	6.5	6.5	6.7	6.8	6.8	6.9	7.3	7.3
37 Cosmetics & Soap	3.5	3.6	3.7	3.9	4.3	4.8	5.1	5.6	6.3	6.5
38 Ex-Village	0.2	0.2	0.2	0.2	0.2	0.2	0.2	0.2	0.2	0.2
39 Cooperatives							0.1	1.0	2.0	3.0
40 AGRICULTURE	22.6	24.0	24.8	25.3	25.8	27.8	28.7	29.2	30.2	30.7
41 State-Coop	13.2	14.5	15.2	15.6	16.0	17.6	18.3	18.8	19.5	20.0
42 Coop. & Dining	1.1	1.1	1.1	1.2	1.2	1.3	1.3	1.3	1.4	1.4
43 Ex-Village	8.3	8.4	8.5	8.5	8.6	8.9	9.1	9.1	9.3	9.3

TABLE B4: RETAIL TRADE
           (billion rubles)

	1965	1970	1971	1972	1973	1974	1975	1976	1977	1978	1979	1980
44 PRINTED MATTER	1.1	1.7	1.8	2.0	2.1	2.2	2.3	2.4	2.6	2.8	3.0	3.2
45 OTHER GOODS	0.0	0.0	0.0	0.0	0.0	0.0	0.0	0.0	0.0	0.0	0.0	0.0

Supporting Data on Retail Trade:

	1965	1970	1971	1972	1973	1974	1975	1976	1977	1978	1979	1980
46 ORG. PURCHASES	5.5	7.4	7.9	8.3	8.7	9.4	10.0	10.4	10.7	10.8	11.1	11.3
47 Non-Prod. Orgs.	4.1	5.6	5.9	6.2	6.5	7.0	7.4	7.8	8.0	8.1	8.3	8.5
48 Agriculture	0.3	0.3	0.3	0.4	0.4	0.4	0.4	0.4	0.4	0.4	0.4	0.4
49 Food Ind.	3.1	4.4	4.6	4.8	5.0	5.2	5.4	5.7	5.9	6.0	6.1	6.2
50 Light Ind.	0.3	0.4	0.4	0.4	0.5	0.5	0.5	0.6	0.6	0.6	0.7	0.8
51 Heavy Ind.	0.4	0.5	0.6	0.6	0.6	0.9	1.1	1.1	1.1	1.1	1.1	1.1
52 MBMW	0.1	0.1	0.1	0.1	0.1	0.2	0.2	0.2	0.2	0.2	0.2	0.2
53 Wood & Paper	0.1	0.2	0.2	0.2	0.2	0.3	0.4	0.4	0.4	0.4	0.4	0.4
54 N.E.C. Ind.	0.1	0.1	0.2	0.2	0.2	0.3	0.4	0.4	0.4	0.4	0.4	0.4
55 Printing	0.1	0.1	0.1	0.1	0.1	0.1	0.1	0.1	0.1	0.1	0.1	0.1
56 Production Orgs.	1.4	1.8	2.0	2.1	2.2	2.4	2.6	2.6	2.7	2.7	2.8	2.8
57 Food	0.4	0.6	0.6	0.6	0.7	0.7	0.8	0.8	0.9	0.9	1.0	1.0
58 Heavy Ind.	0.8	1.0	1.1	1.2	1.2	1.4	1.5	1.5	1.5	1.5	1.5	1.5
59 MBMW	0.2	0.2	0.2	0.2	0.2	0.2	0.2	0.2	0.2	0.2	0.2	0.2
60 Fuels	0.3	0.4	0.5	0.5	0.5	0.6	0.7	0.7	0.7	0.7	0.7	0.7
61 Constr. Mater.	0.3	0.4	0.4	0.5	0.5	0.6	0.6	0.6	0.6	0.6	0.6	0.6
62 Ex-Village	0.2	0.2	0.3	0.3	0.3	0.3	0.3	0.3	0.3	0.3	0.3	0.3
63 SECOND HAND GOODS	0.3	0.6	1.0	1.4	1.9	2.2	2.4	2.5	2.7	2.8	2.9	3.1
64 Autos	0.0	0.1	0.3	0.7	1.1	1.4	1.5	1.5	1.6	1.7	1.8	1.9
65 Other MBMW	0.1	0.2	0.3	0.3	0.3	0.3	0.3	0.4	0.4	0.4	0.4	0.4
66 Apparel, etc.	0.2	0.3	0.4	0.4	0.5	0.5	0.6	0.6	0.7	0.7	0.7	0.7
67 Printed Matter											0.1	0.1

TABLE B4: RETAIL TRADE
        (billion rubles)

	1981	1982	1983	1984	1985	1986	1987	1988	1989	1990
44 PRINTED MATTER	3.3	3.7	4.1	4.3	4.6	4.9	5.2	5.6	6.1	6.4
45 OTHER GOODS	0.1	0.1	0.1	0.1	0.1	0.1	0.1	2.7	10.1	9.8

Supporting Data on Retail Trade:

	1981	1982	1983	1984	1985	1986	1987	1988	1989	1990
46 ORG. PURCHASES	11.4	11.7	12.1	12.5	13.3	13.5	13.8	14.7	15.8	16.5
47 Non-Prod. Orgs.	8.6	8.8	9.1	9.5	9.9	10.0	10.2	10.8	11.7	12.2
48 Agriculture	0.4	0.4	0.4	0.4	0.4	0.4	0.4	0.4	0.5	0.5
49 Food Ind.	6.3	6.4	6.5	6.7	6.8	6.9	7.0	7.3	7.5	7.9
50 Light Ind.	0.8	0.8	0.9	0.9	0.9	1.0	1.0	1.2	1.5	1.6
51 Heavy Ind.	1.1	1.2	1.3	1.5	1.7	1.7	1.8	1.9	2.2	2.2
52 MBMW	0.2	0.3	0.3	0.4	0.5	0.5	0.6	0.6	0.7	0.7
53 Wood & Paper	0.4	0.4	0.4	0.5	0.6	0.6	0.6	0.6	0.7	0.7
54 N.E.C. Ind.	0.4	0.4	0.4	0.4	0.4	0.4	0.4	0.4	0.5	0.5
55 Printing	0.1	0.1	0.2	0.2	0.2	0.2	0.2	0.3	0.3	0.3
56 Production Orgs.	2.8	2.9	3.0	3.0	3.4	3.5	3.6	3.9	4.1	4.3
57 Food	1.0	1.0	1.1	1.1	1.2	1.3	1.4	1.7	1.8	1.9
58 Heavy Ind.	1.5	1.6	1.6	1.6	1.9	1.9	1.9	1.9	2.0	2.1
59 MBMW	0.2	0.2	0.2	0.2	0.3	0.3	0.3	0.3	0.3	0.3
60 Fuels	0.7	0.8	0.8	0.8	0.9	0.9	0.9	0.9	0.9	0.9
61 Constr. Mater.	0.6	0.6	0.6	0.6	0.7	0.7	0.7	0.7	0.8	0.9
62 Ex-Village	0.3	0.3	0.3	0.3	0.3	0.3	0.3	0.3	0.3	0.3
63 SECOND HAND GOODS	3.4	3.7	4.0	4.3	4.6	5.5	6.4	8.2	8.8	9.2
64 Autos	2.1	2.4	2.6	2.8	3.0	3.7	4.4	5.7	6.0	6.3
65 Other MBMW	0.5	0.5	0.6	0.7	0.8	1.0	1.1	1.1	1.2	1.2
66 Apparel, etc.	0.7	0.7	0.7	0.7	0.7	0.7	0.8	1.3	1.5	1.6
67 Printed Matter	0.1	0.1	0.1	0.1	0.1	0.1	0.1	0.1	0.1	0.1

TABLE B5: INTEGRATED CAPITAL FLOWS
          (billion rubles)

	1965	1970	1971	1972	1973	1974	1975	1976	1977	1978	1979	1980
1 TOTAL	62.1	103.2	111.9	118.1	123.1	132.5	142.8	148.4	153.6	164.2	167.4	173.2
2 CAPITAL INVEST+	48.5	81.5	88.2	93.4	96.5	103.5	113.0	116.7	119.9	127.8	129.2	132.7
3 price index*	0.85	0.99	1.00	0.99	0.98	0.98	1.00	0.99	0.98	0.99	0.99	0.99
4 New Construct.	32.9	56.0	61.5	64.4	66.3	70.8	75.3	76.6	77.8	81.2	80.3	81.1
5 price index*	0.84	1.00	1.01	0.99	0.99	1.00	1.00	1.00	0.99	0.99	0.99	0.99
6 M & E Invest.	15.3	25.0	26.2	28.5	29.7	32.2	37.1	39.4	41.6	45.8	48.1	50.8
7 const. prices	17.5	25.3	26.6	28.8	31.1	34.1	37.1	40.7	43.0	46.6	48.6	50.8
8 price index*	0.87	0.99	0.99	0.99	0.96	0.95	1.00	0.97	0.97	0.98	0.99	1.00
9 Domestic M & E	13.1	21.7	22.9	24.5	25.1	26.9	29.1	30.2	31.4	33.1	35.4	37.4
10 const. prices	14.8	22.1	23.5	25.2	27.4	30.1	30.1	33.3	35.0	37.0	39.4	41.4
11 price index*	0.89	0.98	0.98	0.97	0.92	0.90	0.97	0.91	0.90	0.90	0.90	0.90
12 Imported M & E	2.2	3.3	3.3	4.0	4.6	5.3	8.1	9.2	10.2	12.7	12.6	13.4
13 const. prices	2.7	3.2	3.1	3.6	3.7	4.0	7.0	7.4	8.0	9.6	9.2	9.4
14 price index*	0.80	1.03	1.06	1.12	1.25	1.32	1.15	1.24	1.28	1.32	1.37	1.42
15 AGRI. CAPITAL	0.7	1.2	1.5	1.3	1.4	1.7	1.0	1.2	1.5	1.7	1.4	1.3
16 BUDGET M & E	0.6	0.9	0.9	1.0	1.0	1.1	1.3	1.3	1.4	1.5	1.6	1.6
17 CAPITAL REPAIR	12.4	19.7	21.3	22.5	24.2	26.1	27.5	29.3	30.7	33.2	35.1	37.6
18 Constr.	7.2	11.6	12.7	13.2	14.6	15.7	16.5	17.8	18.6	20.2	21.5	23.0
19 M & E	5.2	8.1	8.6	9.2	9.6	10.4	11.0	11.5	12.2	13.0	13.6	14.6
20 UNINSTALLED	4.5	6.9	8.3	10.2	4.8	7.5	7.7	10.6	11.8	10.3	11.7	2.0
21 NET INVESTMENT	28.0	51.1	53.7	55.2	60.2	62.0	62.5	59.6	60.1	66.1	62.3	69.4
22 DEPRECIATION	29.6	45.1	50.0	52.7	58.1	63.0	72.6	78.2	81.7	87.7	93.4	101.8

Capital Outlays by Sector of Origin:

	1965	1970	1971	1972	1973	1974	1975	1976	1977	1978	1979	1980
23 Construction	40.0	67.5	74.3	77.4	80.7	86.2	91.5	93.9	95.9	100.8	101.1	103.4
24 Total for M & E	21.0	34.0	35.7	38.7	40.3	43.8	49.4	52.1	55.1	60.4	63.3	67.0
25 MBMW	20.7	33.5	35.2	38.2	39.7	43.2	48.7	51.3	54.3	59.5	62.4	66.0
26 Wood Prod.	0.3	0.5	0.5	0.5	0.6	0.6	0.7	0.8	0.8	0.9	0.9	1.0
27 Agriculture	1.0	1.6	1.9	1.8	1.9	2.2	1.6	1.9	2.1	2.4	2.2	2.1
28 Light Industry	0.1	0.1	0.0	0.2	0.2	0.3	0.3	0.4	0.4	0.5	0.5	0.5
29 Other Prod.	0.0	0.0	0.0	0.0	0.0	0.0	0.0	0.1	0.1	0.1	0.2	0.2

+ including gardening in public farms
* 1969, 1976 & 1984 construction indexes were used in 1965-1974, 1975-1983 & 1984-1990
  1969, 1973 and 1984 M & E price indexes were used in 1965-1974, 1975-1983 & 1984-1990

TABLE B5: INTEGRATED CAPITAL FLOWS
     (billion rubles)

	1981	1982	1983	1984	1985	1986	1987	1988	1989	1990
1 TOTAL	179.9	192.9	202.6	221.3	230.4	249.0	262.9	278.0	283.4	285.7
2 CAPITAL INVEST+	137.2	147.1	153.8	167.5	173.4	189.0	199.6	212.0	216.1	216.4
3 price index*	0.99	1.02	1.01	0.96	0.97	0.97	0.97	0.97	0.95	0.94
4 New Construct.	83.0	89.5	92.1	102.4	104.3	114.5	121.7	129.8	132.7	137.2
5 price index*	0.98	1.03	1.01	0.94	0.94	0.94	0.93	0.93	0.91	0.92
6 M & E Invest.	53.4	56.8	60.9	64.3	68.3	73.7	77.1	81.4	82.7	78.5
7 const. prices	53.4	55.9	59.7	64.3	67.6	72.3	73.3	77.1	77.5	72.1
8 price index*	1.00	1.02	1.02	1.00	1.01	1.02	1.05	1.06	1.07	1.09
9 Domestic M & E	39.4	40.8	42.7	43.3	45.7	51.5	53.3	58.0	58.2	54.7
10 const. prices	43.8	45.2	47.7	43.3	45.7	51.5	52.1	56.9	57.1	53.1
11 price index*	0.90	0.90	0.90	1.00	1.00	1.00	1.02	1.02	1.02	1.03
12 Imported M & E	14.0	16.0	18.2	21.0	22.6	22.2	23.8	23.4	24.5	23.8
13 const. prices	9.6	10.7	12.0	21.0	21.9	20.8	21.3	20.2	20.4	19.0
14 price index*	1.46	1.49	1.52	1.00	1.03	1.07	1.12	1.16	1.20	1.25
15 AGRI. CAPITAL	1.4	1.0	1.7	2.3	2.2	2.5	2.7	3.0	3.0	2.6
16 BUDGET M & E	1.8	1.9	1.9	2.0	2.2	2.3	2.5	2.5	2.6	2.6
17 CAPITAL REPAIR	39.6	43.0	45.3	49.5	52.5	55.2	58.1	60.5	61.6	64.0
18 Constr.	24.1	26.5	28.1	30.8	32.9	34.3	35.4	36.9	37.1	38.0
19 M & E	15.5	16.5	17.1	18.7	19.7	21.0	22.7	23.6	24.5	26.0
20 UNINSTALLED	5.2	8.1	6.1	9.9	10.9	13.9	9.0	22.2	30.0	30.8
21 NET INVESTMENT	65.9	69.9	73.7	80.6	80.2	87.0	94.1	89.5	79.9	72.6
22 DEPRECIATION	108.8	114.9	122.8	130.8	139.3	148.1	159.8	166.3	173.5	182.3

Capital Outlays by Sector of Origin:

	1981	1982	1983	1984	1985	1986	1987	1988	1989	1990
23 Construction	106.4	115.1	119.3	132.3	136.3	147.9	155.9	165.5	168.6	174.0
24 Total for M & E	70.6	75.1	79.9	85.0	90.2	96.9	102.3	107.5	109.8	107.1
25 MBMW	69.5	73.9	78.7	83.7	88.8	95.4	100.7	105.8	108.1	105.5
26 Wood Prod.	1.1	1.2	1.2	1.3	1.4	1.5	1.6	1.7	1.7	1.6
27 Agriculture	2.2	1.8	2.5	3.1	3.0	3.3	3.5	3.8	3.8	3.4
28 Light Industry	0.5	0.6	0.6	0.6	0.6	0.6	0.8	0.8	0.8	0.8
29 Other Prod.	0.2	0.3	0.3	0.3	0.3	0.3	0.4	0.4	0.4	0.4

+ including gardening in public farms
* 1969, 1976 & 1984 construction indexes were used in 1965-1974, 1975-1983 & 1984-1990
  1969, 1973 and 1984 M & E price indexes were used in 1965-1974, 1975-1983 & 1984-1990

TABLE B5: INTEGRATED CAPITAL FLOWS
        (billion rubles)

	1965	1970	1971	1972	1973	1974	1975	1976	1977	1978	1979	1980
30 PROD. SECTORS	41.9	69.8	76.2	80.6	84.9	92.5	100.5	104.9	107.7	115.8	117.9	122.4
31 CAPITAL INVEST.	32.0	53.8	58.9	62.5	65.4	71.4	79.0	81.7	83.4	89.3	90.2	92.8
32 New Constr.	17.8	30.4	34.1	35.6	37.5	41.1	44.1	44.7	44.3	46.3	45.2	45.7
33 M & E	14.0	22.8	24.1	26.1	27.2	29.6	34.2	36.3	38.3	42.1	43.9	46.1
34 Gardening**	0.3	0.6	0.7	0.7	0.6	0.7	0.7	0.7	0.8	0.9	1.0	0.9
35 LIVESTOCK	0.7	1.0	1.2	1.1	1.3	1.5	0.9	1.2	1.3	1.5	1.2	1.2
36 Public Farms	0.5	0.9	1.0	0.9	1.2	1.2	0.8	1.5	1.3	1.5	1.2	1.1
37 Private	0.2	0.1	0.2	0.2	0.1	0.3	0.1	-0.3	0.0	0.0	0.0	0.1
38 CAPITAL REPAIR	9.2	15.0	16.1	17.0	18.2	19.6	20.6	22.0	23.0	25.0	26.5	28.4
39 Constr.	4.5	7.6	8.3	8.6	9.4	10.1	10.6	11.6	12.0	13.2	14.1	15.2
40 M & E	4.7	7.4	7.8	8.4	8.8	9.5	10.0	10.4	11.0	11.8	12.4	13.2
41 UNINSTALLED	4.0	5.9	7.2	8.7	4.7	6.7	6.8	8.7	10.3	8.5	10.0	1.9
42 Writeoffs	1.6	2.5	2.7	2.4	2.3	2.4	2.6	2.8	2.9	3.3	3.7	3.3
43 State-Coop	2.0	2.9	4.0	5.8	1.9	3.9	3.9	5.7	7.1	4.8	5.8	-1.3
44 Collectives	0.4	0.5	0.5	0.5	0.5	0.4	0.3	0.2	0.3	0.4	0.5	-0.1
45 NET INVESTMENT	17.5	32.1	33.5	34.7	39.0	40.9	40.9	39.3	38.4	43.8	39.9	45.5
46 DEPRECIATION	20.3	31.8	35.5	37.2	41.2	45.0	52.8	56.9	59.0	63.5	68.0	75.0
47 SERVICE SECTORS	20.2	33.4	35.7	37.6	38.2	40.0	42.4	43.5	45.9	48.4	49.5	50.9
48 CAPITAL INVEST.	16.4	27.8	29.6	31.2	31.2	32.3	34.3	35.0	36.7	38.7	39.3	40.1
49 New Constr.	15.1	25.6	27.4	28.8	28.8	29.7	31.4	31.9	33.4	34.9	35.1	35.4
50 M & E Invest.	1.3	2.2	2.2	2.4	2.5	2.7	2.9	3.1	3.3	3.7	4.2	4.7
51 BUDGET M & E	0.6	0.9	0.9	1.0	1.0	1.1	1.3	1.3	1.4	1.5	1.6	1.6
52 All-Republic	0.5	0.7	0.7	0.8	0.8	0.9	1.0	1.0	1.1	1.2	1.3	1.3
53 All-Union	0.1	0.2	0.2	0.2	0.2	0.2	0.3	0.3	0.3	0.3	0.3	0.3
54 CAPITAL REPAIR	3.2	4.7	5.2	5.5	6.0	6.5	6.9	7.3	7.7	8.2	8.6	9.2
55 Constr.	2.8	4.0	4.4	4.6	5.2	5.6	5.9	6.2	6.6	7.0	7.4	7.8
56 M & E	0.5	0.7	0.8	0.8	0.8	0.9	1.0	1.1	1.2	1.2	1.2	1.4
57 UNINSTALLED	0.5	1.0	1.1	1.5	0.1	0.9	0.9	1.9	1.5	1.8	1.7	0.1
58 State-Coop	0.5	1.0	1.0	1.5	0.1	0.8	1.0	1.7	1.3	1.7	1.6	0.0
59 Coll. & Priv.	0.0	0.0	0.1	0.0	0.0	0.1	-0.1	0.2	0.2	0.1	0.1	0.1
60 NET INVESTMENT	10.4	19.0	20.2	20.5	21.2	21.1	21.7	20.3	21.7	22.3	22.4	23.9
61 DEPRECIATION	9.2	13.3	14.5	15.6	16.9	18.0	19.8	21.3	22.7	24.3	25.4	26.9
62 Services	8.6	12.5	13.5	14.7	15.9	17.0	18.7	20.2	21.5	23.1	24.2	25.6
63 Defense	0.6	0.8	0.9	0.9	1.0	1.0	1.1	1.1	1.2	1.2	1.2	1.3

Supporting data on capital investment in constant prices:

	1965	1970	1971	1972	1973	1974	1975	1976	1977	1978	1979	1980
64 PROD SECTORS*	39.7	57.2	61.6	66.8	71.3	77.6	83.0	87.2	90.2	96.1	97.2	99.7
65 Constr.	23.4	33.7	36.8	39.9	42.3	45.7	48.2	49.0	50.0	52.6	52.0	52.9
66 M & E	16.0	23.1	24.4	26.4	28.5	31.4	34.2	37.5	39.6	42.8	44.4	46.0
67 Gardening	0.3	0.4	0.4	0.5	0.5	0.5	0.6	0.7	0.6	0.7	0.8	0.8
68 NONPROD SECTORS	17.3	24.8	26.4	27.5	27.4	28.1	30.0	30.8	32.1	33.1	33.4	34.0
69 Constr.	15.8	22.6	24.2	25.1	24.8	25.4	27.1	27.6	28.7	29.3	29.2	29.2
70 M & E	1.5	2.2	2.2	2.4	2.6	2.7	2.9	3.2	3.4	3.8	4.2	4.8
71 Nonprod T & C	1.9	2.7	2.9	3.3	3.5	3.8	4.3	4.5	4.7	5.5	5.5	5.5

*including investment in nonproductive T&C sectors

TABLE B5: INTEGRATED CAPITAL FLOWS
(billion rubles)

	1981	1982	1983	1984	1985	1986	1987	1988	1989	1990
30 PROD. SECTORS	126.6	136.6	141.9	154.1	163.8	177.4	186.3	193.0	200.0	204.0
31 CAPITAL INVEST.	95.3	103.0	105.9	113.9	120.0	131.4	137.9	142.7	149.4	151.8
32 New Constr.	46.1	50.9	49.9	54.5	56.9	62.8	66.4	67.4	72.9	80.3
33 M & E	48.3	51.3	55.0	58.3	62.1	67.4	70.3	74.1	75.1	70.3
34 Gardening**	0.9	0.9	1.0	1.1	1.0	1.2	1.2	1.3	1.4	1.2
35 LIVESTOCK	1.3	0.9	1.5	2.0	2.0	2.1	2.3	2.5	2.4	2.2
36 Public Farms	1.2	0.8	1.2	1.7	1.7	1.8	2.0	2.1	2.0	1.9
37 Private	0.1	0.1	0.3	0.3	0.3	0.3	0.3	0.4	0.4	0.3
38 CAPITAL REPAIR	30.0	32.7	34.5	38.2	41.8	43.9	46.1	47.8	48.2	50.0
39 Constr.	15.9	17.7	18.9	21.1	23.8	24.7	25.4	26.2	25.8	26.2
40 M & E	14.1	15.0	15.6	17.1	18.0	19.2	20.7	21.6	22.4	23.8
41 UNINSTALLED	4.4	6.9	5.1	8.0	9.8	12.0	8.3	18.7	27.5	28.8
42 Writeoffs	4.2	4.8	4.2	4.7	5.1	6.8	7.5	6.9	7.0	7.5
43 State-Coop	0.2	2.0	0.8	3.2	4.6	5.1	0.8	11.2	20.0	20.8
44 Collectives	0.0	0.1	0.1	0.1	0.1	0.1	0.0	0.6	0.5	0.5
45 NET INVESTMENT	41.6	44.6	45.4	48.4	46.1	50.4	53.5	45.5	38.2	33.6
46 DEPRECIATION	80.6	85.1	91.4	97.7	107.9	115.0	124.5	128.8	134.3	141.6
47 SERVICE SECTORS	53.3	56.3	60.7	67.2	66.6	71.7	76.6	84.9	83.4	81.7
48 CAPITAL INVEST.	42.0	44.1	48.1	53.9	53.6	58.1	62.1	69.7	67.3	65.1
49 New Constr.	36.9	38.6	42.2	47.9	47.5	51.8	55.3	62.4	59.8	56.9
50 M & E Invest.	5.1	5.5	5.9	6.0	6.2	6.3	6.8	7.3	7.6	8.2
51 BUDGET M & E	1.8	1.9	1.9	2.0	2.2	2.3	2.5	2.5	2.6	2.6
52 All-Republic	1.4	1.5	1.5	1.6	1.7	1.8	2.0	2.0	2.1	2.1
53 All-Union	0.4	0.4	0.4	0.4	0.5	0.5	0.5	0.5	0.5	0.5
54 CAPITAL REPAIR	9.6	10.3	10.8	11.3	10.7	11.3	12.0	12.7	13.4	14.0
55 Constr.	8.2	8.8	9.2	9.7	9.1	9.6	10.0	10.7	11.3	11.8
56 M & E	1.4	1.5	1.5	1.6	1.7	1.8	2.0	2.0	2.1	2.2
57 UNINSTALLED	0.8	1.2	1.0	1.9	1.1	1.9	0.7	3.5	2.5	2.0
58 State-Coop	0.7	1.0	0.8	1.8	1.0	1.8	0.7	3.4	2.4	2.0
59 Coll. & Priv.	0.1	0.2	0.2	0.1	0.1	0.1	0.0	0.1	0.1	0.0
60 NET INVESTMENT	24.3	25.3	28.3	32.2	34.1	36.6	40.6	44.0	41.7	39.0
61 DEPRECIATION	28.2	29.8	31.4	33.1	31.4	33.2	35.3	37.4	39.2	40.7
62 Services	26.9	28.4	30.0	31.6	29.8	31.5	33.6	35.6	37.3	38.7
63 Defense	1.3	1.4	1.4	1.5	1.5	1.6	1.7	1.8	1.9	2.0

Supporting data on capital investment in constant prices:

	1981	1982	1983	1984	1985	1986	1987	1988	1989	1990
64 PROD SECTORS*	103.2	106.6	112.2	126.5	129.8	139.5	145.2	154.2	161.4	160.0
65 Constr.	54.2	55.4	57.6	67.4	67.5	72.7	77.7	83.2	85.3	87.2
66 M & E	48.2	50.4	53.8	58.3	61.5	66.0	66.7	70.2	75.3	72.0
67 Gardening	0.8	0.8	0.8	0.8	0.8	0.8	0.8	0.8	0.8	0.8
68 NONPROD SECTORS	35.6	37.2	39.8	47.8	49.7	54.9	60.2	64.0	67.1	70.0
69 Constr.	30.4	31.7	33.9	41.8	43.6	48.6	53.6	57.1	60.0	62.5
70 M & E	5.2	5.5	5.9	6.0	6.1	6.3	6.6	6.9	7.1	7.5
71 Nonprod T & C	5.7	6.0	6.4	7.7	7.4	7.8	8.2	8.5	8.8	9.5

*including investment in nonproductive T&C sectors

127

TABLE B6: CAPITAL INVESTMENT BY SECTOR IN CONSTANT PRICES
        (billion rubles)

		1965	1970	1971	1972	1973	1974	1975	1976	1977	1978	1979	1980
1	TOTAL	57.0	82.0	88.0	94.3	98.7	105.7	112.9	118.0	122.3	129.1	130.6	133.7
2	Construction	39.1	56.1	60.8	64.9	66.9	70.9	75.0	76.4	78.5	81.6	81.0	81.9
3	Assembly Works	35.8	50.3	54.7	58.2	59.6	62.9	66.2	67.0	68.4	69.5	69.1	69.0
4	Design Works	0.7	1.5	1.5	1.8	1.9	2.1	2.3	2.3	2.5	2.5	2.6	2.7
5	Oil & Gas Expl	0.8	1.2	1.3	1.4	1.5	1.6	1.7	1.7	1.9	2.0	2.2	2.4
6	Land Reclam.	0.2	0.2	0.3	0.3	0.5	0.6	0.5	0.3	0.9	0.7	0.4	0.3
7	Other Expend.	1.6	2.9	3.0	3.2	3.4	3.7	4.3	5.1	4.8	6.9	6.7	7.5
8	M & E	17.5	25.3	26.6	28.8	31.1	34.1	37.1	40.7	43.0	46.6	48.6	50.8
9	Gardening	0.3	0.4	0.4	0.4	0.5	0.5	0.6	0.7	0.6	0.7	0.8	0.8
10	Light Industry	0.1	0.2	0.2	0.2	0.2	0.2	0.2	0.2	0.2	0.2	0.2	0.2
11	PROD. SECTORS	39.7	57.2	61.9	67.2	71.6	77.6	83.0	87.2	90.2	96.1	97.2	99.7
12	Assembly Works	21.0	29.1	32.2	35.0	36.7	39.3	41.2	41.3	41.7	42.4	42.1	42.0
13	M & E	16.0	23.1	24.4	26.4	28.5	31.4	34.1	37.5	39.6	42.8	44.4	46.1
14	Other	2.7	5.0	5.3	5.8	6.4	6.9	7.7	8.4	8.9	10.9	10.7	11.6
15	INDUSTRY	21.1	29.3	30.9	33.1	34.9	37.6	39.8	41.6	43.5	45.7	45.7	47.6
16	Assembly Works	11.1	14.4	15.4	16.5	17.1	18.2	18.8	18.4	18.6	18.5	17.6	17.4
17	M & E	8.2	11.7	12.1	12.9	13.7	15.0	16.1	17.8	19.2	20.9	21.4	22.4
18	Other	1.8	3.2	3.4	3.7	4.1	4.4	4.9	5.4	5.7	6.3	6.7	7.8
19	AGRICULTURE	9.5	14.4	16.5	18.1	20.0	21.7	23.3	24.3	24.9	25.8	26.3	26.9
20	Assembly Works	5.0	8.0	9.7	10.6	11.6	12.5	13.3	13.4	13.4	13.6	13.6	13.7
21	M & E	3.8	5.0	5.3	5.8	6.6	7.3	8.0	8.8	9.3	9.9	10.3	10.7
22	Other	0.7	1.4	1.5	1.7	1.8	1.9	2.0	2.1	2.2	2.3	2.4	2.5
23	PROFIT T & C	5.5	7.8	8.5	9.6	10.3	11.3	12.7	13.3	13.9	16.2	16.3	16.1
24	Assembly Works	2.7	3.4	3.8	4.4	4.6	5.0	5.6	5.8	5.9	6.3	6.7	6.6
25	M & E	2.6	4.1	4.4	4.9	5.3	5.9	6.5	6.8	7.2	7.8	8.2	8.4
26	Other	0.2	0.3	0.3	0.3	0.4	0.4	0.6	0.7	0.8	2.1	1.4	1.1
27	CONSTRUCTION	1.5	3.0	3.4	3.6	3.7	4.0	4.3	4.9	4.7	4.9	5.2	5.4
28	Assembly Works	0.4	1.1	1.2	1.2	1.2	1.1	1.1	1.2	1.2	1.2	1.2	1.3
29	M & E	1.1	1.9	2.2	2.4	2.5	2.8	3.1	3.6	3.4	3.6	3.9	4.0
30	Other	0.0	0.0	0.0	0.0	0.0	0.1	0.1	0.1	0.1	0.1	0.1	0.1
31	T & D	1.9	2.4	2.3	2.5	2.4	2.7	2.6	2.8	2.9	3.1	3.3	3.3
32	Assembly Works	1.6	1.9	1.8	2.0	1.9	2.2	2.1	2.2	2.3	2.4	2.6	2.6
33	M & E	0.3	0.4	0.4	0.4	0.4	0.4	0.4	0.5	0.5	0.6	0.6	0.6
34	Other	0.0	0.1	0.1	0.1	0.1	0.1	0.1	0.1	0.1	0.1	0.1	0.1
35	OTHER PROD	0.2	0.3	0.3	0.3	0.3	0.3	0.3	0.3	0.3	0.4	0.4	0.4

*1969, 1973 (1976) and 1984 prices were used in 1970-1974, 1975-1983 and 1984-1990

TABLE B6: CAPITAL INVESTMENT BY SECTOR IN CONSTANT PRICES
     (billion rubles)

	1981	1982	1983	1984	1985	1986	1987	1988	1989	1990
1 TOTAL	138.8	143.8	152.0	174.3	179.5	194.4	205.4	218.2	228.5	230.0
2 Construction	84.4	86.9	91.3	109.0	110.9	120.9	130.9	139.9	144.9	149.3
3 Assembly Works	70.3	71.7	74.7	90.5	91.7	99.0	104.5	109.7	111.0	113.5
4 Design Works	2.4	2.4	2.4	2.5	2.6	2.9	3.1	3.4	3.7	4.2
5 Oil & Gas Expl	2.5	2.7	2.8	2.9	3.0	3.3	3.7	4.4	5.0	6.0
6 Land Reclam.	0.3	0.8	0.7	0.9	0.9	0.9	0.9	0.9	0.9	0.9
7 Other Expend.	8.9	9.3	10.7	12.2	12.7	14.8	18.7	21.5	24.3	24.7
8 M & E	53.4	55.9	59.7	64.3	67.6	72.3	73.3	77.1	82.4	79.5
9 Gardening	0.8	0.8	0.8	0.8	0.8	0.9	0.9	0.9	0.9	0.9
10 Light Industry	0.2	0.2	0.2	0.2	0.2	0.3	0.3	0.3	0.3	0.3
11 PROD. SECTORS	103.2	106.6	112.2	126.5	129.8	139.5	145.2	154.2	161.4	160.0
12 Assembly Works	42.3	43.3	44.3	52.5	52.3	54.4	55.5	57.8	56.4	56.7
13 M & E	48.2	50.3	53.6	58.2	61.3	66.0	66.7	70.2	75.3	72.0
14 Other	12.7	13.0	14.3	15.8	16.2	19.1	23.0	26.2	29.7	31.3
15 INDUSTRY	49.5	50.9	53.7	61.9	65.5	71.0	75.0	79.5	85.7	84.5
16 Assembly Works	17.3	17.6	17.9	22.8	23.3	23.9	23.9	24.5	24.7	23.5
17 M & E	23.6	24.5	26.2	28.5	31.1	33.4	33.8	35.0	38.0	36.7
18 Other	8.6	8.8	9.6	10.6	11.1	13.7	17.3	20.0	23.0	24.3
19 AGRICULTURE	27.6	28.0	29.0	31.0	31.5	33.5	34.4	36.5	38.4	39.0
20 Assembly Works	13.8	13.9	14.2	15.7	15.8	16.3	17.1	18.2	18.6	20.3
21 M & E	11.2	11.4	11.7	12.0	12.5	13.8	13.8	14.6	16.0	14.7
22 Other	2.6	2.7	3.1	3.3	3.2	3.4	3.5	3.7	3.8	4.0
23 PROFIT T & C	16.7	17.6	18.9	22.6	21.9	22.8	24.0	25.1	21.6	21.0
24 Assembly Works	7.0	7.6	7.8	9.2	8.8	9.0	9.4	9.7	7.6	7.0
25 M & E	8.5	8.8	9.8	11.9	11.6	12.2	12.8	13.4	11.7	11.6
26 Other	1.2	1.2	1.3	1.5	1.5	1.6	1.8	2.0	2.3	2.4
27 CONSTRUCTION	5.4	5.9	5.9	6.0	6.1	6.8	6.9	8.3	10.6	10.2
28 Assembly Works	1.1	1.0	0.9	1.1	1.0	1.3	1.5	1.9	2.1	2.0
29 M & E	4.2	4.8	4.9	4.8	5.0	5.4	5.3	6.2	8.3	8.0
30 Other	0.1	0.1	0.1	0.1	0.1	0.1	0.1	0.2	0.2	0.2
31 T & D	3.6	3.8	4.2	4.5	4.3	4.9	4.4	4.3	4.6	4.8
32 Assembly Works	2.7	2.8	3.0	3.2	2.9	3.4	3.1	3.0	2.9	3.4
33 M & E	0.7	0.8	1.0	1.0	1.1	1.2	1.0	1.0	1.3	1.0
34 Other	0.2	0.2	0.2	0.3	0.3	0.3	0.3	0.3	0.4	0.4
35 OTHER PROD	0.4	0.4	0.5	0.5	0.5	0.5	0.5	0.5	0.5	0.5

*1969, 1973 (1976) and 1984 prices were used in 1970-1974, 1975-1983 and 1984-1990

TABLE B6: CAPITAL INVESTMENT BY SECTOR IN CONSTANT PRICES
(billion rubles)

	1965	1970	1971	1972	1973	1974	1975	1976	1977	1978	1979	1980
36 NONPROD SECTORS	17.3	24.8	26.1	27.1	27.1	28.1	29.9	30.8	32.1	33.0	33.4	34.0
37 Assembly Works	14.8	21.2	22.5	23.2	22.9	23.6	25.0	25.7	26.7	27.1	27.0	27.0
38 M & E	1.5	2.2	2.2	2.4	2.6	2.7	3.0	3.2	3.4	3.8	4.2	4.7
39 Other	1.0	1.4	1.4	1.5	1.6	1.8	1.9	1.9	2.0	2.1	2.2	2.3
40 HOUSING	9.6	13.4	14.1	14.6	15.1	15.6	16.3	16.5	17.0	17.4	17.3	17.9
41 Assembly Works	8.7	12.3	13.0	13.4	13.8	14.1	14.7	14.8	15.3	15.6	15.4	15.8
42 M & E	0.1	0.1	0.1	0.1	0.1	0.1	0.1	0.2	0.2	0.2	0.2	0.3
43 Other	0.8	1.0	1.0	1.1	1.2	1.4	1.5	1.5	1.5	1.6	1.7	1.8
44 COMMUNAL, etc.	2.0	3.2	3.3	3.4	3.5	3.7	4.0	4.3	4.5	4.7	4.8	4.9
45 Assembly Works	1.8	2.8	2.9	3.0	3.1	3.3	3.5	3.8	4.0	4.2	4.2	4.4
46 M & E	0.1	0.2	0.2	0.2	0.2	0.3	0.3	0.3	0.3	0.3	0.3	0.3
47 Other	0.1	0.2	0.2	0.2	0.2	0.2	0.2	0.2	0.2	0.2	0.2	0.2
48 EDUCATION, etc.	3.4	4.6	4.7	4.9	4.2	4.5	4.7	4.4	4.4	4.5	4.4	3.9
49 Education	2.0	2.9	2.7	2.9	2.7	3.0	3.1	3.0	3.0	3.0	2.9	2.5
50 Culture & Arts	0.6	0.6	0.6	0.5	0.5	0.6	0.7	0.7	0.7	0.7	0.7	0.7
51 Health, etc.	0.7	1.2	1.4	1.5	1.0	0.9	0.9	0.7	0.7	0.8	0.7	0.7
52 Assembly Works	3.2	4.4	4.4	4.6	3.8	4.1	4.3	4.0	4.0	4.1	4.0	3.4
53 M & E	0.2	0.3	0.3	0.3	0.4	0.4	0.4	0.4	0.4	0.4	0.4	0.5
54 SCIENCE	0.9	1.0	1.5	1.6	1.7	1.8	2.1	2.4	2.7	2.7	2.9	3.4
55 Assembly Works	0.5	0.5	1.0	1.0	1.0	1.1	1.3	1.6	1.8	1.7	1.8	2.1
56 M & E	0.4	0.5	0.5	0.6	0.7	0.7	0.8	0.8	0.9	1.0	1.1	1.3
57 BUDGET T & C	0.5	1.0	1.0	1.0	1.0	1.1	1.2	1.4	1.6	1.7	1.8	1.8
58 Assembly Works	0.2	0.4	0.4	0.4	0.4	0.4	0.5	0.6	0.6	0.6	0.6	0.6
59 M & E	0.2	0.4	0.4	0.4	0.4	0.5	0.5	0.6	0.7	0.8	0.9	0.9
60 Other	0.1	0.2	0.2	0.2	0.2	0.2	0.2	0.2	0.3	0.3	0.3	0.3
61 RESIDUAL	0.9	1.5	1.5	1.6	1.6	1.4	1.6	1.8	1.9	2.0	2.3	2.1
62 Assembly Works	0.5	0.8	0.8	0.8	0.8	0.6	0.7	0.9	1.0	0.9	1.0	0.7
63 M & E	0.5	0.7	0.7	0.7	0.8	0.8	0.9	0.9	0.9	1.1	1.3	1.4

Supporting Data on Education, Culture & Arts, and Science:

	1965	1970	1971	1972	1973	1974	1975	1976	1977	1978	1979	1980
64 49 + 50 + 54	3.5	4.5	4.9	5.0	4.9	5.3	5.9	6.1	6.4	6.4	6.6	6.6
65 40 + 44 + 48												

Supporting data on capital investment in household services in current prices:

	1965	1970	1971	1972	1973	1974	1975	1976	1977	1978	1979	1980
66 Education	1.6	2.7	2.6	2.8	2.6	3.0	3.1	3.0	2.9	2.9	2.8	2.5
67 Culture & Arts	0.5	0.6	0.6	0.5	0.5	0.6	0.7	0.7	0.7	0.7	0.7	0.7
68 Health, etc.	0.6	1.1	1.3	1.5	1.0	0.9	0.9	0.7	0.7	0.8	0.7	0.7

*1969, 1973 (1976) and 1984 prices were used in 1970-1974, 1975-1983 and 1984-1990

TABLE B6: CAPITAL INVESTMENT BY SECTOR IN CONSTANT PRICES
   (billion rubles)

Page 4

	1981	1982	1983	1984	1985	1986	1987	1988	1989	1990
36 NONPROD SECTORS	35.6	37.2	39.8	47.8	49.7	54.9	60.2	64.0	67.1	70.0
37 Assembly Works	28.0	28.4	30.4	38.0	39.4	44.6	49.0	51.9	54.6	56.8
38 M & E	5.2	5.6	6.1	6.1	6.3	6.3	6.6	6.9	7.1	7.5
39 Other	2.4	3.2	3.3	3.7	4.0	4.0	4.6	5.2	5.4	5.7
40 HOUSING	19.0	20.3	21.9	26.9	28.1	30.9	33.5	35.6	37.7	41.2
41 Assembly Works	16.8	17.4	18.9	23.5	24.4	27.2	29.2	30.7	32.8	36.1
42 M & E	0.3	0.3	0.3	0.4	0.4	0.4	0.5	0.5	0.5	0.5
43 Other	1.9	2.6	2.7	3.0	3.3	3.3	3.8	4.4	4.4	4.6
44 COMMUNAL, etc.	5.2	5.6	6.0	7.4	7.7	8.5	9.2	9.8	10.8	11.3
45 Assembly Works	4.6	4.9	5.2	6.4	6.7	7.5	8.1	8.7	9.5	9.7
46 M & E	0.4	0.4	0.5	0.6	0.6	0.6	0.6	0.6	0.6	0.8
47 Other	0.2	0.3	0.3	0.4	0.4	0.4	0.5	0.5	0.7	0.8
48 EDUCATION, etc.	3.8	3.9	4.0	4.9	5.7	6.3	8.1	9.7	9.3	8.9
49 Education	2.5	2.6	2.7	3.1	3.7	4.0	5.3	6.4	6.1	5.9
50 Culture & Arts	0.7	0.7	0.7	0.9	1.0	1.1	1.3	1.5	1.4	1.3
51 Health, etc.	0.7	0.7	0.7	0.9	1.0	1.2	1.5	1.8	1.8	1.7
52 Assembly Works	3.3	3.4	3.4	4.3	5.2	5.2	5.2	5.2	5.2	5.2
53 M & E	0.5	0.5	0.6	0.6	0.6	0.7	0.8	0.8	0.9	1.1
54 SCIENCE	3.4	3.3	3.9	4.0	3.8	4.5	4.7	4.2	3.6	3.7
55 Assembly Works	1.8	1.5	1.8	2.3	1.9	2.7	2.9	2.1	1.9	1.7
56 M & E	1.6	1.8	2.1	1.7	1.9	1.8	1.8	2.1	1.7	2.0
57 BUDGET T & C	1.8	1.8	1.7	2.1	2.1	2.3	2.3	2.3	2.6	2.7
58 Assembly Works	0.6	0.6	0.5	0.7	0.7	1.0	1.0	1.0	1.0	1.0
59 M & E	0.9	0.9	0.9	1.1	1.1	1.0	1.0	1.0	1.3	1.4
60 Other	0.3	0.3	0.3	0.3	0.3	0.3	0.3	0.3	0.3	0.3
61 RESIDUAL	2.3	2.2	2.2	2.5	2.3	2.4	2.4	2.4	3.1	2.2
62 Assembly Works	0.8	0.6	0.5	0.8	0.6	0.6	0.5	0.5	1.0	0.5
63 M & E	1.5	1.7	1.7	1.7	1.7	1.8	1.9	1.9	2.1	1.7

Supporting Data on Education, Culture & Arts, and Science:

	1981	1982	1983	1984	1985	1986	1987	1988	1989	1990
64 49 + 50 + 54	6.6	6.6	7.3	8.0	8.5	9.6	11.3	12.1	11.1	10.8
65 40 + 44 + 48					41.6		50.8	55.1	58.9	

Supporting data on capital investment in household services in current prices:

	1981	1982	1983	1984	1985	1986	1987	1988	1989	
66 Education	2.5	2.6	2.7	3.1	3.6	3.9	5.3	6.3	6.0	
67 Culture & Arts	0.7	0.7	0.7	0.9	1.0	1.1	1.3	1.5	1.4	.
68 Health, etc.	0.7	0.7	0.7	0.9	1.0	1.2	1.5	1.8	1.8	.

*1969, 1973 (1976) and 1984 prices were used in 1970-1974, 1975-1983 and 1984-1990

131

TABLE B7: ADDITIONS TO FIXED CAPITAL STOCK BY SECTOR
        (billion rubles)

	1965	1970	1971	1972	1973	1974	1975	1976	1977	1978	1979	1980
1 TOTAL (2-3)	45.2	76.3	82.2	85.0	94.1	98.8	107.2	108.2	111.0	120.5	120.4	133.1
2 Investment	49.7	83.2	90.5	95.2	98.9	106.3	114.9	118.8	122.8	130.8	132.1	135.1
3 Unfinished	4.5	6.9	8.3	10.2	4.8	7.5	7.7	10.6	11.8	10.3	11.7	2.0
4 PROD SECTORS	30.6	51.8	56.1	58.7	66.5	70.6	76.9	78.1	79.5	88.1	87.5	98.0
5 Investment	34.6	57.7	63.3	67.4	71.2	77.3	83.7	86.8	89.8	96.6	97.5	99.9
6 Unfinished	4.0	5.9	7.2	8.7	4.7	6.7	6.8	8.7	10.3	8.5	10.0	1.9
7 INDUSTRY	15.1	24.7	25.6	26.6	30.5	31.8	34.7	34.5	35.1	38.7	37.7	44.8
8 Investment	18.0	29.1	30.9	32.8	34.0	36.7	39.8	41.0	42.6	45.2	45.2	47.3
9 Unfinished	2.9	4.4	5.3	6.2	3.5	4.9	5.1	6.5	7.5	6.4	7.5	2.5
10 AGRICULTURE	8.3	14.6	17.1	17.7	20.5	22.2	23.7	24.4	25.3	28.1	27.3	28.9
11 Investment	9.0	15.5	18.2	19.1	21.3	23.2	24.6	25.5	26.7	29.2	28.7	28.6
12 Unfinished	0.7	0.9	1.1	1.4	0.8	1.0	0.9	1.1	1.4	1.1	1.4	-0.3
13 PROFIT T & C	4.3	7.1	7.8	8.6	9.4	10.2	11.7	11.8	12.1	13.4	14.2	15.1
14 Investment	4.5	7.4	8.2	9.2	9.6	10.6	12.1	12.4	12.8	13.9	14.7	14.9
15 Unfinished	0.2	0.3	0.4	0.6	0.2	0.4	0.4	0.6	0.7	0.5	0.6	-0.1
16 CONSTRUCTION	1.2	2.8	3.2	3.3	3.5	3.7	4.1	4.5	4.2	4.6	4.9	5.4
17 Investment	1.3	3.0	3.4	3.6	3.6	3.8	4.3	4.8	4.6	4.8	5.1	5.4
18 Unfinished	0.1	0.1	0.2	0.3	0.1	0.2	0.2	0.3	0.4	0.2	0.3	-0.1
19 TRADE, etc.	1.7	2.5	2.4	2.5	2.6	2.8	2.7	2.8	2.8	3.2	3.4	3.7
20 Investment	1.8	2.7	2.6	2.8	2.7	3.0	2.9	3.1	3.2	3.5	3.7	3.7
21 Unfinished	0.1	0.1	0.2	0.3	0.1	0.2	0.2	0.3	0.4	0.2	0.3	-0.1
22 NONPROD SECTORS	14.6	24.5	26.2	26.3	27.7	28.2	30.3	30.1	31.5	32.4	33.0	35.2
23 Investment	15.1	25.5	27.2	27.8	27.8	29.0	31.2	32.0	33.0	34.2	34.7	35.3
24 Unfinished	0.5	1.0	1.1	1.5	0.1	0.9	0.9	1.9	1.5	1.8	1.7	0.1
25 HOUSING	7.8	12.8	13.7	13.7	14.9	15.1	15.9	15.6	16.0	16.4	16.3	17.6
26 Investment	8.0	13.3	14.2	14.5	14.9	15.5	16.3	16.5	16.8	17.3	17.1	17.7
27 Unfinished	0.3	0.5	0.5	0.8	0.1	0.4	0.5	1.0	0.8	0.9	0.9	0.1
28 COMMUNAL, etc.	1.6	3.1	3.2	3.2	3.5	3.6	3.9	4.1	4.3	4.5	4.5	4.9
29 Investment	1.7	3.2	3.3	3.4	3.5	3.7	4.0	4.3	4.5	4.7	4.7	4.9
30 Unfinished	0.1	0.1	0.1	0.2	0.0	0.1	0.1	0.2	0.2	0.2	0.2	0.0
31 EDUCATION, etc.	3.2	5.2	5.4	5.5	5.0	5.3	5.6	5.2	5.3	5.5	5.4	5.2
32 Investment	3.3	5.4	5.5	5.7	5.0	5.4	5.8	5.5	5.5	5.7	5.7	5.2
33 Unfinished	0.1	0.2	0.2	0.2	0.0	0.1	0.1	0.3	0.2	0.3	0.3	0.0
34 SCIENCE	0.8	1.0	1.6	1.6	1.8	1.8	2.2	2.4	2.7	2.7	2.9	3.6
35 Investment	0.8	1.1	1.7	1.7	1.8	1.9	2.3	2.5	2.9	2.9	3.1	3.6
36 Unfinished	0.1	0.1	0.1	0.2	0.0	0.1	0.1	0.2	0.2	0.2	0.2	0.0
37 OTHER NONPROD	1.2	2.3	2.4	2.3	2.5	2.3	2.7	2.9	3.2	3.4	3.8	3.8
38 Investment	1.2	2.5	2.5	2.5	2.5	2.4	2.8	3.2	3.4	3.6	4.0	3.9
39 Unfinished	0.1	0.2	0.2	0.2	0.0	0.1	0.1	0.3	0.2	0.3	0.3	0.0

132

TABLE B7: ADDITIONS TO FIXED CAPITAL STOCK BY SECTOR
      (billion rubles)

		1981	1982	1983	1984	1985	1986	1987	1988	1989	1990
1	TOTAL (2-3)	134.8	140.6	150.6	160.9	165.7	178.5	194.5	194.4	191.6	199.4
2	Investment	140.0	148.7	156.7	174.7	181.0	197.3	207.3	219.7	227.1	238.7
3	Unfinished	5.2	8.1	6.1	13.8	15.3	18.8	12.8	25.3	35.5	39.3
4	PROD SECTORS	98.9	101.8	109.6	116.3	117.3	125.7	137.6	134.8	127.4	130.3
5	Investment	103.3	108.7	114.7	127.3	131.5	142.6	148.3	157.1	160.4	166.6
6	Unfinished	4.4	6.9	5.1	11.0	14.2	16.9	10.7	22.3	33.0	36.3
7	INDUSTRY	44.7	45.9	49.7	52.8	54.8	58.7	65.8	62.7	57.8	57.6
8	Investment	49.0	52.0	54.4	61.6	65.8	71.8	75.4	79.7	81.9	84.0
9	Unfinished	4.3	6.1	4.7	8.7	11.0	13.2	9.6	17.0	24.1	26.4
10	AGRICULTURE	29.6	29.2	31.5	32.5	32.5	34.8	37.5	37.4	35.4	37.3
11	Investment	29.6	29.6	31.7	33.5	33.9	36.5	38.0	40.2	39.7	42.0
12	Unfinished	0.0	0.4	0.2	1.0	1.5	1.7	0.5	2.8	4.3	4.7
13	PROFIT T & C	15.3	16.5	17.8	20.6	19.8	20.7	22.5	22.3	22.3	23.6
14	Investment	15.4	16.7	17.8	21.3	20.7	21.8	22.8	23.7	24.9	26.5
15	Unfinished	0.0	0.2	0.1	0.6	0.9	1.0	0.3	1.5	2.6	2.8
16	CONSTRUCTION	5.4	5.9	6.0	5.9	6.0	6.7	7.2	8.0	7.6	7.6
17	Investment	5.4	6.0	6.0	6.2	6.4	7.2	7.3	8.8	8.9	9.0
18	Unfinished	0.0	0.1	0.0	0.3	0.5	0.5	0.2	0.7	1.3	1.4
19	TRADE, etc.	3.9	4.2	4.7	4.5	4.2	4.8	4.6	4.4	4.2	4.2
20	Investment	3.9	4.3	4.7	4.8	4.7	5.3	4.7	4.6	5.0	5.1
21	Unfinished	0.0	0.1	0.0	0.3	0.5	0.5	0.2	0.2	0.8	0.9
22	NONPROD SECTORS	35.9	38.8	41.0	44.6	48.4	52.8	56.9	59.6	64.2	69.2
23	Investment	36.7	40.0	42.0	47.4	49.5	54.7	59.0	62.6	66.7	72.2
24	Unfinished	0.8	1.2	1.0	2.8	1.1	1.9	2.1	3.0	2.5	3.0
25	HOUSING	18.2	20.3	21.5	23.8	25.8	28.2	30.0	31.7	33.8	37.2
26	Investment	18.6	20.9	22.0	25.2	26.4	29.2	31.1	32.9	35.1	38.7
27	Unfinished	0.4	0.6	0.5	1.4	0.6	1.0	1.1	1.2	1.3	1.5
28	COMMUNAL, etc.	5.0	5.6	6.0	6.7	7.1	7.8	8.3	8.8	9.4	10.3
29	Investment	5.1	5.7	6.1	6.9	7.3	8.0	8.5	9.1	9.6	10.6
30	Unfinished	0.1	0.1	0.1	0.3	0.1	0.2	0.2	0.3	0.3	0.3
31	EDUCATION, etc.	5.1	5.5	5.5	5.9	7.1	7.6	9.2	10.2	11.5	12.3
32	Investment	5.3	5.6	5.7	6.3	7.2	7.9	9.5	11.1	12.2	12.7
33	Unfinished	0.1	0.2	0.2	0.4	0.2	0.3	0.3	1.0	0.7	0.5
34	SCIENCE	3.6	3.6	4.2	4.0	4.1	4.7	4.8	4.4	4.5	4.7
35	Investment	3.7	3.7	4.3	4.3	4.2	4.8	5.0	4.6	4.8	5.0
36	Unfinished	0.1	0.1	0.1	0.3	0.1	0.2	0.2	0.2	0.3	0.3
37	OTHER NONPROD	3.9	3.9	3.8	4.2	4.3	4.6	4.5	4.5	4.7	4.7
38	Investment	4.1	4.1	3.9	4.7	4.5	4.8	4.9	4.9	5.0	5.1
39	Unfinished	0.1	0.2	0.2	0.4	0.2	0.3	0.3	0.4	0.4	0.5

TABLE B8: STOCKS OF INVENTORIES BY SECTOR OF USE
(billion rubles; year-end)

	1965	1970	1971	1972	1973	1974	1975	1976	1977	1978	1979	1980
1 TOTAL	120.1	179.7	191.1	199.6	216.2	228.9	239.4	254.0	268.7	284.0	297.9	316.5
2 -/- Additions	6.6	15.5	11.4	8.5	16.6	12.7	10.5	14.6	14.7	15.3	13.9	18.6
3 State-Coop*	107.3	160.5	170.2	177.4	191.7	202.7	212.7	225.6	238.8	252.3	264.6	282.3
4 Collectives	12.8	19.2	20.9	22.2	24.5	26.2	26.7	28.4	29.9	31.7	33.3	34.2
5 INDUSTRY	45.0	64.4	67.6	70.5	74.1	78.9	85.6	90.3	97.2	103.1	110.1	117.8
6 -/-Additions	3.4	4.9	3.2	2.9	3.6	4.8	6.7	4.7	6.9	5.9	7.0	7.7
7 State-Coop	44.3	63.3	66.4	69.2	72.8	77.5	84.2	88.8	95.6	101.4	108.3	115.9
8 Collectives	0.7	1.1	1.2	1.3	1.3	1.4	1.4	1.5	1.6	1.7	1.8	1.9
9 AGRICULTURE	22.3	36.6	40.3	42.3	47.3	50.2	50.9	55.4	58.7	62.3	65.5	68.0
10 -/-Additions	1.3	4.1	3.7	2.0	5.0	2.9	0.7	4.5	3.3	3.6	3.2	2.5
11 State Farms	10.2	18.5	20.6	21.4	24.1	25.4	25.6	28.5	30.4	32.3	34.0	35.7
12 Collectives	12.1	18.1	19.7	20.9	23.2	24.8	25.3	26.9	28.3	30.0	31.5	32.3
13 T & C	1.6	2.6	2.7	2.8	3.0	3.1	3.4	3.7	4.0	4.2	4.6	5.1
14 -/- Additions	0.1	0.4	0.1	0.1	0.2	0.1	0.3	0.3	0.3	0.2	0.4	0.5
15 CONSTRUCTION	5.0	9.0	9.6	9.9	10.4	11.0	12.0	13.2	14.4	15.6	16.3	17.3
16 -/-Additions	0.4	0.4	0.6	0.3	0.5	0.6	1.0	1.2	1.2	1.2	0.7	1.0
17 Construction	4.2	7.8	8.3	8.6	9.1	9.7	10.6	11.8	13.0	14.1	14.8	15.8
18 Oil & Gas Expl	0.8	1.2	1.3	1.3	1.3	1.3	1.4	1.4	1.4	1.5	1.5	1.5
19 STATE TRADE	32.4	43.7	47.5	49.1	52.0	54.8	57.0	58.8	61.5	61.6	64.3	69.0
20 -/-Additions	0.8	3.9	3.8	1.6	2.9	2.8	2.2	1.8	2.7	0.1	2.7	4.7
21 PROCUREMENT	6.0	9.2	8.5	9.1	12.7	11.7	9.4	11.4	10.1	12.7	10.4	9.7
22 -/-Additions	0.2	0.8	-0.7	0.6	3.6	-1.0	-2.3	2.0	-1.3	2.6	-2.3	-0.7
23 SUPPLY	5.6	9.3	9.7	10.4	10.8	11.6	12.4	13.1	13.9	14.8	15.7	16.1
24 -/-Additions	0.3	0.5	0.4	0.7	0.4	0.8	0.8	0.7	0.8	0.9	0.9	0.4
25 OTHER SECTORS	2.2	4.9	5.2	5.5	5.9	7.6	8.7	8.1	8.9	9.7	11.0	13.5
26 Personal Serv.	1.2	2.3	2.4	2.5	2.7	3.1	3.3	3.4	3.6	3.7	3.8	3.9
27 Public Orgs.	0.3	0.5	0.5	0.5	0.5	0.5	0.6	0.6	0.6	0.6	0.6	0.7
28 Communal Serv.	0.4	0.7	0.7	0.8	0.8	0.9	1.0	1.0	1.1	1.1	1.1	1.2
29 Trade Coop.	0.3	1.4	1.6	1.7	1.9	3.1	3.8	3.1	3.6	4.3	5.5	7.7
30 Agrosupply												
31 Agrobusiness												
32 Other (Defense)												

Supporting Data on Inventories in Construction:

	1965	1970	1971	1972	1973	1974	1975	1976	1977	1978	1979	1980
33 Published Total	4.6	10.7	13.2	18.4	22.9	28.0	32.2	38.3	43.5	48.3	52.8	55.4
34 Unfin. Product.	0.4	2.9	4.9	9.8	13.8	18.3	21.6	26.5	30.5	34.2	38.0	39.6

* excluding unfinished production in construction

TABLE B8: STOCKS OF INVENTORIES BY SECTOR OF USE
(billion rubles; year-end)

| | Appreciated Value | | | | | | | | | Page 2 | |
	1981	1981	1982	1983	1984	1985	1986	1987	1988	1989	1990
1 TOTAL	331.0	350.4	376.7	407.4	429.6	450.0	464.0	466.0	474.4	483.5	488.5
2 -/- Additions	14.5		26.3	30.7	22.2	20.4	14.0	2.0	8.4	9.0	5.0
3 State-Coop*	295.0	313.9	338.6	366.6	385.2	402.4	415.0	416.9	427.0	435.5	441.0
4 Collectives	36.0	36.5	38.1	40.8	44.4	47.6	49.0	49.1	47.4	48.0	47.5
5 INDUSTRY	123.5	132.6	140.1	152.7	159.5	165.2	164.2	165.8	169.0	175.3	176.3
6 -/-Additions	5.7		7.5	12.7	6.8	5.7	-1.0	1.6	3.2	6.3	1.0
7 State-Coop	121.5	130.6	138.0	150.5	157.2	162.9	161.9	163.6	166.7	173.0	174.0
8 Collectives	2.0	2.0	2.1	2.2	2.3	2.3	2.3	2.2	2.3	2.3	2.3
9 AGRICULTURE	71.6	72.6	76.9	83.6	93.8	100.1	107.3	108.4	99.4	101.7	104.2
10 -/-Additions	3.6		4.3	6.6	10.2	6.3	7.2	1.1	-9.0	2.3	2.5
11 State Farms	37.6	38.1	40.9	45.0	51.7	54.8	60.6	61.5	54.3	56.0	59.0
12 Collectives	34.0	34.5	36.0	38.6	42.1	45.3	46.7	46.9	45.1	45.7	45.2
13 T & C	5.5	6.3	6.8	7.3	7.4	7.8	7.9	7.6	8.0	8.5	8.5
14 -/- Additions	0.4		0.5	0.5	0.1	0.4	0.1	-0.3	0.4	0.5	0.0
15 CONSTRUCTION	18.2	19.1	19.8	20.6	21.3	21.9	23.2	24.1	24.7	24.7	24.4
16 -/-Additions	0.9		0.7	0.8	0.7	0.7	1.2	0.9	0.6	0.0	-0.3
17 Construction	16.7	17.5	18.2	19.0	19.7	20.3	21.5	22.4	23.0	23.0	22.7
18 Oil & Gas Expl	1.5	1.6	1.6	1.6	1.6	1.6	1.7	1.7	1.7	1.7	1.7
19 STATE TRADE	72.0	77.9	86.9	93.4	95.6	100.2	91.9	87.1	85.9	85.0	85.0
20 -/-Additions	3.0		9.0	6.5	2.2	4.6	-8.3	-4.8	-1.2	-0.9	0.0
21 PROCUREMENT	9.4	9.7	10.5	11.6	10.8	11.1	11.2	12.2	10.5	10.5	10.5
22 -/-Additions	-0.3		0.8	1.1	-0.8	0.3	0.1	1.0	-1.7	0.0	0.0
23 SUPPLY	16.6	17.7	18.7	20.3	21.6	21.9	22.9	11.7	11.7	11.5	11.5
24 -/-Additions	0.5		1.0	1.6	1.3	0.3	1.0	-11.2	0.0	-0.2	0.0
25 OTHER SECTORS	14.2	14.5	17.0	17.9	19.6	21.8	35.4	49.1	65.2	66.3	68.2
26 Personal Serv.	4.1	4.1	4.3	4.3	4.4	4.5	4.6	4.6	4.6	4.6	4.5
27 Public Orgs.	0.7	0.7	0.8	0.8	0.8	0.8	0.8	0.8	0.8	0.8	0.8
28 Communal Serv.	1.2	1.2	1.3	1.3	1.3	1.4	1.4	1.5	1.5	1.4	1.4
29 Trade Coop.	8.2	8.5	10.6	11.5	13.1	15.1	16.7	18.6	20.2	19.7	20.0
30 Agrosupply								11.1	11.4	11.5	11.4
31 Agrobusiness								13.7	14.0	14.0	
32 Other (Defense)							11.9	12.5	13.0	14.3	16.2

Supporting Data on Inventories in Construction:

33 Published Total	58.5	64.0	69.8	73.4	78.0	81.2	27.3	25.3	26.7	27.0	27.0
34 Unfin. Product.	41.8	46.5	51.6	54.4	58.3	60.9	5.8	2.9	3.7	4.0	4.3

* excluding unfinished production in construction

TABLE B9: STOCKS OF INVENTORIES BY TYPES
        (billion rubles)

	1965	1970	1971	1972	1973	1974	1975	1976	1977	1978	1979	1980
1 STATE-COOP*	107.7	163.4	175.1	187.2	205.5	221.0	234.3	252.1	269.3	286.5	302.6	321.9
2 PROD. ASSETS	43.9	69.0	73.4	76.8	82.0	86.6	92.1	99.1	106.4	112.9	119.2	126.8
3 INDUSTRY	65.2	98.9	104.0	109.1	115.4	122.0	132.6	142.0	151.5	161.0	170.1	180.7
4 Materials	27.6	42.1	44.2	46.2	49.0	51.9	56.7	60.9	64.8	68.9	72.8	77.1
5 a) Industrial	25.7	39.6	41.6	43.6	46.0	48.7	53.3	57.4	60.9	64.9	68.6	72.6
6 b) Agricultural	1.9	2.5	2.6	2.6	3.0	3.2	3.4	3.5	3.9	4.0	4.2	4.5
7 Fuels	1.1	1.6	1.8	1.9	1.9	2.0	2.1	2.0	2.0	2.1	2.2	2.4
8 Repair & Instr.	8.4	12.3	13.3	14.0	14.6	15.2	16.6	17.5	18.9	20.1	21.2	22.9
9 Other Prod. Inv	0.5	0.8	0.5	0.7	0.8	0.9	0.5	0.8	1.1	1.1	1.2	1.3
10 AGRICULTURE	6.4	12.2	13.5	13.9	15.6	16.6	16.2	18.0	19.7	20.8	21.9	23.2
11 Agri Materials	2.4	3.9	4.2	4.4	5.1	5.3	5.2	6.0	6.5	6.8	7.0	7.4
12 Young Livestock	4.0	8.2	9.3	9.5	10.5	11.3	11.0	12.0	13.2	14.0	14.8	15.8
13 UNFIN. PROD.	11.5	20.1	23.6	28.6	34.4	40.2	45.5	52.0	58.1	61.7	67.1	70.5
14 Unfinished Ind.	9.0	13.7	14.5	14.8	16.0	17.0	18.5	19.9	21.3	22.9	24.2	25.8
15 Unfinished Agri	1.7	2.8	3.2	3.0	3.3	3.5	3.7	3.8	4.0	4.2	4.5	4.8
16 Unfinished Cons	0.8	3.6	6.0	10.9	15.1	19.7	23.2	28.3	32.7	34.5	38.3	39.9
17 COMMODITIES	43.3	63.6	66.5	69.6	75.8	80.0	81.8	85.6	88.4	94.7	98.1	105.2
18 Trade Inventory	35.7	45.7	48.3	49.7	52.2	56.0	58.1	58.9	61.1	60.3	62.0	67.1
19 a) Trade & Ind.	31.2	42.3	45.4	47.0	49.7	52.7	55.0	56.6	59.2	59.1	61.7	66.8
20 -/-Trade	30.0	40.6	44.1	45.6	48.2	50.7	52.7	54.3	56.6	56.4	59.0	63.9
21 -/-Industry	1.2	1.7	1.3	1.4	1.5	2.0	2.3	2.3	2.6	2.7	2.7	2.9
22 b) Other	4.5	3.4	2.9	2.7	2.5	3.3	3.1	2.3	1.9	1.2	0.3	0.3
23 Supplies	7.6	17.9	18.2	19.9	23.6	24.0	23.7	26.7	27.3	34.4	36.1	38.1
24 OTHER ASSETS	8.9	10.8	11.6	12.2	13.2	14.1	15.0	15.4	16.4	17.4	18.2	19.4
25 Ready Goods	6.5	7.5	8.1	8.6	9.0	9.7	10.3	10.6	11.3	12.0	12.7	13.5
26 Future Exp. etc	2.5	3.3	3.5	3.6	4.2	4.4	4.7	4.8	5.1	5.3	5.5	5.9
27 COLLECTIVES	13.1	19.6	21.3	22.6	24.9	26.7	27.2	29.0	30.5	32.4	34.0	34.9
28 Young Livestock	6.9	9.2	10.1	10.8	12.1	12.9	13.0	13.8	14.4	15.2	16.2	16.5
29 Agri Materials	2.6	4.4	4.7	5.0	5.5	5.8	5.9	6.2	6.6	6.8	7.0	7.2
30 Ind. Materials	1.0	1.6	1.9	2.0	2.2	2.4	2.5	2.9	3.1	3.6	3.9	4.1
31 Unfinished Prod	1.4	2.6	2.8	2.9	3.2	3.6	3.8	4.0	4.1	4.3	4.4	4.5
32 Other Agri.	0.5	0.7	0.6	0.6	0.6	0.6	0.6	0.6	0.7	0.7	0.7	0.7
33 Unfin. Constr.	0.7	1.1	1.2	1.3	1.3	1.4	1.4	1.5	1.6	1.8	1.8	1.9

Supporting Data on the State-Coop Sector:

	1965	1970	1971	1972	1973	1974	1975	1976	1977	1978	1979	1980
34 2/1*100	40.8	42.2	41.9	41.0	39.9	39.2	39.3	39.3	39.5	39.4	39.4	39.4
35 4/1*100	25.6	25.8	25.2	24.7	23.9	23.5	24.2	24.1	24.1	24.0	24.0	24.0
36 7/1*100	1.0	1.0	1.1	1.0	0.9	0.9	0.9	0.8	0.7	0.7	0.7	0.8
37 8/1*100	7.8	7.5	7.6	7.5	7.1	6.9	7.1	6.9	7.0	7.0	7.0	7.1
38 11/1*100	2.2	2.3	2.4	2.4	2.5	2.4	2.2	2.4	2.4	2.4	2.4	2.3
39 12/1*100	3.7	5.0	5.3	5.1	5.1	5.1	4.7	4.8	4.9	4.9	4.9	4.9
40 9/1*100	0.5	0.5	0.3	0.4	0.4	0.4	0.2	0.3	0.4	0.4	0.4	0.4
41 14/1*100	8.4	8.4	8.3	7.9	7.8	7.7	7.9	7.9	7.9	8.0	8.0	8.0
42 15/1*100	1.6	1.7	1.8	1.6	1.6	1.6	1.6	1.5	1.5	1.5	1.5	1.5
43 16/1*100	0.7	2.2	3.4	5.8	7.4	8.9	9.9	11.2	12.2	12.1	12.7	12.4
44 25/1*100	6.0	4.6	4.6	4.6	4.4	4.4	4.4	4.2	4.2	4.2	4.2	4.2
45 17/1*100	40.2	38.9	38.0	37.2	36.9	36.2	34.9	34.0	32.8	33.0	32.4	32.7
46 26/1*100	2.3	2.0	2.0	1.9	2.0	2.0	2.0	1.9	1.9	1.9	1.8	1.8

*including unfinished production in construction organizations

136

TABLE B9: STOCKS OF INVENTORIES BY TYPES
    (billion rubles)

| | | Appreciated Value | | | | | | | | | Page 2 |
	1981	1981	1982	1983	1984	1985	1986	1987	1988	1989	1990
1 STATE-COOP*	336.8	360.4	390.2	421.0	443.5	463.3	420.8	419.8	430.7	435.5	441.0
2 PROD. ASSETS	133.0	142.4	150.9	164.2	174.5	181.3	194.9	195.5	209.1	214.4	226.7
3 INDUSTRY	189.7	117.8	124.5	134.6	142.0	147.1	157.4	157.7	171.3	176.2	189.5
4 Materials	80.9	99.1	104.7	113.8	120.3	124.1	132.6	132.3	143.5	147.3	158.4
5 a) Industrial	76.2	94.3	99.6	108.3	114.5	118.1	126.3	126.1	137.4	141.2	152.4
6 b) Agricultural	4.7	4.8	5.1	5.5	5.8	6.0	6.3	6.2	6.1	6.1	6.0
7 Fuels	2.6	3.0	3.1	3.1	3.1	3.2	3.2	3.3	3.3	3.3	3.1
8 Repair & Instr.	23.9	14.5	15.3	16.2	17.1	18.2	20.0	20.5	22.8	23.9	26.3
9 Other Prod. Inv	1.3	1.3	1.4	1.5	1.5	1.6	1.6	1.6	1.7	1.7	1.7
10 AGRICULTURE	24.2	24.5	26.4	29.6	32.5	34.2	37.5	37.8	37.8	38.2	37.2
11 Agri Materials	7.7	7.8	8.4	9.6	11.0	11.7	12.9	13.1	13.5	13.7	13.2
12 Young Livestock	16.5	16.7	18.0	20.0	21.5	22.5	24.6	24.7	24.3	24.5	24.0
13 UNFIN. PROD.	73.8	80.2	87.3	93.3	99.6	104.0	49.3	69.2	70.6	72.1	74.8
14 Unfinished Ind.	26.9	28.6	30.2	32.9	34.4	35.7	35.4	59.0	60.3	61.1	63.5
15 Unfinished Agri	5.1	5.1	5.5	6.0	6.9	7.4	8.1	7.3	6.6	7.0	7.0
16 Unfinished Cons	41.8	46.5	51.6	54.4	58.3	60.9	5.8	2.9	3.7	4.0	4.3
17 COMMODITIES	109.6	115.1	127.9	137.6	142.1	149.4	145.8	122.2	97.4	94.6	84.8
18 Trade Inventory	70.0	75.3	84.7	91.0	93.1	98.1	90.0	84.2	82.4	80.9	73.3
19 a) Trade & Ind.	69.7	75.0	84.4	90.7	92.8	97.8	89.7	83.9	82.1	80.6	77.5
20 -/-Trade	66.6	71.9	81.0	87.1	89.1	94.0	85.8	80.1	78.5	77.0	74.0
21 -/-Industry	3.1	3.1	3.4	3.6	3.7	3.8	3.9	3.8	3.6	3.6	3.5
22 b) Other	0.3	0.3	0.3	0.3	0.3	0.3	0.3	0.3	0.3	0.3	0.3
23 Supplies	38.8	39.8	43.2	46.6	49.0	51.3	55.8	38.0	15.0	13.7	11.5
24 OTHER ASSETS	21.1	22.7	24.1	25.9	27.3	28.6	30.8	32.9	53.6	54.4	54.7
25 Ready Goods	14.1	15.2	16.1	17.5	18.3	19.0	20.5	21.3	40.3	40.9	41.1
26 Future Exp. etc	7.0	7.5	8.0	8.4	9.0	9.6	10.3	11.6	13.3	13.5	13.6
27 COLLECTIVES	38.0	38.5	40.2	43.0	46.7	49.9	49.0	49.1	47.4	47.0	46.5
28 Young Livestock	19.3	19.4	20.3	21.8	24.0	25.8	26.5	26.3	24.9	24.6	25.4
29 Agri Materials	7.6	7.6	8.0	8.6	9.4	10.1	10.4	10.4	10.1	10.0	9.9
30 Ind. Materials	4.3	4.6	4.8	5.1	5.5	6.0	6.2	6.4	6.5	6.4	6.3
31 Unfinished Prod	4.2	4.2	4.3	4.5	4.6	4.8	5.0	5.1	5.0	5.1	5.0
32 Other Agri.	0.7	0.7	0.7	0.8	0.9	0.9	0.9	0.9	0.9	0.9	0.9
33 Unfin. Constr.	2.0	2.0	2.1	2.2	2.3	2.3					

*including unfinished production in construction organizations

137

TABLE B10: LOSSES OF NATIONAL INCOME
(billion rubles)

	1965	1970	1971	1972	1973	1974	1975	1976	1977	1978	1979	1980
1 NMP	193.5	289.9	305.0	313.6	337.8	353.7	363.3	385.7	405.6	426.4	440.7	462.3
2 NATIONAL INCOME	190.5	285.5	300.1	310.7	334.6	348.4	363.0	383.0	399.4	420.6	432.9	454.1
3 FOREIGN EARNINGS	0.2	1.3	1.7	-1.3	0.4	2.2	-4.8	-1.2	2.6	0.9	3.2	3.1
4 TOTAL LOSSES	2.8	3.1	3.2	4.2	2.8	3.1	5.1	3.9	3.6	4.9	4.6	5.1
5 Industry												
6 Apparel												
7 Other												
8 Construction	1.6	2.5	2.7	2.4	2.3	2.4	2.6	2.8	2.9	3.3	3.7	3.3
9 Agriculture	1.2	0.6	0.5	1.8	0.5	0.7	2.5	1.1	0.7	1.6	0.9	1.8

TABLE B11: OTHER (RESIDUAL) USES OF NATIONAL INCOME
(billion rubles)

	1965	1970	1971	1972	1973	1974	1975	1976	1977	1978	1979	1980
1 PUBLISHED TOTAL	22.3	33.1	33.4	30.1	37.4	36.1	34.1	43.7	46.8	46.6	47.0	39.2
2 Inventories	6.6	15.5	11.4	8.5	16.6	12.7	10.5	14.6	14.7	15.3	13.9	18.6
3 Unfinished Constr.	2.9	4.4	5.6	7.8	2.5	5.1	5.1	7.8	8.9	7.0	8.0	-1.3
4 State-Coop.	2.5	3.9	5.0	7.3	2.0	4.6	4.9	7.4	8.4	6.5	7.4	-1.3
5 Collective Farms	0.4	0.5	0.6	0.5	0.5	0.5	0.2	0.4	0.5	0.5	0.6	0.0
6 Other Uses	12.8	13.2	16.5	13.8	18.3	18.3	18.5	21.3	23.2	24.3	25.1	21.9
7 Current Expend.+	11.5	14.6	15.3	16.3	16.8	17.6	18.8	19.5	20.5	21.7	22.7	24.4
8 Depreciation	0.6	0.8	0.9	0.9	1.0	1.0	1.1	1.1	1.2	1.2	1.2	1.3
9 Agri. Reserves*	0.6	-2.2	0.2	-3.4	0.5	-0.3	-1.4	0.7	1.5	1.4	1.2	-3.8

+defense output included in material product accounts
*estimates of annual changes in strategic agricultural reserves are speculative
  and contain a significant margin of error

TABLE B10: LOSSES OF NATIONAL INCOME
       (billion rubles)

	1981	1982	1983	1984	1985	1986	1987	1988	1989	1990
1 NMP	486.7	523.4	548.1	569.6	578.5	587.4	599.6	630.8	657.0	676.0
2 NATIONAL INCOME	477.9	512.9	536.4	559.0	568.7	576.0	585.8	619.1	652.0	669.5
3 FOREIGN EARNINGS	2.6	4.2	5.0	5.4	2.1	4.0	5.0	1.5	-5.0	-5.0
4 TOTAL LOSSES	6.2	6.3	6.7	5.2	7.7	7.4	8.8	10.3	10.0	11.5
5 Industry								1.4	1.5	1.5
6 Apparel								1.1	1.2	1.2
7 Other								0.3	0.3	0.3
8 Construction	4.2	4.8	4.2	4.7	5.1	6.4	7.5	6.9	7.0	7.5
9 Agriculture	2.0	1.5	2.5	0.5	2.6	1.0	1.3	2.0	1.5	2.5

TABLE B11: OTHER (RESIDUAL) USES OF NATIONAL INCOME
       (billion rubles)

Page 2

	1981	1982	1983	1984	1985	1986	1987	1988	1989	1990
1 PUBLISHED TOTAL	47.1	64.5	69.7	71.2	70.1	61.4	49.8	63.9	70.4	68.7
2 Inventories	14.5	26.3	30.7	22.2	20.4	14.0	2.0	8.4	9.0	5.0
3 Unfinished Constr.	1.0	3.3	1.9	5.2	5.8	7.1	1.5	15.3	23.0	23.3
4 State-Coop.	0.9	3.0	1.6	5.0	5.6	6.9	1.5	14.6	22.4	22.8
5 Collective Farms	0.1	0.3	0.3	0.2	0.2	0.2	0.0	0.7	0.6	0.5
6 Other Uses	31.6	34.9	37.1	43.8	43.9	40.3	46.3	40.2	38.4	40.4
7 Current Expend.+	28.2	30.6	33.8	39.6	42.4	42.7	40.7	44.6	37.2	38.2
8 Depreciation	1.3	1.4	1.4	1.5	1.5	1.6	1.7	1.8	1.9	2.0
9 Agri. Reserves*	2.1	2.9	1.9	2.7	0.0	-4.0	3.9	-6.2	-0.7	0.2

+defense output included in material product accounts
*estimates of annual changes in strategic agricultural reserves are speculative
 and contain a significant margin of error

Part C

Foreign Trade

TABLE C1: EXPORTS IN FOREIGN AND DOMESTIC PRICES
        (billion rubles)   Page 1

		1965	1970	1971	1972	1973	1974	1975	1976	1977	1978	1979	1980
1	TOTAL f.p.	7.4	11.5	12.4	12.7	15.8	20.8	24.0	28.0	33.3	35.7	42.4	49.6
2	TOTAL d.p.	8.3	16.6	17.3	17.5	19.6	23.1	22.9	24.6	27.0	28.2	29.7	31.0
3	Coefficient	1.12	1.44	1.39	1.38	1.24	1.11	0.95	0.88	0.81	0.79	0.70	0.63
4	POWER f.p.	0.0	0.1	0.1	0.1	0.1	0.1	0.1	0.2	0.2	0.3	0.3	0.3
5	POWER d.p.	0.0	0.1	0.1	0.1	0.1	0.1	0.1	0.2	0.2	0.2	0.2	0.2
6	FUELS f.p.	1.3	1.7	2.0	2.1	2.9	5.2	7.4	9.4	11.5	12.4	17.6	23.0
7	FUELS d.p.	1.3	2.4	2.6	2.7	2.9	3.1	3.4	3.9	4.3	4.4	4.5	4.6
8	growth rate	0.70	1.00	1.08	1.12	1.23	1.28	1.42	1.64	1.79	1.83	1.88	1.93
9	physic. units	116.7	167.0	180.5	187.6	205.7	214.1	237.7	273.5	298.7	306.3	314.1	321.5
10	Coefficient	1.00	1.38	1.28	1.27	1.00	0.59	0.46	0.42	0.37	0.35	0.26	0.20
11	METALS f.p.	1.6	2.3	2.4	2.6	2.7	3.0	3.4	3.7	3.7	3.7	3.9	4.4
12	METALS d.p.	1.3	3.1	3.2	3.3	3.4	3.5	3.4	3.5	3.4	3.4	3.4	3.7
13	Coefficient	0.80	1.35	1.33	1.30	1.25	1.15	1.00	0.95	0.93	0.91	0.87	0.84
14	CHEMICALS f.p.	0.2	0.5	0.5	0.5	0.5	0.7	0.8	0.8	0.9	1.1	1.2	1.6
15	CHEMICALS d.p.	0.4	1.0	1.0	1.0	1.1	1.3	1.3	1.3	1.4	1.5	1.7	2.2
16	Coefficient	1.80	2.05	2.05	2.00	1.95	1.85	1.70	1.60	1.50	1.40	1.40	1.40
17	M & E f.p.	1.5	2.5	2.7	3.0	3.4	4.0	4.4	5.4	6.2	7.0	7.4	7.9
18	M & E d.p.	1.2	2.4	2.5	2.7	3.1	3.6	3.7	4.4	4.8	5.4	5.7	6.0
19	Coefficient	0.80	0.96	0.92	0.92	0.91	0.90	0.85	0.82	0.78	0.77	0.77	0.76
20	WOOD f.p.	0.5	0.8	0.8	0.8	1.0	1.3	1.4	1.5	1.7	1.6	1.8	2.0
21	WOOD d.p.	0.5	1.3	1.2	1.2	1.4	1.8	1.8	1.8	2.0	1.9	2.2	2.4
22	Coefficient	1.00	1.60	1.55	1.50	1.40	1.35	1.25	1.20	1.20	1.20	1.20	1.20
23	CONSTR. MATER.	0.1	0.1	0.1	0.1	0.1	0.1	0.1	0.1	0.1	0.1	0.1	0.1
24	TEXTILES f.p.	0.4	0.5	0.5	0.6	0.6	0.7	0.8	0.9	1.0	1.1	1.1	1.1
25	TEXTILES d.p.	0.9	1.5	1.6	1.8	1.9	2.1	2.3	2.8	2.9	3.2	3.3	3.3
26	Coefficient	2.30	3.00	3.00	3.00	3.00	3.00	3.00	3.00	3.00	3.00	3.00	3.00
27	FOOD f.p.	0.4	0.5	0.5	0.4	0.4	0.7	0.6	0.5	0.5	0.6	0.6	0.5
28	FOOD d.p.	1.3	2.0	2.0	1.7	1.7	2.7	2.1	1.7	1.5	1.8	1.7	1.4
29	Coefficient	3.20	4.00	4.00	4.00	4.00	3.80	3.50	3.20	3.20	3.15	3.05	3.00
30	AGRI f.p.	0.2	0.4	0.5	0.4	0.5	0.8	0.6	0.3	0.5	0.2	0.5	0.4
31	AGRI d.p.	0.3	0.7	0.8	0.7	0.8	1.3	0.8	0.4	0.7	0.3	0.6	0.6
32	Coefficient	1.70	1.70	1.65	1.70	1.65	1.60	1.40	1.40	1.40	1.40	1.40	1.40
33	Consumer f.p.	0.2	0.4	0.4	0.4	0.5	0.6	0.6	0.7	0.9	1.0	0.9	1.0
34	Consumer d.p.	0.2	0.4	0.4	0.4	0.5	0.6	0.6	0.7	0.9	1.0	0.9	1.0
35	MBMW	0.1	0.1	0.1	0.1	0.2	0.2	0.2	0.2	0.3	0.4	0.4	0.4
36	N.E.C. Ind.	0.1	0.2	0.3	0.3	0.3	0.3	0.3	0.4	0.5	0.6	0.4	0.5
37	Apparel	0.0	0.0	0.0	0.0	0.0	0.1	0.1	0.1	0.1	0.1	0.1	0.1
38	Other f.p.	0.0	0.2	0.2	0.3	0.4	0.6	0.6	0.7	0.8	0.9	1.0	1.0
39	Other d.p.	0.0	0.1	0.1	0.2	0.3	0.4	0.4	0.5	0.6	0.6	0.7	0.7
40	Residual f.p.	1.0	1.5	1.6	1.6	2.6	3.0	3.2	3.7	5.3	5.8	6.1	6.3
41	Residual d.p.	0.8	1.4	1.5	1.4	2.3	2.7	2.7	3.1	4.1	4.5	4.7	4.8

TABLE C1: EXPORTS IN FOREIGN AND DOMESTIC PRICES
        (billion rubles)   Page 2

		1981	1982	1983	1984	1985	1986	1987	1988	1989	1990
1	TOTAL f.p.	57.1	63.2	67.9	74.4	72.7	68.3	68.1	67.1	68.2	69.0
2	TOTAL d.p.	33.4	39.4	40.4	43.3	42.1	44.3	44.4	46.2	47.1	48.0
3	Coefficient	0.59	0.62	0.60	0.58	0.58	0.65	0.65	0.69	0.69	0.70
4	POWER f.p.	0.4	0.5	0.6	0.7	0.7	0.7	0.7	0.9	0.9	0.9
5	POWER d.p.	0.3	0.5	0.6	0.6	0.6	0.6	0.6	0.8	0.8	0.8
6	FUELS f.p.	28.3	32.6	35.9	39.8	37.8	31.6	31.0	27.4	27.6	28.5
7	FUELS d.p.	4.6	8.1	8.6	8.9	8.5	10.6	11.2	11.5	11.6	12.0
8	growth rate	1.94	2.03	2.15	2.23	2.11	2.37	2.50	2.67	2.69	2.73
9	physic. units	323.2	339.7	358.7	371.8	352.2	396.0	418.0	446.5	450.0	456.0
10	Coefficient	0.16	0.25	0.24	0.22	0.22	0.34	0.36	0.42	0.42	0.42
11	METALS f.p.	4.6	4.7	5.1	5.4	5.4	5.7	5.8	6.4	6.5	6.5
12	METALS d.p.	3.9	4.7	4.9	5.1	5.1	5.6	5.7	5.7	5.9	5.9
13	Coefficient	0.84	1.00	0.97	0.96	0.94	0.98	0.98	0.90	0.90	0.90
14	CHEMICALS f.p.	2.0	2.0	2.1	2.6	2.8	2.4	2.3	2.7	2.8	2.8
15	CHEMICALS d.p.	2.7	2.9	3.1	3.8	4.1	3.5	3.4	4.0	4.1	4.1
16	Coefficient	1.40	1.47	1.47	1.47	1.47	1.47	1.47	1.47	1.47	1.47
17	M & E f.p.	7.8	8.2	8.5	9.3	9.9	10.2	10.6	10.9	11.3	11.9
18	M & E d.p.	5.9	6.0	6.0	6.4	6.4	6.4	6.4	6.1	6.3	6.8
19	Coefficient	0.76	0.74	0.71	0.69	0.65	0.63	0.60	0.56	0.56	0.57
20	WOOD f.p.	1.9	1.8	1.9	2.1	2.2	2.3	2.3	2.4	2.3	2.3
21	WOOD d.p.	2.3	2.4	2.6	2.8	3.0	3.1	3.2	3.2	3.1	3.2
22	Coefficient	1.20	1.35	1.35	1.35	1.35	1.35	1.35	1.35	1.36	1.38
23	CONSTR. MATER.	0.1	0.1	0.1	0.1	0.2	0.2	0.2	0.2	0.2	0.2
24	TEXTILES f.p.	1.2	1.2	1.2	1.3	1.2	1.0	1.0	1.1	1.1	1.1
25	TEXTILES d.p.	3.6	4.0	3.9	4.4	3.9	3.2	3.4	3.6	3.6	3.8
26	Coefficient	3.00	3.35	3.35	3.35	3.35	3.35	3.35	3.40	3.40	3.40
27	FOOD f.p.	0.6	0.6	0.7	0.8	0.8	0.8	0.7	0.9	0.9	0.9
28	FOOD d.p.	1.7	1.8	1.9	2.2	2.0	2.0	1.6	2.3	2.4	2.5
29	Coefficient	2.80	2.80	2.75	2.70	2.65	2.60	2.45	2.45	2.75	2.80
30	AGRI f.p.	0.5	0.4	0.3	0.3	0.3	0.3	0.2	0.2	0.2	0.2
31	AGRI d.p.	0.7	0.6	0.5	0.5	0.4	0.4	0.2	0.3	0.3	0.3
32	Coefficient	1.40	1.40	1.50	1.50	1.45	1.45	1.40	1.40	1.45	1.50
33	Consumer f.p.	1.0	1.1	1.1	1.2	1.2	1.6	1.9	1.8	2.0	2.1
34	Consumer d.p.	1.0	1.0	1.0	1.1	1.2	1.5	1.7	1.5	1.6	1.7
35	MBMW	0.4	0.4	0.4	0.4	0.4	0.5	0.6	0.6	0.7	0.8
36	N.E.C. Ind.	0.5	0.5	0.5	0.5	0.6	0.7	0.8	0.6	0.6	0.6
37	Apparel	0.1	0.2	0.2	0.2	0.2	0.3	0.3	0.3	0.3	0.3
38	Other f.p.	1.2	1.2	1.3	1.3	1.4	1.3	1.4	1.7	1.6	1.7
39	Other d.p.	0.8	0.8	0.8	0.8	0.9	0.8	0.9	1.1	1.0	1.1
40	Residual f.p.	7.5	8.9	9.2	9.6	8.8	10.3	10.0	10.6	10.9	9.9
41	Residual d.p.	5.7	6.6	6.5	6.6	5.7	6.3	6.0	5.9	6.1	5.6

143

TABLE C2: IMPORTS IN FOREIGN AND DOMESTIC PRICES
        (billion rubles)

	1965	1970	1971	1972	1973	1974	1975	1976	1977	1978	1979	1980
1 TOTAL f.p.	7.2	10.6	11.2	13.3	15.5	18.8	26.7	28.7	30.1	34.6	37.9	44.6
2 TOTAL d.p.	14.8	26.9	28.0	31.8	34.2	36.7	46.5	49.5	51.2	56.3	60.5	71.4
3 Coefficient	2.06	2.54	2.50	2.39	2.21	1.95	1.74	1.72	1.71	1.63	1.60	1.60
4 FUELS f.p.	0.2	0.2	0.3	0.4	0.6	0.7	1.1	1.1	1.1	1.3	1.3	1.3
5 FUELS d.p.	0.2	0.4	0.6	0.7	0.9	0.9	1.1	1.0	0.9	1.0	0.8	0.7
6 Coefficient	0.91	1.90	1.75	1.75	1.55	1.29	1.00	0.91	0.82	0.79	0.63	0.54
7 METALS f.p.	0.7	1.0	1.1	1.2	1.5	2.6	3.1	3.1	2.8	3.4	4.2	4.8
8 METALS d.p.	0.7	1.1	1.1	1.2	1.4	2.4	2.8	2.6	2.4	2.8	3.4	3.9
9 Coefficient	1.00	1.05	1.05	1.00	0.95	0.92	0.89	0.86	0.85	0.84	0.81	0.81
10 CHEMICALS f.p.	0.4	0.6	0.6	0.7	0.7	1.2	1.3	1.2	1.3	1.4	1.8	2.4
11 CHEMICALS d.p.	1.1	1.5	1.5	1.6	1.6	2.4	2.4	2.3	2.4	2.6	3.2	4.1
12 Coefficient	2.46	2.48	2.48	2.46	2.40	2.03	1.91	1.86	1.85	1.85	1.80	1.75
13 M & E f.p.	2.5	3.8	3.8	4.6	5.3	6.1	9.2	10.5	11.7	14.6	14.5	15.2
14 M & E d.p.	2.6	4.2	4.1	4.8	5.3	6.1	9.2	10.5	11.7	14.6	14.5	15.2
15 Coefficient	1.04	1.11	1.07	1.04	1.00	1.00	1.00	1.00	1.00	1.00	1.00	1.00
16 WOOD f.p.	0.1	0.2	0.2	0.2	0.3	0.3	0.6	0.5	0.5	0.5	0.6	0.9
17 WOOD d.p.	0.3	0.6	0.6	0.7	0.8	0.9	1.3	1.0	1.0	1.0	1.2	1.6
18 CON. MAT. f.p.	0.1	0.1	0.1	0.1	0.1	0.2	0.2	0.2	0.2	0.2	0.2	0.2
19 CON. MAT. d.p.	0.1	0.3	0.3	0.3	0.4	0.4	0.4	0.5	0.5	0.6	0.6	0.6
20 Coefficient	2.80	3.10	3.10	3.00	2.85	2.65	2.24	2.08	2.00	1.92	1.85	1.78
21 TEXTILES f.p.	0.3	0.5	0.5	0.4	0.6	0.6	0.6	0.7	0.7	0.7	0.7	1.0
22 TEXTILES d.p.	1.1	1.9	1.7	1.6	1.5	1.5	1.6	1.7	1.8	1.8	1.8	2.2
23 Coefficient	3.67	3.73	3.69	3.62	2.67	2.59	2.55	2.54	2.54	2.54	2.47	2.27
24 FOOD f.p.	1.0	1.1	1.0	1.0	1.5	2.0	3.3	3.6	4.3	4.2	4.7	6.6
25 FOOD d.p.	2.9	4.5	4.1	4.1	4.8	5.0	6.3	6.8	8.1	8.0	8.9	12.5
26 Coefficient	2.93	4.15	4.10	4.00	3.14	2.50	1.94	1.92	1.91	1.90	1.90	1.90
27 AGRI f.p.	0.5	0.6	0.7	1.3	1.6	1.2	2.8	3.0	2.0	2.4	3.6	4.2
28 AGRI d.p.	0.9	1.4	1.7	3.2	3.9	2.8	5.8	6.0	4.5	5.1	6.8	8.0
29 Coefficient	1.90	2.33	2.34	2.43	2.44	2.26	2.07	2.00	2.25	2.13	1.90	1.90
30 CONSUMER f.p.	1.1	2.0	2.3	2.4	2.4	2.8	3.4	3.6	3.9	4.1	4.3	5.4
31 CONSUMER d.p.	4.1	9.9	10.9	11.9	12.0	12.7	14.0	15.1	15.8	16.4	16.8	19.4
32 Coefficient	3.89	4.88	4.86	4.86	4.90	4.54	4.12	4.17	4.07	4.03	3.95	3.59
33 Chemicals	0.0	0.3	0.3	0.3	0.3	0.2	0.2	0.2	0.2	0.2	0.2	0.2
34 Rubber Footware	0.2	0.3	0.3	0.3	0.2	0.2	0.2	0.2	0.2	0.2	0.2	0.2
35 MBMW Products	0.3	0.5	0.4	0.5	0.4	0.5	0.7	0.9	0.9	0.8	0.8	0.8
36 Furniture	0.3	0.5	0.6	0.6	0.6	0.6	0.6	0.7	0.7	0.8	0.8	0.9
37 N.E.C. Prod.	0.1	0.2	0.2	0.3	0.3	0.3	0.4	0.4	0.4	0.5	0.5	0.6
38 Soap & Cosmetic	0.2	0.1	0.1	0.2	0.2	0.2	0.2	0.3	0.3	0.3	0.3	0.5
39 Apparel, etc.	3.1	8.0	8.9	9.8	10.0	10.7	11.6	12.4	13.0	13.6	13.9	16.1

TABLE C2: IMPORTS IN FOREIGN AND DOMESTIC PRICES
     (billion rubles)

	1981	1982	1983	1984	1985	1986	1987	1988	1989	1990
1 TOTAL f.p.	52.6	56.4	59.6	65.3	69.4	62.6	60.7	65.0	70.2	70.0
2 TOTAL d.p.	80.8	88.3	92.2	97.0	102.8	98.5	88.9	96.6	107.4	106.5
3 Coefficient	1.54	1.57	1.55	1.49	1.48	1.57	1.46	1.49	1.53	1.52
4 FUELS f.p.	1.9	2.6	3.3	4.0	3.7	2.9	2.4	2.9	3.1	3.1
5 FUELS d.p.	0.8	1.4	1.6	1.7	1.5	1.2	0.7	0.7	0.8	0.8
6 Coefficient	0.42	0.54	0.48	0.43	0.41	0.41	0.29	0.24	0.26	0.26
7 METALS f.p.	5.3	5.6	5.2	5.4	5.8	5.2	4.9	5.2	5.5	5.5
8 METALS d.p.	4.2	5.3	4.9	5.0	5.3	5.0	4.5	4.8	5.0	5.0
9 Coefficient	0.79	0.95	0.94	0.92	0.91	0.96	0.92	0.92	0.91	0.91
10 CHEMICALS f.p.	2.7	2.5	2.7	2.9	3.4	3.2	3.2	3.3	3.2	3.2
11 CHEMICALS d.p.	4.8	4.6	5.1	5.4	6.3	5.9	5.9	6.0	5.9	5.9
12 Coefficient	1.75	1.85	1.85	1.85	1.85	1.85	1.85	1.85	1.85	1.85
13 M & E f.p.	16.2	19.6	23.0	24.2	25.9	25.5	25.1	26.6	27.0	27.5
14 M & E d.p.	16.2	19.6	23.0	24.2	25.9	25.5	25.1	26.6	27.0	27.5
15 Coefficient	1.00	1.00	1.00	1.00	1.00	1.00	1.00	1.00	1.00	1.00
16 WOOD f.p.	1.0	0.8	0.8	0.8	0.9	0.8	0.7	0.8	0.8	0.8
17 WOOD d.p.	1.6	1.5	1.5	1.4	1.6	1.4	1.2	1.3	1.3	1.3
18 CON. MAT. f.p.	0.2	0.3	0.4	0.4	0.4	0.4	0.4	0.5	0.5	0.5
19 CON. MAT. d.p.	0.6	0.8	0.9	0.9	0.9	0.9	0.9	0.9	0.9	0.9
20 Coefficient	1.68	1.88	1.83	1.79	1.78	1.76	1.71	1.68	1.63	1.63
21 TEXTILES f.p.	0.9	0.9	1.2	1.0	1.2	0.6	0.8	1.0	1.0	1.0
22 TEXTILES d.p.	2.0	2.3	3.0	2.6	2.8	1.7	2.0	2.5	2.5	2.5
23 Coefficient	2.20	2.47	2.42	2.50	2.37	2.83	2.50	2.50	2.50	2.50
24 FOOD f.p.	8.6	7.6	7.1	9.0	8.7	7.1	6.7	6.3	9.5	9.3
25 FOOD d.p.	16.0	16.2	15.6	18.0	17.6	16.4	13.7	13.1	19.5	18.6
26 Coefficient	1.87	2.14	2.20	2.00	2.02	2.31	2.04	2.08	2.05	2.00
27 AGRI f.p.	6.0	5.8	5.1	5.7	5.9	3.5	3.1	4.0	4.1	4.5
28 AGRI d.p.	9.7	9.3	9.2	9.3	9.8	7.6	5.8	8.6	8.5	9.5
29 Coefficient	1.62	1.60	1.80	1.63	1.66	2.17	1.87	2.14	2.07	2.11
30 CONSUMER f.p.	6.8	7.2	6.9	7.5	8.6	8.4	7.9	8.3	9.2	9.0
31 CONSUMER d.p.	21.2	23.1	22.9	23.7	25.9	27.4	23.6	25.8	29.3	28.1
32 Coefficient	3.13	3.20	3.31	3.15	3.00	3.26	2.99	3.11	3.18	3.12
33 Chemicals	0.2	0.3	0.3	0.3	0.3	0.2	0.2	0.2	0.3	0.3
34 Rubber Footware	0.2	0.2	0.3	0.3	0.3	0.3	0.3	0.3	0.4	0.4
35 MBMW Products	0.9	1.1	1.1	1.1	1.1	1.2	1.1	1.2	1.2	1.3
36 Furniture	1.0	1.1	1.1	1.1	1.1	1.1	1.1	1.1	1.1	1.0
37 N.E.C. Prod.	0.9	1.0	1.0	1.1	1.2	1.2	1.2	1.2	1.3	1.3
38 Soap & Cosmetic	0.8	0.9	0.9	0.8	0.8	1.0	1.1	1.4	1.5	1.8
39 Apparel, etc.	17.2	18.5	18.2	19.0	21.1	22.4	18.6	20.4	23.5	22.0

TABLE C2: IMPORTS IN FOREIGN AND DOMESTIC PRICES
(billion rubles)

	1965	1970	1971	1972	1973	1974	1975	1976	1977	1978	1979	1980
40 RESIDUAL f.p.	0.4	0.5	0.6	0.8	0.9	1.1	1.2	1.3	1.6	1.8	2.0	2.6
41 RESIDUAL d.p.	0.9	1.2	1.3	1.6	1.6	1.6	1.6	1.9	2.2	2.4	2.5	3.2
42 Other Chemicals	0.2	0.4	0.4	0.4	0.4	0.4	0.4	0.5	0.6	0.6	0.6	0.7
43 Other MBMW	0.6	0.7	0.8	1.0	1.0	1.0	1.0	1.2	1.4	1.6	1.7	2.2
44 Other Products	0.1	0.1	0.1	0.2	0.2	0.2	0.2	0.2	0.2	0.2	0.2	0.3

Total Imports for Selected Sectors:

	1965	1970	1971	1972	1973	1974	1975	1976	1977	1978	1979	1980
45 Total Chemicals	1.5	2.5	2.5	2.6	2.5	3.3	3.3	3.2	3.4	3.6	4.2	5.3
46 Total MBMW	3.5	5.4	5.3	6.3	6.7	7.6	10.9	12.6	14.0	17.0	17.0	18.2
47 Total Wood	0.6	1.1	1.3	1.3	1.4	1.5	1.9	1.7	1.7	1.8	2.0	2.5
48 Total Food	3.1	4.6	4.2	4.3	5.0	5.2	6.5	7.1	8.4	8.3	9.2	13.0
49 Total Light	4.2	9.9	10.6	11.4	11.5	12.2	13.2	14.1	14.8	15.4	15.7	18.3

Supporting Data on Consumer Goods in Foreign Trade Prices:

	1965	1970	1971	1972	1973	1974	1975	1976	1977	1978	1979	1980
50 Chemicals	0.0	0.1	0.1	0.1	0.1	0.1	0.1	0.1	0.1	0.2	0.2	0.2
51 MBMW Products	0.1	0.1	0.1	0.2	0.2	0.2	0.3	0.3	0.3	0.3	0.3	0.3
52 Furniture	0.1	0.2	0.2	0.2	0.2	0.3	0.3	0.3	0.4	0.4	0.4	0.4
53 Medical Supply	0.1	0.2	0.2	0.3	0.3	0.3	0.4	0.4	0.4	0.5	0.5	0.6
54 Other N.E.C.	0.1	0.1	0.1	0.1	0.1	0.2	0.2	0.2	0.2	0.2	0.2	0.2
55 Soap & Cosm.	0.1	0.1	0.1	0.1	0.1	0.1	0.1	0.2	0.2	0.2	0.2	0.4
56 Apparel, etc.	0.6	1.2	1.4	1.5	1.5	1.7	2.0	2.1	2.3	2.4	2.5	3.3

TABLE C2: IMPORTS IN FOREIGN AND DOMESTIC PRICES
        (billion rubles)

		1981	1982	1983	1984	1985	1986	1987	1988	1989	1990
41	RESIDUAL f.p.	3.0	3.5	3.8	4.3	4.9	5.1	5.5	6.1	6.3	5.6
42	RESIDUAL d.p.	3.6	4.2	4.6	4.7	5.1	5.5	5.6	6.4	6.7	6.4
43	Other Chemicals	0.7	0.8	0.9	0.9	0.9	1.0	1.0	1.1	1.1	1.1
44	Other MBMW	2.4	2.8	3.0	3.1	3.4	3.5	3.6	3.8	4.0	3.7
45	Other Products	0.5	0.6	0.7	0.7	0.8	1.0	1.0	1.5	1.6	1.6

Total Imports for Selected Sectors:

		1981	1982	1983	1984	1985	1986	1987	1988	1989	1990
51	Total Chemicals	5.9	5.9	6.6	6.9	7.8	7.4	7.4	7.6	7.7	7.7
52	Total MBMW	19.5	23.5	27.1	28.4	30.4	30.2	29.8	31.6	32.2	32.5
53	Total Wood	2.6	2.6	2.6	2.5	2.7	2.5	2.3	2.4	2.4	2.3
54	Total Food	16.8	17.1	16.5	18.8	18.4	17.4	14.8	14.5	21.0	20.4
55	Total Light	19.2	20.8	21.2	21.6	23.9	24.1	20.6	22.9	26.0	24.5

		1981	1982	1983	1984	1985	1986	1987	1988	1989	1990
43	Chemicals	0.2	0.2	0.2	0.2	0.2	0.1	0.1	0.1	0.1	0.1
44	MBMW Products	0.4	0.4	0.4	0.4	0.4	0.4	0.4	0.4	0.4	0.4
45	Furniture	0.5	0.6	0.6	0.6	0.6	0.6	0.6	0.6	0.6	0.6
46	Medical Supply	0.9	1.0	1.0	1.2	1.3	1.3	1.3	1.4	1.5	1.5
47	Other N.E.C.	0.2	0.2	0.2	0.3	0.3	0.3	0.2	0.2	0.3	0.3
48	Soap & Cosm.	0.5	0.4	0.3	0.4	0.4	0.4	0.4	0.5	0.5	0.6
49	Apparel, etc.	4.1	4.4	4.3	4.4	5.4	5.4	4.9	5.1	5.7	5.5

Chapter 4

The Real Size and Structure of the USSR Economy

CHAPTER 4

THE REAL SIZE AND STRUCTURE OF THE USSR ECONOMY

It is often emphasized that one of the major difficulties with analyzing the USSR economy lies in obtaining a reliable and complete set of GNP accounts. Rather than publishing a full set of Soviet GNP accounts, Goskomstat officials are instructed to conceal important Soviet financial, defense and other economic secrets.[1] As a result, they limit the volume of published results to the total size of Soviet GNP and its three major end uses-- consumption, investment and other (see Chart 6 below).[2]

However, even these GNP data are difficult to use because of the absence of detailed notes on how to convert the regularly published Goskomstat's production and financial statistics into the GNP format. In the absence of such notes, doubts remain about the accuracy of the Goskomstat's estimates of the total size and structure of Soviet GNP. This makes it desirable to have alternative estimates.

As was discussed in previous chapters, any alternative estimate must be based on the detailed analysis of how Soviet production and financial methods of national accounts differ from each other as well as from a standard GNP method. It must be noted that this analysis was already performed to some extent in two articles written by three known

Soviet experts on national accounts.[3] In addition, they co-authored the Gosplan-Goskomstat study which contains a description of GNP accounts estimated using production, distribution and end use methods of compiling Soviet national income data.[4] Goskomstat officials also published methodological notes on how to measure Soviet GNP in the 1989 Narkhoz issue.[5]

Whereas Goskomstat officials have presented a fairly complete procedure for estimating GNP for civilian sectors of the USSR economy, they have not resolved all difficult methodological issues that are often encountered in using Soviet national accounts. Moreover, they conspicuously avoided any mention of how the Goskomstat accounts for secret defense and other national security (NS) activities in their published GNP estimates. Thus, their procedure for converting published i-o, production and financial data into a GNP format remains far from complete.

Below I propose to complete the Goskomstat procedure using the integrated approach to analyzing all published Soviet statistics. I describe how the new i-o, production and financial data can be used a) to compile Soviet GNP accounts by sector of origin and end use and b) to improve the precision of estimates performed for particular sectors of the economy. Afterward, I present my estimates of Soviet GNP by sector of origin and end use for 1965, 1970-1990 in current established prices. In the end of this chapter I suggest that the obtained data on GNP by end use lead to a misleading picture of the real structure of the USSR economy. Specifically, I argue that the additional statistics must be brought into light to determine the real Soviet defense burden and to understand the current finacial crisis.

---

[1] The First Deputy Chairman of the Goskomstat I. Pogosov wrote recently that "the implementation of the glasnost policy in the area of economics is undermined by the existence of the continuous ban on the publication of statistical data of the type used abroad, particularly in UN publications." He added that "the absence of such data makes it difficult to conduct economic research in academic institutions, leads to criticism on the part of business circles, and creates mistrust in the official statistics." He suggested that the first priority should be the publication of data on household budgets, foreign currency transactions and military expenditures. I. Pogosov, "Sotsial'no-Ekonomicheskie Preobrazovaniya i Informatsiya." Ekonomika i Matematicheskie Metody, Vol. XXV, N5, 1989, p. 786.

[2] V. Kirichenko, "Natsional'nyi Dokhod: Spravedlivost' Raspredeleniya." Ekonomika i Zhizn', N13, March 1990, p. 4.

[3] Yu. Ivanov and B. Ryabushkin, "Integratsiya Balansa Narodnogo Khozyaistva i Sistemy Natsional'nykh Schetov." Vestnik Statistiki (VS), N9, 1989; and to Yu. Ivanov, B. Ryabushkin and M. Eidel'man, "Ischislenie Valovogo Natsional'nogo Produkta SSSR," VS, N7, 1988.

[4] Goskomstat SSSR and Gosplan SSSR, Metodika Ischisleniya Valovogo Natsional'nogo Produkta SSSR. Moscow, 31 March 1988.

[5] Narkhoz for 1989, pp. 696-699.

Chart 6: New Soviet Data on GNP by End Use
        (billion rubles)

	1985	1986	1987	1988	1989
Total GNP	777.0	798.5	825.0	875.4	924.1
Consumption	421.5	431.7	444.7	472.5	508.5
Households	369.3	377.0	388.2	411.2	444.5
Free Services	52.2	54.7	56.5	61.3	64.0
Investment	248.4	255.5	264.8	276.1	287.6
Other Uses	107.1	111.3	115.5	126.8	128.0

Official GNP by Sector of Origin

The analysis of Goskomstat publications leads to the conclusion that a) it estimates the size of Soviet GNP using data on total labor and non-labor income earned in all sectors of the economy, and b) it does not have data on other end uses--primarily defense--which it derives as a residual.

I attempt to duplicate the Goskomstat estimates of Soviet GNP by sector of origin in Table D1 below. In accord with Soviet national accounting practices, I present estimates for three major groups of sectors: 1) enterprises engaged in material production, 2) "non-material" services, and 3) so-called "other sectors" which are excluded from the above. Following the Goskomstat's estimation methodology, I measure total Soviet GNP as the sum of labor and non-labor income and depreciation.

Labor and non-labor income of material production and "non-material" service sectors as well as depreciation were already estimated within the format of Soviet national accounts in the previous chapter. An attempt was also made in this chapter to measure the net output of "other" sectors using published data on total wages and profits.

At issue here is how to convert the NEB and i-o data in such a way that it would be possible to compare estimation results with those published by the Goskomstat. The issue arises because Goskomstat officials have not revealed all the necessary details on how they have actually derived total GNP nor consumption and investment.

Methodological Issues

Compared to net material and "non-material" products estimated in Soviet NEB and i-o tables, GNP should include:

- labor and non-labor income of "other" sectors of the USSR economy
- privately provided services
- imputed rent for owner-occupied houses
- military subsistence (free food and uniforms), and
- foreign monetary revenues.

Similarly, GNP should exclude:

- the unamortized value of scrapped fixed capital
- all losses of national income
- employees' business trip and other business-type expenditures, and
- all purchases of business services by enterprises.

The problem is that Goskomstat officials have not clarified its methodology for estimating GNP with respect to the above revenues and outlays.

"Other" sectors are never mentioned in Goskomstat publications, except in employment reports. Despite this fact, Goskomstat officials are well aware of this sector when they compile the wage, profit and social security data.

Private services are mentioned in some publications but are omitted from other. Soviet economists reveal that Goskomstat experts make no independent estimates of private services and instead rely on Gosplan's extensive research results performed by Tatyana Koryagina using a sociological approach.[6]

While imputed rent and foreign revenues are mentioned in all Goskomstat's publications, there is no evidence that its experts actually estimate these components of Soviet GNP by sector of origin. Moreover, given the existing compartmentalization of information flows between Soviet government bureaucracies, it is doubtful whether Goskomstat even has sufficient data for measuring the real size and structure of foreign monetary revenues. The same apparently applies to military subsistence.

It is unclear which losses are excluded and which are included by Goskomstat officials in deriving the official figure for total GNP. As was discussed in the previous chapter, there are three types of losses in Soviet national income accounts: a) the unamortized value of scrapped fixed capital, b) losses of inventories and reserves in industry (since 1988) and in agriculture, c) and losses in construction (investment writeoffs and other losses which are unspecified).

It appears from Goskomstat's publications that the official GNP excludes the unamortized value.[7] In my independent estimates of the official GNP I assume that the same applies to losses of inventories and reserves and industry and agriculture.

As presented in Goskomstat's publications, both labor and non-labor incomes are reduced by the amount of business expenditures in preparing official GNP estimates. It remains unclear, however, what procedure was employed to measure business expenditures. According to one senior Goskomstat official, business expenditures which are treated as labor income are estimated using the i-o data on wages and other earnings in material production sectors and as 4 percent of "non-material" sectors' wages.[8] In addition, purchases of goods supposedly account for 25 percent of all business expenditures treated as labor income.

What remains unknown is the amount of wages received by employees of material production sectors that are treated as business expenditures. Another unknown is why these expenditures comprise 4 percent of "non-material" sectors' wages when other earnings comprise only 3 percent of the same wages in i-o tables. As expected, there is no information on "other" sectors' business expenditures. In sum, given the existing lack of information on the subject, it is difficult to avoid making three assumptions:

1) other earnings in all sectors of the economy which are listed in Goskomstat's tables on the household budget (see Chart 3) approximately equal total business expenditures;

2) the same as above applies to material production and "non-material" sectors; and

3) the already mentioned 3 percent rate applies to "other" sectors as well.

The validity of the second assumption depends to a large extent on the compatibility of published NEB, i-o and financial data on other earnings. However, such a compatibility can be observed only till 1985. As estimated in NEB and i-o tables, other earnings in material production sectors decreased from 6.9 to 6.3 b.r. in 1986-1988 apparently as a result of changes in the industrial wage accounting procedures (see reconstructed estimates in Table A4).

If the trend in the growth of other earnings continued uninterrupted, then other earnings in material production sectors would have equaled around 8.2 b.r. rather than 6.3 b.r. in 1988. I thus assume that business expenditures treated as labor income in material production sectors also equaled 8.2 b.r., of which 1.9 b.r. were registered as industrial wages.

Purchases of individual business services by enterprises were derived in the previous chapter as the difference between the total volume of services and the sum of purchases made by the state budget and households. Included in the list of estimated business services are free (subsidized) housing-communal, education, cultural and health services provided by enterprises to their employees. Also included in the list are business purchases of passenger transportation, research, banking, insurance, and administrative services.

---

[6] For example, refer to her recent book T. Koryagina, Platnye Uslugi v SSSR, Moscow: Ekonomika, 1990, pp. 40-50.

[7] See Goskomstat and Gosplan, Metodika..., op. cit., pp. 9 and 12.

[8] B. Ryabushkin, "Statisticheskoe Izuchenie Urovnya Zhizni," VS, N4, 1988, p. 22.

Excluded from this list are government expenditures connected with housing depreciation which in Soviet national income accounts are treated as consumption of material wealth by households. In theory, housing depreciation is included in GNP when housing services are operated as a business enterprise. In the context of the USSR economy, however, housing services are primarily run by the state or occupied by owners. This poses a difficult methodological question as to how to account for housing depreciation.

Two valid approaches can be proposed as solutions: 1) to exclude housing depreciation altogether or 2) to include it twice as part of capital depreciation and as business expenditures. The advantage of the second approach is that it enables to integrate GNP accounts with capital flows as well as to measure net national product (NNP) as GNP less total depreciation.

I arbitrarily assumed that most purchases of business services are made by material production services since a large majority of "non-material" sectors do not earn profit from their operation.

Estimating GNP by Sector of Origin

I estimated the official GNP in Table D1 below in three ways as the sum of household and public sectors' incomes, as value added in material production, "non-material" services and "other" sectors, and as the sum of net national product and depreciation.

Most of my estimates for material production sectors were based on data presented in Tables A6 and A10. Total value added in these sectors (row 4) equals the sum of labor income (row 5), revenues (row 8), revenues from foreign trade (row 13) and total depreciation of capital operated in these sectors (row 14). Labor income equals the difference between total labor income (row 6) and business expenditures (row 7).

Revenues (row 8) equal the difference between gross revenues (row 9) and the sum of three types of expenditures on business services (row 10), housing depreciation (row 11) and losses of fixed and working capital (row 12). As was noted above, the latter exclude losses in construction.

I estimated total value added in "non-material"

service and "other" sectors in a way similar to that in material production sectors. The derived data on "non-material" sectors were based primarily on estimates presented in Tables A5 and A9. As was discussed above, data on private service incomes were based on Gosplan's estimates performed by Koryagina.

I assumed that regular unreported wages of "other" sectors (row 26) equal the difference between total officially reported wages (row 33) and the sum total of published material and "non-material" sectors' wages (row 37).[9] I also assumed that revenues of "other" sectors equal social security deductions derived as 10 percent of unreported wages.

Estimates of total value added in "other" sectors contain the margin of error whose size however is quite small. Estimation errors cannot be avoided for two reasons.

First, it remains unclear to what extent total reported wages (row 34) correspond to the total wage bill. Goskomstat's table on household budgetary income (see Chart 3) contains data on other monetary income received from the state. This income has grown significantly during the 1980s largely as a result of a significant increase in payments to households from the so-called "operational fund" of enterprises. Some of these payments may be treated as labor income.

Second, there is no information on profits and bonus wages of "other" sectors. In preparing the 1988 i-o data I assumed that "other" sectors' profits can be derived similar to unreported wages as a residual. Thus, total profits amounted to 281 b.r. of which published sectors accounted for 279.3 b.r.[10] I therefore assumed that "other" sectors earned around 1.7 b.r. as profits in 1988 and that bonus wages comprised more than half of this amount (1 b.r.).

---

[9] Total wages (row 33) was derived as the sum of monetary and non-monetary wages. Monetary wages were already discussed in the previous chapter (see Charts 3 and 4). Non-monetary wages are known to equal around 6 percent of collective farmers' labor income.

[10] As reported in Narkhoz for 1988, p. 615, total net profit amounted to 260 b.r. As known, net profit excludes other revenues which, according to the 1988 table, equaled 21 b.r.

I propose to measure the total estimation error for 1985-1989 as the difference between the officially reported GNP and the sum of labor and non-labor income that can be accounted for. As presented in Table D1 (row 28), the estimation error is surprisingly insiginificant--around 2-3 b.r. I assumed that this amount represents other labor income paid from the "operational expenditure" fund. As I discuss below, this assumption is supported by other available evidence.

The Need for Alternative Estimates

At this point, it is necessary to address the issue as to what extent the officially reported GNP accurately measures the real size of the USSR economy in rubles.

I believe that although the Goskomstat's estimate can be significantly improved in theory, especially with respect to imputed income, any alternative approach to measuring total GNP will probably lead to quite hypothetical results.

First of all, there is a lack of alternative sources of information on value added in "other" sectors. It would be useful to have alternative sources of information because it is difficult to accept wild fluctuations in Goskomstat's estimates of unreported wages (see Table D1, row 26).

In my previous work on the real size of Soviet military expenditure (SME), I discussed a hypothesis that "other" sectors include the military, civilians working for the Defense Ministry, the police, as well as employees of defense industries specializing in assembly, repair and maintenance of weapons systems.[11] The hypothesis was based on estimation results which indicated that the military and the police alone cannot account for the exclusion of a large number of employees and their wages from published statistics on production and service activities.[12]

If the hypothesis can be proven, then it becomes possible to resolve the controversial issue whether

the cost of weapons intended for domestic use is fully incorporated in published aggregates for material production and national income. The issue has arisen because doubts exist whether secretive defense agencies share enough information on defense activities with departments of Goskomstat and Gosplan responsible for compiling Soviet national accounts. Unless the issue is resolved it would be difficult for the public to have confidence in any estimate of the real size and structure of Soviet GNP. This applies not only to estimates published by the CIA but also by the Soviet government.

Soviet officials have recently revealed in private conversations that "other" sectors indeed include some defense industries which do not submit statistical reports to Goskomstat similar to the military and the police. However, they have indicated that "other" sectors also include some functionaries of party and other political organizations.

In 1989, wages paid by the Defense Ministry, MVD and KGB amounted to 14.2 b.r. (see Chart 7 below).[13] Given the overall increase in average wage rates in 1987-1989, it can be assumed that the total wage bill of the above three organizations was around 12-13 b.r. in 1987-1988. It also can be assumed that party functionaries received wages in the amount not exceeding 1-2 b.r. Thus, unreported wages of defense industries doubled from 7 to 14 b.r. in 1987-1988. This jump is unbelievable unless one assumes that a) Goskomstat officials unwittingly perform inconsistent estimates for published sectors' wages each year or b) a large part of defense wages has been paid from the "operational expenditures" fund.

Is it possible that Goskomstat officials have been kept in the dark not only about the size of the defense economy but also about the total wage bill?

Given the restricted flow of financial information in the USSR, it can be speculated that a large part of labor income registered as "operational expenditures" may contain defense

---

[11] Dmitri Steinberg, "Estimating Total Soviet Military Expenditures: An Alternative Approach Based on Reconstructed Soviet National Accounts", in The Soviet Defense Enigma, edited by C.G. Jacobsen. Oxford: Oxford University Press, 1987.
[12] ibid, p. 39.

[13] For data on the official defense budget refer to Pravitel'stvennyi Vestnik, N45, 1990. Military personnel received 7.2 b.r. (or 9.5 b.r. with pensions), while civilians--2.5 b.r. The combined police budget was reported as 9 b.r., of which wages apparently equaled around 4.5 b.r.

CHART 7: OFFICIAL 1989 U.S.S.R. DEFENSE BUDGET

Column groups: [6]–[7] = CENTRAL SUPPORT AND ADMINISTRATION; [9]–[11] = MILITARY ASSISTANCE

Line Item	STRATE-GIC FORCES [1]	LAND FORCES [2]	NAVAL FORCES [3]	AIR FORCES [4]	OTHER COMBAT FORCES [5]	Support [6]	Command [7]	PARA-MILITARY FORCES [8]	Home Territory [9]	Abroad [10]	UN Peace-keeping [11]	UNDIS-TRIBUTED [12]	TOTAL MILITARY EXPEN-DITURES [13]	CIVIL DEFENSE [14]
1.0 OPERATING COSTS	0	9,950	2,737	2,226	2,540	2,506	121	1,113	0	0	0	2,239	23,432	114
1.1 PERSONNEL	0	4,943	1,241	1,296	1,626	0	94	537	0	0	0	2,239	11,976	61
1.1.1 Conscripts	0	3,393	953	1,092	1,392	0	82	336	0	0	0	0	9,487	53
1.1.2 Other Military Personnel Including Reserves	0	1,550	288	204	234	0	12	201	0	0	0	0	2,489	0
1.1.3 Civilian Personnel	0	0	0	0	0	0	0	0	0	0	0	0	0	0
1.2 OPERATIONS & MAINTENANCE	0	5,007	1,496	930	914	2,506	27	576	0	0	0	0	11,456	53
1.2.1 Materials for Current Use	0	2,824	450	144	229	1,632	0	201	0	0	0	0	6,553	0
1.2.2 Maintenance & Repair	0	1,158	931	732	486	0	6	268	0	0	0	0	3,665	0
1.2.3 Purchased Services	0	0	0	0	0	0	6	0	0	0	0	0	0	0
1.2.4 Rent Costs	0	1,025	115	54	199	874	15	52	0	0	0	0	2,334	0
1.2.5 Other	0	0	0	0	0	0	0	0	0	0	0	0	4	0
2.0 PROCUREMENT AND CONSTRUCTION	0	10,093	7,193	7,360	6,771	5,521	0	740	0	0	0	2,031	39,709	0
2.1 PROCUREMENT	0	9,273	6,531	6,941	5,465	4,361	0	637	0	0	0	2,031	35,239	0
2.1.1 Aircraft & Engines	0	0	210	1,845	949	0	0	94	0	0	0	0	3,098	0
2.1.2 Missiles and conventional warheads	0	995	695	662	1,621	0	0	0	0	0	0	0	3,973	0
2.1.3 Nuclear Warheads and Bombs	0	0	0	0	0	0	0	0	0	0	0	2,031	2,031	0
2.1.4 Ships and Boats	0	0	2,993	0	0	0	0	205	0	0	0	0	3,198	0
2.1.5 Armored Vehicles	0	2,137	0	0	0	0	0	30	0	0	0	0	2,169	0
2.1.6 Artillery	0	418	12	0	0	2	0	1	0	0	0	0	431	0
2.1.7 Other Ordnance and Ground Force Weapons	0	819	0	0	0	0	0	17	0	0	0	0	872	0
2.1.8 Ammunition	0	2,256	487	757	0	36	0	30	0	0	0	0	3,530	0
2.1.9 Electronics & Communications	0	779	495	927	767	2,591	0	185	0	0	0	0	5,744	0
2.1.10 Non-Armored Vehicles	0	0	0	0	0	0	0	60	0	0	0	0	60	0
2.1.11 Other	0	1,869	1,639	2,850	2,128	1,732	0	25	0	0	0	0	10,243	0
2.2 CONSTRUCTION	0	820	662	419	1,306	1,160	0	103	0	0	0	0	4,470	0
2.2.1 Air Bases & Air Fields	0	0	0	0	0	0	0	0	0	0	0	0	0	0
2.2.2 Missile Sites	0	0	0	0	0	0	0	0	0	0	0	0	0	0
2.2.3 Naval Bases & Facilities	0	0	0	0	0	0	0	0	0	0	0	0	0	0
2.2.4 Electronics, etc.	0	0	0	0	0	0	0	0	0	0	0	0	0	0
2.2.5 Personnel Facilities	0	0	0	0	0	0	0	0	0	0	0	0	0	0
2.2.6 Medical Facilities	0	0	0	0	0	0	0	0	0	0	0	0	0	0
2.2.7 Training Facilities	0	0	0	0	0	0	0	0	0	0	0	0	0	0
2.2.8 Warehouses, Depots, etc	0	0	0	0	0	0	0	0	0	0	0	0	0	0
2.2.9 Command & Admin Facilities	0	0	0	0	0	0	0	0	0	0	0	0	0	0
2.2.10 Fortifications	0	0	0	0	0	0	0	0	0	0	0	0	0	0
2.2.11 Shelters	0	0	0	0	0	0	0	0	0	0	0	0	0	0
2.2.12 Land	0	0	0	0	0	0	0	0	0	0	0	0	0	0
2.2.13 Other	0	0	0	0	0	0	0	0	0	0	0	0	0	0
3.0 RESEARCH & DEVELOPMENT	0	966	2,160	2,734	7,484	140	394	9	0	0	0	250	14,137	9
3.1 BASIC & APPLIED RESEARCH	0	0	0	0	0	0	0	0	0	0	0	0	0	0
3.2 DEVELOPMENT, TESTING & EVALUATION	0	0	0	0	0	0	0	0	0	0	0	0	0	0
TOTAL EXPENDITURES >>	0	21,009	12,090	12,320	16,795	8,167	515	1,862	0	673	0	4,520	77,951	123

industrial wages that were paid in secret and thus were excluded from the reported statistics on the total state wage bill. The possible existence of these unaccounted wages may explain why the wage gap has exhibited such unexpected fluctuations.

It appears that the decision to exclude statistics on some defense industrial output from all published national economic aggregates was not widely advertised within the state apparatus and that until recently most Goskomstat officials were not even informed of the decision. What supports this speculative view is that the above decision made it possible to preserve the "great Soviet defense economic secret" without violating the inner logic of Soviet national accounts.

After some production of weapons for domestic use was classified as a "non-material" activity similar to household and government administrative services, its full size could no longer be found within the published boundaries of material production activities. Instead, economists could find only a part of this production in the official statistics--the value of defense output less labor and non-labor income of secret defense enterprises which have not submitted their statistical reports to Goskomstat.

The reduced value of defense production has become popularly known as strategic reserves. In the course of the last three decades these strategic reserves--identified in the last chapter as other uses of national income--have been predictably mistaken for the full value of weapons produced for domestic use. Thus, a large part of value added in secret defense industries was hidden beyond the reach of even the most inquisitive readers of the official statistics.

Adjusting the Official GNP Estimate

Before more information on "other" sectors becomes available it would be possible to make only a partial adjustment in the official GNP estimate. Specifically, the official total should be increased by imputed rent for owner-occupied housing, foreign monetary revenues and military subsistence, and should be reduced by losses in construction.

No information exists on how Goskomstat officials might eventually decide to measure imputed rent. Public housing is heavily subsidized in the USSR. As was determined in the previous

chapter, rent payments account for one third of all costs connected with mainitaining housing facilities (see Table A10). At issue is whether imputed rent should be based on the subsidized or total housing costs. Undoubtedly, if private housing were made public, then owners would have paid subsidized rents. However, if the USSR housing market were privatized, then rents would far exceed current unsubsidized costs.

Before the introduction of the full-fledged housing market it can be assumed that for the same level of provided amenities imputed rent must be similar in size per one square meter of living space as state rent. Private housing comprises around 65 percent of public housing in the USSR.[14] Given differences in levels of housing amenities, I assume that imputed rent comprises less than 50 percent of total rent paid residents of public dwellings. As rent payments amounted to 6.3 b.r., imputed rent was around 3.0 b.r. in 1988.[15]

Foreign monetary revenues--which account for the difference between GNP and gross domestic product (GDP)-- are estimated as the net flow of financial capital in and from the USSR. These revenues are also estimated as net state budgetary earnings connected with foreign monetary transactions. These cover sale of monetary gold, tourism, net private earnings, foreign economic aid, and net debt payments. In 1988, the difference between state budgetary revenues and outlays connected with foreign monetary transactions, excluding foreign trade earnings, amounted to -1.4 (14.8-16.2) b.r.[16]

Soviet military subsistence has been regularly estimated by the CIA using intelligence data on purchases of food and uniforms for the military. As estimated by CIA analysts, it supposedly equaled 3.5 b.r. in 1982.[17] If CIA analysts are on the mark, then it should have grown to more than 4 b.r. by 1988. However, as presented by the Soviet military, total purchases of food and uniform did not exceed 3 b.r.[18] The nature of the discrepancy between the

---

[14] *Narkhoz* for 1989, p. 165.

[15] See Table A10, row 21.

[16] These data were kindly provided by a Soviet economist.

[17] CIA, *Measures of Soviet Gross National Product in 1982 Prices*, Washington, D.C.: U.S. GPO, 1991, p. 96.

[18] This was reported as part of the detailed data

two estimates is unclear.

Total unaccounted revenues thus amounted to around 5-6 b.r. in 1988. Losses in construction reached 7 b.r. during the same year. I thus determine net adjustment to the official GNP estimate as around -1 to -2 b.r. Since this is a relatively small amount, I tentatively conclude that the official GNP estimate approximately measures the real size of the USSR economy in rubles in 1988. For the preceding years the accuracy of the official estimates depends on the extent to which Goskomstat officials had underestimated the total wage bill.

Official GNP by End Use

As was noted above, the official estimate of total GNP by sector of origin is used as a control total for analyzing the allocation of resources in the USSR economy. Goskomstat officials make independent estimates for consumption and investment and then derive other uses as a residual. I follow the same approach bbelow in trying to duplicate published Goskomstat's estimates.

Household Consumption

I expected to duplicate the Goskomstat's estimate of household consumption by adding:

a) individual household purchases of goods and services (without purchases of producer and second hand goods, producer services and business trip expenditures);

b) net insurance payments and membership dues;

c) agricultural and other consumption-in-kind; and

d) purchases of privately provided services.

As presented in the 1988 household budget, all

individual purchases equaled 394.4 b.r.[19] Of this amount, purchases of producer and second-hand goods, producer services and business expenditures equaled

$$34.8 \ (12.8+8.2+2.6+11.2) \text{ b.r.}[20]$$

Net insurance payments and membership dues equaled 7.7 (3.3+4.4) b.r.[21] Agricultural and other consumption-in-kind equaled 33.8 b.r.[22] As was dscussed above, privately provided services were estimated as 10.2 b.r.

The Goskomstat's estimate of the total household consumption can be successfully duplicated as:

$$394.4-34.8+7.7+33.8+10.2 = 411.2 \text{ b.r.}$$

I thus conclude that the Goskomstat's estimate of the total household consumption is compatible with published statistics on household budgets.

In theory, it also should be possible to duplicate the Goskomstat's estimate using published national income and i-o data on consumption of material wealth and financial data on purchases of services. In 1988, consumption of material wealth equaled around 380 b.r. (see Table B3), while household purchases of "non-material" services from the state and individuals--around 43 b.r. (see Table A10) and 10.2 b.r. respectively. Total household consumption then equaled:

$$380+43+10.2-11.2 = 422 \text{ b.r.}$$

I thus conclude that consumption of material wealth as defined in NEB and i-o tables exceeds actual household purchases of goods by around 11 b.r. Since the NEB and i-o data are compatible with the published retail trade data (see Tables B3 and B4), I further conclude that the discovered discrepancy in the amount of 11 b.r. represents retail trade purchases made by the military and the police as institutions.

---

on the official USSR defense budget by senior Soviet military officers at the CSBM seminar held in Vienna in January of 1990. See C. Wilkinson, "Perestroika and the Soviet Defense Sector: Some Implications for NATO," The paper presented at the IV World Congress for Soviet and East European Studies, Harrogate, England, July 21-26, 1990, p. 11.

[19] Narkhoz for 1989, p. 76.
[20] See Table B4, rows 23, 45, 61 and 63. Purchases of business services (primarily transportation and agricultural services offered to private producers) were derived using published data on purchases of other services. See Narkhoz for 1988, p. 143.
[21] See Table A10, rows, 46 and 50.
[22] See Table B3, row 4.

According to the official defense budgetary statistics (see Chart 7 above), the military purchased goods in the amount of 9.1 b.r. in 1989, including materials for current use--5.6 b.r. and materials for maintenance and current repair purposes--3.6 b.r. There was little change in defense budgetary appropriations on operations and maintenance of the armed forces in 1988-1989. As a result, it appears that the military made most of their purchases of goods through what Goskomstat considers as the retail trade network. I hence conclude that the total retail trade turnover has grossly exaggerated in the official Soviet statistics to conceal the size and structure of the defense budget.

Free Services

In addition to household consumption, the GNP consumption component also includes state budgetary outlays on transportation and communications (media), education and culture, health and sports offered to households for free. I derived budgetary financing of transportation services in Table A10 as the difference between total and paid services: 5.1 = 23.5 - 18.4. As listed in the 1988 Narkhoz issue, state budgetary outlays on all other free services (without stipends--2.8 b.r.) were as follows:[23]

Education	35.3
Culture and arts	2.1
Media	2.7
Health and sports	21.9
Total	62.1 (b.r.)

They derived total includes net fixed investment (NFI) in the above service sectors, which must be excluded in estimating GNP. NFI can be estimated as the difference between total capital outlays and depreciaion. I propose to obtain data on capital outlays in education, culture, arts and media (12.2 b.r.) and on health and sports (3.4 b.r.) by comparing two sets of data on the services listed in the Narkhoz. The two sets are presented with and without capital outlays.[24] Depreciation in two groups of sectors (6.5 and 2.2 b.r.) were derived in Table A3 using the i-o data. NFI in these sectors then equaled 5.7 and 1.2 b.r. I assume that the state budget financed around 85 percent of NFI in

[23] Narkhoz for 1988, pp. 83 and 626-627.
[24] ibid, pp. 83 and 626.

education and culture sectors or 4.8 b.r.--in proportion to its overall financing of these sectors.

Total state budgetary outlays on free services then equaled

61.3 (5.1+62.1-4.8-1.2) b.r.

I thus conclude that the Goskomstat's GNP estimate of free services is compatible with the published NEB, i-o and financial data on free services.

Investment

The GNP investment component is estimated as gross fixed investment (GFI) minus the unamortized value (4.2 b.r.) plus annual additions of inventories (8.4 b.r.) plus annual changes in reserves whose size is unknown.[25] I estimated GFI (278 b.r.) in Table B5 as the sum of NFI (89.5 b.r.), depreciation (166.3 b.r.) and "uninstalled" capital (22.2 b.r.). The latter equals additions to unfinished construction (15.3 b.r.) plus losses in construction (6.9 b.r.).

The GNP investment component for 1988 then can be estimated without changes in reserves as:

278-4.2+8.4=282.2 b.r.

This exceeds the Goskomstat estimate by 6.2 b.r., which I assume equals depletion of reserves plus the unknown amount of error. The question that must be explored is whether the proposed alternative estimate was derived using the methodology which differs from that used by Goskomstat.

One way to answer this question is to examine the extent to which the error is repeated for other year for which Goskomstat estimates are available, i.e., for 1985-1989. I compare Goskomstat and alternative statistics on total investment in Table D2 below together with other estimates of GNP by end use. As evident from the table, the residual row 21--changes in reserves plus errors--equaled 0 in 1985, -4 b.r. in 1986, 3.9 b.r. in 1987 and -0.7 b.r. in

[25] Total additions of inventories amounted to 9.3 b.r. See Narkhoz for 1988, p. 623. To avoid double counting of construction works I reduced total additions of inventories by additions of unfinished construction works (0.9 b.r.).

1989. It is clear from these results that the discrepancy between the two estimates is caused not by differences in methodology but by the fact that I did not to take into account changes in reserves. What is surprising is that changes in reserves have fluctuated to such an extent in 1986-1988.

Other Uses

As was noted above, Goskomstat officials appear to lack data on total defense outlays and thus are forced to derived other uses as a residual. In Soviet GNP accounts, the category "other uses" represents a) state budgetary outlays on NS (national security--defense and police), "science" and administration and b) net exports. The "science" sector in turn represents a group of sectors engaged in basic and applied R&D work, geological exploration, agricultural services and space.

According to Goskomstat estimates, other uses amounted to 126.8 b.r. in 1988. In comparison, I estimated other uses as 116 b.r. in Table D2. The discrepancy between the two estimates in the amount of 10.8 b.r. was caused by using published retail trade data to estimate household consumption. As I demonstrated above, this approach results in the overestimation of household consumption by the amount of retail trade purchases made by the military and the police.

In 1988, state budgetary outlays on "science" and administration equaled 30.5 and 3 b.r. respectively (see Table A10), while net exports equaled 1.4 b.r. (see Table A7). Other uses intended for meeting NS objectives then can be estimated as a residual:

116-30.5-3-1.4=81 b.r.

What do these other uses consist of? The derived residual can be divided into three parts:

1) total output of "other" sectors (67.7 b.r.) which was estimated as the sum of other uses and "unreported" value added in the i-o column 35 (see Chart 3);

2) the unidentified uses of machinery and chemical products (11.3 b.r.) hidden with additions to working capital--inventories and reserves--in the i-o column 34 (see Chart 3);[26] and

3) state budgetary outlays on other communications (2.1 b.r.) which were estimated in Table A10;

As estimates presented in Table D2 indicate, there was a significant 15 b.r. increase in other other uses from 66 b.r. in 1987 to 81 b.r. in 1988. Some of this increase was caused by the fact that military purchases made through the retail trade network dropped by 2 b.r. duringthe same period (see Table D2, row 6). However, most of the increase in the amount of 13 b.r. cannot be explained by the existing evidence on trends in defense allocations under Gorbachev. As Gorbachev himself noted, the USSR defense budget was frozen on the same level in 1987-1988.

A careful search for the possible source of error causing the unexplained jump in other uses led me to examine again published data on the total wage bill. To remind the reader, while estimating value added in "other" sectors I discovered that these sectors' labor income (derived as residual wages) jumped unexpectedly by 10 b.r. in 1987-1988. The correlation found between fluctuations in the official sector-of-origin and end-use data further led me to consider two hypotheses:

a) Goskomstat officials have underestimated both the total wage bill and the official GNP prior to 1988; or

b) Goskomstat officials have overestimated both the total wage bill and the official GNP starting in 1988.

It must be emphasized that 1988 was an important year in the Goskomstat history: as known it was during that year that the Soviet government decided for the first time to begin estimating GNP. Is it possible that as part of government's decree to measure total Soviet GNP Goskomstat obtained for the first time the correct information on total wages and employment in the economy? If this was indeed the case then one must accept the first hypothesis and at the same time reject Goskomstat's pre-1988 estimates of value added in "other" sectors as unreliable.

---

26 Total for the column equals 13.5 b.r. or 19.7

b.r. after accounting for the depletion of reserves estimated as 6.2 b.r. Other uses hidden as additions to working capital were estimated as: 19.7-8.4=11.3 b.r.

Measuring the Total NS Budget

In Table D2, I disaggregate the official GNP into four end uses: consumption, investment, NS and other government expenditures. This disaggregation required the approximate estimation of secret budgetary allocations on NS sectors hidden with total allocations on free services, fixed investment, R&D works, space and state administration.

In estimating the total NS budget I made the following assumptions:

● free transportation services financed by the state budget are provided primarily to NS sectors;

● NS sectors purchase around 5 percent of education, cultural and health services;

● NS sectors accounts for 60 and 3 percent of total fixed investment in "science" and service sectors respectively;[27]

● NS sectors account for all fixed investment in "other" sectors;

● NS sectors account for 55 percent of total budgetary financing of "science" and administration sectors.[28]

After accounting for the above allocations on NS sectors, it is now possible to measure the defense burden in established prices that were used in the USSR economy in 1988. Total NS expenditures amounted to 127.7 b.r. (see Table D2, row 33). On the one hand, this amount includes allocations on civilian police (around 4.8 b.r.) which are excluded from the total defense budget.[29] On the other hand, this amount excludes military pensions which I believe equaled at least 3 b.r. in

1988.[30] I thus estimate total defense budget as:

$$127.7 - 4.8 + 3 = 126 \text{ b.r.}$$

The defense burden in established prices hence equaled 14.6 (126:875.4*100) percent.

Total defense budget had the following structure:[31]

O&M Expenditure	24.5
Weapons and charges	60.5
Military R&D	14.7
Military Space	4.6
Free Services	8.0
Communications	2.1
Installations	8.6
Pensions	3.0

How do these figures compare with published Soviet defense budgetary numbers? It must be emphasized that Soviet officials have so far published only the entire 1988 official defense budget without outlays on space--82.5 b.r.[32] The actual breakdown of the official defense budget has been reported for 1989-1990 (see Chart below):

CHART   Official Defense Budget for 1989-1990

O&M	20.2	19.3
Weapons	32.6	31.0
Military R&D	15.3	13.2
Construction	4.6	3.7
Nuclear Charges	2.3	1.3
Pensions	2.3	2.4
Total	77.3	70.9

---

[27] As was published in the Soviet press, defense industries also receive around 40 percent of total investment in machinery and chemical sectors. These allocations were not, however, counted as part of total NS budget.

[28] This assumption is based on reported data on military R&D works. For example, see Pravda, June 8, 1989, pp. 3-4.

[29] Total police budget equaled 8 b.r. in 1988. See V. Panskov, "U Opasnoi Cherty," Ekonomika i Zhizn', N15, 1990, p. 7. I assumed that allocations on the military police account for 40 percent of the total police budget.

[30] Military pensions were reported as 2.4 b.r. in the official budget. I believe this amount covers pensions paid to the retired military personnel and hence excludes pensions received by veterans and survivors.

[31] O&M expenditures were estimated by adding the official budget (20.2 b.r.) and the budget of the militarized police. Outlays on weapons (56.7 b.r.) and nuclear charges (2.3 b.r.) were derived as a residual after accounting for all other defense budgetary items.

[32] See fn. 29.

I assume that defense outlays on O&M, construction and pensions remained the same in 1988-1989 and that outlays on military R&D equaled around 14.7 b.r.. Outlays on weapons and nuclear charges thus equaled the remaining 40 b.r.[33]

Compared to the independently derived GNP data, the official defense budget excludes outlays on most services provides to the military as well as on miltary police and space. In addition, the official defense budget contains a much smaller figure for weapons procurement and construction of military-related installations.

In the case of construction, it appears that the authors of the official estimate fail to account for construction activities in the defense area not directly supervised by the military. The reason for the understatement of the total weapons procurement bill is still a subject of heated controversy among specialists on Soviet defense industries.

As I argued elsewhere, total SME in estabished prices can be estimated in two ways depending on whether producer or purchasers' prices are used to evaluate various defense-related activities.[34] The new official budgetary figure was apparently derived using artificial prices which contain various subsidies that reduce the recorded value of various defense-related activities, primarily production of weapon systems. The mechanism of subsidized purchases is similar in defense and agricultural sectors. In both sectors producer prices far exceed purchasers' prices, whereby the difference between the two price levels is financed by the state budget.

Thus, I believe that the existence of hidden subsidies explain why the figure for weapons procurement estimated within the framework of GNP accounts exceeds the official budgetary figure by around 20 b.r. These subsidies are apparently classified by Soviet financial authorities as budgetary items apart from reported allocations on

defense. In this way, the total weapons procurement bill is continued to be concealed not only from the general public but also from the highest legislative authority in the country--the USSR Supreme Soviet.

The Real Defense Burden

So far, the focus of discussion was on the defense burden in established prices which are used in the economy. While the analysis of subsidies may help integrate the defense sector within GNP accounts, it only provides the first step to measuring the real Soviet defense burden. The reason for this is that even unsubsidized prices cannot be used as a reliable substitute for measuring real resource costs in the Soviet defense sector.

As evident from the reconstructed UB table of the USSR economy, unsubsidized prices on defense goods consist of five standard components:

● purchases of material inputs for current use
● depreciation of fixed capital operated in defense sectors
● wages and other earnings of defense employees
● social security deductions, and
● gross revenues.

Each of the above defense price components is kept by Soviet planning and financial authorities at the artificially low level to conceal the country's enormous defense burden.

During the past year Soviet economists began to unveil various secrets about the distorted price formation in Soviet defense sectors. All these distortions can be legitimately classified as hidden defense costs:

Hidden material costs. The most rampant manifestation of such costs has been established in Soviet technologically advanced industries which suffer large production losses that occur due to poor management and lack of know how. For example, the privileged position that manufacturers of armaments have enjoyed in the Soviet economy has enabled them to select electronic components of the best available quality without having to pay any extra cost of producing these components.

All extra costs are absorbed by civilian industries

---

[33] Total budgetary outlays on R&D works equaled 16.7 b.r. in 1988 (see Narkhoz for 1988, p. 627). I assume that outlays on civilian research equaled around around 2 b.r.

[34] See Dmitri Steinberg, "Soviet Defense Allocations Under Gorbachev: Analyzing New Data for 1988-1990," in Roy Allison, ed., Radical Reform in Soviet Military Policy, London: Macmillan, 1991.

which assemble producer and consumer durables.[35] As a result, Soviet civilian users purchase low quality electronic products at exorbitant prices. In any market economy extra costs connected with production of higher quality electronic components would be passed to defense industries which in turn would increase prices on weapons. The opposite takes place in the Soviet economy where the cost of technological advancement i the defense area is paid by consumers. In this connection, it can be speculated that price increases that are registered in consumer industries far outpace those in defense industries.

Hidden capital costs. The process of fixed capital formation in defense industries provides another vivid example of how defense costs are concealed in the Soviet economy. All civilian enterprises are required to make deposits on their depreciation accounts in the amount that is believed to be sufficient for repairing and replacing fixed capital.

For various reasons, including quick wear and tear of capital due to its poor quality and maintenance, civilian enterprises pay the enormous capital repair bill. In comparision, defense industry's capital repair bill is much smaller than in civilian industries due to higher capital replacement rates. In addition, it is a well known fact that financing of fixed capital formation at defense enterprises takes place by means of direct state budget allocations. The cost of this fixed capital is not added, however, to prices on defense goods and services.

Hidden inventory costs. A similar situation can be observed with respect to inventory buildup in defense industries which is known to be financed by the state budget as so-called "additions to strategic reserves". It appears that the cost for purchasing and storing these reserves has also been excluded from prices on defense goods. Moreover, the privilege position enjoyed by defense industries has enabled then to avoid incurring at least three extra costs connected with the problem of procuring scarce supplies of inventories which is so rampant in the Soviet economy.

First, civilian enterprises must pay for hoarding extra supplies of inventories which may not be needed for the current production. Second, they must continue paying wages and other bills during frequent interruptions in the production process when their suppliers fail to deliver promised products. Third, they cannot avoid large losses when their suppliers deliver shoddy products.

Hidden labor costs. For decades defense industries were granted the privilege to hire the most skilled labor available in the country. The opportunity to earn higher wages has been one but not the major incentive for attracting talented engineers and scientists. In fact, according to Soviet defense officials, wages in defense industries have increased at a much slower pace than in civilian industries.[36]

In addition to higher wages, defense employees have enjoyed significant privileges and perks, such as better housing and working conditions, access to scarce consumer goods at subsidized prices and free services. As opposed to civilian industries, most defense industries have used state budgetary subsidies to pay for these perks. Similar to hidden capital costs, these hidden labor costs have also been excluded from prices on defense goods.

Savings from small profits. According to Moiseev and other high level Soviet defense officials, defense industries began to operate on a contractual basis and earn sizable profits from producing weapons in 1989.[37] This caused a large increase in prices on many weapons systems. Even though he probably referred to a long-term trend when he mentioned that prices increased by 2-3 times, his still made his point quite clearly: the military had been used to making substantial savings from purchasing relatively cheap defense goods and services prior to the 1989 price reforms. Moiseev's statement was corroborated by the deputy chief of the air forces main staff whose published figures imply that price increases exceeded 7 percent in 1989 and 14 percent in 1990.

It must be emphasized that the military had indeed enjoyed multiple savings because of small profit margins that existed in the entire defense industry. The reason for this is that prices on finished military goods must theoretically contain profits received not only by head enterprises where

---

[35] See Ye. Kuznetsov and F. Shirokov, "Naukoyomkie Proizvodstva i Konversiya," Kommunist, N10, 1989, p. 17.

[36] Pravda, August 18, 1989, p. 4.

[37] Izvestiya, February 22, 1990, p. 3. Also see Kommunist Vooruzhenykh Sil, N1, January 1990.

numerous suppliers of components and other materials used in weapons production. In an attempt to compensate for losses of budgetary revenues resulting from small profit margins in defense industries, Soviet financial and planning authorities had traditionally pursued a policy of sustaining artificially high prices on apparel and other consumer items. The burden of defense was carried in this hidden way by Soviet consumers, resulting in lowest living standards among developed countries.

As evident from the above analysis, officially published estimates of SME exclude not only subsidies but also a number of hidden costs that must be taken into account in measuring the total Soviet defense burden. Some economists believe that there are numerous indirect defense allocations which also must be accounted for.[38] These include outlays on:

● decentralized civil defense measures
● construction and maintenance of transportation and communications facilities and equipment that serve both civilian and military needs
● wages paid by civilian enterprises to draftees and reserves, and
● military aid to Cuba, Vietnam and other Soviet client states.

Following the tradition established by U.S. intelligence agencies, it is desirable to consider both lower and upper bound estimates of SME with the upper bound covering all indirect defense allocations. The estimate of 200 b.r. derived by experts of the USSR Supreme Soviet is widely assumed to be within the upper bound of SME. Since SME equaled around 125 b.r. in unsubsidized prices in 1988, these experts seem to believe that the sum of various hidden costs and indirect defense allocations amounted to as much as 75 b.r.

Some Western economists, for example, Igor Birman, believe that 200 b.r. is a realistic number confirming numerous impressionistic accounts about the militaristic nature of the Soviet economy.[39] It is difficult to see however how 75

b.r., which is an enormous sum of money for the Soviet economy, was derived. Instead, I believe it is more realistic to assume that all hidden defense costs and indirect defense allocations amounted to no more than 30-40 b.r. a year.

Hidden costs can be hypothetically divided into two parts:

1) preferential price discounts received by defense sectors when they made purchases of materials used in developing, producing, repairing and operating weapons systems; and

· 2) low profit margins which make it impossible for defense sectors to finance own social programs for their employees and technological innovation.

I assume that discounts comprise from 25 to 30 percent of producer prices or 13-15 b.r. I also assume that a "normal" profit rate for Soviet defense producers should be around 15 percent or 10 b.r. Total hidden costs then equaled 23-25 b.r. Indirect defense allocations amounted to another 10-13 b.r. These include various costs incurred by civilian enterprises (around 3-5 b.r.) and military aid (7-8 b.r.).

In sum, the lower bound of SME equaled 145-150 b.r., while the upper bound--no more than 165 b.r. in 1988. This suggests that the real defense burden in 1988 ranged from 18 to 20 percent as compared to 14 percent in established prices. It must be emphasized that after increasing the defense burden from 14 to 18 percent the share of consumption in total GNP drops from 53 to 49 percent.

I believe that after accounting for all other price distortions in the economy, such as excessive taxes and profits contained in prices on consumer goods, the share of consumption would drop further to 46-47 percent. Likewise, the share of investment and civilian government expenditures would increase by the same amount.

Even if the officially announced reduction of SME by 10 percent has indeed taken place in 1988-1991, the size of the real defense burden is still around 15 percent. It appears that the Soviet leadership has still a long way to go toward de-

[38] For a recent Soviet discussion on the subject refer to V. Lopatin, "Armiya i Ekonomika," Voprosy Ekonomiki, N10, 1990, pp. 8-9.
[39] Igor Birman, "The size of Soviet military expenditures: methodological aspects," A paper presented at the AEI Conference on the Comparison of Soviet and American Economies, Virginia, April 19-22, 1990, p. 35.

militarizing the economy and saving it from the collapse. By all accounts, the USSR economy cannot carry such an enormous defense burden for much longer.

The Financial Crisis

The most devastating effect of the immense defense burden on the civilian economy has been in the financial area. Unfortunately, this effect is difficult to analyze within the constraints of Soviet GNP accounts. As a reliable indicator, GNP can be used only for the economy with a stable financial system. However, as it has become increasingly evident during the past year, the Soviet financial system is experiencing the most severe crisis since the end of the war. For this reason, some discussion of this crisis seems necessary to understand the structure of the USSR economy, even though the analysis of Soviet financial accounts is outside the focus of this study.

Five aspects of the financial crisis must appear particularly troublesome to the Soviet leadership:

1. The state budgetary deficit sky-rocketed from 14 to 81 b.r. in 1985-1989. Altogether financial authorities borrowed 92 b.r. to cover All-Union state budgetary outlays.[40] In effect, these funds were borrowed to pay for military buildup. As a result of this borrowing spree, the internal debt has grown from 142 to 398 b.r. in 1985-1989.[41] By Spring of 1991 the internal debt was expected to reach 500 b.r. or more than 50 percent of GNP.

2. The country's external debt has grown with the same pace as the internal debt reaching 55 b.r.[42] Most hard currency earnings are expected to be used for making debt payments. In 1991, enterprises are required to pay taxes on foreign currency earnings at the rate of 40 percent to help pay for the debt. This is expected to dampen the interest of enterprises in increasing exports.[43]

3. By the beginning of 1991 households have deposited as much as 450 b.r. in savings and other

deposits in the State Bank.[44] This constitutes 80 percent of total household monetary revenues. Most of these deposits have been used to finance the debt as well as construction and other loans to enterprises.

4. Around 40 percent of household savings are "forced"--individuals simply cannot spend their money because of shortages of goods and services. As estimated by Goskomstat, "forced" savings doubled from 79 to 166 b.r. in 1985-1989 as the rapid growth of income has outstripped the growth of goods and services.[45] If this growth trend would continue, then by the beginning of 1991 households would be ready to withdraw around 200 b.r. for purchases, providing there was something to buy.

In addition, by the beginning of 1991 households accumulated near 130 b.r. in cash savings which they were also ready to cash. Given the scope of the crisis, it is understandable why the Soviet government is in a desperate position to begin financial reforms aimed at reducing its financial debt to households. Some drrastic measures will be required, such as freezing all bank deposits and introducing new monetary notes to substantially reduce the size of cash savings.

5. The impressive increase in total investment in 1985-1990 has been largely due to the phenomenal rise of unfinished construction and inventories of unsold goods. By 1989 unfinished construction reached 90 percent of total capital investment, a full 10 percent increase since 1985.[46] The same trend continued in 1990. Since a large part of unfinished construction probably will be never completed, it appears that most of the increase in total fixed investment in production sectors during Gorbachev's reign must be considered as losses and

---

[40] Narkhoz for 1989, p. 612.

[41] ibid, p. 614.

[42] A. Volkov, "Finansovyi Krizis v SSSR," A Paper presented at the Stanford-Berkeley Seminar, December 24, 1990, p. 13.

[43] ibid, p. 14.

[44] This is based on data listed in Sh. Sverdlik, "The Balance of Monetary Turnover in the USSR," A Paper presented at the Stanford-Berkeley Seminar, December 24, 1990, Tables 13-15. By the end of 1989 households deposited 338 b.r. in savings accounts (Narkhoz for 1989, p. 92), 21 b.r. in life insurance policies, and 27 b.r. in old and new bond. According to Sverdlik, more than half of household savings has been directly used to finance the internal debt (data are from his Table 15).

[45] Goskomstat SSSR, Metodika Ischisleniya Normal'nykh, Izbytochnykh Sberezheniy i Neudovletvorennogo Sprosa, Moscow, 1990, p. 10.

[46] Narkhoz for 1989, p. 547.

hence excluded from GNP. The same applies to excessive stocks of inventories which eventually will be either sold at a discounted price or disposed of as wasted resources.

Conclusion

In this chapter I proposed an alternative procedure for converting Soviet national accounts into a GNP format. Specifically, I examined whether the new Soviet GNP and financial data are compatible. I also compared these new data with the i-o data and with the regularly published production and national income data. I found these four sets of data to be surprisingly compatible with respect to material production and "non-material" service sectors. However, I discovered the major incompatibility between the new GNP and financial data with respect to "other" sectors for the pre-1988 period. I suggested that both the total wage bill and total GNP were underestimated by around 10 b.r. in 1987 due to the inability of Goskomstat officials to account for all wages of "other" sectors.

I also analyzed the structure of value added and major uses of GNP in current established prices. I paid particular attention to the way the defense sector is integrated in Soviet GNP accounts. In conclusion, I proposed a procedure for converting SME from established prices to prices that reflect real resource costs in the economy. As a result of following this procedure, I proposed alternative estimates of the real Soviet defense burden. In the end, I summarized major aspects of the current financial crisis.

Part D

GNP and Established Prices

TABLE D1: OFFICIAL SOVIET GNP BY SECTOR OF ORIGIN
     (billion rubles)

	1965	1970	1971	1972	1973	1974	1975	1976	1977	1978	1979	1980
1 OFFICIAL GNP	235.3	366.9	387.7	400.4	432.8	456.5	475.2	504.6	530.5	558.8	581.2	613.2
2 HOUSEHOLDS	129.6	184.8	195.1	203.1	215.1	229.0	239.6	252.2	263.6	276.2	287.1	299.6
3 PUBLIC SECTOR	105.7	182.1	192.7	197.2	217.8	227.5	235.6	252.4	266.9	282.6	294.2	313.6
4 MATERIAL PRODUCTION	199.0	299.9	316.1	324.3	350.9	368.3	381.0	406.4	426.3	447.6	464.6	490.1
5 Labor Income	98.8	138.8	146.3	151.9	160.4	169.9	177.1	188.1	195.7	204.0	211.1	219.5
6 Total Income	100.7	141.6	149.4	155.4	164.1	173.9	181.4	192.7	200.5	209.1	216.4	225.0
7 Business Expend.*	1.9	2.9	3.1	3.5	3.7	3.9	4.3	4.6	4.9	5.0	5.3	5.5
8 Revenues	73.1	118.5	122.5	122.2	134.4	137.6	132.3	137.7	144.9	151.2	151.4	152.2
9 Gross Revenues	86.0	136.9	143.2	145.2	158.7	164.3	163.1	169.3	178.3	188.2	190.3	193.8
10 Business Services+	5.9	9.7	11.0	12.1	14.3	16.1	17.7	19.7	21.5	23.9	26.1	27.6
11 Housing Depreciation	4.7	6.4	6.7	7.1	7.4	7.7	8.0	8.2	8.5	8.8	9.0	9.3
12 Losses of Capital	2.3	2.3	2.9	3.7	2.6	2.9	5.1	3.7	3.4	4.4	3.8	4.7
13 Foreign Trade	6.8	11.4	12.4	13.0	15.0	15.8	18.8	23.7	26.8	29.0	34.0	43.5
14 Depreciation	20.3	31.2	34.8	37.2	41.2	45.0	52.8	56.9	58.9	63.5	68.0	75.0
15 SERVICES	32.3	59.2	63.2	67.7	73.2	78.1	82.6	86.9	91.5	97.7	102.3	108.0
16 Labor Income	27.7	39.5	41.7	44.3	47.6	50.6	52.7	54.7	57.2	60.8	63.7	67.2
17 Total Wages	25.0	35.9	37.9	40.3	43.3	46.0	47.9	49.7	52.0	55.2	57.8	60.9
18 Business Expend.*	0.6	1.1	1.2	1.3	1.4	1.4	1.5	1.6	1.6	1.7	1.7	1.7
19 Private Income	3.3	4.7	5.0	5.3	5.7	6.1	6.3	6.6	6.9	7.3	7.6	8.0
20 Revenues	4.6	7.2	8.0	8.8	9.7	10.5	11.2	12.1	12.8	13.9	14.4	15.2
21 Gross Revenues	4.8	7.5	8.3	9.1	10.1	10.9	11.6	12.6	13.4	14.4	15.0	15.9
22 Business Services+	0.2	0.3	0.3	0.4	0.4	0.4	0.5	0.5	0.5	0.6	0.6	0.6
23 Depreciation**	8.6	12.5	13.5	14.7	15.9	17.0	18.7	20.2	21.5	23.1	24.2	25.6
24 UNPUBLISHED SECTORS	3.9	7.8	8.5	8.3	8.7	10.2	11.5	11.3	12.7	13.5	14.3	15.1
25 Labor Income	3.1	6.5	7.1	6.9	7.2	8.5	9.7	9.5	10.7	11.4	12.2	12.9
26 Unreported Wages	3.2	6.8	7.4	7.2	7.5	8.9	10.1	9.9	11.1	11.9	12.6	13.3
27 Business Expend.*	0.2	0.3	0.3	0.3	0.4	0.4	0.4	0.4	0.4	0.4	0.4	0.4
28 Other Labor Income												
29 Revenues	0.2	0.5	0.5	0.5	0.5	0.6	0.7	0.7	0.8	0.8	0.9	0.9
30 Depreciation	0.6	0.8	0.9	0.9	1.0	1.0	1.1	1.1	1.2	1.2	1.2	1.3
31 DEPRECIATION	29.6	44.5	49.3	52.7	58.1	63.0	72.6	78.2	81.6	87.7	93.4	101.8
32 NNP	205.7	322.4	338.5	347.6	374.7	393.6	402.6	426.4	449.0	471.1	487.8	511.4

Supporting Data on Total Labor Income (including payments for business trips):

	1965	1970	1971	1972	1973	1974	1975	1976	1977	1978	1979	1980
33 Total Official Wages	107.0	158.4	167.6	177.2	187.1	199.8	211.1	221.9	232.3	243.7	253.4	265.6
34 Monetary Wages	106.0	157.3	166.4	175.9	185.7	198.4	209.7	220.4	230.7	242.0	251.7	263.8
35 Annual Growth	108.9	106.9	105.8	105.7	105.6	106.8	105.7	105.1	104.7	104.9	104.0	104.8
36 Non-Monetary Wages	1.0	1.1	1.2	1.3	1.3	1.4	1.5	1.5	1.6	1.7	1.7	1.8
37 Published Wages	103.8	151.6	160.2	169.9	179.5	190.9	201.0	212.0	221.2	231.8	240.8	252.3

* including compensation for travel and other business expenditures paid as labor income
+ excluding housing and other subsidies to service sectors
**including housing depreciation

TABLE D1: OFFICIAL SOVIET GNP BY SECTOR OF ORIGIN
        (billion rubles)

	1981	1982	1983	1984	1985	1986	1987	1988	1989	1990
1 OFFICIAL GNP	647.7	693.2	726.3	758.0	777.0	798.5	825.0	875.4	924.1	965.3
2 HOUSEHOLDS	310.2	321.6	334.1	345.8	358.9	373.1	387.0	417.0	465.3	499.5
3 PUBLIC SECTOR	337.5	371.6	392.1	412.2	418.1	425.4	438.0	458.4	458.8	465.8
4 MATERIAL PRODUCTION	518.2	558.0	586.0	613.0	629.0	644.0	660.7	692.4	717.6	741.8
5 Labor Income	225.9	236.3	246.1	255.2	264.6	274.7	282.1	296.4	325.0	344.6
6 Total Income	231.6	242.4	252.4	261.7	271.4	282.2	289.8	304.6	334.3	354.6
7 Business Expend.*	5.7	6.0	6.3	6.5	6.9	7.5	7.7	8.2	9.3	10.0
8 Revenues	161.6	183.4	191.7	202.3	193.7	196.1	204.6	215.3	202.9	202.2
9 Gross Revenues	205.1	227.9	238.9	248.8	244.3	247.0	260.3	274.4	266.8	267.9
10 Business Services+	28.9	30.8	32.2	32.9	34.5	34.5	38.1	38.6	43.2	44.7
11 Housing Depreciation	9.6	9.9	10.3	10.7	11.1	12.0	12.5	13.1	13.7	14.3
12 Losses of Capital	4.9	3.8	4.7	2.9	4.9	4.4	5.2	7.4	7.0	6.8
13 Foreign Trade	50.0	53.1	56.8	57.8	62.8	58.2	49.5	51.8	55.3	53.5
14 Depreciation	80.6	85.1	91.4	97.7	107.9	115.0	124.5	128.8	134.3	141.6
15 SERVICES	112.5	119.1	122.7	127.0	128.4	133.8	141.8	149.4	159.9	169.3
16 Labor Income	69.6	71.9	73.4	75.6	77.9	81.0	86.0	91.8	99.4	106.5
17 Total Wages	63.1	65.3	66.7	68.6	70.8	73.4	77.9	83.3	90.1	96.7
18 Business Expend.*	1.8	2.0	2.0	2.1	2.1	2.1	2.2	2.4	2.6	3.0
19 Private Income	8.3	8.6	8.7	9.1	9.2	9.7	10.3	11.0	11.9	12.8
20 Revenues	16.0	18.8	19.3	19.8	20.7	21.3	22.1	21.9	23.3	24.1
21 Gross Revenues	16.7	19.6	20.1	20.6	21.6	22.1	23.1	22.8	24.2	25.1
22 Business Services+	0.7	0.8	0.8	0.8	0.9	0.9	0.9	0.9	1.0	1.0
23 Depreciation**	26.9	28.4	30.0	31.6	29.8	31.5	33.6	35.6	37.3	38.7
24 UNPUBLISHED SECTORS	17.0	16.1	17.5	18.1	19.6	20.7	22.5	33.6	46.6	54.2
25 Labor Income	14.6	13.3	14.6	15.1	16.4	17.4	18.9	28.7	40.9	48.5
26 Unreported Wages	15.0	13.8	15.1	15.6	17.0	15.8	18.2	27.7	38.5	45.5
27 Business Expend.*	0.5	0.5	0.5	0.5	0.5	0.5	0.5	0.6	1.0	1.1
28 Other Labor Income						2.1	1.2	1.6	3.4	4.1
29 Revenues	1.1	1.4	1.5	1.6	1.6	1.7	1.9	3.1	3.8	3.8
30 Depreciation	1.3	1.4	1.4	1.5	1.5	1.6	1.7	1.8	1.9	2.0
31 DEPRECIATION	108.8	114.9	122.8	130.8	139.3	148.1	159.9	166.3	173.5	182.3
32 NNP	538.9	578.3	603.4	627.3	637.7	650.3	665.1	709.1	750.7	783.1

Supporting Data on Total Labor Income (including payments for business trips):

	1981	1982	1983	1984	1985	1986	1987	1988	1989	1990
33 Total Official Wages	275.8	285.4	294.7	304.2	315.1	325.2	336.9	364.2	411.0	443.1
34 Monetary Wages	273.9	283.5	292.7	302.1	313.0	323.0	334.6	361.8	408.5	440.4
35 Annual Growth	103.8	103.5	103.3	103.2	103.6	103.2	103.6	108.2	113.1	107.8
36 Non-Monetary Wages	1.9	1.9	2.0	2.1	2.2	2.3	2.3	2.4	2.5	2.7
37 Published Wages	260.7	271.6	279.6	288.6	298.2	309.4	318.8	336.5	372.5	397.6

* including compensation for travel and other business expenditures paid as labor income
+ excluding housing and other subsidies to service sectors
**including housing depreciation

TABLE D2: OFFICIAL SOVIET GNP BY END USE
   (billion rubles)

	1965	1970	1971	1972	1973	1974	1975	1976	1977	1978	1979	1980
1 OFFICIAL GNP	235.3	366.9	387.7	400.4	432.8	456.5	475.2	504.6	530.5	558.8	581.2	613.2
2 CONSUMPTION	147.8	213.0	224.6	237.0	249.7	263.3	279.1	293.9	307.3	323.5	338.9	361.4
3 Households	128.7	186.3	196.8	207.6	217.9	230.1	244.3	257.6	269.5	283.4	297.5	317.8
4 Goods	114.5	163.7	172.7	182.1	190.3	200.4	212.5	224.2	234.0	245.7	258.4	276.5
5 Family Budgets	108.0	155.8	164.7	174.0	181.9	192.1	204.0	213.6	223.2	234.7	247.4	265.3
6 National Security	6.5	7.9	8.0	8.1	8.4	8.4	8.6	10.7	10.8	11.0	11.1	11.3
7 Paid Services	14.2	22.6	24.2	25.5	27.6	29.7	31.8	33.4	35.5	37.7	39.1	41.3
8 "Material"	3.7	5.1	5.5	5.8	6.1	6.5	7.0	7.7	8.2	8.7	9.1	9.7
9 "Non-material"	7.7	13.1	14.0	14.8	16.2	17.5	18.9	19.6	21.0	22.3	22.9	24.1
10 "Private"	2.8	4.4	4.7	4.9	5.3	5.6	5.9	6.1	6.3	6.7	7.1	7.5
11 Free Services	19.1	26.6	27.7	29.3	31.7	33.2	34.8	36.2	37.8	40.0	41.4	43.5
12 Family Budgets	17.0	23.5	24.4	25.8	27.6	29.0	30.3	31.4	32.8	34.9	36.0	37.9
13 National Security	2.0	3.1	3.3	3.5	4.1	4.2	4.5	4.8	4.9	5.1	5.4	5.6
14 INVESTMENT	68.2	114.7	121.0	121.2	138.0	142.6	149.2	161.0	167.0	178.0	179.6	185.0
15 Fixed Capital	61.0	101.4	109.4	116.1	120.9	130.2	140.1	145.7	150.8	161.3	164.5	170.2
16 New Fixed Invest.	47.6	81.7	88.1	93.6	96.7	104.1	112.6	116.4	120.0	128.1	129.4	132.6
17 Civilian	44.4	77.2	83.0	88.4	91.3	98.8	107.2	110.8	114.2	122.0	123.8	126.7
18 Defense	3.2	4.5	5.1	5.2	5.4	5.3	5.4	5.6	5.8	6.1	5.6	5.9
19 Capital Repair	12.4	19.7	21.3	22.5	24.2	26.1	27.5	29.3	30.7	33.2	35.1	37.6
20 Inventories	6.6	15.5	11.4	8.5	16.6	12.7	10.5	14.6	14.7	15.3	13.9	18.6
21 Reserves and Errors	0.6	-2.2	0.2	-3.4	0.5	-0.3	-1.4	0.7	1.5	1.4	1.2	-3.8
22 GOVERNMENT	19.3	39.3	42.1	42.2	45.1	50.7	46.8	49.7	56.3	57.4	62.7	66.8
23 R&D & Admin.	9.3	13.4	14.4	15.4	16.0	17.0	17.7	17.9	18.8	19.7	20.6	22.2
24 Civilian	3.8	6.0	6.5	6.9	7.2	7.6	8.3	8.2	8.8	9.0	9.5	10.3
25 Defense	5.5	7.4	7.9	8.5	8.8	9.4	9.4	9.7	10.0	10.7	11.2	11.9
26 Net Exports	0.2	1.3	1.7	-1.3	0.4	2.2	-4.8	-1.2	2.6	0.9	3.2	3.1
27 National Security	9.8	24.6	26.0	28.1	28.7	31.5	33.9	33.0	34.9	36.8	38.9	41.5
28 CIVIL CONSUMPTION	139.2	201.9	213.3	225.3	237.2	250.6	266.0	278.4	291.5	307.3	322.4	344.4
29 percent of total	59.2	55.0	55.0	56.3	54.8	54.9	56.0	55.2	55.0	55.0	55.5	56.2
30 CIVIL INVESTMENT	65.0	110.2	115.9	116.0	132.6	137.3	143.8	155.4	161.2	171.9	174.0	179.1
31 percent of total	27.6	30.0	29.9	29.0	30.6	30.1	30.3	30.8	30.4	30.8	29.9	29.2
32 GOVERNMENT, etc.	31.1	54.8	58.5	59.0	63.0	68.6	65.3	70.7	77.8	79.7	84.8	89.6
33 National Security	27.1	47.5	50.3	53.4	55.3	58.7	61.8	63.7	66.4	69.8	72.1	76.2
34 percent of total	11.5	12.9	13.0	13.3	12.8	12.9	13.0	12.6	12.5	12.5	12.4	12.4
35 Other (24+26)	4.0	7.3	8.2	5.6	7.6	9.8	3.5	7.0	11.4	9.9	12.7	13.4

TABLE D2: OFFICIAL SOVIET GNP BY END USE
(billion rubles)

	1981	1982	1983	1984	1985	1986	1987	1988	1989	1990
1 OFFICIAL GNP	647.7	693.2	726.3	758.0	777.0	798.5	825.0	875.4	924.1	965.3
2 CONSUMPTION	380.8	394.7	408.0	422.5	434.2	444.5	457.7	483.3	518.0	551.8
3 Households	335.8	347.0	359.5	372.3	381.9	389.7	401.2	422.0	453.9	484.7
4 Goods	292.9	301.8	312.3	323.2	330.7	336.1	343.6	361.5	390.8	417.9
5 Family Budgets	281.5	289.8	300.2	310.8	318.1	323.3	330.7	350.7	381.4	408.5
6 National Security	11.4	12.0	12.2	12.4	12.7	12.8	13.0	10.8	9.5	9.5
7 Paid Services	42.9	45.2	47.2	49.1	51.1	53.6	57.6	60.5	63.1	66.8
8 "Material"	10.1	10.9	11.4	11.9	12.6	13.5	14.7	15.5	16.5	17.5
9 "Non-material"	25.1	26.5	27.7	28.9	30.0	31.2	33.3	34.8	35.7	37.6
10 "Private"	7.7	7.8	8.0	8.4	8.5	8.9	9.5	10.2	10.9	11.7
11 Free Services	45.0	47.6	48.5	50.1	52.2	54.7	56.5	61.3	64.0	67.0
12 Family Budgets	39.1	41.3	42.1	43.5	45.5	47.7	49.3	53.3	55.8	58.5
13 National Security	5.9	6.3	6.4	6.6	6.8	7.0	7.2	8.0	8.2	8.5
14 INVESTMENT	193.5	219.7	232.9	243.7	248.4	255.5	264.8	276.1	287.6	286.5
15 Fixed Capital	176.9	190.5	200.3	218.8	228.0	245.5	258.9	273.9	279.3	281.3
16 New Fixed Invest.	137.3	147.5	155.1	169.3	175.5	190.3	200.8	213.4	217.7	217.3
17 Civilian	130.6	139.7	147.4	161.3	166.6	181.0	191.2	204.8	210.2	210.2
18 Defense	6.7	7.8	7.7	8.0	8.9	9.3	9.6	8.6	7.5	7.1
19 Capital Repair	39.6	43.0	45.3	49.5	52.5	55.2	58.1	60.5	61.6	64.0
20 Inventories	14.5	26.3	30.7	22.2	20.4	14.0	2.0	8.4	9.0	5.0
21 Reserves and Errors	2.1	2.9	1.9	2.7	0.0	-4.0	3.9	-6.2	-0.7	0.2
22 GOVERNMENT	73.3	78.9	85.3	91.9	94.4	98.5	102.4	116.0	118.6	127.1
23 R&D & Admin.	23.4	25.3	26.6	27.6	28.6	29.4	31.5	33.5	36.9	40.4
24 Civilian	10.7	11.2	11.9	12.3	12.8	12.9	13.3	14.3	15.1	16.8
25 Defense	12.7	14.1	14.7	15.3	15.8	16.5	18.2	19.2	21.8	23.6
26 Net Exports	2.6	4.2	5.0	5.4	2.1	4.0	5.0	1.4	-5.0	-5.0
27 National Security	47.3	49.4	53.7	58.9	63.7	65.1	65.9	81.1	86.7	91.7
28 CIVIL CONSUMPTION	363.5	376.3	389.4	403.4	414.6	424.6	437.5	464.5	500.2	533.7
29 percent of total	56.1	54.3	53.6	53.2	53.4	53.2	53.0	53.1	54.1	55.3
30 CIVIL INVESTMENT	186.8	211.9	225.2	235.7	239.5	246.2	255.2	267.5	280.1	279.4
31 percent of total	28.8	30.6	31.0	31.1	30.8	30.8	30.9	30.6	30.3	28.9
32 GOVERNMENT, etc.	97.3	104.9	111.6	118.8	122.8	127.6	132.2	143.4	143.8	152.2
33 National Security	83.9	89.5	94.7	101.2	107.9	110.7	113.9	127.7	133.7	140.4
34 percent of total	13.0	12.9	13.0	13.3	13.9	13.9	13.8	14.6	14.5	14.5
35 Other (24+26)	13.3	15.4	16.9	17.7	14.9	16.9	18.3	15.7	10.1	11.8

Chapter 5

The Long Term Soviet Economic Performance: An Alternative Assessment

CHAPTER 5:

LONG-TERM SOVIET ECONOMIC PERFORMANCE:
AN ALTERNATIVE ASSESSMENT

This chapter has the following major objectives:

● substantiate the need for alternative estimates

● outline the problem of measuring inflation in the Soviet economy

● evaluate Goskomstat's estimates of the Soviet economic performance, particularly for 1989

● summarize my major findings on long-term Soviet economic growth paying particular attention to the systemic causes of the current Soviet economic crisis

● evaluate existing approaches to estimating Soviet economic growth

● present the new method of integrated national accounts as it is applicable to measuring economic growth

● describe the data base on the Soviet economic growth

● compare in detail my and CIA estimates of Soviet GNP by sector of origin and end use, and

● outline unresolved problems as subjects of future research.

Why Alternative Estimates?

According to official reports on the Soviet economic performance prepared by Goskomstat, the average annual growth rates of Soviet GNP dropped from an impressive 5.5 to a still respectable 3.7 percent during the 1970s and 1980s. This robust long-term growth was allegedly sustained with an incredulously low rate of inflation which decreased from less than 1 percent a year before 1985 to near zero since Gorbachev took office.[1] Goskomstat

reports also advertize a spectacular growth of household consumption during the 1980s (4.5 percent a year), but contain no data on the growth of military expenditures.[2]

It is known that official Soviet estimates of growth rates are based on deflating the value of output in current prices with the so-called "comparable price indexes." Numerous deficiencies that undermine the reliability of these indexes have been discussed in detail on a number of occasions.

For some unknown reason, however, Goskomstat officials find it convenient to continue compiling these indexes in such a way as to ignore evidence about hidden inflation, changes in the structure of production, and the growth of fictitious output and illicitly provided services.

The Goskomstat estimates the NMP in comparable prices as the difference between gross value of output and the sum of material inputs and depreciation--all estimated in comparable prices. The NMP is used as as measure of total net output of material production sectors which exclude those sectors that are not involved directly in the production of material wealth.

What is known is that Goskomstat officials do not compare indicators of output in value and physical terms and instead rely on comparable price lists which change only in the aftermath of announced price reforms. Furthermore, in measuring the growth of NMP Goskomstat officials do not take into account changes in the structure of material inputs. As a result, they significantly overestimate the growth of NMP compared to the growth of material wealth that is available for end use.

Comparison of the recently published official data on the growth of Soviet GNP with traditional estimates of the NMP growth suggests that Goskomstat officials simply fail to deflate the net output of service sectors.

Goskomstat reports have been continuously contradicted by impressionistic accounts which have

---

[1] Goskomstat SSSR, Narodnoe Khozyaistvo SSSR v 1988 Godu. Moscow: Finansy i Statistika, 1989, pp. 5-7. This annual Soviet statistical manual is referred to here as Narkhoz. For 1989 estimates refer to Pravda, 28 January, p.1.

[2] Narkhoz for 1988, p. 17.

indicated that Soviet society has been in a big economic slump at least since the mid-1970s but has found sufficient resources to support one of the largest military buildups in peacetime history. The issue that has preoccupied economists for decades is how to reconcile official reports with impressionistic accounts. This issue has attracted considerable attention because the official statistics are the only available original source of systematically collected information on the Soviet national economy.

Of all independent estimates challenging the official statistics, those prepared by CIA analysts in their annual reports to the U.S. Congress are considered to be the most reliable. As opposed to other estimates, the CIA data on the Soviet civilian economy are derived using a commonly accepted methodology based on the systematic reconstruction of GNP accounts. Similarly, CIA estimates of Soviet military expenditures constitute a detailed assessment based on technological means of collecting intelligence.

According to CIA reports, the average annual growth rates of Soviet GNP estimated by sector of origin fell from 2.8 to 1.7 percent during the 1970s and the 1980s, thus indicating a change in hidden inflation from 2.7 to 2.0 percent.[3] CIA reports also indicate that hidden inflation in the Soviet civilian economy decreased from 2.3 to 1.7 percent but remained the same 5 percent in defense production.[4]

Even though CIA reports provide a much more pessimistic assessment of the Soviet civilian economic performance than the official statistics, some economists believe that this assessment is still too optimistic in light of the mounting evidence about the current Soviet economic crisis.[5] In recent years the accuracy of some CIA estimates was questioned increasingly in connection with the published work of the maverick Soviet economist Khanin.[6]

As estimated by Khanin, Soviet production activities grew on the average by 2 percent a year in the 1970s and by 1 percent in the 1980s.[7] He derived these surprising results using several unconventional methods of counting various physical outputs as a proxy for the overall national economic growth.

However, most of Khanin's methods cannot be tested for reliability using conventional tools of statistical analysis. In this respect, Khanin's work must be thoroughly tested for reliability using specially designed testing methods.[8] Before such tests are performed his work can be only tentatively accepted as a viable alternative to published

[3] The Soviet Economy in 1988: Gorbachev Changes Course. A Report by the CIA and the DIA Presented to the Subcommittee on National Security Economics of the Joint Economic Committee. April 14, 1989, p. 41; and CIA, USSR Measures of Economic Growth and Development, 1950-1980. Studies Prepared for the Use of the Joint Economic Committee of the U.S. Congress. Washington, D.C.: U.S. GVO, 1982, p. 15. Cited CIA estimates for the 1970s were derived as the average of the two estimates contained in the above publications.

[4] The rate of hidden inflation is estimated as the difference between the independent and official estimate of inflation. According to CIA estimates the actual rate of inflation averaged 2 percent a year for the civilan economy. No information exists on the official estimate of inflation in the defense economy, though official estimates of total GNP discussed below indicate that the Goskomstat makes no such estimate. For the CIA data on the Soviet defense economy refer to CIA, "A Guide to Monetary Measures of Soviet Defense Activities," A Reference

Aid, SOV 87-10069, November 1987, pp. 12-13.

[5] For the discussion of some of these reports, which are largely impressionist, refer to Anders Aslund, "How Small is the Soviet National Income?" A Paper Presented at the RAND-Hoover Conference on the Defense Sector in the Soviet Economy held in March, 1988.

[6] Khanin's work first came to be widely known with the publication of his article (written together with Selyunin) in monthly Soviet journal Novyi Mir, N2, 1987. Khanin's estimates and their significance in understanding how the oficial statistics distort Soviet economic reality were discussed extensively in Richard E. Ericson, The Soviet Statistical Debate: Khanin vs. TsSU." A Paper Presented at the RAND-Hoover Conference on the Defense Sector in the Soviet Economy held in March, 1988. Also refer to Boris Rumer, "Soviet Estimates of the Rate of Inflation, Soviet Studies, Vol. XLI, N2, April 1989, pp. 306-307.

[7] G. Khanin, "Ekonomicheskiy Rost: Al'ternativnaya Otsenka," Kommunist, N17, 1988, p. 85. Khanin's estimates are limited to material production sectors of the Soviet economy which account for more than 80 percent of GNP by sector of origin.

[8] For an excessively unfair criticism of Khanin's work see Vestnik Statistiki, N4, 1987, pp. 3-12, N6, 1987, pp. 53-60.

Goskomstat and CIA reports.

As opposed to Khanin's work, CIA estimates are based on the accepted GNP methodology and thus can be tested for reliability. Such a test is especially desirable for those CIA estimates that are based on Goskomstat's production and comparable price indexes.

In addition, Soviet GNP accounts compiled by the CIA contain large "statistical discrepancies" which need to be removed in order to make these accounts internally consistent and useful for deriving independent estimates of military expenditures. Thus, there is a need to have alternative estimates for:

● testing the validity of all Goskomstat's indexes
● solving the problem of statistical discrepancies
● determining how the new methodology affects estimates of Soviet GNP.

After performing the detailed analysis of hidden inflation in the Soviet economy I concluded that none of published Goskomstat's indexes compiled in comparable prices is reliable. This conclusion led to a different outlook on the Soviet civilian and defense economic growth since 1970, particularly on the growth of consumer industries and services, freight transportation, construction, investment into producer durables, and weapons procurement.

According to alternative estimates presented in this paper, the average annual price increase in Soviet consumer, investment and defense sectors was 3.2, 4.4 and around 3 percent respectively in 1970-1989. These estimates indicate that average annual growth rates of Soviet GNP declined from 2.4 percent during the 1970s to 1.3 percent during the 1980s.

The annual growth of the Soviet civilian and defense economy averaged 2.2 and above 3 percent during the 1970s and 1.2 and under 3 percent during the 1980s. Compared to CIA reports, alternative estimates suggest that the average annual growth of the Soviet civilian economy was slower by around 0.7-0.8 percent and that of the Soviet defense economy was faster by around 0.5-0.6 percent.

At the same time, CIA estimates of the total level of Soviet defense burden as well as the structure of Soviet military expenditures are fully corroborated by alternative estimates.

While the Soviet GNP data presented in this study are readily comparable with the CIA data, the comparison with Khanin's estimates is complicated by the fact that he did not account for service and defense activities that are not registered in Soviet material product accounts (MPA). After converting GNP into the MPA format I arrived at estimates which are notably higher, particularly for the late 1970s and early 1980s, than Khanin's.

Further research must determine why my alternative estimates differ from that of Khanin's. Specifically, it should be explored whether in some way Khanin was able to account indirectly for the impact of changes in quality on real growth rates. As known, the analysis of this impact is excluded from most independent estimates of the Soviet economic growth.

Measuring Open and Hidden Inflation

The driving force behind alternative estimates of Soviet economic growth has been the widespread belief that official estimates published by the Goskomstat are based on an inherently flawed methodology that is designed to advertise Soviet economic achievements rather than measure real rates of inflation.

There are two commonly accepted types of hidden inflation that are not registered by Goskomstat officials. Hidden inflation occurs when: 1) price increases are not accompanied by similar increases in the quality of goods and services, and 2) prices remain the same but the quality of goods and services deteriorates.

In addition, there are several other types of hidden inflation which are usually not taken into account. These types occur as a result of increases in fictitious output and money supply far in excess of the supply of goods and services. Fictitious output is a widespread Soviet economic phenomenon when enterprises grossly exaggerate their performance by fudging their output reports, especially those concerning the production of intermediate goods.

Some fictitious output, such as excessive inventories of soft goods and durables whose low quality does not meet consumer expectations, must be taken into account in addition to hidden inflation. The same applies to excessive unfinished construction--another form of fictitious output that can be analyzed as the second type of hidden inflation.

Certain difficulties also arise in measuring unregistered price rises in service sectors. For example, the commonly accepted ton-kilometer index conceals the widespread cheating ('pripiski') and inefficiencies in the traveled distances which are recorded each year by freight transportation services.

Hidden deterioration in the quality of many household services is uniquely manifested in the growth of employment that exceeds significantly the growth of capital and supplies. As known, the latter exert a crucial impact on productivity trends in service sectors.

As a rule, the dearth of available data on changes in quality makes it difficult to analyze the second type of hidden inflation. In response to mounting criticism, Goskomstat officials recently made an effort to increase collection and evaluation of data on changes in quality of consumer goods and services. Hopefully, they will publish sufficient data on changes in quality in the near future that will make it possible to incorporate the analysis of the second type of hidden inflation into estimates of Soviet economic growth.

Despite the dearth of data on changes in quality, there has been enough data published to analyze other types of hidden inflation by comparing indicators of the same output in value and physical terms. The underlying assumption behind such a comparison is that when there is no inflation the growth of output in value terms should not grow faster than the output measured in tons, square meters, horse power units, etc.

There are three measures of output in value terms: gross value of output in producer prices (GVO), GNP by sector of origin and GNP by end use. While GVO measures the total value of output of particular sectors, GNP measures that part of GVO which corresponds to value added in these sector.

Total GVO exceeds GNP in established prices by the difference between:

a) production inputs that equal intermediate purchases of goods and services plus losses of inventories and reserves, and

b) turnover, foreign trade and other indirect taxes less subsidies plus certain hidden revenues of the second economy.

In order to measure the real growth of GVO it is necessary to compare the gross output in value and physical terms for each year of the observed and for as many products with the same technological parameters as possible. The price index based on this comparison is referred to as the Paasche index.

The growth of value added by sector of origin depends not only on changes in GVO but also on changes in production inputs. The method for measuring the growth of value added that takes into account changes in both GVO and inputs is referred to as the double deflation method.

This method is considered more accurate than the commonly applied procedure based on the dubious assumption that the growth of inputs does not differ much from that of GVO. Such a procedure is usually applied in conjunction with the Laspeyres-type index which is compiled by multiplying base-year prices and the growth of physical output.

There are two ways to estimate the growth of end uses. One way is to convert GVO for each sector of the economy from producer to purchasers' prices which include net taxes, the delivery charge and imports. The other way is to compare as many types of end uses as possible in both value and physical terms. If all the above estimates are performed successfully, than hidden inflation can be estimated as the difference between real and officially reported rates of growth.

The above discussion focused on measures of growth using prices established during economic transactions. The problem is that these prices distort the real cost of producing goods and services. Hence, these prices must be adjusted for the purpose of assigning a correct price "weight", i.e., the true contribution of each sector to economic growth.

Prices adjusted in this way are referred to as factor cost prices. In market economies, prices are usually adjusted for indirect taxes and subsidies. In the case of the Soviet economy where most prices are fixed by the government fiat, it is also necessary to adjust profits in proportion to employed factors of production.

Evaluating Goskomstat Estimates

As estimated by the Goskomstat, the growth of Soviet GNP under Gorbachev was 2.3 percent in 1985, around 3 percent in 1986-1987 and 1989, and 5.5 percent in 1988.[9] This performance can be hardly cited as evidence of a long-term economic slowdown. The comparison of Goskomstat's estimates of GNP in current established and comparable prices indicates, however, that the above growth rates were derived using the highly improbable assumption that inflation in the Soviet economy was almost non-existent under Gorbachev.

Under mounting criticism for publishing distorted data, the Soviet government decided in 1989 to change the Goskomsat's top leadership which was publicly instructed by Gorbachev himself to introduce glasnost in its official statistical reports.[10] As of the end of 1990 the Goskomstat made several albeit modest steps in making its reports more objective.

Most significantly, it published some previously secret input-output and budgetary data and compiled 1989 price indexes for 650 key commodities without taking into considration changes in quality. The general price index increased by 2 percent while the unsatisfied consumer demand for goods and services grew by 5.5 percent, thus resulting in an overall inflation rate of 7.5 percent.[11]

As reported by the new Goskomstat Chairman Kirichenko, the average consumer price index increased by more than 40 percent in 1971-1988 or by 2.3 percent a year. In 1986-1988 alone prices rose 7.2 percent, including food items--11.7 percent and non-food items--3.2 percent. He also noted that if sale prices were balanced with the unsatisfied household demand for goods and services (165 billion rubles in 1989), then prices would increase by at least 40 percent.[12]

Despite Goskomstat's increased openness, it still

remains unclear how new price indexes were used for measuring the reported growth of Soviet GNP in 1989 or why these indexes were not utilized for reassessing the long-term growth. In addition, the Goskomstat has not yet revealed how it accounts for the growth of civilian and defense sectors financed by the state budget as well as for the growth of private production and service sectors. Intil Kirichenko and his associates decide to reveal more about their new methodology, published Goskomstat's statistics will continue to lack credibility. The recently published Goskomstat report for 1989 confirms this observation.[13]

According to this report, the growth in most production sectors ranged from minus 2 to plus 2 percent, including 1.7 percent in industry, 0.8 percent in agriculture, 0.5 percent for capital investment (including imports), minus 2 percent in freight transportation (the ton index), and 1 percent in passenger transportation. This explains why net product estimated by the Goskomstat for domestic material production sectors increased by only 1.5 percent.[14]

Soviet government officials also expected that the growth of weapons production declined by 4.5 percent in 1989.[15] The conversion of defense industries made it possible to increase production of consumer goods at these industries by 11 percent in comparable prices.[16]

There was negligible growth in the derived number of employed in state service sectors. In sum, total GVO measured by the Goskomstat should have grown in 1989 by no more than 1 percent, which was notably smaller than the reported GNP growth of 3 percent. How is it possible to reconcile this gross discrepancy?

Soviet economist Bolotin suggested that the explanation for the noted discrepancy lies in the possibility that that there was a sharp decline in intermediate purchases.[17] This economist noted himself, however, that intermediate purchases of goods by material production sectors declined by no more than 1 percent in 1989.[18] Moreover, a part of

[9] Narkhoz for 1988, p. 7.

[10] For a summary of Soviet economists' critical assessment of the official statistics refer to three informative articles published in Ekonomika i Matematicheskie Metody, Vol. XXV, N5, 1989.

[11] N. Zhelnorova, "Statistika Stanovitsya Ob'yektivnoi," Argumenty i Fakty, N4, 1990, p. 1.

[12] V. Kirichenko, "Vernut' Doverie Statistike," Kommunist, N3, 1990, pp. 27-28.

[13] Pravda, 28 January, 1990, pp. 1-2.

[14] Zhelnorova, op. cit., p. 1.

[15] Izvestiya, August, 1989, p. 4.

[16] Pravda, 28 January, 1990, p. 1.

[17] Moscow News, N6, 1990, p. 10.

[18] This was apparently based on the comparison of

the decline was offset by the growth of intermediate purchases of services. This suggests that, excluding foreign trade revenues, the growth of GNP estimated using the Goskomstat methodology should have been no more than 1.5-1.7 percent.

The most likely explanation for the noted discrepancy appears to be rather a sizable 8 percent growth of revenues from sales of imports and services provided to households in established prices which the Goskomstat apparently fails to deflate. In addition, there was a spectacular growth of budgetary revenues resulting from the 25 percent increase in the sale of alcohol.[19] There was also a sizable growth of revenues from private production and service activities which were previously rendered illicitly. There is no valid reason, however, why the growth of most of the noted budgetary and private revenues should be counted as contributing to the growth of GNP in factor cost prices.

The issue that must be explored is whether even the above estimated 1.5-1.7 percent growth of Soviet GNP is at all accurate. A large part of this estimate is based on the official statistics compiled in comparable prices which must be tested for reliability. This applies particularly to comparable prices on the machinery and construction output.

As reported by the Goskomstat, there was on the average a 1 percent decline in physical output in the combined power and fuels sector as well as in basic industrial materials sectors producing metals, wood and paper, construction materials, and chemicals.

The GVO of food (without alcoholic beverages), textiles, apparel and footware industries grew by around 1.5 percent in comparable prices. There was at least 1.5-2 percent decline in the reported number of produced machinery and equipment. While the number of produced automobiles and agricultural equiment declined by 4 and 6 percent respectively, the output of consumer electronics industry increased by around 3-4 percent.

Given the noted conversion of defense industries, it appears that there was a sizable decline in the output of the entire machinery sector. The real size

of this decline is probably impossible to determine due to the lack of published data on hidden inflation in Soviet computer and other electronic industries.

Overall, it appears that the industrial GVO measured in factor cost prices declined by at least 0.5-1 percent in 1989, thus suggesting that hidden inflation in industry was around 2-2.5 percent. This notable decline explains why intermediate purchases of goods fell after rising sharply for decades.

Similar to comparable industrial prices, comparable construction prices are known to be designed so as to conceal significant hidden price increases. The most precise way of estimating the growth of construction output is to compare data in value and physical terms for as many types of buildings and installations as published in the official statistics.

Since the detailed data on construction activities will be reported for 1989 later in the year, at this time it is possible to make only speculative observations using published data on produced construction materials and metals and continuing hidden inflation trends established during the past two decades.

The negligible growth of reported construction works corresponds closely to the near zero growth of produced materials that are used in construction. Hidden price increases estimated by Soviet planners have averaged around 1.2 percent a year in construction.[20] These estimates, however, fail to take into acount the enormous waste and fictitious output which are manifested first of all in the spectacular increase of unfinished construction which brings no return on investment.

Given peculiarly inefficient Soviet investment practices, it seems that the annual increase in unfinished construction ought to be excluded from GNP growth. In 1989 alone this increase exceeded 20 billion rubles or 18 percent of new construction works.[21] As reported by the Goskomstat, installed capital measured in comparable prices declined in 1989 by 2 percent. The decline measured in physical units amounted to 4 percent in the housing sector, 9 percent in education and as much as 16 percent in

reported rates of growth of GVOs by sector and net material product (2.4 percent).

[19] Y. Gaidar, "Trudnyi Vybor. Ekonomicheskoe Obozrenie Po Itogam 1989 Goda." Kommunist, N2, 1990, p.27.

[20] This can be deduced by comparing new construction prices introduced every 15 years. See Narkhoz for 1980, p. 364, and Narkhoz for 1985, p. 334.

[21] Gaidar, op. cit., p. 27.

the health sector.[22] The construction GVO had to decline by at least 3.5-4 percent.

The brief review of the Goskomstat report for 1989 indicates that alcoholic beverage and consumer electronics industries were star performers in the Soviet economy followed by sugar, confectionaries and apparel industries. A large part of household purchases of sugar was used for manufacturing moonshine, while a 2 percent growth of the produced apparel was estimated by the Goskomstat in comparable prices which were probably not adjusted for hidden inflation. All other sectors of the Soviet economy reported negligible or negative growth.

In sum, the Goskomstat report indicates that the growth of consumer sectors was around 1 percent without taking into account changes in quality. Combined investment and defense uses fell by around 3.5-4 percent. Considering that household consumption accounts for less than 50 percent of total end uses measured in factor cost prices, it can be tentatively concluded that Soviet GNP actually declined by 1-1.5 percent in 1989. This was a disastrous performance even in comparison with the stagnation 1987-1988 period when the Soviet GNP grew by 1.3-1.5 percent.[23]

Preliminary estimates thus suggest that the Soviet economy was indeed in the midst of a deep crisis already in 1989 with the population enjoying no increase in living standards and increasingly addicted to alcohol and sugar "blues". It will be possible to make more precise estimates after the publication of the Narkhoz edition for 1989 later in the year and after the Goskomstat clarifies how it deflates various GNP components both by sector of origin and end use.

Major Findings

Growth of GNP

As evident from Chart 8 below, my estimates of the growth of Soviet GNP by sector of origin for 1966-1970 are identical with those reported by the CIA. The two estimates differ, however, for the post-1970 period. My results indicate that the Soviet economy grew on the average by 1.6 percent in 1970-1985. This is notably smaller than 2.4 percent that is

---

[22] ibid, p. 27.
[23] CIA and DIA (1989), p. 41.

quoted in CIA reports. Since the Soviet population grew by 1.1 percent a year during this period, my results suggest that the average per capita growth in the Soviet economy was around 0.5 percent a year.

This slow growth must be undoubtedly viewed as a sign of a prolonged Soviet economic slump. A more detailed analysis of estimation results further reveals that the current Soviet socio-economic crisis did not occur suddenly; it evolved gradually over two decades and thus could have been predicted more than a decade ago.

Its first signs were already manifested in the early 1970s which witnessed a notable slowdown in the growth of consumption and fixed investment. Starting in the mid-1970s the slowdown was replaced with a negative per capita growth which by the mid-1980s evolved into negative absolute growth.

Initially, the general economic slowdown resulted from the poor performance of "light" industry (textiles, apparel and footware), food processing, wood-making, agricultural and construction sectors. The robust performance of utilities, chemical, machinery and transportation sectors was not good enough to reverse the general slowdown in the economy.

The Soviet economy came to an inevitable halt in the mid-1970s when Soviet planners lost their ability to maintain the extensive type of economic growth, which was fueled during the preceding decades primarily by the rapid expansion of production facilities, particularly in industries that mine and process basic materials. These include fuels, ferrous and nonferrous metals, wood and construction materials.

It appears that planners' ability to continue the expansion of production capacities was gradually undermined by increasing defense allocations and inefficient investment policies in consumer and agricultural sectors. By the late 1970s planners could no longer find sufficient financial and material resources to meet sky-rocketing construction and operating costs in mining and processing industries.

The absolute decline in the Soviet economy was triggered in the 1980s by the inevitable decline that occured in sectors producing basic industrial materials. The author of this report believes that this decline combined with the disastrous harvests of 1979-1982 finally crippled the growth of the

181

CHART 8: DS AND CIA ESTIMATES OF AVERAGE ANNUAL RATES OF SOVIET ECONOMIC GROWTH IN 1982 PRICES
(base year is the year prior to the stated period)

	1966-1970		1971-1975		1976-1980		1981-1985		1971-1985	
	DS	CIA	DS	CIA	DS	CIA	DS	CIA	DS	CIA
GNP by Sector of Origin*	4.8	4.8	2.1	3.1	1.6	2.3	1.0	1.9	1.6	2.4
Industrial Production	3.9	5.3	4.5	6.0	2.8	2.8	1.4	1.8	2.9	3.5
Machinery & Metal-Working	6.8	5.8	5.6	7.4	2.7	4.0	3.0	1.3	3.8	4.2
Agricultural Production	5.7	3.6	-1.2	-2.2	-0.3	0.0	2.2	1.2	0.2	-0.3
Construction	2.5	4.6	0.8	6.3	-1.6	3.5	-2.4	3.3	-1.1	4.4
Transportation	8.8	8.2	5.8	7.6	2.2	3.9	1.6	2.5	3.0	4.7
Services	4.4	4.6	2.2	3.7	1.3	2.9	0.3	2.3	1.3	3.0
GNP by End Use**	4.5	5.6	2.3	4.0	1.8	3.0	1.0	2.2	1.7	3.1
GNP by Civilian End Use	4.6	5.8	2.1	4.2	1.6	3.1	0.9	2.3	1.5	3.2
Per Capita Consumption+	3.2	5.0	1.5	3.0	0.0	2.0	0.0	0.8	0.5	1.9
Fixed Investment	5.0	6.0	4.6	5.2	0.9	3.3	0.7	3.5	2.0	4.0
Direct Defense Allocations	4.2	4.0	2.7	3.0	3.3	2.0	2.9	2.0	3.0	2.3

*in 1982 factor cost prices
**CIA estimates of consumption and fixed investment are in established prices
+both estimates are based on the assumption that there were no changes in quality

manufacturing and household service sectors.

One way to test the reliability of the above assessment of Soviet economic development during the 1970s and 1980s is to compare estimates of GNP growth by sector of origin and end use. The method of integrated national accounts has made it possible to attain relatively consistent estimates. In comparison, CIA estimates of the growth of Soviet GNP by end use are 25 percent higher than its estimates of the growth of Soviet GNP by sector of origin.

This inconsistency can be explained by the fact that CIA estimates are based on the traditional methodology which does not call for the integration of interindustry flows, foreign trade, defense production and other hidden economic activities within the broad framework of GNP accounts.

Growth of Per Capita Consumption

My estimation results indicate that the average annual growth of per capita consumption declined from 3 percent in the 1960s and early 1970s to zero in 1975-1985 without even taking into account the dramatic decrease in quality of goods and services consumed by Soviet households. Furthermore, my results indicate that from 1975 to 1985 there appears to be little increase in the total physical quantity of many categories of consumed goods and state provided services. In fact, transportation and illicitly provided services (second economy) were the only two consumer sectors that displayed any significant per capita growth.

To measure the real index in living standards it is necessary to account for changes in quality. Unfortunately, I did not have sufficient resources at my disposal to collect a reliable data base on changes in quality parameters of goods and services consumed by Soviet households. However, I made some tentative estimates based on interviews with recent emigres and tourists, news reports and published Soviet statistics. The collected evidence points overwhelmingly to a significant decline in the quality index ranging from 0.5-1.5 percent a year for many food items, soft goods and consumer durables and to 1-2 percent a year for many services, including trade and health.

By accounting for changes in quality I determined that living standards of Soviet people have declined

by around 1 percent a year since the early 1970s. This decline was further exacerbated by ecological disasters that have made a number of regions in the Soviet Union uninhabitable. In sum, it appears that Soviet people were at least 15 percent worse off in 1985 than they were in the early 1970s.

This relatively large decline in living standards still fails to reveal the actual collapse of the Soviet consumer sector in the 1970s and 1980s. In order to develop the full picture of the collapse I measured the growth of monetary incomes of Soviet households and the share of consumption in total GNP. According to my estimates, the total monetary income of households increased from 195 b.r. in 1970 to 415 b.r. in 1985, thus indicating an average annual increase of around 6 percent. During the same period total household savings increased from 73 to 314 b.r. or twice as fast as monetary incomes. The author of this report calculated that in 1985 Soviet people kept 221 b.r. in state banks, 82 b.r. in unorganized savings (rubles stashed at home) and 11 b.r. in bonds and other deposits.

Most of these savings occured because Soviet people were unable to make purchases of goods and state provided services of their choice. Despite this large increase in total household savings, stocks of inventories of unsold finished consumer goods and other commodities more than doubled during the same period from 64 to 150 b.r.

If it is assumed that the Soviet economy absorbed excess savings through price increases, then the comparison of the growth of monetary incomes and living standards would indicate an average annual inflation rate in the consumer sector of around 7-8 percent in 1970-1985. According to the official Soviet statistics, this rate was 0.5 percent a year. The comparison of real and official inflation thus points to the fact that hidden inflation in the Soviet consumer sector was around 7 percent a year and that the ruble has lost 50 percent of its purchasing power.

Actual purchases of goods now account for only 26 percent of total GNP in factor cost prices. Given the fact that services account for 19 percent, the share of total consumption is only 45 percent. It is difficult to find a country in the world where the share of consumption is below 55 percent. For example, in the U.S. this share has fluctuated between 60 and 65 percent.

Considering the fact that living standards in the U.S. have grown by 2-3 percent a year, today Soviet people appear twice as poor relative to American people than they were in 1970. According to some of the more optimistic assessments, Soviet living standards were one third of the American level in the mid-1970s. It can be concluded therefore that today Soviet people enjoy living standards which are no more than 20-25 percent of the American level.

Growth of Investment

The gross inefficiency of the Soviet economy is manifested first of all by the fact that it expends as much as 35 percent of all allocated resources on investment into civilian and defense sectors. Given this gross inefficiency, it is not surprising that any decline in the growth of production capacities had such grave consequences for the rest of the Soviet economy. The average annual growth of total new fixed investment declined dramatically from 5 percent in the 1960s and early 1970s to 0.8 percent in 1975-1985.

Furthermore, the growth of investment in housing, household services, science and defense sectors substantially exceeded investment into civilian production facilities and services. The actual growth of production-type fixed investment was less than 0.5 percent a year. My estimate is somewhat larger than that derived by the maverick Soviet economist Khanin who believes that there was an absolute negative growth in fixed investment into production sectors. If Khanin's estimates are correct, then the author of this report probably underestimated fictitious output of construction and machinery sectors.

However, despite some differences between my and Khanin's estimates, they are quite close compared with CIA and official Soviet estimates, which indicate that the average annual growth of fixed investment was 3.4 and 3.7 percent respectively. It appears that CIA analysts decided to accept the questionable assumption made by Harvard Professor Bergson that Soviet officials fully account for cost increases in compiling comparable price series for construction works and installed machinery.

The disappointing performance of Soviet civilian construction and machinery sectors stands in stark contrast to the robust increase in the production of goods that were stored as inventories and reserves and that were delivered to defense sectors. Annual additions to inventories and reserves doubled in the 1975-1985 period alone, thus pointing to the fact that the gross inefficiency with which production resources are utilized in the Soviet economy has reached catastrophic proportions.

In this light, it appears that a large percent of accumulated inventories should not be even counted as economic growth.

Growth of Defense Activities

The evidence discussed in this study confirms two well known facts that a) defense sectors were star performers in the Soviet economy during the past two decades and b) Soviet leaders significantly underestimate the total defense claim on allocated resources. Calculations of Soviet defense activities were performed in both established and factor cost prices. In 1965-1985 direct defense allocations grew on the average by 4.5 percent a year in established prices and by 3.4 percent in factor cost prices.

It is unclear, however, to what extent factor cost prices correctly reflect short-term fluctuations in growth rates of defense expenditures. The advantage of using factor cost prices lies instead in evaluating real long-term changes in the Soviet defense burden.

During the observed period the Soviet defense burden increased from 13 to 15 percent of total GNP in constant established prices and from 17 to 19 percent in constant factor cost prices that exclude net budgetary revenues from foreign trade. The defense burden equaled 18 percent if one assumes that GNP in factor cost prices include foreign trade revenues.

This high level of defense allocations can be appreciated fully only after comparing military and consumer purchases of goods and services in the USSR and the USA. While the ratio between total military and consumer purchases has been 40-45 percent in the USSR, it has been approximately 10 percent in the USA. Herein lies one of the major reasons why Soviet living standards have declined so dramatically compared to those in the U.S.

Growth of Foreign Trade Revenues

In 1975-1985, foreign trade revenues increased by as much as 250 percent, thus exerting a considerable influence on the growth of Soviet GNP. In fact, the IDS estimated that as much as one quarter of Soviet GNP growth in established prices was due to the increase in foreign trade revenues during the above period. During 1965-1985 foreign trade revenues increased by almost ten times from 7 to 65 b.r. in current prices. The growth record of the Soviet domestic economy was one half that of foreign trade revenues. In order to put the phenomenal growth of these revenues in a proper perspective it is necessary to analyze how they are earned and whether they should be counted as part of the overall economic growth.

Foreign trade revenues equal net earnings of the state budget from foreign trade. These earnings are estimated as the difference between budgetary revenues and subsidies connected with foreign trade activities. The state budget collects revenues when imports in domestic (established) prices exceed imports in foreign trade prices and when exports in foreign trade prices exceed exports in domestic prices.

In effect, foreign trade revenues occur when the state takes advantage of its monopolistic control over trade to capitalize on differences between price levels in the domestic economy and in the world markets. The growth of these revenues thus provides a distorted picture of Soviet production capabilities. For this reason, the author of this report believes that foreign trade revenues should be excluded from estimates of Soviet GNP growth in factor cost prices.

Alternative Approach

To estimate the growth of Soviet GNP necessary to perform a three-step procedure entailing the conversion of integrated accounts from:

1) established prices into constant producers' prices adjusted for hidden inflation,

2) constant producer prices into constant purchasers' prices which are used for measuring uses of output, and

3) established prices into adjusted factor cost prices that do not distort the relative scarcity value of produced goods and services.

In addition to uncovering numerous Soviet economic secrets, the method of integrated national accounts makes it possible the first time to apply for three superior methodological tools for measuring hidden inflation in the Soviet economy.

First, I was able to use the "Paasche" (current-year weighted) price index. Economists consider this index to be superior to the commonly applied "Laspeyres" (base-year weighted) price index which is based on two questionable assumptions: a) the base-year ratio between the net and gross output remains constant, and b) base-year relative prices are not significantly different from those of the current year.[24]

Second, I was able to apply the so-called "double deflation" procedure of converting value added of production sectors from current to constant prices. This procedure, which is based on deflating the GVO on the one hand and material inputs and business services on the other, is considered to be more precise than the common practice of applying GVO deflators as a substitute for the value added index.[25]

Third, I attempted to enhance measures of growth of civilian and defense service sectors by combining data on the growth of labor, material and capital inputs in proportion to the share of these inputs in each sector's GVO measured in current prices. While trends in employment are often considered sufficient for measuring growth of service sectors in market economies, the analysis of growth of Soviet consumer services must take into account the fact that annual changes in labor productivity depend to a great extent on the availability of material and capital resources. As will be discussed below, the dramatic decline in the growth of Soviet service sectors occured primarily as a result of acute shortages in material supplies and adequate facilities.

Estimating Producers' Price Indexes

Because of the lack of reliable data on annual changes in the quality of Soviet goods and services, it is usually assumed that Soviet economic growth can

24 Refer to CIA, "USSR Measures of Economic Growth and Development, 1950-1980." Studies Prepared for the Use of the Joint Economic Committee of the U.S. Congress. Washington, D.C.: U.S. GVO, 1982, pp. 42-43.

25 ibid, pp. 43-45.

be best measured in terms of the growth of physical output by sector of origin and end use. I relied on the same assumption in deriving price indexes for all sectors of the Soviet economy with the exception of the housing construction sector where the annual increase of 1.5 percent in quality was assumed.

In order to use the Paasche index it is necesary to have an extensive set of data on outputs of technologically homogeneous sectors in both value and physical terms for each year of the observed period. The ratio between the output of a particular sector in value and physical terms equals the price of one unit of commodity produced in this sector.

Thus, I tried to compile data on GVO and inputs for as many sectors as possible in both value and physical terms. I complied such data in these five major estimation steps:

1. I compiled the GVO data in established prices for thirty aggregate production and service sectors.

2. I disaggregated the GVO data into around 300 sectors using published official statistics on input-output flows, capital investment, growth rates and price changes.

3. I compared GVO in value and physical terms for around 230 production sectors which resulted in the same number of producer price indexes.

4. I aggregated the latter into 20 indexes one for each major production sector of the economy using the derived price weights.

5. I compiled indexes for major service sectors by combining the derived data on the growth of labor, material and capital inputs weighted in proportion to their share in the total value of these services in established prices.

Altogether I compiled an extensive data base on the output in both value and physical terms for around 165 industrial, 35 agricultural, 15 construction and 10 other production sectors of the Soviet economy.

Various sources of evidence were used to derive producer price indexes for disaggregated production sectors. These include published Soviet statistics on: production, input-output and financial flows, end uses of national income, GVO by sector in current and comparable established prices, growth rates by sector, official price indexes, and output in physical terms.

Published Soviet statistics contain a sufficient amount of information for estimating real annual price changes for most sectors with a technologically homogeneous structure of production output. Groups of sectors producing furniture and apparel were notable exceptions to the rule.

To estimate hidden inflation in these sectors, I utilized the derived price indexes for wood-works and six types of fabrics. This "substitution" procedure probably leads to fairly reliable results for two reasons: 1) the cost of production materials comprises the predominant part of producer prices in these sectors, and 2) the structure of production is quite similar within each of these sectors.

In the case of construction, I rejected the procedure based on trends in the production of construction materials in physical units which is used by the CIA. I believe that this procedure results in a significant overestimation of the growth of the construction GVO for two major reasons.

First, construction materials comprise less than half (45-47 percent) of material outlays of the construction sector. Prices on other material inputs-- fuels, lumber, metal structure and metal ware--rose much faster than prices on construction materials. At the same time, an ever increasing share of the construction materials GVO is used for intermediate deliveries within this sector.

Second, the structure of production and hence factors that trigger price increases are quite different in the construction sector than in sectors manufacturing construction materials. In the 1970s and 1980s construction enterprises had an incentive to use more rather than less expensive construction materials in order to fulfill the plan and to increase wage rates. This fact combined with frequent work stoppages due to interruptions in the supply of material inputs has resulted in sky-rocketing increases in the cost of one square meter of completed floor space.

In view of the above, it appears that the published Soviet data on completed industrial and housing floor space and on a number of other physical indicators of construction activities provide a more reliable measure of physical growth than the growth of produced construction materials. CIA analysts, however, could not use published data on physical output of construction sectors because they made no attempt to disaggregate the construction GVO by end use.

The advantage of using the integrated data base is that it contains a detailed information on construction activities performed in 36 sectors of the Soviet economy. This information was utilized first to compare the construction output in both value and physical terms and then to derive the annual growth of the construction GVO within the format of the Paasche price index.

As was expected, major problems arise in estimating the growth of machinery production sectors--technologically the most heterogeneous sectors in the economy. CIA analysts assert that the Soviet official statistics contain a sufficient amount of data on machinery output in physical terms for compiling reliable machinery price indexes. Contrary to this assertion, I found the published statistics to be quite inadequate for measuring the real growth of the machinery production sector and, as a result made an effort to develop an alternative procedure.

The latter is based on combining all data on the machinery sector in both value and physical terms that can be collected from open Soviet publications. These include the results of the extensive research on the subject performed primarily by four Soviet academic economists: Kheinman, Palterovich, Faltsman and Kornev.[26]

In brief, the alternative procedure consists of the following nine major estimation steps:

1. Disaggregating the GVO of the MBMW (machine-building, metal-working and machinery repairs) sector for 1972 into 125 major sectors using published input-output, financial and national income data as well as ministry-type data which are reported by Soviet authors.

2. Aggregating the output of major MBMW sectors for 1972 into five NMP-type production groups: a)

---

26 See S.A. Kheinman and D.M. Palterovich, (eds.), Naucho-Tekhnicheskiy Progress i Struktura Proizvodstva Sredstv Proizvodstva. Moscow: Nauka, 1982, pp. 25-54; V.K. Faltsman, Proizvodstvennyi Potentsial SSSR: Voprosy Prognozirovaniya. Moscow: Ekonomika, 1987; and A. Kornev, "Kak Rastut Tseny Na Machiny i Oborudovanie." Voprosy Ekonomiki, N6 1988.

civilian and defense interindustry components, b) producer durables, c) consumer durables, d) components used by nonproduction service sectors (science, passenger transportation, etc.); and e) capital and current repair works.

3. Combining the 1972 input-output data with published growth rates to derive the GVO for 125 sectors in comparable 1972 prices.

4. Combining national income, financial and foreign trade data to derive the total value of producer durables in established and comparable producer prices.

5. Estimating the official 1972 price index for the MBMW GVO and its major components, including producer and consumer durables, metal and repair works.

6. Converting the GVO for 125 sectors from comparable 1972 into current prices and then aggregating the detailed data into five NMP-type production groups.

7. Deriving real price indexes for selected machinery producer and consumer sectors for which data are published in physical units.

8. Deflating the derived GVO of producer and consumer durable sectors using a) derived price indexes for sectors whose output is published in physical units, and b) price indexes estimated by Soviet academic economists.

9. Determining the rate of hidden inflation (the difference between real and official price indexes) for a selected number of producer and consumer durables sectors and extrapolating the same rate on other components of MBMW GVO.

General Price Indexes

So far, the focus of the discussion was on the procedure for estimating producer price indexes which register changes in prices on outputs that are charged by enterprises. In comparison, general price indexes register changes in prices paid by purchasers of output. In order to estimate a general price index it is necessary to convert the value of supplied goods from producer to purchasers prices.

The integrated data base enclosed in the end of

the chapter includes consolidated accounts of supply and uses of goods and services produced in 14 material production sectors of the Soviet economy in both current and constant 1973 prices. It must be emphasized that purchasers' prices exceed producer prices by the sum of net taxes, T&C and T&D charges, and imports in domestic prices.

I assumed that net taxes and supply charges grow in tandem with the physical growth of output by sector. I estimated the value of imported goods in both current and constant by combining published data on import price indexes with the derived data on domestic production price indexes.[27]

Producers' price indexes hence proved useful in estimating general price indexes for production, household and government service sectors. General production price indexes for T&C and T&D services were compiled by aggregating supply charges for each sector of industry and agriculture in constant prices. The freight transportation index derived in this way appears to be much more reliable than the officially reported "ton/kilometer" data used by the CIA.

Unfortunately, there is no alternative procedure to using the officially reported "passenger/kilometer" data for converting the GVO of transportation sectors serving households from current to constant prices. Price indexes for all other services were compiled by combining the independently derived data on material and capital inputs into these sectors in constant prices with published data on employment trends.

General price indexes are needed to estimate the growth of supplied goods and services that are used as:

● inputs into production of finished goods and services
● household consumption
● government administrative services
● new fixed investment and capital repair
● annual changes in inventories and reserves
● defense goods and services, and
● exports of goods.

I applied general price indexes to convert the

above uses from current to constant prices following the format of input-output tables. This made it possible to estimate Soviet GNP by end use in constant prices as the sum total of all end uses in constant prices. I used the derived data on intermediate uses in estimating total Soviet GNP by sector of origin in constant prices using the double delation method.

Factor Cost Prices

Although the available methodological tools help measure the rate of hidden inflation in the Soviet economy, they are not sufficient for estimating reliable rates of Soviet economic growth. The reason for this is that established prices (even those adjusted for hidden inflation) distort the contribution of each production and service sector to the overall growth of the national economy.

In order to determine the correct contribution of each sector to GNP growth it is necessary to estimate what economists call "price weights". This entail developing a reliable procedure for converting established prices into factor cost prices.

As was discussed in the first chapter, the traditional methodology, which was originally developed by Bergson, is based on adjusting net revenues according to the relative size of stocks of fixed capital and inventories operated by each sector of the economy. I cited new evidence suggesting the need for a new adjusted factor cost (AFC) procedure.

Rather than using an arbitrary rate of return on capital, I decided to distribute net revenues earned in the whole Soviet economy into sectors according to the size of the total cost included in these sectors' value added. This cost consists of adjusted wages, social security deductions and depreciation on capital operated in profit-seeking and budget-supported sectors.

The new AFC procedure thus takes into account labor inputs as a factor of production as well as differences in capital productivity. Unfortunately, it remains difficult to determine to what extent the new procedure can be considered as an improvement over Bergson's methodology. The reason for this is that there is the dearth of published data on how wage and depreciation rates actually reflect the quality of labor and capital inputs by sector.

27 V. Sel'tsovskiy, "Sovershenstvovanie Analiza Effektivnosti Vneshney Torgovli SSSR." Vestnik Statistiki, N6, 1983.

CHART 9:   DS AND CIA ESTIMATES OF THE STRUCTURE OF SOVIET GNP BY SECTOR OF ORIGIN
           (percent; current prices)

	Established Prices				Factor Cost Prices			
	1970		1982		1970		1982	
	DS	CIA	DS	CIA	DS	CIA	DS	CIA
Total GNP	100	100	100	100	100	100	100	100
Industry*	42.4	45.8	42.5	50.0	29.6	32.0	31.4	33.4
Metals	3.3	2.9	3.2	n.a.	2.7	3.6	2.6	3.4
Fuel	4.2	3.9	5.7	n.a.	2.6	3.1	2.8	3.5
Power	1.8	1.5	1.8	n.a.	1.6	2.2	1.3	2.5
MBMW	11.4	11.9	12.3	n.a.	10.6	10.1	12.5	11.2
Chemicals	2.2	2.6	2.2	n.a.	1.7	2.0	2.3	2.6
Wood and Paper	2.5	2.9	1.9	n.a.	2.6	2.4	2.2	2.0
Construction Materials	1.8	2.0	1.7	n.a.	1.9	2.1	1.9	2.0
Light Industry	7.0	8.7	6.0	n.a.	3.0	2.5	2.6	2.3
Food Industry	6.9	7.4	5.9	n.a.	2.7	3.0	2.5	2.6
Other Industry	1.5	1.7	1.8	n.a.	0.6	0.9	0.7	1.3
Agriculture	17.4	20.5	12.8	15.4	22.1	21.1	18.4	20.2
Construction	8.0	7.4	7.2	6.7	10.8	7.3	10.6	8.0
T & C	7.8	8.1	8.2	9.3	7.5	9.6	8.8	10.8
Trade	5.0	4.8	4.9	6.7	4.9	7.3	5.1	6.5
Services	10.5	10.7	10.5	10.6	16.1	20.2	11.3	18.5
Other Legal Sectors	7.6	2.1	12.1	2.6	9.0	2.2	12.5	2.6
a) Foreign Trade**	3.2	n.a.	7.9	n.a.	0.5	n.a.	0.8	n.a.
b) Other Civilian Production	0.8	0.4	0.8	0.6	1.2	0.4	1.1	0.6
b) Hidden Defense Sectors	3.7	1.8	3.4	2.0	5.8	1.9	5.3	2.0
Second Economy+	1.3	0.7	1.8	0.7	1.3	0.3	1.9	0.3

* agricultural subsidies reduce value added in industry in established prices
**in CIA estimates, foreign trade revenues are distributed among production sectors
+ illicit private incomes less fictitious output (CIA estimates exclude fictitious output)

Data Base on Soviet Economic Growth

The data base enclosed in the end of this chapter consists of three parts. The first two parts consist of summary tables on Soviet economic performance, each containing estimates in established and factor cost prices. Both sets of tables contain estimates of the growth and structure of Soviet GNP by sector of origin and end use in both current and constant prices. In addition, the first set contains estimates of the growth of total NMP, GVO by sector and price deflators by sector of origin and end use. The second set also contains data on factor cost adjustment coefficients.

Tables on the growth of Soviet GNP in established prices were compiled using 1973 production price indexes. This year was chosen with the intent of making alternative estimates in established prices compatible with the official Soviet data on growth rates. Tables on the growth of Soviet GNP in factor cost prices were compiled using both 1973 and 1982 price indexes. The 1982 prices were chosen to make alternative estimates compatible with CIA estimates.

Summary tables on Soviet GNP and GVO were based on detailed calculations performed in part three of the data base. Total Soviet GNP equals the sum of value added in economic sectors that operate legally plus value added in the second economy minus fictitious output of uninstalled capital and excessive inventories of soft goods that are not available for end use because of their poor quality. Data on the second economy and fictitious output were compiled using recent Soviet publications.[28]

Part three consists of detailed data on supply and uses of goods and services and on price indexes compiled for approximately 300 production and services sectors which are aggregated into 30 major sectors. Data on each sector of the economy are presented in current and constant 1973 prices and in physical units. The classification of sectors

corresponds to that used in compiling original Soviet planning tables. Thus, sectors are divided into three groups: material production, "non-material" services and hidden NS activitie.

Several aggregagate sectors--MBMW, light industry, food processing, agricultural production and construction--consist of a large number of small sectors which require detailed coverage. Overall, I followed a standard format. Each table consists of four sections: a) the unified accounts of supplies and uses in current and constant prices; b) derivation of 1973 price indexes; c) GVO in current producer prices by sector with a homogeneous technological output; d) supporting data on domestic output in physical units; and e) supporting data on imports in both value and physical units (when appropriate).

Analysis of Soviet GNP by Sector of Origin

Established and Factor Cost Prices

The comparison of CIA and my alternative estimates of the GNP structure in established and factor cost prices is presented in Chart 9 below. As a result of misreading its own finding on hidden wages of defense sectors, the CIA appears to have misrepresented the actual structure of Soviet GNP by sector of origin. A comparison of estimates indicates that instead of listing value added in hidden defense sectors as a separate GNP component, the CIA distributed it among production sectors.

The comparison also indicates that the same distribution procedure was employed with respect to hidden revenues from foreign trade. Since these revenues are earned by the entire Soviet economy, they must be listed as a separate component of GNP. There are a number of other differences between DS and CIA estimates in established prices that are evident from the chart and that require further investigation.

The comparison of CIA and my estimates of relative price weights for 1970 and 1982 points to the existence of significant discrepancies which range from 10 percent for the MBMW sector to 30 percent for service sectors and to as much as 50 percent for electrical power. Moreover, the discrepancy is much greater for 1982 than it is for 1970. This can be explained by the fact that CIA estimates for 1970 are much closer than those for 1982 to the published 1972 input-output data which were extensively used

---

28 For the most recent authoritative estimate of the second economy refer to T.I. Koryagina, "Uslugi Tenevyei Legal'nye." EKO, N2, 1989, pp.60-65; for uninstalled equipment refer to Ekonomicheskaya Gazeta, N10, 1988, p. 8, and for excessive inventories of soft goods to P.A. Lokshin, Spros, Proizvodstvo, Torgovlya. Moscow: Ekonomika, 1975, p.178, and Ekonomicheskaya Gazeta, N3, 1988, p.10.

by the CIA in deriving 1970 estimates.

Comparison of estimation results further indicates that the size of the discrepancy between CIA and my estimates in factor cost prices depends as much on different assessments of the GNP structure in established prices as on differences in methodology. I converted established into factor cost prices using a procedure that takes into account both the size and quality of labor and capital inputs. In comparison, the CIA relied on the AFC procedure which takes into account only the size of capital inputs. As a result, the discrepancy between the two estimates depends on the extent to which labor/capital and capital depreciation rates for a given sector diverge from the mean rates that exist in the whole economy.

Compared to CIA estimates based on Bergson's AFC procedure, the alternative estimates point to relatively smaller price weights for such capital-intensive sectors of the Soviet economy as electric power, fuels, ferrous and non-ferrous metallurgy, transportation, trade and housing. At the same time, alternative estimates point to relatively higher price weights for such labor-intensive sectors as MBMW, wood and paper, light industry, capital construction, agriculture and other production sectors. The two sets of estimates are close only with respect to construction materials, chemical and food production sectors.

The observed discrepancies between the two estimates of price weights for the construction sector on the one hand and T&C, trade and other service sectors on the other are too large to be dismissed simply as results of applying a different AFC procedure. The very existence of these discrepancies demonstrates better than any theoretical formulation why applying Bergson's methodology leads to unreliable estimates of Soviet GNP in factor cost prices.

The fact that CIA analysts assign relatively large price weights to service sectors is the result of not taking into account labor inputs as well as sectoral differences in depreciation rates, which serve as the best substitute for capital productivity. It is known that depreciation rates in production sectors exceed those in service sectors by more than 50 percent because most machinery is operated in production sectors. For the same reason, depreciation rates in construction are highest among production sectors.

Large stocks of inventories held by trade sectors also undoubtedly contributed to the relatively large price weight assigned to these sectors by the CIA. It is known, however, that Soviet trade organizations are forced to accept a large percentage of products which cannot be sold due to their intolerably poor quality. Moreover, they are even forced to pay turnover taxes for these products, which often will never be sold, to the state budget. At the same time, another large percentage of stored products cannot be sold because they are kept in reserve in case of unexpected production failures. In view of the above, it appears that current stocks of trade inventories serve as a poor indicator of any hypothetical return on capital.

Value Added Indexes

The comparative analysis of value added indexes further illustrates the advantage of using the AFC procedure that takes into account the quality of both labor and capital inputs. As presented in Chart 10 below, alternative estimates of the growth of value added in 1965-1985 diverge considerably from those of the CIA for all sectors of the Soviet economy. The discrepancy in estimates is particularly large with respect to such labor-intensive sectors with a poor growth record as wood-making and light industries and construction.

I estimated that the cumulative share of these sectors in total GNP declined from 16.4 to 15.4 percent in 1970-1982. According to CIA reports, this share remained around 12.2 percent during the same period. As a result of the price weight effect, the poor performance of the above sectors had a much smaller impact on CIA estimates of the slowdown in growth rates of total GNP than it had on my estimates.

The reverse impact can be observed with respect to such capital-intensive sectors with an impressive overall growth record as electric power, fuels, chemicals, T&C, trade and other services. According to CIA reports, the cumulative share of these sectors in total Soviet GNP remained 44.4 percent. In comparison, I estimated that the same share decreased from 34.4 to 31.6 percent. In sum, the price weight effect helps explain at least some of the observed dicrepancy between CIA and my estimates of the growth of Soviet GNP by sector of origin. To determine all sources of this discrepancy it is necessary to compare estimates of price trends for each of the five-year periods.

CHART 10: DS AND CIA ESTIMATES OF SOVIET GNP GROWTH BY SECTOR OF ORIGIN
           (in 1982 factor cost prices)*

	1966-1970		1971-1975		1976-1980		1981-1985	
	DS	CIA	DS	CIA	DS	CIA	DS	CIA
Total GNP	1.28	1.28	1.11	1.16	1.08	1.12	1.05	1.09
Industry**	1.22	1.35	1.22	1.30	1.13	1.15	1.07	1.09
Metals	1.06	1.35	1.31	1.25	1.20	1.06	0.92	1.07
Fuel	0.94	1.30	1.18	1.29	1.19	1.17	1.15	1.05
Power	1.10	1.47	1.49	1.41	1.34	1.24	1.14	1.19
MBMW	1.52	1.41	1.28	1.37	1.14	1.20	1.14	1.10
Chemicals	1.59	1.49	1.62	1.49	1.20	1.16	1.32	1.30
Wood and Paper	1.28	1.37	1.00	1.13	0.93	0.97	0.91	1.11
Construction Materials	1.27	1.35	1.27	1.28	1.17	1.07	0.97	1.07
Light Industry	1.28	1.37	0.91	1.13	1.10	1.13	0.90	1.02
Food Industry	1.25	1.32	1.31	1.16	1.10	1.13	1.14	1.10
Other Industry	1.43	1.35	1.74	1.30	1.15	1.15	1.07	1.07
Agriculture	1.39	1.18	0.94	0.89	0.99	1.01	1.11	1.06
Construction	1.15	1.30	1.04	1.31	0.92	1.14	0.88	1.11
T & C	1.43	1.43	1.29	1.37	1.11	1.20	1.08	1.13
Trade	1.52	1.41	1.20	1.26	1.07	1.14	1.12	1.09
Services	1.25	1.24	1.17	1.18	1.10	1.14	1.02	1.12
Defense Sectors	1.32	1.22	1.22	1.12	1.09	1.09	1.10	1.03

*base year is the year prior to the stated period
**excluding value added in sectors producing weapons for domestic use

CHART 11: COMPARISON OF DS AND CIA ESTIMATES OF THE STRUCTURE OF SOVIET GNP BY END USE
(percent; current prices)

	Established Prices					Factor Cost Prices			
	1970		1982			1970		1982	
	DS	CIA	DS	CIA		DS	CIA	DS	CIA
Total GNP	100	100	100	100		100	100	100	100
Consumption	56.1	55.1	53.7	52.7		50.7	54.2	44.5	52.7
Consumer Goods	43.0	43.4	40.1	41.4		30.7	34.7	25.7	34.7
Food	26.3	28.2	22.0	24.7		23.9	25.5	18.1	24.6
Soft and Durable Goods	16.7	15.3	18.2	16.7		6.8	9.3	7.6	10.1
Consumer Services	11.1	11.2	11.2	11.0		18.0	19.8	16.0	18.4
Second Economy	2.0	0.4	2.6	0.4		2.0	0.4	2.8	0.5
Fixed Investment	26.7	28.5	25.9	27.9		30.4	28.2	31.6	31.2
New Fixed Investment	21.8	23.5	20.4	21.7		23.8	23.4	24.8	24.3
Machinery and Equipment	7.0	6.8	8.8	8.6		7.2	6.7	9.2	9.3
Construction and Other	13.7	15.6	11.2	13.0		15.3	15.6	15.8	14.7
Net Additions to Livestock	1.1	1.1	0.5	0.2		1.3	1.2	0.6	0.3
Capital Repair	4.9	5.0	5.6	6.2		6.6	4.8	7.0	6.9
Other End Uses	17.2	16.4	20.4	19.4		18.9	17.6	23.9	16.1
Inventories and Reserves*	2.0	4.0	3.5	n.a.		1.7	n.a.	3.6	n.a.
Government Services	1.3	2.4	1.2	n.a.		1.6	n.a.	1.4	n.a.
Research and Development	2.3	2.7	2.5	n.a.		2.6	n.a.	2.8	n.a.
Defense Expenditures**	11.1	7.1	12.5	n.a.		12.3	n.a.	15.4	n.a.
Net Exports	0.5	0.2	0.7	n.a.		0.7	n.a.	0.8	n.a.

* DS estimates include changes in agricultural reserves less fictitious output
**weapons procurement, defense construction and O&M (including defense-related services, except for R&D)

193

CHART 12:   DS AND CIA ESTIMATES OF AVERAGE ANNUAL GROWTH OF PER-CAPITA CONSUMPTION IN THE USSR
(established prices)

	1966-1970		1971-1975		1976-1980		1981-1985	
	DS	CIA	DS	CIA	DS	CIA	DS	CIA
Total Consumption*	3.8	5.0	2.0	3.0	1.2	2.0	0.0	0.8
Food & Alcohol	3.3	4.2	1.7	2.1	1.2	1.1	-1.4	-0.4
Food Items	1.7	n.a.	0.5	n.a.	-0.2	n.a.	-0.2	n.a.
Alcohol	9.4	n.a.	5.5	n.a.	4.7	n.a.	-4.0	n.a.
Soft Goods	4.1	7.2	0.3	2.7	0.9	2.8	1.3	1.4
Durables	9.4	9.5	4.9	9.7	2.8	5.4	2.4	3.0
Services	4.4	4.1	3.8	3.0	1.7	2.0	0.3	1.7
Housing and Utilities	2.5	3.7	2.7	3.4	2.1	2.5	1.3	2.4
Transportation	8.4	8.2	6.4	6.4	3.0	2.2	2.2	1.7
Repair and Personal Care	4.8	6.4	4.9	4.4	3.1	4.1	1.6	3.1
Education	2.8	3.0	1.3	1.5	0.6	1.4	-0.4	0.9
Health, etc.	4.7	3.2	0.7	3.2	0.0	0.9	-0.3	0.9
Recreation and Other	10.0	2.6	5.0	4.1	-2.0	1.2	0.1	1.0

* excluding illicit incomes from privately provided services

The two sets of estimates are identical with respect to total GNP for 1966-1970. However, this similarity conceals substantial differences in estimates of the growth of value added by sector. While CIA estimates indicate that value added in industry and construction increased by 7 and 6 percent per year respectively, my estimates point instead to 4.4 and 3 percent. The opposite can be observed with respect to agriculture, trade and defense sectors.

Large differences in estimates also exist with respect to particular sectors of industry. My estimates indicate a spectacular growth of 10 percent a year in MBMW, chemical and other industries, a negligable growth in capital-intensive industries, and a growth of 6 percent a year in labor-intensive industries. In comparison, CIA estimates point to an average growth rate of 7-8 percent throughout the industrial sector.

While my estimates indicate that Soviet economic growth in 1971-1975 was reduced to 2 percent a year, CIA estimates point to 3 percent a year. The two estimates diverge with respect to total industry, agriculture, construction and defense sectors in a way similar to the previous period. However, the difference between the two estimates is not as large for agriculture and increased substantially for construction. In addition, DS estimates are significantly smaller than CIA estimates for T&C and trade sectors. As was the case for the previous period, the two sets of estimates indicate a similar rate of growth for service sectors.

Although CIA and my estimates for 1976-1980 are more similar for total industry, agriculture and defense sectors than for previous periods, I still continued to find a greater rate of hidden inflation in most sectors of the Soviet economy. This rate was particularly significant in construction, T&C, trade and other services. As a result, I determined that the overall growth of GNP for this period was 1.6 percent rather than 2.4 percent as derived by the CIA.

My estimates indicate that the overall Soviet economic growth further declined during 1981-1985 averaging only 1 percent a year. A notable slowdown occurred in all sectors of the economy, except for MBMW, chemical, food and defense industries. There was also a small improvement in the growth of value added in agriculture and trade sectors.

The poor performance of labor-intensive industries, metallurgy, construction, T&C and household services, however, reached crisis proportions. Average rates of growth were below zero in many industries and services, and even reached -2.4 percent in construction. In comparison, CIA estimates indicate that there was some growth in every sector of the Soviet economy, except for light industry. This explains why the CIA has provided a relatively optimistic assessment of the Soviet economic performance for this period.

How is it possible that CIA and alternative estimates provide such a divergent outlook on hidden inflation in the Soviet economy, given the fact that the two estimates are based on essentially the same pool of published Soviet data? The full answer to this question requires an extensive analysis of how particular changes in methodology influence the procedure for estimating value added in each sector of the Soviet economy. While such an analysis is outside the scope of the present report, the above comparison of DS and CIA estimates warrant two preliminary observations.

First, double deflation and "base-year value added" methods lead to quite different assessments of long-term trends in hidden inflation for each sector of the Soviet economy. The double deflation method is believed to be far more precise because it enables one to take into account changes in the structure of material inputs measured in constant prices. The reason why it is necessary to account for these changes stems from the fact that my estimates performed for 1965-1985 reveal the existence of significant sectoral variations on national and intra-industry levels in a) annual rate of hidden inflation and b) ratios between value added and GVO.

For example, hidden price increases were found to be larger in producer and consumer durables sectors rather than in sectors manufacturing interindustry machinery components, which constitute a large share of material inputs in many sectors of the economy. The author of this report found the same pattern in wood-making, apparel, agriculture and some other sectors. As a result, the growth of material inputs for most sectors of the Soviet economy far outstripped the GVO growth. The latter, in turn, was much larger than the growth of value added, thus negating the major assumption of the "base-year value added" method.

Second, the approach based on the comparison of output in value and physical terms always leads to a rate of inflation which is much higher than that estimated using published "comparable prices" and physical indexes. The above applies particularly to machinery, construction and T&C sectors, where the hidden inflation was especially rampant because of planners' difficulties in monitoring unjustified increases in prices and fictitious output. Hidden inflation in household service sectors has accelerated since the late 1970s as a result of the slowdown in the growth of material and capital inputs.

GNP by End Use

Growth of Per-Capita Consumption

A general consumption index is an aggregate of many indexes measuring the growth of various goods and services consumed by households. In theory, there are two procedures for estimating trends in consumption of goods. One procedure, which I followed in this study, is based on compiling purchasers' price indexes adjusted for hidden inflation for major categories of consumer goods and services. It must be emphasized that this procedure is indeed quite cumbersome because it requires the availability of data on producer, T&C, trade and import price indexes.

The other procedure, which is used by the CIA, is based on combining published data on per capita consumption of selected items that are reported in physical units with official price indexes. Although the CIA procedure is much less cumbersome than the DS procedure, there is evidence indicating that official price indexes are quite unreliable. In addition, there are large gaps in the availability of data on per capita consumption in physical units that make it difficult to measure annual changes in consumption for 1970s and 1980s using these data.

Beginning in 1986 the Goskomstat began to release annual reports on per capita consumption of 10 major food items, alcohol, tobacco, textiles, knitwear, hosiery and footwear. In addition, official statistical publications include data in physical units on 12 major types of consumer durables. The publication of consumption data in physical units on a regular basis now makes it possible for the IDS to integrate these data with its independently derived purchasers' price indexes for the post-1985 period.

I also followed new procedures to estimate the growth of the following consumer services:

Utilities. These services consist of productive-type utilities (electricity, gas and water) and housing. Data on production-type utilities are available in physical units. The CIA estimated the housing index using published data on the mid-year housing stock. I rejected this approach because it does not take into account changes in quality. Instead, I estimated the index for the entire Soviet housing-comunal sector by combining independently estimated indexes for labor, material and capital inputs into this sector.

Transportation. The CIA procedure is based on combining published data on fares and passenger-kilometers for nine transportation sectors. I could not use this procedure in connection with the Paasche index because of the lack of data on the total volume of each transportation service in current rubles. Instead, I estimated the total volume of all passenger transportation services in current rubles and than compared this volume with published data on passenger-kilometers. The advantage of using the CIA procedure is that it takes into account changes in the general transportation index that are caused by the unequal growth of various component indexes. The advantage of using the alternative procedure is that its reliability is not affected by unannounced increases in the transportation fare.

Communications. I could not find any suitable measure of growth in physical units for postal, telephone and broadcasting services and thus decided to use the derived general T&C index as a substitute. In comparison, the CIA relied on various published physical indexes, such as the number of telephones, TV sets, mailed letters, etc.

Repair and Personal Care. These services encompass what Soviets call "production and nonproduction everyday services." I believe that the officially reported data on these services in constant prices, which are used by the CIA, conceal a considerable rate of hidden inflation. Thus, I decided to compile the alternative index that is based on the independently derived production indexes for textiles and apparel, footware and consumer durables. The derived index for housing-communal sectors encompasses everyday services in accord with the published statistics on labor, material and capital inputs.

Communal Services. The CIA assumed that the index

for communal services is a weighted aggregate of separate CIA indexes for education and health.

Education and Health. I compiled indexes for these sectors in the same way as for housing-communal services. In comparison, the CIA made an estimate that is based on a weighted composite of two sub-indexes--personnel costs and other current purchases (primarily food, utilities and supply). The growth of these services is also affected by the growth and quality of capital inputs which can be analyzed conditionally using depreciation accounts.

Recreation and Other. These sectors consist of arts, sports, tourism, banking and insurance, and other small services. The CIA used various physical indexes for measuring the growth of recreational services but did not estimate indexes for banking and insurance sectors, even though the latter earn substantial profits from serving households. I based the index for all these sectors on the growth of inputs.

The comparison of DS and CIA estimates of average annual growth of per-capita consumption in established prices is presented in Chart 12 above. As evident from the chart, the two sets of estimates are quite different, except for consumption of food and soft goods in 1981-1985. I believes that the discrepancy between the two estimates of soft goods in the 1970s results primarily from the fact that the CIA did not adjust official price indexes for apparel sectors where hidden inflation was particularly rampant. The same applies to a number of sectors manufacturing consumer durables.

In contrast, the detected discrepancy between the two estimates of services primarily results from using different methodologies. A significant slowdown in the growth of health, education and other services can be explained by a zero growth in the volume of material and capital inputs into these sectors beginning in the mid-1970s.

Soviet consumption trends evaluated in factor cost prices exhibit a much slower rate of growth than trends evaluated in established prices (see Chart 8 above). This can be explained by two factors: 1) a decrease in the price weight of sectors with a relatively high growth rate--alcoholic beverages, tobacco, consumer durables and T&C services; and 2) an increase in the price weight of sectors that were in a slump--many food items, education and health services.

Growth of Fixed Investment

Fixed investment consists of four components: 1) new construction works, 2) commissioning of M&E, 3) additions of agricultural capital, and 4) capital repair of fixed capital stock (buildings, installations, machinery, etc.).

Table 0.06 on machinery production sectors consists of four parts: a) unified accounts; b) derivation of 1973 production price indexes; c) total MBMW GVO disaggregated into 100 major sectors in current producer prices; and d) supporting data on selected machinery output. Unified accounts include data on supply and uses of machinery output in current and 1973 prices and on components of the general MBMW index.

The general index was estimated for all uses as a composite of four weighted sub-indexes for the output in producer (enterprise) prices, turnover tax and subsidies, imported machinery and TC and TD charges. The general index for investment into machinery excludes a sub-index for turnover tax but includes a sub-index for subsidies on agricultural machinery. The most influential component of the general index is a sub-index for producer prices, which is derived by aggregating individual indexes compiled for 16 machinery sectors.

As was discussed in the methodological section of the report, these individual producer price indexes are based on combining official data on the machinery reported in physical units with data on hidden price increases derived by Soviet academic economists. Hidden price increases that were first found for the total producer durable sector were then extrapolated for intermediate, consumer and defense goods.

The extrapolation was performed on the basis of comparable price indexes that were derived for 100 MBMW sectors in Table A9. Specifically, it was assumed that hidden inflation is higher in producer durable sectors by the difference between comparable price indexes for these and other MBMW sectors.

As estimated in Table 0.06, the combined producer price index of published MBMW sectors increased on the average by 2.3 percent a year. The largest price increases occured in sectors that manufacture producer and consumer durables--3.9 and 3.2 percent a year respectively. The smallest price increase--1.2 percent a year--occurred in sectors that manufacture complex electronic component parts for civilian and defense machinery.

This estimation result, which seems paradoxical at first sight, can be explained by the fact that comparable prices in electronics industries were lowered by 50 percent during the observed period as a result of a steep decline in these industries' profitability ratio. Comparable producer prices on electronic components declined by 10 percent in 1971, by 14 percent in 1973, and by 20 percent in 1975-1977. It appears that planners made a concerted effort to contain cost increases in defense-related industries by manipulating the price structure. This explains why value added in the Soviet electronics industry is much higher in factor cost prices compared to established prices.

Table 0.13 on capital construction has two parts: a) output by 30 sectors in current prices, and b) output by 20 sectors in constant 1973 prices adjusted for hidden inflation. Most data in current prices were compiled by converting published investment series in comparable prices with the derived comparable price index. Data on some service sectors were obtained from the statistics on "public consumption funds" (financing of free services) and on the state budget. Data on capital repair were obtained by applying published repair rates to the derived value of fixed capital stock.

Many construction activities in physical units are evaluated in terms of square meters of completed floor space. The official statistics include data on three types of floor space that are used by industrial, housing and service facilities. Other construction activities are evaluated in various physical units which are reported for electrical power stations, oil and gas exploration, agricultural facilities, land and water projects, oil and gas pipelines, freeways, trade and education facilities.

When viewed as a whole, published physical units represent most of the construction works performed during a given year. This makes it possible to determine annual changes in the cost of all construction activities using the ratio between particular types of output in current prices and in physical units.

Table 0.13 also contains data on annual changes in the cost of one unit of 10 types of performed construction works. It was assumed that some

increase (1.5 percent a year) in the cost of housing construction was the result of improvements in quality. Another assumption that had to be made concerns the problem with the availability of data on square meters of completed industrial floor after 1980. Thus, it was assumed that price increases in industry were the same with respect to price increases in housing after 1980s as they were before that year. The general price index was estimated by weighing individul sub-indexes for one year at a time.

Estimation results indicate that hidden inflation was most rampant in oil and gas exploration, agricultural and trade sectors, where the cost of construction more than doubled during the observed period. The cost of construction in manufacturing sectors increased by 5 percent a year, and in housing, highway construction and services--by around 3.0 percent a year.

The general construction index, which increased by 4.8 percent a year, was only slightly higher than the comparable price index for 1965-1970 but was much higher for 1970-1985. The rate of hidden inflation increased from 3.5 percent a year in the 1970s to 4.5 percent a year in the 1980s. As a result of hidden inflation, there has been no real increase in construction activities since the mid-1970s.

Growth of Defense Activities

Total Soviet defense expenditures in current and constant established prices are estimated in Table 0.18 following both GNP and Soviet national accounting formats. According to the traditional GNP format, procurement of weapons is evaluated in purchasers' prices that contain the transportation cost. Similarly, the O&M budget covers the cost of defense-related services. In Soviet planning tables, the production of weapons and space technology is evaluated in producer prices, while outlays on military administration (a Soviet term for O&M) do not cover purchases of producer durables (dual-use technologies) nor the cost of defense-related services.

Unfortunately, the official statistics are presented in a way which makes the actual estimation of total defense expenditures possible only within the format of Soviet planning tables. All Soviet defense activities, including production of weapons and space technology, are evaluated in prices that exclude profit. Therefore, in order to account for all defense resource costs, total defense expenditures estimated according to the Soviet format must be increased by the cost of defense-related services and investment into defense sectors.

In order to estimate the real growth of Soviet defense activities, it is necessary to determine price indexes for defense output. Given the dearth of data on Soviet defense price trends, it is not surprising that the issue of hidden inflation in Soviet defense sectors has not been sufficiently explored. The proper analysis of this issue requires a separate large-scale effort which extends far beyond the objectives of the present study. However, the collected evidence on hidden inflation in machinery and other sectors of the Soviet economy appears to be sufficiently complete for making two preliminary observations.

First of all, it appears that the CIA has significantly underestimated the growth of Soviet defense production sectors because it has assumed that inflation in these sectors has averaged above 5 percent a year. This assumption contradicts the above discussed evidence that price reforms in defense electronic sectors during the 1970s and 1980s succeeded in keeping the rate of inflation at a low pace in these sectors compared to civilian machinery sectors.

Sophisticated defense electronic components comprise at least 30 percent of the total weapons production cost. Material supplies and other components comprise another 30 percent, while value added--the remaining 40 percent. Given this price structure, there is a high probability that the noted price reforms had a strong impact on the total weapons production cost. What remains unclear is the extent to which these price reforms counterbalanced a considerable rise in prices on material supplies and defense capital.

In the absence of more reliable information, it can be safely assumed that the rate of inflation in defense industries was not higher than in the entire machinery production sector where it averaged 2.5 percent a year. This leads to the conclusion that the average annual growth of weapons production was approximately 4.5 percent a year during the 1970s and 1980s.

Price increases in other Soviet defense sectors were measured by combining data on labor, material and capital inputs into these sectors. As presented in Table 0.18, the average annual growth of total Soviet military expenditures was almost 4 percent a year,

while the general price index increased by 3 percent a year.

Conclusion: Suggestions for Further Research

It is now recognized that the rapid disintegration of the Soviet economy in the late 1980s has caught most experts by surprise. During the last decade some lonely Cassandras made prophetic predictions but their voices were barely heard beneath the chorus of relatively optimistic assessments of the Soviet economic performance presented by academic and government experts. The chorus quieted down and changed its tune after Soviet leaders themselves recognized publicly that the Soviet economy entered a period of negligible growth already in the mid-1970s.

Today when many leading Soviet economists echo Cassandras' warnings about the imminent collapse of the Soviet socio-economic order their words are accepted as an unconditional truth. Soon, the past optimistic assessments will be forgotten as are other unsuccessul forecasts. However, the subject of this chapter was not the discussion of the history of various unsuccessful forecasts.[29] Instead, the subject was an alternative methodology and estimates that can be used to improve the current assessment of the long-term Soviet economic performance.

The proposed methodology is based on reconstructing Soviet national account as the first step in compiling Soviet GNP and national product accounts. It was demonstrated that the format of the official statistics is not amenable to the direct estimation of Soviet GNP by sector of origin and end use and that reliable estimates can be made by extracting relevant data from a complete set of reconstructed Soviet planning tables.

In addition, it was demonstrated that reliable estimates of Soviet economic growth depend to a large extent on the availability of data in current established prices. These data are needed not only for estimating annual price changes but also for deriving reliable price weights in factor cost prices.

As was emphasized in the report, the major flaw

---

29 For such a discussion refer to Igor Birman, "Rosefielde and My Cumulative Disequilibrium Hypothesis: A Comment." Soviet Studies, January 1989.

of the widely accepted Bergson-CIA methodology stems from the fact that it was never designed to analyze Soviet national accounts in current established prices as an integrated system. As a result, the CIA and other users of its methodology have had great difficulties in interpreting the official statistics for the purpose of compiling Soviet GNP accounts.

These difficulties, in turn, have led to the development of inferior analytical tools for a) measuring Soviet civilian and defense economic activities in current prices, and b) converting current prices into constant prices adjusted for hidden inflation. The comparison of IDS and CIA estimates presented above indicated that using inferior analytical tools leads to the overestimation of Soviet civilian economic performance and to the underestimation of the size and growth of Soviet defense activities.

It must be emphasized that it was not the objective of this study to solve all methodological problems that arise in measuring Soviet economic growth. In fact, this study must be viewed as the initial step to develop an alternative methodology and estimates of Soviet civilian and defense economic might. This step entailed the estimation of Soviet output in established and factor cost prices which were adjusted for hidden inflation using published data in physical units. The completion of further steps requires a large-scale effort that must focus on the following four issues:

> 1) collection of reliable data on quality changes in Soviet consumer, investment, material supply and defense sectors;

> 2) development of methodology designed to integrate physical and quality indexes for particular sectors of the economy;

> 3) evaluation of the impact of financial factors on hidden inflation trends; and

> 4) development of a methodology for measuring the type of factor cost prices that better approximate the relative scarcity value of Soviet goods and services than factor cost prices derived using alternative methodologies.

Unfortunately, the latter issue cannot be addressed within the constraints imposed by the inconvertible Soviet domestic currency. What is required is the systematic re-evaluation of all Soviet goods and services in some domestic and world market-clearing prices.

Soviet Data Base

II. Annual Growth Rates and Price Changes

Part A

GNP Growth in Established Prices

TABLE 1A: GROWTH OF SOVIET GNP BY SECTOR OF ORIGIN
        (percent)

	1965	1970	1971	1972	1973	1974	1975	1976	1977	1978	1979	1980	1981	1982	1983	1984	1985
1 TOTAL GNP	77.8	100.0	103.7	104.9	110.9	113.7	114.6	119.3	122.8	124.6	124.9	127.8	130.4	133.3	135.9	137.0	136.4
2 annual growth			3.7	1.2	6.0	2.7	0.9	4.7	3.4	1.8	0.3	2.9	2.7	2.8	2.6	1.1	-0.5
3 per capita	81.9	100.0	102.9	103.0	107.9	109.7	109.6	112.8	115.2	116.0	115.4	117.0	118.3	120.0	121.2	121.1	119.5
4 PROFIT SECTORS	77.6	100.0	103.8	104.8	110.9	113.8	114.5	119.6	123.1	124.9	125.1	128.0	130.8	133.8	136.7	137.8	137.1
5 annual growth			3.8	1.0	6.2	2.9	0.7	5.1	3.5	1.8	0.2	2.9	2.8	3.0	2.9	1.1	-0.7
6 TOTAL INDUSTRY	77.3	100.0	106.5	110.2	112.8	118.8	123.8	128.0	132.3	135.0	138.2	142.9	147.4	145.8	146.6	150.0	149.9
7 annual growth			6.5	3.7	2.6	6.0	4.9	4.3	4.3	2.6	3.3	4.7	4.5	-1.6	0.7	3.4	-0.1
8 POWER	65.3	100.0	105.6	119.8	127.3	136.3	142.2	155.3	158.8	170.9	178.8	181.6	190.8	199.0	201.0	208.2	215.6
9 annual growth			5.6	14.2	7.4	9.0	5.9	13.0	3.6	12.0	8.0	2.7	9.2	8.2	2.0	7.2	7.4
10 FUELS	72.7	100.0	107.0	112.9	120.1	124.7	130.0	138.3	147.1	154.7	160.8	166.7	167.7	167.4	171.4	175.9	180.7
11 annual growth			7.1	5.9	7.2	4.6	5.3	8.3	8.9	7.5	6.2	5.9	1.0	-0.3	4.0	4.5	4.8
12 METALS	66.2	100.0	101.5	104.6	111.3	120.3	123.9	126.0	124.9	131.1	128.4	129.7	131.8	126.1	136.6	141.6	147.4
13 annual growth			1.5	3.1	6.7	8.9	3.6	2.1	-1.0	6.2	-2.7	1.3	2.1	-5.7	10.5	5.0	5.8
14 CHEMICALS	58.8	100.0	113.8	125.3	143.6	167.0	179.8	188.7	190.7	197.8	188.6	204.1	215.1	204.3	211.6	221.4	225.3
15 annual growth			13.8	11.5	18.3	23.4	12.9	8.8	2.0	7.1	-9.2	15.5	11.0	-10.8	7.4	9.7	4.0
16 TOTAL MBMW	66.9	100.0	105.9	115.1	118.4	124.1	129.4	137.3	141.9	145.1	152.0	153.7	157.0	163.7	165.8	171.0	174.1
17 annual growth			5.9	9.2	3.3	5.7	5.3	7.9	4.6	3.2	6.9	1.7	3.3	6.7	2.1	5.2	3.1
18 WEAPONS & SPACE	75.3	100.0	108.5	115.7	126.0	131.4	136.1	144.4	148.1	152.5	156.0	161.6	165.4	171.3	173.6	177.4	178.8
19 annual growth			8.5	7.1	10.3	5.4	4.7	8.2	3.8	4.4	3.5	5.7	3.8	5.9	2.3	3.8	1.4
20 WOOD & PAPER	95.0	100.0	99.7	99.6	98.4	95.2	96.0	94.8	91.7	87.4	82.6	84.3	86.1	90.2	89.2	89.2	91.0
21 annual growth			-0.3	-0.1	-1.2	-3.2	0.8	-1.2	-3.2	-4.2	-4.8	1.7	1.8	4.1	-1.0	-0.1	1.9
22 CONSTR. MATER.	71.7	100.0	113.7	119.4	127.0	128.3	131.1	135.5	142.8	143.9	137.9	137.6	142.0	135.5	144.3	150.0	154.3
23 annual growth			13.7	5.7	7.6	1.3	2.9	4.4	7.3	1.1	-6.1	-0.2	4.3	-6.5	8.8	5.7	4.3
24 N.E.C. INDUSTRY	75.3	100.0	114.2	128.1	142.3	156.4	182.7	192.7	212.8	222.7	220.7	240.9	242.8	238.5	236.2	231.9	227.8
25 annual growth			14.2	14.0	14.2	14.1	26.3	10.0	20.1	9.9	-2.0	20.2	1.9	-4.3	-2.3	-4.4	-4.0
26 LIGHT INDUSTRY	86.5	100.0	102.9	103.6	91.0	89.0	89.0	90.7	91.4	91.0	93.7	98.7	102.2	89.2	86.0	87.4	89.2
27 annual growth			2.9	0.7	-12.6	-1.9	-0.1	1.7	0.8	-0.5	2.7	5.0	3.5	-13.0	-3.1	1.3	1.8
28 FOOD INDUSTRY	91.1	100.0	108.9	106.4	111.9	122.9	128.2	127.4	134.0	133.1	138.8	144.4	152.3	149.1	143.8	144.9	131.3
29 annual growth			8.9	-2.5	5.4	11.0	5.4	-0.8	6.6	-0.9	5.7	5.6	7.9	-3.2	-5.3	1.1	-13.7
30 AGRICULTURE	79.2	100.0	94.5	79.1	103.3	93.1	75.2	85.7	87.6	90.4	80.3	72.0	67.6	81.6	89.5	87.3	82.1
31 annual growth			-5.5	-15.5	24.2	-10.1	-17.9	10.5	1.9	2.8	-10.1	-8.3	-4.4	14.0	8.0	-2.3	-5.2
32 TRANSP. & COM.	69.5	100.0	106.7	111.2	116.0	122.1	128.1	132.2	133.5	134.7	134.4	136.2	137.0	140.7	144.6	148.2	146.2
33 annual growth			6.7	4.5	4.8	6.1	6.0	4.1	1.3	1.2	-0.3	1.8	0.8	3.6	3.9	3.6	-1.9
34 CONSTRUCTION	84.1	100.0	101.7	103.0	101.5	108.6	111.1	109.5	110.6	104.7	99.2	98.6	96.9	100.3	101.9	91.9	90.3
35 annual growth			1.7	1.4	-1.5	7.1	2.5	-1.7	1.1	-5.8	-5.6	-0.6	-1.7	3.3	1.7	-10.0	-1.7

TABLE 1A: GROWTH OF SOVIET GNP BY SECTOR OF ORIGIN
   (percent)

	1965	1970	1971	1972	1973	1974	1975	1976	1977	1978	1979	1980	1981	1982	1983	1984	1985
36 TRADE, etc.	68.4	100.0	104.3	105.4	113.9	118.5	121.7	123.0	127.9	129.4	128.2	128.4	127.6	129.2	136.9	139.4	138.9
37 annual growth			4.4	1.0	8.5	4.6	3.3	1.3	4.9	1.5	-1.2	0.3	-0.8	1.6	7.6	2.5	-0.5
38 OTHER PROD.	76.5	100.0	99.1	101.3	105.7	113.7	113.7	119.5	122.4	128.3	122.4	122.4	125.4	122.4	125.4	128.3	131.2
39 annual growth			-0.9	2.2	4.4	8.0	0.0	5.8	2.9	5.8	-5.8	0.0	2.9	-2.9	2.9	2.9	2.9
40 FOREIGN TRADE	81.3	100.0	111.0	116.6	120.1	108.7	103.9	124.8	130.8	137.3	162.3	197.8	228.3	252.8	255.9	257.0	263.5
41 annual growth			11.0	5.6	3.6	-11.4	-4.8	20.9	6.0	6.4	25.1	35.4	30.5	24.5	3.1	1.1	6.5
42 OTHER N-PROD	74.5	100.0	105.8	113.5	118.5	123.0	129.5	134.8	141.3	144.3	151.0	154.1	156.4	157.2	163.2	167.9	170.5
43 annual growth			5.8	7.7	5.0	4.5	6.5	5.3	6.5	3.0	6.7	3.1	2.3	0.8	6.0	4.7	2.5
44 BUDGET SECTORS	80.1	100.0	103.2	106.4	111.0	112.3	115.1	116.4	119.2	121.2	122.2	125.8	127.1	127.7	128.2	128.8	130.3
45 annual growth			3.2	3.2	4.6	1.3	2.8	1.3	2.8	2.0	1.0	3.6	1.4	0.5	0.5	0.6	1.5
46 EDUCATION	84.8	100.0	102.0	103.0	109.3	110.7	112.1	114.0	115.1	115.9	117.5	120.6	119.5	120.4	120.5	121.4	124.8
47 annual growth			2.0	1.1	6.3	1.4	1.5	1.9	1.1	0.8	1.6	3.1	-1.1	0.9	0.1	0.9	3.4
48 CULTURE & ART	72.5	100.0	102.1	102.9	104.7	106.6	110.6	119.3	127.0	128.9	131.1	135.4	131.8	132.9	133.5	139.2	146.8
49 annual growth			2.1	0.8	1.8	1.9	4.0	8.6	7.7	1.9	2.2	4.3	-3.6	1.1	0.6	5.7	7.7
50 HEALTH, etc.	78.4	100.0	104.2	107.6	104.3	105.0	104.6	103.5	106.5	109.7	110.8	112.3	111.6	112.6	112.8	114.0	116.1
51 annual growth			4.2	3.5	-3.3	0.6	-0.4	-1.0	2.9	3.3	1.0	1.5	-0.7	1.0	0.2	1.2	2.2
52 SCIENCE	79.0	100.0	107.5	115.1	124.9	126.6	133.8	133.7	139.9	143.5	148.8	158.8	164.8	162.9	163.7	157.4	154.4
53 annual growth			7.5	7.6	9.7	1.7	7.2	-0.1	6.2	3.6	5.3	10.0	6.0	-1.9	0.8	-6.3	-3.0
54 STATE ADMIN.	77.1	100.0	101.6	106.1	112.0	115.1	117.5	119.4	122.8	126.0	126.4	131.4	137.8	136.7	136.0	141.4	139.3
55 annual growth			1.6	4.5	5.9	3.1	2.4	1.9	3.4	3.3	0.3	5.0	6.4	-1.1	-0.7	5.4	-2.1
56 MILITARY ADMIN.	76.7	100.0	100.6	103.3	109.1	109.6	114.9	116.3	116.2	116.7	111.0	109.7	113.4	116.1	118.4	120.8	122.3
57 annual growth			0.6	2.7	5.8	0.5	5.3	1.5	-0.1	0.5	-5.7	-1.2	3.7	2.7	2.3	2.4	1.4

TABLE 1B: SOVIET GNP BY SECTOR OF ORIGIN IN 1973 PRICES
(billion rubles)

Page 1

	1965	1970	1971	1972	1973	1974	1975	1976	1977	1978	1979	1980	1981	1982	1983	1984	1985
1 TOTAL	314.9	404.7	419.8	424.5	449.0	460.1	463.7	482.8	496.8	504.1	505.3	517.1	527.9	539.3	550.0	554.2	552.1
2 Legal Economy	309.9	399.1	414.4	418.6	443.1	454.1	457.5	476.7	490.3	497.5	498.8	510.3	521.1	532.2	542.7	546.8	543.9
3 Second Economy	7.0	8.2	8.3	8.8	9.0	9.2	9.4	9.6	9.9	10.1	10.3	10.6	10.8	11.1	11.3	11.5	11.7
4 Fictitious Output	-1.9	-2.6	-2.8	-2.9	-3.1	-3.2	-3.1	-3.4	-3.4	-3.5	-3.7	-3.7	-3.9	-3.9	-4.0	-4.0	-3.5
5 PROFIT SECTORS	285.0	367.3	381.2	384.8	407.5	418.1	420.6	439.3	452.2	458.8	459.6	470.1	480.4	491.5	502.0	506.1	503.4
6 TOTAL INDUSTRY	146.8	189.4	200.9	208.9	214.0	225.5	234.5	242.7	250.6	255.4	261.3	269.9	278.3	274.9	276.0	282.2	281.5
7 "NMP" Industry	140.5	181.0	191.7	199.2	203.4	214.4	223.0	230.5	238.2	242.5	248.1	256.3	264.4	260.4	261.3	267.2	266.4
8 Weapons & Space	6.4	8.5	9.2	9.8	10.7	11.1	11.5	12.2	12.5	12.9	13.2	13.7	14.0	14.5	14.7	15.0	15.1
9 POWER	4.1	6.3	6.7	7.6	8.0	8.6	9.0	9.8	10.0	10.8	11.3	11.5	12.1	12.6	12.7	13.1	13.6
10 GVO	8.4	11.9	12.6	14.0	14.8	15.7	16.5	17.7	18.2	19.1	19.8	20.4	21.1	22.3	22.8	23.7	24.4
11 Intermediate	4.3	5.6	5.9	6.4	6.8	7.1	7.5	7.9	8.2	8.3	8.5	8.9	9.0	9.7	10.1	10.6	10.8
12 Materials	4.1	5.3	5.6	6.1	6.4	6.7	7.1	7.5	7.7	7.8	8.0	8.4	8.5	9.2	9.5	9.9	10.1
13 Losses	0.1	0.1	0.2	0.1	0.2	0.2	0.2	0.2	0.2	0.2	0.2	0.2	0.2	0.2	0.2	0.2	0.3
14 Services	0.1	0.2	0.2	0.2	0.2	0.2	0.2	0.2	0.3	0.3	0.3	0.3	0.3	0.3	0.4	0.4	0.4
15 FUELS	11.9	16.4	17.5	18.5	19.7	20.4	21.3	22.7	24.1	25.3	26.3	27.3	27.5	27.4	28.1	28.8	29.6
16 GVO	22.3	28.1	29.7	31.2	32.8	34.6	36.5	38.3	40.1	41.6	42.6	44.0	44.7	46.3	47.3	48.2	49.4
17 Intermediate	10.4	11.7	12.2	12.7	13.1	14.2	15.2	15.7	16.0	16.3	16.3	16.7	17.2	18.9	19.2	19.4	19.8
18 Materials	10.1	11.3	11.7	12.2	12.6	13.6	14.6	15.0	15.3	15.5	15.5	15.9	16.4	18.0	18.3	18.4	18.7
19 Losses	0.1	0.1	0.1	0.1	0.1	0.1	0.1	0.2	0.2	0.2	0.2	0.2	0.2	0.2	0.2	0.2	0.2
20 Services	0.2	0.3	0.3	0.4	0.4	0.4	0.5	0.5	0.5	0.6	0.6	0.6	0.7	0.7	0.7	0.8	0.9
21 METALS	8.6	12.9	13.2	13.6	14.4	15.6	16.1	16.4	16.2	17.0	16.7	16.9	17.1	16.4	17.8	18.4	19.2
22 GVO	27.7	37.1	38.6	40.7	42.8	45.2	47.5	48.6	49.9	51.4	50.8	50.2	50.2	49.3	50.7	51.8	53.6
23 Intermediate	19.1	24.2	25.4	27.1	28.4	29.6	31.4	32.2	33.7	34.4	34.1	33.3	33.1	32.9	32.9	33.4	34.4
24 Materials	18.8	23.9	25.0	26.7	28.0	29.1	30.9	31.7	33.1	33.8	33.5	32.7	32.4	32.2	32.2	32.6	33.6
25 Losses	0.1	0.1	0.1	0.1	0.1	0.1	0.1	0.2	0.2	0.2	0.2	0.2	0.2	0.2	0.2	0.2	0.2
26 Services	0.2	0.3	0.3	0.3	0.3	0.3	0.4	0.4	0.4	0.4	0.4	0.5	0.5	0.5	0.6	0.6	0.6
27 CHEMICALS	4.4	7.4	8.5	9.4	10.8	12.5	13.5	14.2	14.3	14.8	14.1	15.3	16.1	15.3	15.9	16.6	16.9
28 GVO	13.7	22.2	24.5	26.6	28.9	32.0	34.0	35.7	36.7	37.9	37.5	39.4	40.9	41.6	43.2	44.4	44.6
29 Intermediate	9.3	14.8	16.0	17.2	18.1	19.5	20.5	21.6	22.4	23.1	23.4	24.1	24.8	26.3	27.3	27.8	27.7
30 Materials	9.0	14.4	15.5	16.7	17.6	18.9	19.9	20.9	21.7	22.3	22.6	23.3	24.0	25.4	26.4	26.8	26.7
31 Losses	0.1	0.1	0.1	0.1	0.1	0.1	0.1	0.2	0.2	0.2	0.2	0.2	0.2	0.2	0.2	0.2	0.2
32 Services	0.2	0.3	0.3	0.4	0.4	0.4	0.5	0.5	0.5	0.6	0.6	0.6	0.6	0.7	0.7	0.8	0.8
33 TOTAL MBMW	29.2	43.6	46.1	50.1	51.5	54.2	56.5	60.0	62.0	63.4	66.4	67.1	68.6	71.5	72.4	74.7	76.0
34 GVO	60.8	97.1	103.7	112.5	117.3	124.9	132.7	136.6	140.9	147.4	152.5	157.4	160.6	163.2	167.4	172.5	176.0
35 Intermediate	31.5	53.6	57.6	62.4	65.8	70.7	76.2	76.6	78.9	84.0	86.1	90.3	92.1	91.7	94.9	97.8	100.0
36 Materials	29.6	50.7	54.5	59.2	62.4	67.2	72.5	72.7	74.8	79.5	81.7	85.7	87.4	86.7	89.7	92.1	94.2
37 Losses	0.2	0.4	0.5	0.4	0.5	0.5	0.6	0.6	0.6	0.6	0.6	0.7	0.7	0.7	0.7	0.8	0.8
38 Services	1.7	2.5	2.6	2.8	2.9	3.0	3.1	3.3	3.5	3.9	3.8	3.9	4.0	4.3	4.5	4.9	5.0
39 WOOD & PAPER	9.8	10.3	10.2	10.2	10.1	9.8	9.9	9.7	9.4	9.0	8.5	8.7	8.8	9.3	9.2	9.1	9.3
40 GVO	19.2	21.7	22.0	22.3	22.3	22.4	22.5	22.3	21.9	21.6	20.7	20.9	21.0	21.9	22.0	22.3	22.6
41 Intermediate	9.4	11.4	11.8	12.1	12.2	12.6	12.6	12.6	12.5	12.6	12.2	12.2	12.2	12.6	12.8	13.2	13.3
42 Materials	9.3	11.2	11.5	11.8	11.9	12.3	12.3	12.2	12.1	12.2	11.8	11.8	11.7	12.2	12.4	12.7	12.8
43 Losses	0.0	0.1	0.1	0.1	0.1	0.1	0.1	0.1	0.1	0.1	0.1	0.1	0.1	0.1	0.1	0.2	0.2
44 Services	0.1	0.2	0.2	0.2	0.2	0.2	0.2	0.3	0.3	0.3	0.3	0.3	0.3	0.3	0.3	0.3	0.3

206

TABLE 1B: SOVIET GNP BY SECTOR OF ORIGIN IN 1973 PRICES
    (billion rubles)

	1965	1970	1971	1972	1973	1974	1975	1976	1977	1978	1979	1980	1981	1982	1983	1984	1985
45 CONSTR. MATER.	4.7	6.5	7.4	7.8	8.3	8.4	8.6	8.9	9.3	9.4	9.0	9.0	9.3	8.9	9.4	9.8	10.1
46 GVO	10.6	15.1	16.3	17.0	18.0	18.8	19.8	20.5	20.8	21.2	20.6	20.8	21.1	21.5	22.1	22.7	23.2
47 Intermediate	5.9	8.6	8.9	9.2	9.7	10.4	11.2	11.7	11.5	11.8	11.6	11.8	11.8	12.7	12.7	12.9	13.1
48 Materials	5.8	8.4	8.7	9.0	9.5	10.2	11.0	11.4	11.2	11.5	11.3	11.5	11.5	12.3	12.3	12.5	12.7
49 Services	0.1	0.2	0.2	0.2	0.2	0.2	0.2	0.3	0.3	0.3	0.3	0.3	0.3	0.4	0.4	0.4	0.4
50 N.E.C. INDUSTRY	3.7	4.9	5.6	6.3	7.0	7.7	9.0	9.5	10.5	10.9	10.9	11.8	11.9	11.7	11.6	11.4	11.2
51 GVO	9.0	12.2	13.4	14.3	16.4	17.4	19.3	20.6	22.0	23.0	23.2	24.2	24.4	24.9	25.3	25.4	25.6
52 Intermediate	5.3	7.3	7.8	8.0	9.4	9.7	10.3	11.1	11.5	12.1	12.3	12.4	12.5	13.2	13.7	14.0	14.4
53 Materials	5.3	7.2	7.7	7.9	9.3	9.6	10.2	11.0	11.4	11.9	12.2	12.2	12.3	13.0	13.5	13.8	14.2
54 Services	0.0	0.1	0.1	0.1	0.1	0.1	0.1	0.1	0.1	0.2	0.1	0.2	0.2	0.2	0.2	0.2	0.2
55 LIGHT INDUSTRY	28.8	33.3	34.3	34.5	30.3	29.7	29.6	30.2	30.5	30.3	31.2	32.9	34.0	29.7	28.7	29.1	29.7
56 GVO	67.7	88.1	91.8	91.7	91.9	93.2	95.2	97.9	99.2	100.6	99.2	100.6	102.1	99.6	99.5	99.5	100.9
57 Intermediate	38.9	54.8	57.5	57.2	61.6	63.5	65.6	67.7	68.7	70.3	68.0	67.7	68.1	69.9	70.8	70.4	71.2
58 Materials	38.6	54.5	57.2	56.8	61.2	63.1	65.1	67.2	68.2	69.7	67.4	67.1	67.4	69.2	70.1	69.6	70.4
59 Services	0.3	0.3	0.3	0.4	0.4	0.4	0.5	0.5	0.5	0.6	0.6	0.6	0.7	0.7	0.7	0.8	0.8
60 FOOD INDUSTRY	35.3	39.5	42.2	41.2	43.3	47.6	49.6	49.3	51.9	51.5	53.7	55.9	58.9	57.7	55.7	56.1	50.8
61 GVO	88.3	108.8	114.8	114.1	119.3	126.9	131.0	129.4	134.1	134.9	135.6	135.7	137.2	138.8	144.8	146.3	140.2
62 Intermediate	53.0	69.3	72.6	72.9	76.0	79.3	81.4	80.1	82.2	83.4	81.9	79.8	78.3	81.1	89.1	90.2	89.4
63 Materials	52.8	69.0	72.3	72.5	75.6	78.9	80.9	79.6	81.7	82.8	81.3	79.2	77.6	80.4	88.4	89.4	88.6
64 Services	0.2	0.3	0.3	0.4	0.4	0.4	0.5	0.5	0.5	0.6	0.6	0.6	0.7	0.7	0.7	0.8	0.8
65 AGRICULTURE	44.8	56.6	53.5	44.8	58.4	52.7	42.6	48.5	49.6	51.2	45.5	40.7	38.3	46.2	50.7	49.4	46.5
66 GVO	86.1	109.2	110.3	104.9	121.9	118.2	111.0	119.5	121.4	127.4	121.8	119.5	117.4	125.1	132.1	132.3	130.9
67 Net Subsidies	8.6	12.9	13.5	13.3	14.4	14.8	14.9	15.1	15.4	16.1	15.4	14.9	14.4	15.4	17.6	17.5	17.5
68 Intermediate	32.7	39.7	43.3	46.8	49.1	50.7	53.5	55.9	56.4	60.1	60.9	63.9	64.7	63.5	63.8	65.4	66.9
69 Materials	28.7	36.1	39.3	41.5	45.7	46.9	47.9	51.6	52.3	54.7	55.6	58.1	57.9	58.2	57.6	59.5	59.0
70 Losses	2.4	1.8	2.2	3.3	1.3	1.6	3.3	1.9	1.6	2.6	2.6	3.0	3.8	2.1	2.8	2.3	4.1
71 Services	1.6	1.8	1.8	2.0	2.1	2.2	2.3	2.4	2.5	2.8	2.7	2.8	3.0	3.2	3.4	3.6	3.8
72 TOTAL TRANSP.	19.8	28.3	30.2	31.4	32.7	34.4	36.0	37.0	37.3	37.4	37.1	37.5	37.6	38.8	39.7	40.7	40.0
73 GVO	26.3	36.3	38.6	40.2	42.4	44.6	46.6	48.2	49.3	50.8	51.3	52.2	52.8	54.4	56.0	57.3	57.9
74 Intermediate	6.5	8.0	8.4	8.8	9.8	10.2	10.6	11.2	12.0	13.3	14.3	14.8	15.2	15.6	16.3	16.6	17.8
75 Materials	6.2	7.4	7.8	8.1	9.0	9.5	9.9	10.4	11.1	12.4	13.4	13.8	14.2	14.5	15.1	15.4	16.5
76 Losses	0.0	0.1	0.1	0.1	0.1	0.1	0.1	0.2	0.2	0.2	0.2	0.2	0.2	0.2	0.2	0.2	0.2
77 Services	0.3	0.5	0.5	0.6	0.6	0.6	0.6	0.7	0.7	0.7	0.7	0.8	0.8	0.9	1.0	1.0	1.1
78 TOTAL COMMUNIC.	1.7	2.8	3.0	3.2	3.4	3.5	3.8	4.1	4.2	4.4	4.7	4.9	4.9	4.9	5.2	5.4	5.4
79 GVO	2.3	3.5	3.7	3.9	4.1	4.3	4.7	4.9	5.1	5.3	5.6	5.8	5.9	6.0	6.3	6.6	6.6
80 Intermediate	0.6	0.7	0.7	0.7	0.7	0.8	0.9	0.9	0.9	0.9	0.9	1.0	1.0	1.1	1.1	1.2	1.2
81 Materials	0.6	0.6	0.6	0.7	0.7	0.7	0.8	0.8	0.8	0.8	0.8	0.9	0.9	0.9	0.9	1.0	1.0
82 Services	0.0	0.0	0.0	0.0	0.0	0.1	0.1	0.1	0.1	0.1	0.1	0.1	0.1	0.2	0.2	0.2	0.2
83 CONSTRUCTION	30.2	35.9	36.5	37.0	36.5	39.0	39.9	39.3	39.7	37.6	35.6	35.4	34.8	36.0	36.6	33.0	32.4
84 GVO	57.7	74.8	79.6	80.1	81.0	84.8	87.5	86.7	86.5	84.7	81.8	81.8	81.4	83.6	84.9	83.5	83.7
85 Intermediate	27.5	38.9	43.1	43.1	44.5	45.8	47.6	47.4	46.8	47.1	46.2	46.4	46.6	47.6	48.3	50.5	51.3
86 Materials	25.4	35.5	39.5	39.8	40.8	42.0	43.6	43.1	42.3	42.4	41.6	41.6	41.6	40.8	41.5	44.4	44.7
87 Losses	1.1	2.0	2.2	1.8	2.2	2.2	2.3	2.5	2.6	2.7	2.6	2.7	2.8	4.5	4.4	3.5	3.9
88 Services	1.0	1.4	1.4	1.5	1.5	1.6	1.7	1.8	1.9	2.0	2.0	2.1	2.2	2.3	2.4	2.6	2.7

TABLE 1B: SOVIET GNP BY SECTOR OF ORIGIN IN 1973 PRICES
(billion rubles)

		1965	1970	1971	1972	1973	1974	1975	1976	1977	1978	1979	1980	1981	1982	1983	1984	1985
89	TRADE, etc.	13.7	20.0	20.9	21.1	22.8	23.7	24.4	24.6	25.6	25.9	25.7	25.7	25.6	25.9	27.4	27.9	27.8
90	Total GVO	17.7	23.5	24.6	25.1	26.6	27.8	28.5	28.9	29.7	30.3	30.0	30.1	30.1	30.5	32.0	32.5	32.9
91	GVO	17.7	23.4	24.5	24.9	26.4	27.6	28.2	28.6	29.3	29.9	29.6	29.6	29.6	29.9	31.4	31.9	32.2
92	Consignment	0.0	0.1	0.1	0.2	0.2	0.2	0.3	0.4	0.4	0.4	0.4	0.5	0.5	0.6	0.6	0.6	0.7
93	Intermediate	4.0	3.5	3.7	4.0	3.8	4.1	4.1	4.3	4.1	4.4	4.3	4.4	4.5	4.6	4.6	4.6	5.1
94	Materials	3.5	2.8	2.9	3.2	2.9	3.1	3.1	3.2	3.0	3.2	3.1	3.1	3.2	3.2	3.2	3.1	3.6
95	Losses	0.0	0.1	0.2	0.1	0.2	0.2	0.2	0.2	0.2	0.2	0.2	0.2	0.2	0.2	0.2	0.2	0.2
96	Services	0.5	0.6	0.6	0.7	0.7	0.8	0.8	0.9	0.9	1.0	1.0	1.1	1.1	1.2	1.2	1.3	1.3
97	OTHER PROD.	2.6	3.4	3.4	3.5	3.6	3.9	3.9	4.1	4.2	4.4	4.2	4.2	4.3	4.2	4.3	4.4	4.5
98	GVO	3.4	4.3	4.4	4.5	4.6	4.9	4.9	5.1	5.2	5.4	5.2	5.2	5.3	5.3	5.4	5.5	5.7
99	Intermediate	0.8	0.9	1.0	1.0	1.0	1.0	1.0	1.0	1.0	1.0	1.0	1.0	1.0	1.1	1.1	1.1	1.2
100	TOTAL FOREIGN	12.2	15.0	16.6	17.5	18.0	16.3	15.6	18.7	19.6	20.5	24.3	29.6	34.2	37.8	38.3	38.5	39.4
101	Trade Revenue	10.2	12.5	12.5	16.2	14.9	11.8	18.1	18.2	16.4	17.9	20.9	26.5	31.2	34.3	34.5	34.4	37.3
102	Monetary Revenue	2.0	2.5	4.1	1.3	3.1	4.5	-2.5	0.5	3.2	2.6	3.4	3.1	3.0	3.5	3.8	4.1	2.1
103	OTHER N-PROD	8.0	10.3	10.8	11.6	12.4	13.0	13.7	14.1	14.9	15.3	14.8	15.3	15.5	15.7	16.6	17.3	17.5
104	Hous.-Com., etc.	6.0	8.0	8.5	9.1	9.5	9.9	10.4	10.7	11.2	11.5	12.1	12.2	12.4	12.5	13.0	13.4	13.6
105	Banks & Insur.	2.1	2.2	2.3	2.5	2.9	3.1	3.3	3.4	3.7	3.8	2.7	3.1	3.0	3.2	3.6	3.8	3.9
106	BUDGET SECTORS	29.9	37.4	38.6	39.8	41.5	42.0	43.1	43.5	44.6	45.3	45.7	47.0	47.6	47.7	47.9	48.2	48.7
107	Education	11.0	13.0	13.2	13.4	14.2	14.4	14.5	14.8	14.9	15.0	15.2	15.6	15.5	15.6	15.6	15.7	16.2
108	Culture & Arts	1.5	2.1	2.2	2.2	2.2	2.2	2.3	2.5	2.7	2.7	2.8	2.9	2.8	2.8	2.8	2.9	3.1
109	Health, etc.	5.9	7.5	7.8	8.1	7.8	7.9	7.9	7.8	8.0	8.2	8.3	8.4	8.4	8.5	8.5	8.6	8.7
110	Science	5.1	6.4	6.9	7.4	8.0	8.1	8.6	8.6	9.0	9.2	9.5	10.2	10.6	10.4	10.5	10.1	9.9
111	Administration	6.5	8.4	8.5	8.8	9.3	9.4	9.8	9.9	10.0	10.1	9.8	9.9	10.3	10.4	10.5	10.8	10.8
112	State	2.5	3.2	3.3	3.4	3.6	3.7	3.8	3.9	4.0	4.1	4.1	4.2	4.5	4.4	4.4	4.6	4.5
113	Military	4.0	5.2	5.2	5.4	5.7	5.7	6.0	6.0	6.0	6.0	5.8	5.7	5.9	6.0	6.1	6.3	6.3

TABLE 1C: SOVIET GNP BY SECTOR OF ORIGIN IN CURRENT ESTABLISHED PRICES
(billion rubles)

	1965	1970	1971	1972	1973	1974	1975	1976	1977	1978	1979	1980	1981	1982	1983	1984	1985
1 TOTAL	251.5	376.0	398.9	413.8	449.0	473.2	491.2	523.5	552.7	582.3	605.5	637.6	670.9	717.7	752.5	785.8	805.0
2 Legal Economy	248.3	371.1	394.0	408.1	443.0	466.8	484.2	516.2	544.3	573.4	595.8	626.7	659.2	704.6	738.6	770.9	788.0
3 Second Economy	4.9	7.4	7.7	8.5	9.0	9.8	10.4	11.2	12.4	13.2	14.4	15.6	16.8	18.5	19.7	20.9	22.3
4 Fictitious Outpu	-1.6	-2.4	-2.7	-2.8	-3.1	-3.3	-3.4	-3.8	-3.9	-4.2	-4.6	-4.7	-5.1	-5.4	-5.7	-5.9	-5.3
5 PROFIT SECTORS	227.4	341.6	362.7	375.3	407.5	429.3	445.4	475.9	502.9	529.4	549.9	579.0	609.9	653.8	686.8	717.6	733.7
6 TOTAL INDUSTRY	116.1	181.3	193.9	203.0	214.0	230.0	243.0	254.8	267.1	283.4	294.4	307.7	322.9	347.2	363.3	374.7	380.9
7 "NMP" Industry	109.8	172.4	184.5	193.1	203.4	218.7	231.0	242.1	253.9	269.5	279.9	292.6	307.3	330.6	346.2	357.0	362.6
8 Weapons & Space	6.4	9.0	9.5	10.0	10.7	11.3	12.1	12.7	13.3	13.9	14.5	15.2	15.7	16.7	17.2	17.8	18.4
9 POWER	2.9	6.5	7.0	7.5	8.0	8.7	9.2	9.2	9.5	10.1	10.5	11.0	11.4	13.0	13.6	14.3	15.0
10 GVO	6.4	12.0	12.9	14.0	14.8	15.7	16.6	17.2	17.7	18.6	19.2	20.0	20.6	26.6	27.2	28.5	29.2
11 Intermediate	3.5	5.5	5.9	6.5	6.8	7.0	7.4	8.0	8.3	8.5	8.7	9.0	9.2	13.6	13.6	14.2	14.2
12 Materials	3.3	5.2	5.6	6.2	6.4	6.6	7.0	7.6	7.8	8.0	8.2	8.5	8.7	13.0	13.0	13.5	13.5
13 Losses	0.1	0.1	0.2	0.1	0.2	0.2	0.2	0.2	0.2	0.2	0.2	0.2	0.2	0.2	0.2	0.2	0.3
14 Services	0.1	0.2	0.2	0.2	0.2	0.2	0.2	0.3	0.3	0.3	0.3	0.3	0.3	0.4	0.4	0.4	0.4
15 FUELS	8.5	16.0	17.4	18.4	19.7	20.6	21.5	23.7	25.0	28.0	29.6	31.0	31.1	41.2	43.5	44.6	45.7
16 GVO	16.5	27.3	29.0	30.8	32.8	34.9	37.2	40.0	42.3	46.2	48.3	50.5	51.2	74.4	77.8	79.1	81.0
17 Intermediate	8.0	11.3	11.6	12.4	13.1	14.3	15.7	16.4	17.3	18.2	18.7	19.5	20.1	33.2	34.3	34.5	35.3
18 Materials	7.7	10.9	11.1	11.9	12.6	13.7	15.1	15.7	16.6	17.4	17.9	18.7	19.3	32.3	33.4	33.4	34.1
19 Losses	0.1	0.1	0.1	0.1	0.1	0.1	0.1	0.2	0.2	0.2	0.2	0.2	0.2	0.2	0.2	0.2	0.2
20 Services	0.2	0.3	0.3	0.4	0.4	0.4	0.5	0.5	0.5	0.6	0.6	0.6	0.7	0.7	0.7	0.9	1.0
21 METALS	7.1	12.8	13.3	13.7	14.5	15.6	16.0	17.3	17.4	17.7	17.4	17.4	17.8	23.1	24.7	25.5	26.0
22 GVO	21.6	36.4	38.1	40.5	42.8	45.3	47.9	50.8	53.3	55.0	55.0	54.9	55.0	67.7	70.0	72.0	74.3
23 Intermediate	14.5	23.6	24.8	26.8	28.3	29.7	31.9	33.6	35.9	37.3	37.6	37.5	37.2	44.6	45.3	46.5	48.3
24 Materials	14.3	23.3	24.5	26.5	28.0	29.4	31.5	33.2	35.5	36.8	37.2	37.0	36.7	44.1	44.7	45.9	47.7
25 Losses	0.1	0.1	0.1	0.1	0.1	0.1	0.1	0.2	0.2	0.2	0.2	0.2	0.2	0.2	0.2	0.2	0.2
26 Services	0.2	0.3	0.3	0.3	0.3	0.3	0.4	0.4	0.4	0.5	0.4	0.5	0.5	0.5	0.6	0.6	0.6
27 CHEMICALS	4.3	8.0	8.8	9.6	10.8	12.0	13.2	14.1	15.0	15.7	16.3	17.1	18.3	16.0	16.8	18.2	18.8
28 GVO	12.1	22.6	24.6	26.7	28.9	31.6	33.9	36.3	38.7	40.7	42.1	44.3	46.8	48.8	51.0	53.6	54.8
29 Intermediate	7.8	14.6	15.8	17.1	18.1	19.6	20.7	22.3	23.7	25.0	25.8	27.2	28.5	32.8	34.2	35.4	36.0
30 Materials	7.5	14.2	15.3	16.6	17.6	19.0	20.1	21.6	23.0	24.2	25.0	26.4	27.7	31.9	33.3	34.4	34.9
31 Losses	0.1	0.1	0.1	0.1	0.1	0.1	0.1	0.2	0.2	0.2	0.2	0.2	0.2	0.2	0.2	0.2	0.2
32 Services	0.2	0.3	0.3	0.4	0.4	0.4	0.5	0.5	0.5	0.6	0.6	0.6	0.7	0.7	0.7	0.8	0.9
33 TOTAL MBMW	26.7	42.8	45.2	50.2	51.6	57.2	61.3	64.7	69.4	73.9	79.9	83.2	88.8	88.1	93.9	99.9	106.5
34 GVO	53.9	96.1	102.4	112.5	117.3	128.8	140.5	145.4	154.8	166.7	176.4	186.7	196.6	204.4	215.7	227.4	239.3
35 Intermediate	27.1	53.2	57.1	62.1	65.6	71.7	79.2	80.7	85.4	92.7	96.5	103.5	107.8	116.2	121.7	127.3	132.7
36 Materials	25.5	50.6	54.2	59.1	62.5	68.1	75.4	76.7	81.0	87.9	91.7	98.5	102.7	110.8	116.1	121.3	126.4
37 Losses	0.2	0.4	0.5	0.4	0.5	0.5	0.6	0.6	0.6	0.6	0.6	0.7	0.7	0.7	0.7	0.8	0.8
38 Services	1.4	2.4	2.5	2.8	2.8	3.0	3.2	3.4	3.8	4.2	4.1	4.3	4.4	4.7	5.0	5.4	5.6
39 WOOD & PAPER	5.4	9.1	9.3	9.7	10.1	10.2	10.6	10.6	10.8	10.9	10.9	11.2	11.5	14.3	15.1	15.6	16.2
40 GVO	12.1	19.7	20.5	21.5	22.3	23.1	24.0	24.7	25.0	25.6	25.6	26.3	27.1	34.3	35.7	36.8	38.1
41 Intermediate	6.7	10.6	11.2	11.8	12.2	12.9	13.4	14.1	14.2	14.7	14.7	15.1	15.6	20.0	20.6	21.2	21.9
42 Materials	6.6	10.4	10.9	11.5	11.9	12.6	13.1	13.7	13.8	14.3	14.3	14.7	15.1	19.5	20.1	20.6	21.3
43 Losses	0.0	0.1	0.1	0.1	0.1	0.1	0.1	0.1	0.1	0.1	0.1	0.1	0.1	0.1	0.1	0.2	0.2
44 Services	0.1	0.2	0.2	0.2	0.2	0.2	0.2	0.3	0.3	0.3	0.3	0.3	0.3	0.4	0.4	0.4	0.4

TABLE 1C: SOVIET GNP BY SECTOR OF ORIGIN IN CURRENT ESTABLISHED PRICES
(billion rubles)

		1965	1970	1971	1972	1973	1974	1975	1976	1977	1978	1979	1980	1981	1982	1983	1984	1985
45	CONSTR. MATER.	3.8	6.6	7.1	7.6	8.3	8.9	9.2	9.6	9.8	9.7	9.8	9.7	9.9	12.1	12.8	13.5	14.0
46	GVO	8.7	15.0	15.8	16.7	18.0	19.5	20.9	21.8	22.1	22.5	22.7	23.2	23.5	29.1	30.4	31.4	32.5
47	Intermediate	4.9	8.4	8.7	9.1	9.7	10.6	11.7	12.3	12.3	12.8	12.9	13.5	13.6	17.1	17.6	17.9	18.5
48	Materials	4.8	8.2	8.5	8.9	9.5	10.4	11.5	12.0	12.0	12.5	12.6	13.2	13.3	16.7	17.2	17.5	18.1
49	Services	0.1	0.2	0.2	0.2	0.2	0.2	0.2	0.3	0.3	0.3	0.3	0.3	0.3	0.4	0.4	0.4	0.4
50	N.E.C. INDUSTRY	3.4	5.5	6.0	6.5	7.0	7.6	8.6	9.1	9.6	10.4	11.1	12.8	13.0	13.2	13.9	14.2	14.7
51	GVO	7.9	12.5	13.5	14.4	16.4	17.5	19.4	20.8	22.2	23.6	25.0	26.9	27.8	30.2	31.2	32.2	33.6
52	Intermediate	4.5	7.0	7.5	7.9	9.4	9.9	10.8	11.7	12.6	13.2	13.8	14.1	14.8	17.0	17.3	18.0	18.9
53	Materials	4.5	6.9	7.4	7.8	9.3	9.8	10.7	11.6	12.5	13.0	13.7	13.9	14.6	16.8	17.1	17.8	18.6
54	Services	0.0	0.1	0.1	0.1	0.1	0.1	0.1	0.1	0.1	0.2	0.2	0.2	0.2	0.2	0.2	0.2	0.3
55	LIGHT INDUSTRY	19.4	27.6	29.2	29.9	30.3	31.4	33.0	34.5	36.0	37.0	38.5	41.2	42.7	44.7	45.3	45.7	46.8
56	GVO	52.0	76.5	81.4	83.8	91.9	96.0	101.2	105.4	109.8	113.1	116.6	122.1	125.3	135.6	136.9	137.9	141.5
57	Intermediate	32.6	48.9	52.2	53.9	61.6	64.6	68.2	70.9	73.8	76.1	78.1	80.9	82.6	90.9	91.6	92.2	94.7
58	Materials	32.3	48.6	51.9	53.5	61.2	64.2	67.7	70.4	73.3	75.5	77.5	80.3	81.9	90.2	90.9	91.4	93.8
59	Services	0.3	0.3	0.3	0.4	0.4	0.4	0.5	0.5	0.5	0.6	0.6	0.6	0.7	0.7	0.7	0.8	0.9
60	FOOD INDUSTRY	28.5	37.8	41.4	40.2	43.5	47.0	48.8	49.7	51.8	56.1	56.2	58.4	63.1	65.3	67.1	66.0	59.3
61	GVO	77.0	103.4	110.8	113.1	119.3	128.4	134.6	134.5	140.7	146.8	149.5	152.4	158.6	167.1	174.8	178.6	172.8
62	Intermediate	48.5	65.6	69.4	72.9	75.8	81.4	85.8	84.8	88.9	90.7	93.3	94.0	95.5	101.8	107.7	112.6	113.5
63	Materials	48.3	65.3	69.1	72.5	75.4	81.0	85.3	84.3	88.4	90.1	92.7	93.4	94.8	101.1	107.0	111.8	112.6
64	Services	0.2	0.3	0.3	0.4	0.4	0.4	0.5	0.5	0.5	0.6	0.6	0.6	0.7	0.7	0.7	0.8	0.9
65	AGRICULTURE	41.7	52.4	51.1	45.8	58.4	54.9	47.6	53.8	58.0	58.2	58.1	55.7	58.2	66.1	70.9	74.6	72.1
66	GVO	71.0	103.8	108.1	108.8	121.9	122.1	122.3	132.4	141.7	147.0	151.9	152.6	160.0	170.3	207.9	217.0	219.5
67	Net Subsidies	2.2	13.1	14.6	15.5	14.4	15.2	17.8	18.5	21.0	22.4	23.2	21.6	23.3	25.4	50.6	50.9	52.5
68	Intermediate	27.1	38.3	42.4	47.5	49.1	51.9	57.0	60.1	62.7	66.4	70.7	75.3	78.5	78.8	86.4	91.5	94.9
69	Materials	23.5	34.9	38.6	42.3	45.7	48.0	51.1	55.5	58.2	60.6	64.9	69.0	71.1	73.0	79.6	84.9	86.0
70	Losses	2.2	1.7	2.1	3.2	1.3	1.7	3.5	2.1	1.8	2.8	2.8	3.2	4.1	2.3	3.1	2.6	4.6
71	Services	1.4	1.7	1.7	2.0	2.1	2.2	2.4	2.5	2.7	3.0	3.0	3.1	3.3	3.5	3.7	4.0	4.3
72	TOTAL TRANSP.	17.7	26.6	28.7	30.2	32.7	35.4	37.9	40.1	41.8	43.3	44.2	46.5	48.3	52.3	55.0	56.8	58.1
73	GVO	22.9	34.3	36.9	39.0	42.4	45.7	49.1	52.0	54.8	58.1	60.5	63.5	66.1	72.3	76.0	78.1	84.4
74	Intermediate	5.2	7.7	8.2	8.8	9.7	10.3	11.2	12.0	13.0	14.8	16.3	17.0	17.8	19.7	21.0	21.3	26.3
75	Materials	5.0	7.1	7.6	8.1	9.0	9.6	10.4	11.1	12.1	13.9	15.4	16.0	16.7	18.8	19.6	19.9	24.8
76	Losses	0.0	0.1	0.1	0.1	0.1	0.1	0.1	0.2	0.2	0.2	0.2	0.2	0.2	0.2	0.2	0.2	0.2
77	Services	0.2	0.5	0.5	0.6	0.6	0.6	0.7	0.7	0.7	0.7	0.7	0.8	0.9	1.0	1.2	1.2	1.3
78	TOTAL COMMUNIC.	1.6	2.7	2.9	3.1	3.4	3.6	4.0	4.4	4.7	5.1	5.6	6.0	6.3	6.6	7.2	7.5	8.0
79	GVO	2.0	3.3	3.5	3.8	4.1	4.5	4.9	5.3	5.7	6.1	6.6	7.1	7.4	8.0	8.6	9.0	9.6
80	Intermediate	0.4	0.6	0.6	0.7	0.7	0.8	0.9	0.9	1.0	1.0	1.0	1.1	1.1	1.4	1.4	1.5	1.6
81	Materials	0.4	0.6	0.6	0.7	0.7	0.7	0.8	0.8	0.9	0.9	0.9	1.0	1.0	1.2	1.2	1.3	1.3
82	Services	0.0	0.0	0.0	0.0	0.0	0.1	0.1	0.1	0.1	0.1	0.1	0.1	0.1	0.2	0.2	0.2	0.3
83	CONSTRUCTION	17.7	29.7	32.9	35.0	36.3	39.3	41.7	43.8	45.1	46.9	47.9	49.0	50.6	51.9	53.9	62.6	64.1
84	GVO	40.0	67.6	74.7	77.4	80.9	86.4	91.7	94.2	96.2	99.6	101.1	103.4	106.4	115.1	119.3	132.3	136.3
85	Intermediate	22.3	37.9	41.8	42.4	44.6	47.1	50.0	50.4	51.1	52.7	53.2	54.4	55.8	63.2	65.4	69.7	72.2
86	Materials	20.5	34.7	38.4	39.2	40.9	43.2	45.9	45.9	46.3	47.5	48.2	49.2	50.3	55.6	57.8	62.8	64.5
87	Losses	0.9	1.9	2.1	1.7	2.2	2.2	2.3	2.6	2.7	2.9	2.8	2.9	3.0	5.0	4.8	3.9	4.5
88	Services	0.9	1.3	1.3	1.5	1.5	1.7	1.8	1.9	2.1	2.3	2.2	2.3	2.5	2.6	2.8	3.0	3.2

TABLE 1C: SOVIET GNP BY SECTOR OF ORIGIN IN CURRENT ESTABLISHED PRICES
       (billion rubles)

	1965	1970	1971	1972	1973	1974	1975	1976	1977	1978	1979	1980	1981	1982	1983	1984	1985
89 TRADE, etc.	12.2	19.1	20.1	21.6	22.8	23.9	25.6	27.0	28.4	30.0	31.4	33.9	34.9	35.0	36.3	37.1	37.9
90 Total GVO	15.2	22.3	23.8	25.5	26.6	28.0	29.9	31.5	33.0	34.7	36.2	38.8	40.0	40.6	42.0	43.0	45.0
91 GVO	15.2	22.2	23.7	25.3	26.4	27.8	29.6	31.1	32.6	34.3	35.8	38.3	39.5	40.0	41.4	42.4	44.3
92 Consignment	0.0	0.1	0.1	0.2	0.2	0.2	0.3	0.4	0.4	0.4	0.4	0.5	0.5	0.6	0.6	0.6	0.7
93 Intermediate	3.0	3.3	3.7	3.9	3.8	4.1	4.3	4.4	4.5	4.7	4.7	4.8	5.0	5.6	5.7	5.8	7.0
94 Materials	2.6	2.6	2.8	3.1	2.9	3.2	3.3	3.4	3.3	3.4	3.5	3.5	3.7	4.2	4.2	4.2	5.3
95 Losses	0.0	0.1	0.2	0.1	0.2	0.2	0.2	0.2	0.2	0.2	0.2	0.2	0.2	0.2	0.2	0.2	0.2
96 Services	0.4	0.6	0.6	0.7	0.7	0.8	0.8	0.9	1.0	1.1	1.0	1.1	1.1	1.2	1.3	1.4	1.5
97 OTHER PROD.	2.3	3.1	3.2	3.4	3.6	4.0	4.1	4.4	4.6	4.9	5.0	5.1	5.4	5.7	6.0	6.2	6.5
98 GVO	3.1	3.9	4.2	4.4	4.6	5.0	5.2	5.5	5.7	6.0	6.1	6.3	6.6	7.1	7.4	7.6	8.1
99 Intermediate	0.8	0.9	1.0	1.0	1.0	1.1	1.1	1.1	1.1	1.1	1.2	1.2	1.2	1.4	1.4	1.4	1.6
100 TOTAL FOREIGN	7.8	11.9	14.4	16.0	17.9	18.7	20.4	25.7	28.9	32.0	37.2	46.5	53.5	56.7	59.7	61.4	66.3
101 Trade Revenue	6.8	11.2	12.6	13.5	15.2	15.8	18.8	23.6	26.7	28.8	35.2	44.8	51.5	55.6	58.7	60.3	64.8
102 Monetary Revenue	1.0	0.6	1.9	2.5	2.7	3.0	1.5	1.9	2.1	3.2	2.0	1.7	2.0	1.1	1.0	1.1	1.5
103 HOUS-COM., etc	5.2	7.8	8.4	9.1	9.5	10.2	10.7	11.4	11.9	12.4	13.2	13.8	14.3	15.0	15.8	16.7	17.5
104 Public Sectors	3.9	6.3	6.9	7.5	7.9	8.5	9.0	9.6	10.1	10.6	11.3	11.9	12.3	13.0	13.7	14.5	15.2
105 Net Imputed Rent	1.3	1.5	1.5	1.6	1.6	1.7	1.7	1.8	1.8	1.8	1.9	1.9	2.0	2.0	2.1	2.2	2.3
106 BANKS & INSURANC	1.8	2.1	2.2	2.4	2.9	3.2	3.5	3.6	4.1	4.4	3.4	3.9	3.9	4.3	4.9	5.3	5.6
107 BUDGET SECTORS	24.1	34.5	36.2	38.5	41.5	43.9	45.8	47.6	49.8	52.9	55.6	58.6	60.9	63.8	65.7	68.3	71.3
108 EDUCATION	8.5	11.4	12.1	12.9	14.2	15.0	15.4	16.2	16.9	17.9	18.4	19.5	19.9	20.8	21.5	22.5	23.8
109 CULTURE & ART	1.3	2.0	2.0	2.1	2.2	2.3	2.4	2.6	2.8	2.9	3.1	3.4	3.6	3.7	3.8	4.0	4.2
110 HEALTH, etc.	4.9	6.9	7.2	7.5	7.9	8.3	8.8	9.2	9.9	10.5	11.1	11.7	12.2	12.7	13.0	13.3	13.9
111 SCIENCE	3.9	6.2	6.7	7.3	8.0	8.6	9.0	9.0	9.4	10.1	10.8	11.5	12.1	12.8	13.1	13.6	14.0
112 ADMINISTRATION	5.6	7.9	8.2	8.7	9.3	9.7	10.2	10.5	10.9	11.5	12.1	12.6	13.2	13.9	14.3	14.8	15.4
113 State Admin.	2.2	3.0	3.2	3.4	3.6	3.8	4.0	4.1	4.3	4.6	5.0	5.3	5.6	5.9	5.9	6.2	6.4
114 Military Admin.	3.5	4.9	5.0	5.3	5.6	5.9	6.2	6.4	6.6	6.9	7.1	7.3	7.6	8.0	8.4	8.6	9.0

TABLE 1D: GROWTH AND STRUCTURE OF SOVIET GNP BY END USE IN 1973 PRICES
(percent; billion rubles)

	1965	1970	1971	1972	1973	1974	1975	1976	1977	1978	1979	1980	1981	1982	1983	1984	1985
Rates of Growth:																	
1 TOTAL GNP	78.1	100.0	103.8	104.7	110.7	113.2	114.3	118.8	122.1	124.2	124.8	128.1	130.6	133.5	136.0	136.8	136.2
2 annual growth			3.8	0.9	6.0	2.5	1.1	4.5	3.3	2.1	0.6	3.2	2.5	2.9	2.5	0.8	-0.6
3 CIVILIAN ECONOMY	77.6	100.0	103.3	104.2	110.7	113.1	113.5	117.8	121.7	123.4	123.8	126.8	129.0	130.9	133.8	134.2	132.9
4 annual growth			3.3	0.9	6.5	2.4	0.4	4.3	3.9	1.6	0.4	3.0	2.2	1.9	2.9	0.4	-1.3
5 per capita	81.6	100.0	102.4	102.3	107.6	109.1	108.5	111.4	114.2	114.9	114.4	116.1	117.0	117.8	119.3	118.6	116.3
6 CONSUMPTION	79.1	100.0	103.7	105.9	108.7	111.3	113.9	116.9	119.3	121.3	123.3	126.8	129.6	128.5	130.0	132.1	132.5
7 annual growth			3.6	2.3	2.7	2.6	2.6	3.0	2.4	2.0	2.0	3.5	2.8	-1.1	1.4	2.1	0.4
8 NET CONSUMPTION*	81.1	100.0	103.4	105.1	107.3	109.3	111.1	113.5	115.1	116.5	117.5	120.2	123.5	123.0	124.5	126.5	130.7
9 annual growth			3.4	1.7	2.2	2.1	1.7	2.5	1.6	1.3	1.1	2.7	3.3	-0.5	1.5	1.9	4.2
10 per capita	85.3	100.0	102.6	103.2	104.3	105.4	106.3	107.4	108.0	108.5	108.6	110.0	112.0	110.7	111.1	111.8	114.4
11 INVESTMENT	73.6	100.0	101.0	101.3	112.8	113.5	115.2	119.9	124.3	125.6	121.3	123.0	124.7	131.9	138.0	134.7	131.5
12 annual growth			1.0	0.4	11.5	0.7	1.6	4.7	4.4	1.3	-4.3	1.7	1.7	7.2	6.1	-3.3	-3.2
13 FIXED INVEST.	74.9	100.0	104.8	110.2	111.3	116.7	122.9	124.6	127.1	129.4	127.3	127.5	128.9	128.9	133.4	132.6	133.0
14 annual growth			4.8	5.4	1.0	5.4	6.2	1.7	2.5	2.4	-2.2	0.2	1.4	0.0	4.5	-0.7	0.3
15 BUDGET SERVICES	81.6	100.0	105.3	107.9	115.2	117.5	121.4	124.7	128.3	131.9	134.9	140.3	142.1	141.7	142.8	144.0	143.4
16 annual growth			5.3	2.6	7.3	2.3	3.8	3.3	3.7	3.6	3.0	5.4	1.8	-0.4	1.1	1.2	-0.5
17 TOTAL DEFENSE	81.9	100.0	107.2	108.3	110.8	113.9	119.7	125.4	124.5	130.0	131.7	136.6	141.3	150.9	151.0	154.8	158.5
18 annual growth			7.2	1.2	2.4	3.2	5.8	5.6	-0.9	5.5	1.7	4.9	4.7	9.6	0.1	3.8	3.7
19 POPULATION	231	243	245	248	250	252	254	257	259	261	263	266	268	270	273	275	278
GNP Structure:																	
20 TOTAL GNP	316.7	405.5	420.8	424.6	449.0	459.0	463.3	481.6	495.0	503.7	506.2	519.3	529.5	541.2	551.5	554.9	552.2
21 percent	100.0	100.0	100.0	100.0	100.0	100.0	100.0	100.0	100.0	100.0	100.0	100.0	100.0	100.0	100.0	100.0	100.0
22 CONSUMPTION	162.2	205.1	212.6	217.3	222.9	228.3	233.6	239.8	244.7	248.8	252.8	260.0	265.8	263.6	266.6	270.9	271.8
23 percent	51.2	50.6	50.5	51.2	49.6	49.7	50.4	49.8	49.4	49.4	49.9	50.1	50.2	48.7	48.3	48.8	49.2
24 INVESTMENT	82.6	112.3	113.4	113.8	126.7	127.5	129.3	134.6	139.6	141.0	136.2	138.1	140.0	148.1	155.0	151.3	147.7
25 percent	26.1	27.7	26.9	26.8	28.2	27.8	27.9	27.9	28.2	28.0	26.9	26.6	26.4	27.4	28.1	27.3	26.7
26 BUDGET SERVICES	26.8	32.8	34.6	35.4	37.8	38.6	39.9	40.9	42.1	43.3	44.3	46.1	46.7	46.5	46.9	47.3	47.1
27 percent	8.5	8.1	8.2	8.3	8.4	8.4	8.6	8.5	8.5	8.6	8.8	8.9	8.8	8.6	8.5	8.5	8.5
28 TOTAL DEFENSE	43.1	52.6	56.4	57.0	58.3	59.9	63.0	66.0	65.5	68.4	69.3	71.9	74.3	79.4	79.4	81.5	83.4
29 percent	13.6	13.0	13.4	13.4	13.0	13.1	13.6	13.7	13.2	13.6	13.7	13.8	14.0	14.7	14.4	14.7	15.1
30 NET EXPORTS	2.0	2.5	4.1	1.3	3.1	4.5	-2.5	0.5	3.2	2.6	3.4	3.1	3.0	3.5	3.8	4.1	2.1
31 percent	0.6	0.6	1.0	0.3	0.7	1.0	-0.5	0.1	0.6	0.5	0.7	0.6	0.6	0.6	0.7	0.7	0.4

*consumption less sales of alcohol

TABLE 1E: SOVIET GNP BY END USE IN CONSTANT 1973 PRICES
(billion rubles)

	1965	1970	1971	1972	1973	1974	1975	1976	1977	1978	1979	1980	1981	1982	1983	1984	1985
1 TOTAL GNP	316.7	405.5	420.8	424.6	449.0	459.0	463.3	481.6	495.0	503.7	506.2	519.3	529.5	541.2	551.5	554.9	552.2
2 percent	100.0	100.0	100.0	100.0	100.0	100.0	100.0	100.0	100.0	100.0	100.0	100.0	100.0	100.0	100.0	100.0	100.0
3 CONSUMPTION	162.2	205.1	212.6	217.3	222.9	228.3	233.6	239.8	244.7	248.8	252.8	260.0	265.8	263.6	266.6	270.9	271.8
4 percent of total	51.2	50.6	50.5	51.2	49.6	49.7	50.4	49.8	49.4	49.4	50.0	50.1	50.2	48.7	48.3	48.8	49.2
5 GOODS	139.1	174.6	180.8	183.7	187.7	191.8	195.4	200.3	204.2	207.6	211.3	217.1	222.1	218.9	220.5	224.2	224.0
6 Food	67.7	77.3	79.1	78.4	82.1	82.9	82.6	84.3	84.1	85.0	85.0	85.6	85.4	85.2	85.3	86.8	88.7
7 Alcohol	16.5	25.5	26.9	28.5	30.1	31.9	34.0	35.7	37.8	39.5	41.6	44.0	43.9	42.6	42.8	43.6	37.0
8 Soft Goods	47.9	60.8	63.4	64.3	62.7	63.6	64.4	65.4	67.1	67.0	68.1	70.4	74.6	73.2	73.5	74.6	78.4
9 Durables	7.1	11.0	11.5	12.5	12.7	13.5	14.3	14.9	15.3	16.2	16.5	17.1	18.3	18.0	18.9	19.2	20.0
10 SERVICES	23.1	30.5	31.8	33.6	35.2	36.5	38.3	39.5	40.4	41.2	41.6	42.9	43.7	44.7	46.1	46.7	47.8
11 Total Utilities	5.4	6.4	6.6	6.9	7.2	7.4	7.6	7.9	8.0	8.3	8.6	8.8	9.0	9.2	9.4	9.6	9.8
12 Utilities	4.0	4.9	5.1	5.3	5.6	5.7	5.9	6.1	6.2	6.5	6.7	6.9	7.0	7.2	7.3	7.5	7.7
13 Net Imputed Rent	1.4	1.5	1.5	1.6	1.6	1.7	1.7	1.8	1.8	1.8	1.9	1.9	2.0	2.0	2.1	2.1	2.1
14 Personal Care	2.3	3.0	3.2	3.4	3.6	3.7	3.9	4.0	4.3	4.4	4.5	4.7	4.9	5.0	5.1	5.2	5.3
15 Transp. & Comm.	4.4	6.6	6.9	7.3	7.8	8.3	9.1	9.4	9.7	10.1	10.5	11.0	11.4	11.7	12.1	12.4	12.8
16 Other Services	4.0	6.3	6.7	7.3	7.6	7.9	8.3	8.5	8.5	8.2	7.7	7.8	7.6	7.7	8.2	8.1	8.2
17 Second Economy	7.0	8.2	8.4	8.7	9.0	9.2	9.4	9.6	9.9	10.1	10.3	10.6	10.8	11.1	11.3	11.5	11.7
18 INVESTMENT	87.7	119.3	122.7	120.8	134.6	136.8	132.7	142.2	149.6	150.9	148.4	147.8	150.6	160.1	165.8	162.9	156.3
19 percent of total	27.7	29.4	29.2	28.5	30.0	29.8	28.6	29.5	30.2	30.0	29.3	28.5	28.4	29.6	30.1	29.4	28.3
20 FIXED CAPITAL	79.7	106.1	110.9	116.9	118.1	124.1	130.2	131.7	134.2	136.7	134.0	134.1	135.7	136.1	140.7	140.4	140.7
21 Civilian Sectors	76.7	101.7	106.0	112.0	113.2	119.2	124.6	125.7	128.0	130.4	127.4	127.4	128.9	129.8	134.1	134.2	134.5
22 Non-Residential	63.9	86.7	90.5	96.1	96.3	102.1	107.1	108.2	109.9	112.2	109.6	108.8	109.6	110.2	113.5	113.7	113.6
23 Residential	12.8	15.0	15.5	16.0	16.9	17.1	17.5	17.5	18.1	18.3	17.8	18.7	19.3	19.6	20.6	20.5	21.0
24 Defense Sectors	3.0	4.4	4.9	4.9	4.9	4.9	5.6	6.0	6.2	6.3	6.6	6.7	6.8	6.3	6.7	6.2	6.2
25 INVENTORIES	7.9	16.1	11.7	8.5	16.6	12.3	10.1	13.4	13.3	13.3	12.1	15.6	12.5	20.2	21.8	15.7	13.7
26 Inventories	7.9	15.8	11.5	8.0	16.4	11.8	9.9	13.0	12.6	12.6	9.9	13.2	12.7	18.0	21.4	15.3	12.9
27 Defense	0.0	0.3	0.2	0.5	0.2	0.5	0.2	0.4	0.7	0.7	2.2	2.4	-0.2	2.2	0.4	0.4	0.8
28 FICTITIOUS OUTPUT	-1.9	-2.6	-2.8	-2.9	-3.1	-3.2	-3.1	-3.4	-3.4	-3.5	-3.7	-3.7	-3.9	-3.9	-4.0	-4.0	-3.5
29 Fixed Capital	-1.2	-1.5	-1.7	-1.9	-1.8	-2.0	-2.1	-2.2	-2.2	-2.4	-2.6	-2.4	-2.6	-2.2	-1.6	-2.7	-2.3
30 Inventories	-0.8	-1.1	-1.1	-1.0	-1.3	-1.2	-1.0	-1.2	-1.2	-1.1	-1.1	-1.3	-1.3	-1.8	-2.3	-1.3	-1.2
31 STRATEGIC RESERVE	0.0	-2.8	-1.2	-3.0	0.0	-0.9	-1.9	0.0	2.3	1.7	2.6	-1.3	3.3	4.2	3.4	6.7	3.2
32 Agricultural	0.0	-2.5	-1.6	-3.5	0.4	-0.2	-1.9	-0.9	2.2	0.9	2.7	1.3	2.0	4.2	3.2	5.6	3.8
33 Other Reserves	0.0	-0.3	0.4	0.5	-0.4	-0.7	0.0	0.9	0.1	0.8	-0.1	-2.6	1.3	0.0	0.2	1.1	-0.6
34 NET EXPORTS	2.0	2.5	4.1	1.3	3.1	4.5	-2.5	0.5	3.2	2.6	3.4	3.1	3.0	3.5	3.8	4.1	2.1
35 GOVERNMENT	66.8	81.0	85.5	86.5	91.5	93.9	97.1	99.6	100.7	103.9	104.9	111.5	113.1	117.4	119.1	121.1	124.1
36 percent of total	21.1	20.0	20.3	20.4	20.4	20.4	20.9	20.7	20.3	20.6	20.7	21.5	21.4	21.7	21.6	21.8	22.5
37 BUDGET SERVICES	34.5	43.3	45.4	46.5	49.4	50.4	52.0	53.3	54.7	56.1	57.6	60.0	60.9	61.1	61.8	62.3	62.3
38 Civilian	26.8	32.8	34.6	35.4	37.8	38.6	39.9	40.9	42.1	43.3	44.3	46.1	46.7	46.5	46.9	47.3	47.1
39 Defense	7.7	10.4	10.8	11.0	11.5	11.8	12.2	12.3	12.6	12.8	13.3	14.0	14.3	14.5	15.0	15.0	15.3
40 DEFENSE	32.3	37.8	40.1	40.1	42.1	43.5	45.0	46.3	45.9	47.8	47.3	51.4	52.2	56.4	57.2	58.8	61.8
41 Weapons & Space	18.1	21.0	22.2	23.3	24.5	24.7	25.6	26.9	27.5	27.6	28.0	29.4	30.2	32.9	34.5	36.7	38.7
42 Military Admin.	9.7	12.0	12.1	12.7	12.9	13.7	14.4	14.6	14.7	15.8	15.8	18.3	18.4	18.1	18.2	18.3	18.8
43 Defense Constr.	4.5	4.7	5.8	4.1	4.8	5.1	5.0	4.9	3.7	4.4	3.5	3.7	3.6	5.4	4.5	3.8	4.3
44 TOTAL DEFENSE	43.1	52.6	56.4	57.0	58.3	59.9	63.0	66.0	65.5	68.4	69.3	71.9	74.3	79.4	79.4	81.5	83.4
45 percent of total	13.6	13.0	13.4	13.4	13.0	13.1	13.6	13.7	13.2	13.6	13.7	13.8	14.0	14.7	14.4	14.7	15.1

TABLE 1F: SOVIET GNP BY END USE IN CURRENT ESTABLISHED PRICES
(billion rubles)

	1965	1970	1971	1972	1973	1974	1975	1976	1977	1978	1979	1980	1981	1982	1983	1984	1985
1 TOTAL GNP	251.5	376.0	398.9	413.8	449.0	473.2	491.2	523.5	552.7	582.3	605.5	637.6	670.9	717.7	752.5	785.8	805.0
2 percent	100.0	100.0	100.0	100.0	100.0	100.0	100.0	100.0	100.0	100.0	100.0	100.0	100.0	100.0	100.0	100.0	100.0
3 CONSUMPTION	132.4	190.1	200.9	212.4	222.9	234.8	248.5	260.4	272.1	284.3	299.2	317.1	335.4	348.0	361.6	374.2	384.2
4 percent of total	52.6	50.6	50.4	51.3	49.6	49.6	50.6	49.7	49.2	48.8	49.4	49.7	50.0	48.5	48.0	47.6	47.7
5 PURCHASED GOODS	114.4	161.7	171.0	179.8	187.7	196.8	207.7	217.1	226.3	236.2	248.7	263.4	279.0	287.8	297.6	307.6	314.2
6 Food	58.0	73.8	76.9	79.3	82.1	84.4	86.8	89.7	91.5	94.1	97.5	100.7	104.8	107.4	112.3	116.5	120.9
7 Alcohol	16.0	24.9	26.9	28.5	30.1	31.9	34.0	35.7	37.8	39.9	42.1	44.4	46.5	50.3	51.5	52.8	47.7
8 Soft Goods	35.1	53.4	56.9	60.2	62.7	66.1	71.1	74.9	79.4	83.2	88.9	96.3	103.2	105.2	107.1	110.4	115.8
9 Durables	5.4	9.5	10.3	11.8	12.7	14.4	15.8	16.7	17.7	19.0	20.2	22.0	24.6	24.9	26.8	27.9	29.8
10 PAID SERVICES	17.9	28.4	29.9	32.6	35.2	38.0	40.8	43.3	45.8	48.1	50.5	53.7	56.3	60.3	63.9	66.6	70.0
11 Total Utilities	4.5	6.2	6.4	6.9	7.2	7.5	7.7	8.0	8.2	8.5	8.9	9.1	9.5	10.3	10.7	11.1	11.6
12 Utilities	3.2	4.7	4.9	5.3	5.6	5.8	6.0	6.2	6.4	6.7	7.0	7.2	7.5	8.3	8.6	8.9	9.3
13 Net Imputed Rent	1.3	1.5	1.5	1.6	1.6	1.7	1.7	1.8	1.8	1.8	1.9	1.9	2.0	2.0	2.1	2.2	2.3
14 Personal Care	1.9	2.8	3.0	3.3	3.6	3.8	4.0	4.3	4.7	5.0	5.3	5.7	6.0	6.2	6.5	6.7	7.2
15 Transp. & Comm.	3.6	6.4	6.6	6.9	7.8	8.5	9.4	10.1	10.6	11.0	11.6	12.3	12.8	13.3	14.1	14.4	14.6
16 Other Services	3.1	5.7	6.3	7.0	7.6	8.5	9.3	9.8	10.0	10.4	10.3	11.0	11.2	12.0	12.9	13.6	14.2
17 Second Economy	4.9	7.4	7.7	8.5	9.0	9.8	10.4	11.2	12.4	13.2	14.4	15.6	16.8	18.5	19.7	20.9	22.3
18 INVESTMENT	64.6	110.0	116.2	117.0	134.7	141.3	139.9	155.9	170.2	179.3	185.4	189.5	196.9	218.2	234.4	247.6	246.9
19 percent of total	25.7	29.3	29.1	28.3	30.0	29.9	28.5	29.8	30.8	30.8	30.6	29.7	29.3	30.4	31.1	31.5	30.7
20 FIXED CAPITAL	58.4	97.6	105.2	113.4	118.1	127.7	138.3	144.4	152.1	162.1	165.7	170.2	177.9	186.3	198.4	214.3	222.7
21 Civilian Sectors	56.0	93.5	100.6	108.6	113.2	122.6	132.3	137.8	145.1	154.5	157.4	161.6	168.9	177.5	188.8	204.7	212.8
22 Non-Residential	47.7	79.4	85.5	92.8	96.3	105.2	114.2	119.4	125.7	134.6	137.1	140.0	146.0	152.9	162.5	175.2	181.6
23 Residential	8.4	14.1	15.0	15.8	16.9	17.4	18.1	18.5	19.3	19.9	20.3	21.6	22.9	24.6	26.3	29.5	31.2
24 Defense Sectors	2.4	4.1	4.6	4.8	4.9	5.1	6.0	6.6	7.0	7.6	8.3	8.6	9.0	8.8	9.6	9.6	9.9
25 INVENTORIES	6.6	15.5	11.3	8.4	16.6	12.7	10.5	14.6	14.6	15.3	15.8	20.5	15.1	26.3	30.5	22.2	20.4
26 Civilian Sectors	6.6	15.2	11.1	7.9	16.4	12.2	10.3	14.2	13.9	14.6	13.3	17.8	15.3	23.4	30.0	21.7	19.3
27 Defense Sectors	0.0	0.3	0.2	0.5	0.2	0.5	0.2	0.4	0.7	0.7	2.5	2.7	-0.2	2.9	0.5	0.5	1.1
28 FICTITIOUS OUTPUT	-1.6	-2.4	-2.7	-2.8	-3.1	-3.3	-3.4	-3.8	-3.9	-4.2	-4.6	-4.7	-5.1	-5.4	-5.7	-5.9	-5.3
29 Fixed Capital	-0.5	-0.7	-0.9	-1.2	-1.3	-1.5	-1.9	-2.1	-2.2	-2.2	-2.4	-2.3	-2.6	-2.6	-2.5	-2.9	-2.7
30 Inventories	-1.1	-1.7	-1.8	-1.6	-1.8	-1.8	-1.5	-1.7	-1.7	-2.0	-2.2	-2.4	-2.5	-2.8	-3.2	-3.0	-2.6
31 STRATEGIC RESERVE	0.0	-2.7	-1.1	-3.1	0.0	-1.0	-2.1	0.0	2.7	2.0	3.3	-1.4	4.3	5.7	5.3	10.5	5.5
32 Agricultural	0.0	-2.4	-1.5	-3.6	0.4	-0.3	-2.1	-1.0	2.6	1.1	3.4	1.6	2.7	5.7	5.1	9.1	6.3
33 Other Reserves	0.0	-0.3	0.4	0.5	-0.4	-0.7	0.0	1.0	0.1	0.9	-0.1	-2.9	1.6	0.0	0.2	1.4	-0.8
34 NET EXPORTS	1.3	2.0	3.6	1.2	3.1	5.2	-3.3	0.7	4.7	4.1	5.2	4.9	4.7	5.3	5.9	6.5	3.5
35 GOVERNMENT	54.5	75.9	81.8	84.3	91.4	97.1	102.7	107.2	110.4	118.7	120.9	131.0	138.6	151.5	156.6	164.0	174.0
36 percent of total	21.7	20.2	20.5	20.4	20.4	20.5	20.9	20.5	20.0	20.4	20.0	20.5	20.7	21.1	20.8	20.9	21.6
37 BUDGET SERVICES	27.5	39.8	42.9	44.9	49.3	52.0	55.0	57.4	60.2	64.0	65.6	70.0	75.1	79.6	82.0	84.2	87.6
38 Civilian Services	21.2	30.1	32.7	34.2	37.8	39.8	42.1	44.0	46.3	49.3	49.9	53.0	57.4	60.7	62.1	63.6	66.1
39 Defense Services	6.2	9.7	10.2	10.7	11.5	12.3	12.9	13.3	13.9	14.7	15.7	17.0	17.7	18.9	19.9	20.6	21.5
40 DEFENSE	27.0	36.1	38.9	39.4	42.1	45.1	47.7	49.8	50.2	54.6	55.3	61.0	63.5	71.9	74.6	79.8	86.3
41 Weapons & Space	15.4	20.8	21.8	23.1	24.4	25.6	27.3	28.7	30.0	31.2	32.2	34.0	36.0	40.9	44.1	48.9	53.2
42 Military Admin.	7.8	11.2	11.4	12.3	12.9	14.3	15.2	15.7	16.1	18.2	18.7	22.2	22.8	23.5	24.2	25.1	26.5
43 Defense Constr.	3.7	4.2	5.6	4.0	4.8	5.2	5.2	5.4	4.1	5.2	4.4	4.7	4.7	7.5	6.2	5.8	6.6
44 TOTAL DEFENSE	35.7	49.9	54.3	55.9	58.3	62.2	66.8	71.1	71.9	78.5	81.7	86.4	91.6	102.4	104.7	111.9	118.1
45 percent of total	14.2	13.3	13.6	13.5	13.0	13.1	13.6	13.6	13.0	13.5	13.5	13.5	13.7	14.3	13.9	14.2	14.7

TABLE 1G: SOVIET GNP DEFLATORS BY SECTOR OF ORIGIN AND END USE

	1965	1970	1971	1972	1973	1974	1975	1976	1977	1978	1979	1980	1981	1982	1983	1984	1985
GNP by Sector of Origin:																	
1 TOTAL	0.80	0.93	0.95	0.97	1.00	1.03	1.06	1.08	1.11	1.16	1.20	1.23	1.27	1.33	1.37	1.42	1.46
2 Legal Economy	0.80	0.93	0.95	0.97	1.00	1.03	1.06	1.08	1.11	1.15	1.19	1.23	1.27	1.32	1.36	1.41	1.45
3 Second Economy	0.70	0.90	0.92	0.97	1.00	1.06	1.11	1.16	1.25	1.30	1.40	1.48	1.56	1.67	1.74	1.82	1.90
4 Fictitious Outpu	0.82	0.94	0.96	0.98	1.00	1.05	1.09	1.11	1.16	1.21	1.24	1.28	1.32	1.37	1.43	1.46	1.53
5 PROFIT SECTORS	0.80	0.93	0.95	0.98	1.00	1.03	1.06	1.08	1.11	1.15	1.20	1.23	1.27	1.33	1.37	1.42	1.46
6 TOTAL INDUSTRY	0.79	0.96	0.97	0.97	1.00	1.02	1.04	1.05	1.07	1.11	1.13	1.14	1.16	1.26	1.32	1.33	1.35
7 "NMP" Industry	0.78	0.95	0.96	0.97	1.00	1.02	1.04	1.05	1.07	1.11	1.13	1.14	1.16	1.27	1.32	1.34	1.36
8 Weapons & Space	1.00	1.06	1.03	1.02	1.00	1.02	1.05	1.04	1.06	1.08	1.10	1.11	1.12	1.15	1.17	1.18	1.21
9 POWER	0.71	1.03	1.04	0.99	1.00	1.01	1.02	0.93	0.94	0.94	0.93	0.96	0.94	1.04	1.07	1.09	1.10
10 FUELS	0.71	0.98	0.99	0.99	1.00	1.01	1.01	1.04	1.04	1.11	1.13	1.14	1.13	1.50	1.55	1.55	1.55
11 METALS	0.82	1.00	1.01	1.01	1.01	1.00	1.00	1.05	1.07	1.04	1.04	1.03	1.04	1.41	1.39	1.39	1.35
12 CHEMICALS	0.98	1.08	1.04	1.02	1.00	0.96	0.98	0.99	1.05	1.06	1.16	1.12	1.13	1.05	1.06	1.10	1.12
13 TOTAL MBMW	0.91	0.98	0.98	1.00	1.00	1.06	1.09	1.08	1.12	1.17	1.20	1.24	1.30	1.23	1.30	1.34	1.40
14 WOOD & PAPER	0.55	0.88	0.91	0.95	1.00	1.04	1.07	1.09	1.15	1.21	1.28	1.29	1.31	1.55	1.65	1.71	1.74
15 CONSTR. MATER.	0.81	1.02	0.96	0.97	1.00	1.06	1.07	1.08	1.05	1.03	1.09	1.08	1.06	1.36	1.36	1.38	1.39
16 N.E.C. INDUSTRY	0.92	1.12	1.07	1.03	1.01	0.99	0.96	0.96	0.91	0.95	1.02	1.08	1.09	1.13	1.20	1.25	1.31
17 LIGHT INDUSTRY	0.67	0.83	0.85	0.87	1.00	1.06	1.11	1.14	1.18	1.22	1.23	1.25	1.26	1.51	1.58	1.57	1.58
18 FOOD INDUSTRY	0.81	0.96	0.98	0.98	1.00	0.99	0.98	1.01	1.00	1.09	1.05	1.04	1.07	1.13	1.20	1.18	1.17
19 AGRICULTURE	0.93	0.93	0.95	1.02	1.00	1.04	1.12	1.11	1.17	1.14	1.28	1.37	1.52	1.43	1.40	1.51	1.55
20 TOTAL T & C	0.89	0.94	0.95	0.96	1.00	1.03	1.05	1.08	1.12	1.16	1.19	1.24	1.28	1.35	1.38	1.40	1.45
21 CONSTRUCTION	0.59	0.83	0.90	0.95	1.00	1.01	1.05	1.12	1.14	1.25	1.34	1.38	1.46	1.44	1.47	1.90	1.98
22 TRADE, etc.	0.89	0.95	0.96	1.02	1.00	1.01	1.05	1.10	1.11	1.16	1.22	1.32	1.37	1.35	1.32	1.33	1.36
23 OTHER PROD.	0.89	0.90	0.94	0.97	1.00	1.01	1.05	1.07	1.10	1.10	1.18	1.22	1.25	1.36	1.40	1.41	1.44
24 TOTAL FOREIGN	0.64	0.79	0.87	0.92	1.00	1.15	1.31	1.37	1.48	1.56	1.53	1.57	1.57	1.50	1.56	1.60	1.68
25 HOUS-COM., etc	0.87	0.97	0.98	1.00	1.00	1.03	1.03	1.06	1.06	1.08	1.09	1.13	1.15	1.20	1.22	1.24	1.29
26 BUDGET SECTORS	0.80	0.92	0.94	0.97	1.00	1.05	1.06	1.09	1.12	1.17	1.22	1.25	1.28	1.34	1.37	1.42	1.46
27 Education	0.77	0.88	0.91	0.96	1.00	1.05	1.06	1.10	1.13	1.19	1.21	1.24	1.28	1.33	1.38	1.43	1.47
28 Culture & Arts	0.82	0.95	0.95	0.97	1.00	1.00	1.01	1.04	1.06	1.08	1.14	1.17	1.28	1.32	1.35	1.37	1.37
29 Health, etc.	0.83	0.92	0.92	0.93	1.00	1.05	1.12	1.18	1.23	1.27	1.33	1.39	1.46	1.50	1.53	1.55	1.59
30 Science	0.77	0.97	0.97	0.99	1.00	1.06	1.05	1.05	1.05	1.10	1.14	1.13	1.15	1.23	1.25	1.35	1.41
31 State Admin.	0.87	0.94	0.97	0.99	1.00	1.03	1.05	1.06	1.09	1.13	1.24	1.27	1.29	1.33	1.36	1.38	1.42

TABLE 1G: SOVIET GNP DEFLATORS BY SECTOR OF ORIGIN AND END USE

	1965	1970	1971	1972	1973	1974	1975	1976	1977	1978	1979	1980	1981	1982	1983	1984	1985
GNP by End Use:																	
32 TOTAL GNP	0.79	0.93	0.95	0.97	1.00	1.03	1.06	1.09	1.12	1.16	1.20	1.23	1.27	1.33	1.36	1.42	1.46
33 CONSUMPTION	0.82	0.93	0.94	0.98	1.00	1.03	1.06	1.09	1.11	1.14	1.18	1.22	1.26	1.32	1.36	1.38	1.41
34 GOODS	0.82	0.93	0.95	0.98	1.00	1.03	1.06	1.08	1.11	1.14	1.18	1.21	1.26	1.31	1.35	1.37	1.40
35 Food	0.86	0.96	0.97	1.01	1.00	1.02	1.05	1.06	1.09	1.11	1.15	1.18	1.23	1.26	1.32	1.34	1.36
36 Alcohol	0.97	0.98	1.00	1.00	1.00	1.00	1.00	1.00	1.00	1.01	1.01	1.01	1.06	1.18	1.21	1.21	1.29
37 Soft Goods	0.73	0.88	0.90	0.94	1.00	1.04	1.10	1.15	1.18	1.24	1.31	1.37	1.38	1.44	1.46	1.48	1.48
38 Durables	0.76	0.86	0.90	0.94	1.00	1.07	1.11	1.12	1.15	1.18	1.22	1.28	1.35	1.38	1.42	1.45	1.49
39 SERVICES	0.78	0.93	0.94	0.97	1.00	1.04	1.07	1.10	1.13	1.17	1.22	1.25	1.29	1.35	1.39	1.43	1.46
40 Total Utilities	0.83	0.97	0.96	1.00	1.00	1.01	1.01	1.02	1.02	1.03	1.04	1.03	1.06	1.13	1.14	1.15	1.19
41 Personal Care	0.81	0.94	0.94	0.96	1.00	1.01	1.02	1.07	1.09	1.15	1.19	1.21	1.23	1.23	1.27	1.29	1.35
42 Transp. & Comm.	0.81	0.97	0.96	0.95	1.00	1.02	1.04	1.07	1.09	1.09	1.11	1.12	1.13	1.13	1.17	1.16	1.15
43 Other Services	0.78	0.89	0.94	0.96	1.00	1.08	1.12	1.15	1.18	1.26	1.34	1.41	1.47	1.55	1.58	1.68	1.73
44 SECOND ECONOMY	0.70	0.90	0.91	0.98	1.00	1.06	1.11	1.16	1.25	1.30	1.40	1.48	1.56	1.67	1.74	1.82	1.90
45 INVESTMENT	0.74	0.92	0.95	0.97	1.00	1.03	1.05	1.10	1.14	1.19	1.25	1.28	1.31	1.36	1.41	1.52	1.58
46 FIXED CAPITAL	0.73	0.92	0.95	0.97	1.00	1.03	1.06	1.10	1.13	1.19	1.24	1.27	1.31	1.37	1.41	1.53	1.58
47 Civilian	0.73	0.92	0.95	0.97	1.00	1.03	1.06	1.10	1.13	1.18	1.24	1.27	1.31	1.37	1.41	1.53	1.58
48 Non-Residential	0.75	0.92	0.95	0.97	1.00	1.03	1.07	1.10	1.14	1.20	1.25	1.29	1.33	1.39	1.43	1.54	1.60
49 Construction	0.70	0.89	0.92	0.96	1.00	1.02	1.06	1.09	1.13	1.18	1.23	1.26	1.34	1.38	1.41	1.63	1.67
50 Machinery	0.86	0.95	0.95	0.99	1.00	1.04	1.09	1.12	1.14	1.21	1.27	1.32	1.35	1.42	1.47	1.49	1.54
51 Residential	0.65	0.94	0.97	0.99	1.00	1.02	1.03	1.05	1.07	1.09	1.14	1.16	1.19	1.26	1.28	1.44	1.49
52 Defense	0.78	0.93	0.94	0.97	1.00	1.03	1.07	1.10	1.14	1.20	1.25	1.29	1.33	1.39	1.44	1.55	1.60
53 INVENTORIES	0.83	0.96	0.97	0.99	1.00	1.03	1.04	1.09	1.10	1.15	1.31	1.32	1.21	1.30	1.40	1.41	1.49
54 AGRI. RESERVES	0.89	0.94	0.96	1.01	1.00	1.03	1.08	1.09	1.13	1.12	1.21	1.25	1.32	1.32	1.37	1.44	1.46
55 GOVERNMENT	0.82	0.94	0.96	0.97	1.00	1.03	1.06	1.08	1.10	1.14	1.15	1.18	1.22	1.29	1.31	1.35	1.40
56 BUDGET SERVICES	0.80	0.92	0.94	0.97	1.00	1.03	1.06	1.08	1.10	1.14	1.14	1.17	1.23	1.30	1.33	1.35	1.41
57 Civilian	0.79	0.92	0.94	0.97	1.00	1.03	1.06	1.08	1.10	1.14	1.13	1.15	1.23	1.30	1.32	1.34	1.40
58 Defense	0.81	0.93	0.94	0.97	1.00	1.04	1.06	1.08	1.10	1.15	1.18	1.21	1.24	1.30	1.33	1.37	1.41
59 DEFENSE	0.84	0.96	0.97	0.98	1.00	1.04	1.06	1.07	1.09	1.14	1.17	1.19	1.22	1.27	1.30	1.36	1.40
60 FOREIGN TRADE	0.69	0.81	0.87	0.93	1.00	1.16	1.31	1.39	1.57	1.72	1.66	1.70	1.69	1.62	1.71	1.76	1.84
61 Imports	0.76	0.91	0.93	0.96	1.00	1.06	1.15	1.19	1.24	1.30	1.32	1.38	1.40	1.46	1.51	1.54	1.60
62 Exports	0.83	0.97	0.97	0.99	1.00	1.01	1.02	1.03	1.04	1.05	1.08	1.10	1.12	1.28	1.31	1.34	1.35
63 TOTAL DEFENSE	0.83	0.95	0.96	0.98	1.00	1.04	1.06	1.08	1.10	1.15	1.18	1.20	1.23	1.29	1.32	1.37	1.42

TABLE 1H: GROSS OUTPUT BY SECTOR IN 1973 PRICES AND ITS ANNUAL GROWTH

(billion rubles; percent)

	1965	1970	1971	1972	1973	1974	1975	1976	1977	1978	1979	1980	1981	1982	1983	1984	1985
1 TOTAL GVO	599.3	787.7	827.8	847.1	893.1	924.2	949.9	978.3	1002	1027	1028	1050	1066	1091	1119	1135	1145
2 annual growth	76.1	100.0	105.1	107.5	113.4	117.3	120.6	124.2	127.2	130.4	130.5	133.3	135.3	138.5	142.0	144.1	145.3
3 TOTAL GSP	514.8	682.6	716.4	730.3	770.9	798.6	820.8	845.4	865.6	886.1	885.9	901.3	914.4	936.4	960.4	972.6	978.5
4 annual growth	75.4	100.0	105.0	107.0	112.9	117.0	120.3	123.9	126.8	129.8	129.8	132.1	134.0	137.2	140.7	142.5	143.4
5 TOTAL NMP	242.3	311.4	320.6	321.0	337.8	345.1	343.9	357.1	366.0	372.9	372.2	380.0	388.9	400.6	407.6	411.0	408.8
6 annual growth	77.8	100.0	103.0	103.1	108.5	110.8	110.4	114.7	117.6	119.8	119.5	122.1	124.9	128.7	130.9	132.0	131.3
7 TOTAL INDUSTRY	327.6	442.3	467.8	484.4	504.5	531.0	555.1	567.6	583.8	598.7	602.5	613.5	623.4	629.4	645.0	656.9	660.5
8 annual growth	74.1	100.0	105.8	109.5	114.1	120.1	125.5	128.3	132.0	135.4	136.2	138.7	140.9	142.3	145.8	148.5	149.3
9 POWER	8.4	11.9	12.9	14.0	14.8	15.7	16.5	17.7	18.2	19.1	19.8	20.4	21.1	22.3	22.8	23.7	24.4
10 annual growth	70.5	100.0	108.8	117.7	124.3	131.7	139.0	149.0	153.3	160.8	166.7	171.8	177.7	187.3	191.3	198.9	205.0
11 FUELS	22.3	28.1	29.7	31.2	32.8	34.6	36.5	38.3	40.1	41.6	42.6	44.0	44.7	46.3	47.3	48.2	49.4
12 annual growth	79.4	100.0	105.6	110.9	116.7	123.1	129.8	136.4	142.9	148.2	151.6	156.6	159.1	164.9	168.3	171.7	175.9
13 METALS	27.7	37.1	38.6	40.7	42.8	45.2	47.5	48.6	49.9	51.4	50.8	50.2	50.2	49.3	50.7	51.8	53.6
14 annual growth	74.6	100.0	104.2	109.7	115.4	121.7	128.1	130.9	134.6	138.4	136.8	135.3	135.4	132.8	136.8	139.6	144.4
15 CHEMICALS	13.7	22.2	24.5	26.6	28.9	32.0	34.0	35.7	36.7	37.9	37.5	39.4	40.9	41.6	43.2	44.4	44.6
16 annual growth	61.5	100.0	110.3	119.9	130.3	144.0	153.1	160.8	165.4	170.7	169.1	177.3	184.3	187.6	194.5	199.8	201.0
17 TOTAL MBMW	60.8	97.1	103.7	112.5	117.3	124.9	132.7	136.6	140.9	147.4	152.5	157.4	160.6	163.2	167.4	172.5	176.0
18 annual growth	66.4	106.1	113.3	123.0	128.2	136.5	145.0	149.3	154.0	161.1	166.7	172.0	175.5	178.4	183.0	188.5	192.3
19 WOOD & PAPER	19.2	21.7	22.0	22.3	22.3	22.4	22.5	22.3	21.9	21.6	20.7	20.9	21.0	21.9	22.0	22.3	22.6
20 annual growth	88.3	100.0	101.5	102.8	102.9	103.3	103.7	102.7	100.7	99.5	95.5	96.1	96.6	100.8	101.3	103.0	104.4
21 CONSTR. MATER.	10.6	15.1	16.3	17.0	18.0	18.8	19.8	20.5	20.8	21.2	20.6	20.8	21.1	21.5	22.1	22.7	23.2
22 annual growth	70.2	100.0	107.8	112.6	118.9	124.2	131.2	135.8	137.7	140.2	136.1	138.0	139.6	142.6	146.1	150.1	153.9
23 N.E.C. INDUSTRY	9.0	12.2	13.4	14.3	16.4	17.4	19.3	20.6	22.0	23.0	23.2	24.2	24.4	24.9	25.3	25.4	25.6
24 annual growth	73.5	100.0	109.3	117.0	134.2	142.5	157.6	168.2	179.5	187.8	189.4	197.4	199.3	203.2	206.6	207.8	208.8
25 LIGHT INDUSTRY	67.7	88.1	91.8	91.7	91.9	93.2	95.2	97.9	99.2	100.6	99.2	100.6	102.1	99.6	99.5	99.5	100.9
26 annual growth	76.9	100.0	104.2	104.1	104.3	105.8	108.1	111.1	112.6	114.2	112.6	114.1	115.9	113.0	112.9	113.0	114.6
27 FOOD INDUSTRY	88.3	108.8	114.8	114.1	119.3	126.9	131.0	129.4	134.1	134.9	135.6	135.7	137.2	138.8	144.8	146.3	140.2
28 annual growth	81.2	100.0	105.5	104.8	109.6	116.6	120.4	118.9	123.2	124.0	124.6	124.7	126.1	127.6	133.1	134.5	128.8
29 AGRICULTURE	86.1	109.2	110.3	104.9	121.9	118.2	111.0	119.5	121.4	127.4	121.8	119.5	117.4	125.1	132.1	132.3	130.9
30 annual growth	78.8	100.0	101.0	96.0	111.6	108.3	101.6	109.4	111.1	116.6	111.5	109.5	107.6	114.6	121.0	121.2	119.9
31 AGRI. SUBSIDIES	-8.6	-12.9	-13.5	-13.3	-14.4	-14.8	-14.9	-15.1	-15.4	-16.1	-15.4	-14.9	-14.4	-15.4	-17.6	-17.5	-17.5
32 annual growth	70.7	100.0	110.7	109.0	118.0	121.3	122.1	123.8	126.2	132.0	126.2	122.1	118.0	126.2	144.3	143.4	143.4
33 PROD T & C	20.4	27.4	28.9	30.2	31.8	33.2	34.7	35.7	36.7	37.7	37.5	38.2	38.5	39.9	41.1	41.9	42.9
34 annual growth	74.4	100.0	105.6	110.1	115.9	121.3	126.7	130.4	133.9	137.6	137.0	139.4	140.7	145.7	150.1	153.0	156.7
35 CONSTRUCTION	57.7	74.8	79.6	80.1	81.0	84.8	87.5	86.7	86.5	84.7	81.8	81.8	81.4	83.6	84.9	83.5	83.7
36 annual growth	77.1	100.0	106.4	107.1	108.3	113.4	117.0	115.9	115.6	113.2	109.4	109.4	108.8	111.8	113.5	111.6	111.9

TABLE 1H: GROSS OUTPUT BY SECTOR IN 1973 PRICES AND ITS ANNUAL GROWTH
      (billion rubles; percent)

Page 2

	1965	1970	1971	1972	1973	1974	1975	1976	1977	1978	1979	1980	1981	1982	1983	1984	1985
37 TRADE, etc.	17.7	23.4	24.5	24.9	26.4	27.6	28.2	28.6	29.3	29.9	29.6	29.6	29.9	31.4	31.9	32.2	33.8
38 annual growth	75.6	100.0	104.7	106.4	112.8	117.9	120.5	122.2	125.2	127.8	126.5	126.5	127.8	134.2	136.3	137.6	144.4
39 OTHER PROD.	3.4	4.3	4.4	4.5	4.6	4.9	4.9	5.1	5.2	5.4	5.2	5.2	5.3	5.3	5.4	5.5	5.7
40 annual growth	79.1	100.0	102.3	104.7	107.0	114.0	114.0	118.6	120.9	125.6	120.9	120.9	123.3	123.3	125.6	127.9	132.6
41 FOREIGN TRADE	10.6	14.1	14.5	14.7	15.3	13.7	14.4	17.3	18.1	18.5	23.0	28.5	32.8	37.1	37.7	37.8	38.5
42 annual growth	74.9	100.0	102.4	103.7	107.9	96.7	102.1	122.5	128.3	130.5	162.6	201.4	232.3	262.5	266.5	267.3	272.3
43 TOTAL N-PROD	84.4	105.0	111.3	116.8	122.1	125.5	129.0	132.8	136.3	140.7	141.7	148.6	151.3	154.2	158.2	162.2	166.3
44 annual growth	80.4	100.0	106.0	111.3	116.3	119.5	122.8	126.5	129.9	134.0	134.9	141.5	144.1	146.9	150.7	154.5	158.4
45 N-PROD T & C	8.3	12.4	13.3	14.0	14.8	15.7	16.6	17.4	17.7	18.5	19.6	20.1	20.7	21.1	21.7	22.6	22.9
46 annual growth	67.3	100.0	107.5	113.0	119.6	127.1	134.2	141.1	143.5	149.5	158.3	162.8	167.2	170.6	175.9	182.5	185.0
47 HOUS-COM., etc	7.7	10.3	10.9	11.5	12.3	12.8	13.5	14.0	14.7	15.3	16.1	16.3	16.6	16.8	17.4	17.9	18.1
48 annual growth	74.0	100.0	104.8	110.6	118.3	123.1	129.8	134.6	141.3	147.1	154.8	156.7	159.6	161.5	167.3	172.1	174.0
49 BANKS & INSUR.	2.1	2.2	2.3	2.4	2.9	3.1	3.3	3.4	3.8	4.0	2.8	3.1	3.1	3.3	3.7	3.9	4.1
50 annual growth	95.3	100.0	104.5	108.5	129.7	140.9	150.0	154.5	172.6	183.0	127.3	140.9	140.9	150.0	166.8	177.3	186.4
51 EDUCATION	14.6	17.5	17.9	18.2	18.9	19.1	19.5	19.8	20.2	20.3	20.5	21.0	20.9	21.1	21.1	21.1	21.5
52 annual growth	83.3	100.0	102.1	103.9	108.1	109.1	111.4	113.4	115.4	115.7	117.4	119.9	119.7	120.7	120.7	120.6	123.0
53 CULTURE & ART	1.8	2.4	2.6	2.6	2.7	2.9	3.0	3.3	3.4	3.5	3.5	3.6	3.5	3.6	3.6	3.7	3.8
54 annual growth	75.4	100.0	106.2	107.8	111.6	119.6	124.8	134.4	140.5	145.1	145.2	149.5	142.6	146.8	147.8	151.3	158.0
55 HEALTH, etc.	8.7	11.3	11.7	12.2	12.0	12.1	12.2	12.2	12.4	12.6	12.7	12.8	12.8	12.9	13.0	13.1	13.2
56 annual growth	76.8	100.0	103.9	107.5	106.0	106.8	107.6	107.7	109.6	111.9	112.3	113.5	113.0	114.4	114.8	116.3	117.2
57 SCIENCE	8.7	11.2	12.0	12.9	14.1	14.2	15.0	14.9	15.5	15.7	16.0	17.4	18.2	18.1	18.7	18.5	18.4
58 annual growth	77.9	100.0	107.8	115.9	126.3	127.6	134.0	133.9	138.4	140.6	143.5	156.2	163.4	162.2	167.8	165.4	165.2
59 ADMINISTRATION	3.1	3.9	4.2	4.3	4.5	4.6	4.8	4.9	5.0	5.3	5.3	5.4	5.6	5.6	5.7	5.8	5.8
60 annual growth	79.1	100.0	107.2	110.9	115.0	118.5	122.0	125.0	128.7	134.9	134.4	138.1	142.9	143.9	144.3	148.2	147.0
61 WEAPONS & SPACE	18.1	21.0	22.2	23.3	24.4	24.7	25.6	26.9	27.5	27.6	28.0	29.4	30.2	32.9	34.5	36.7	38.7
62 annual growth	86.0	100.0	105.3	110.9	116.1	117.2	121.8	127.8	130.8	131.3	133.0	139.6	143.6	156.5	164.0	174.3	183.8
63 MILITARY ADMIN.	9.7	12.0	12.1	12.7	12.9	13.7	14.4	14.6	14.7	15.8	15.8	18.3	18.4	18.1	18.2	18.3	18.8
64 annual growth	81.0	100.0	100.6	105.2	106.9	114.1	119.8	121.1	122.0	131.3	131.6	152.4	153.0	150.1	151.6	152.1	156.3
65 FOREIGN REVENUE	1.6	0.8	2.2	2.7	2.7	2.6	1.1	1.4	1.4	2.1	1.3	1.1	1.3	0.7	0.6	0.7	1.0

Supporting Data on NMP Accounts:

	1965	1970	1971	1972	1973	1974	1975	1976	1977	1978	1979	1980	1981	1982	1983	1984	1985
66 Total Materials	246.4	336.7	358.3	370.7	391.7	409.5	427.3	435.8	444.5	457.2	456.3	461.4	463.2	472.5	487.0	496.8	502.9
67 Total Depr.	26.1	34.5	37.5	38.5	41.4	44.0	49.7	52.5	55.0	56.1	57.4	59.9	62.3	63.3	65.8	64.8	66.8

218

Part B

GNP Growth in Factor Cost Prices

TABLE 2A: GROWTH OF SOVIET GNP BY SECTOR OF ORIGIN IN CONSTANT FACTOR COST PRICES
(billion rubles)

	1965	1970	1971	1972	1973	1974	1975	1976	1977	1978	1979	1980	1981	1982	1983	1984	1985
1973 Prices:																	
1 TOTAL GNP*	77.5	100.0	103.6	103.0	110.4	113.4	112.1	117.3	120.7	123.1	120.8	121.7	122.0	127.0	130.7	130.0	128.5
2 annual growth			3.6	-0.6	7.4	3.0	-1.2	5.2	3.3	2.4	-2.3	0.9	0.3	5.0	3.8	-0.7	-1.5
3 per capita	81.6	100.0	102.7	101.1	107.3	109.3	107.3	110.9	113.2	114.6	111.6	111.4	110.6	114.3	116.6	114.9	112.5
4 TOTAL GNP*	310.0	400.0	414.3	412.0	441.4	453.4	448.5	469.2	482.6	492.3	483.2	486.8	488.0	507.9	522.9	520.0	513.8
5 PROFIT SECTORS	259.4	336.2	347.5	346.5	370.9	381.1	383.0	400.1	409.2	418.6	409.4	411.3	412.2	431.0	445.3	441.7	437.1
6 TOTAL INDUSTRY	101.5	127.5	134.2	137.5	146.3	151.3	158.7	168.2	172.6	175.1	176.1	181.3	186.0	189.0	190.5	196.5	196.1
7 POWER	3.7	4.2	4.3	5.0	5.3	5.6	6.1	7.1	7.4	7.9	8.2	8.3	8.8	8.7	8.8	8.9	9.1
8 FUELS	10.6	10.0	10.0	10.0	10.6	10.9	11.8	12.3	13.1	12.9	13.2	14.0	14.8	12.6	12.9	13.8	14.5
9 METALS	9.4	10.0	10.3	10.8	11.6	12.4	13.2	13.3	13.8	14.5	15.0	15.8	16.2	12.7	13.4	14.1	14.6
10 CHEMICALS	3.8	6.0	6.7	7.2	7.9	8.8	9.7	10.3	10.4	11.0	10.6	11.7	12.4	14.6	15.3	15.6	15.4
11 TOTAL MBMW	26.7	40.2	43.7	44.6	48.8	49.4	51.6	56.1	56.9	58.0	57.7	58.6	59.1	68.8	68.4	69.5	67.3
12 WEAPONS & SPACE	9.5	13.1	14.1	14.7	16.2	16.7	16.9	17.9	18.5	19.0	19.2	19.8	20.3	21.0	21.6	22.2	21.9
13 WOOD & PAPER	12.0	11.1	11.0	10.9	11.1	11.0	11.1	11.3	11.0	10.6	10.2	10.3	10.6	9.6	9.4	9.4	9.4
14 CONSTR. MATER.	5.6	7.0	7.8	8.0	8.4	8.5	9.0	9.3	9.8	10.4	10.0	10.5	11.0	9.5	9.9	10.2	10.2
15 N.E.C. INDUSTRY	1.4	2.0	2.3	2.3	2.9	3.2	3.6	3.7	4.2	4.2	4.2	4.1	4.2	4.5	4.4	4.4	4.4
16 LIGHT INDUSTRY	10.4	13.4	13.3	13.3	12.2	12.2	12.2	12.7	12.6	12.7	12.9	13.2	13.5	11.8	11.5	11.9	12.0
17 FOOD INDUSTRY	8.4	10.4	10.5	10.8	11.3	12.6	13.6	14.1	14.9	13.9	14.8	15.1	15.1	15.2	15.0	16.6	17.2
18 AGRICULTURE	63.6	88.8	90.3	84.1	92.9	92.0	83.5	88.3	88.6	95.0	87.3	82.6	77.9	87.4	96.9	93.8	91.5
19 T & C	20.5	29.5	31.4	33.0	35.2	37.0	38.1	40.1	40.7	41.9	41.8	42.3	43.2	44.3	45.1	46.6	45.7
20 CONSTRUCTION	42.5	48.6	48.1	48.1	48.6	50.7	50.7	49.5	51.0	48.2	46.5	46.6	46.2	50.0	50.9	41.1	40.7
21 TRADE, etc.	12.5	18.9	19.6	19.3	21.5	22.3	22.8	23.0	24.1	24.9	24.5	24.3	24.1	25.5	26.6	27.6	27.2
22 OTHER PROD.	4.0	4.9	4.8	4.9	5.1	5.3	5.1	5.3	5.4	5.6	5.3	5.3	5.5	5.3	5.3	5.5	5.4
23 HOUSING	11.1	12.2	12.5	12.8	13.9	14.6	15.7	16.9	17.6	17.9	18.0	18.5	18.8	19.0	19.3	19.5	19.6
24 COMMUNAL, etc.	3.4	5.0	5.5	5.8	6.2	6.7	7.2	7.5	7.9	8.4	8.7	8.9	9.1	9.2	9.3	9.5	9.5
25 BANKS & INSUR.	0.5	1.0	1.1	1.0	1.2	1.3	1.2	1.4	1.4	1.5	1.4	1.4	1.5	1.5	1.5	1.6	1.5
26 BUDGET SECTORS	43.6	55.6	57.2	58.2	61.6	61.7	61.8	62.4	63.6	64.4	63.9	65.5	65.9	66.2	66.6	66.8	66.2
27 EDUCATION	15.8	18.8	19.0	19.1	20.5	20.4	20.3	20.6	20.6	20.7	20.7	21.1	21.0	21.2	21.3	21.4	21.7
28 CULTURE & ART	2.1	3.2	3.3	3.3	3.3	3.4	3.5	3.8	3.9	3.9	3.9	4.0	3.8	3.8	3.8	3.9	4.0
29 HEALTH, etc.	8.6	11.2	11.7	12.0	12.1	11.7	11.3	11.1	11.4	11.8	11.6	11.7	11.5	11.7	11.7	12.0	11.9
30 SCIENCE	7.8	9.8	10.4	10.8	11.8	12.0	12.4	12.5	13.1	13.3	13.6	14.5	14.9	14.7	14.9	14.3	13.8
31 STATE ADMIN.	3.5	4.9	5.0	5.1	5.5	5.6	5.6	5.6	5.9	6.0	5.7	6.0	6.2	6.0	6.0	6.2	6.0
32 MILITARY ADMIN.	5.9	7.8	7.9	7.9	8.4	8.5	8.6	8.8	8.8	8.8	8.3	8.2	8.5	8.6	8.8	9.0	8.8
33 NET EXPORTS	2.0	2.5	4.1	1.3	3.1	4.5	-2.5	0.5	3.2	2.6	3.4	3.1	3.0	3.5	3.8	4.1	2.1
34 NET HIDDEN GNP+	5.0	5.7	5.5	5.9	5.9	6.1	6.2	6.2	6.5	6.7	6.5	6.9	6.9	7.1	7.3	7.4	8.3

*excluding foreign trade tarriffs
+illicit incomes less fictitious output

220

TABLE 2A: GROWTH OF SOVIET GNP BY SECTOR OF ORIGIN IN CONSTANT FACTOR COST PRICES
(billion rubles)

	1965	1970	1971	1972	1973	1974	1975	1976	1977	1978	1979	1980	1981	1982	1983	1984	1985
1982 Prices:																	
35 TOTAL GNP*	77.9	100.0	103.3	102.5	109.7	112.5	110.8	115.8	119.1	121.5	119.0	119.8	119.9	124.5	128.4	127.3	125.9
36 annual growth			3.4	-0.8	7.2	2.8	-1.7	5.0	3.3	2.4	-2.4	0.7	0.1	4.6	3.9	-1.0	-1.5
37 per capita	81.9	100.0	102.5	100.6	106.6	108.5	106.0	109.5	111.7	113.1	110.0	109.6	108.7	112.0	114.5	112.5	110.2
38 TOTAL GNP*	425.1	545.9	564.2	559.6	598.9	614.2	604.8	632.1	650.2	663.1	649.8	653.8	654.6	679.7	700.8	695.2	687.1
39 PROFIT SECTORS	354.2	456.8	470.9	468.4	500.5	513.5	514.3	536.3	548.1	560.5	547.1	548.8	549.2	572.7	592.6	586.0	579.9
40 TOTAL INDUSTRY	133.7	165.5	173.5	177.4	187.9	193.8	202.9	214.7	220.0	223.1	224.2	230.8	236.8	237.9	239.8	247.4	247.1
41 POWER	3.9	4.3	4.5	5.2	5.5	5.8	6.4	7.4	7.7	8.2	8.6	8.6	9.1	9.0	9.1	9.3	9.4
42 FUELS	15.9	15.0	15.1	15.0	15.9	16.3	17.7	18.5	19.7	19.4	19.9	21.0	22.2	18.9	19.4	20.7	21.8
43 METALS	13.2	14.1	14.5	15.2	16.4	17.5	18.5	18.8	19.4	20.4	21.2	22.2	22.8	17.8	18.9	19.8	20.5
44 CHEMICALS	4.0	6.3	7.0	7.6	8.3	9.2	10.2	10.8	11.0	11.5	11.2	12.3	13.0	15.3	16.0	16.3	16.2
45 TOTAL MBMW	32.9	49.6	53.9	54.9	60.2	60.9	63.6	69.1	70.1	71.5	71.1	72.2	72.9	84.8	84.2	85.6	82.9
46 WEAPONS & SPACE	10.9	15.0	16.3	16.9	18.6	19.2	19.5	20.6	21.2	21.9	22.1	22.7	23.3	24.2	24.8	25.5	25.2
47 WOOD & PAPER	18.6	17.2	17.1	16.9	17.1	17.1	17.2	17.5	17.0	16.5	15.8	16.0	16.4	14.9	14.6	14.6	14.6
48 CONSTR. MATER.	7.6	9.6	10.7	10.9	11.5	11.6	12.2	12.6	13.3	14.2	13.6	14.3	14.9	13.0	13.5	13.8	13.9
49 N.E.C. INDUSTRY	1.6	2.3	2.6	2.6	3.3	3.6	4.0	4.2	4.8	4.8	4.7	4.6	4.8	5.0	5.0	4.9	4.9
50 LIGHT INDUSTRY	15.7	20.2	20.1	20.1	18.4	18.3	18.4	19.1	19.1	19.1	19.5	19.9	20.3	17.8	17.4	18.0	18.2
51 FOOD INDUSTRY	9.5	11.8	11.9	12.2	12.7	14.2	15.4	16.0	16.8	15.7	16.7	17.0	17.1	17.2	16.9	18.8	19.5
52 AGRICULTURE	90.9	127.1	129.1	120.3	132.9	131.6	119.6	126.4	126.7	136.0	125.0	118.1	111.4	125.1	138.7	134.2	130.9
53 T & C	27.7	39.8	42.5	44.5	47.5	49.9	51.4	54.2	55.0	56.5	56.4	57.1	58.3	59.8	60.9	62.9	61.7
54 CONSTRUCTION	61.2	70.0	69.3	69.2	70.0	73.0	72.9	71.3	73.5	69.5	66.9	67.1	66.6	72.0	73.3	59.2	58.6
55 TRADE, etc.	16.9	25.6	26.5	26.1	29.0	30.1	30.8	31.0	32.6	33.6	33.0	32.8	32.5	34.4	35.9	37.3	36.7
56 OTHER PROD.	5.4	6.7	6.5	6.6	7.0	7.2	6.9	7.2	7.3	7.6	7.2	7.2	7.5	7.2	7.2	7.4	7.4
57 HOUSING	13.7	15.0	15.4	15.9	17.1	18.0	19.4	20.8	21.7	22.1	22.2	22.9	23.2	23.4	23.8	24.1	24.2
58 COMMUNAL, etc.	4.1	6.0	6.6	7.0	7.5	8.1	8.7	9.0	9.4	10.1	10.4	10.7	11.0	11.0	11.2	11.4	11.4
59 BANKS & INSUR.	0.7	1.3	1.4	1.4	1.6	1.7	1.7	1.8	1.9	2.0	1.8	1.9	1.9	2.0	2.0	2.1	2.0
60 BUDGET SECTORS	58.7	74.9	77.0	78.4	82.8	82.8	82.8	83.6	85.3	86.3	85.5	87.7	88.2	88.5	89.0	89.4	88.7
61 EDUCATION	21.0	25.0	25.3	25.4	27.3	27.2	27.1	27.4	27.4	27.6	27.5	28.1	28.0	28.3	28.4	28.5	28.9
62 CULTURE & ART	2.7	4.2	4.4	4.4	4.4	4.5	4.6	5.0	5.1	5.2	5.2	5.3	5.0	5.1	5.0	5.1	5.3
63 HEALTH, etc.	12.9	16.8	17.5	18.0	18.1	17.5	17.0	16.7	17.1	17.6	17.4	17.5	17.3	17.5	17.6	18.0	17.8
64 SCIENCE	9.6	12.0	12.8	13.3	14.5	14.8	15.3	15.4	16.1	16.3	16.8	17.8	18.3	18.1	18.3	17.6	17.0
65 STATE ADMIN.	4.7	6.5	6.7	6.8	7.3	7.5	7.5	7.4	7.8	8.0	7.6	8.0	8.3	8.0	8.0	8.2	8.0
66 MILITARY ADMIN.	7.8	10.4	10.5	10.6	11.2	11.3	11.4	11.6	11.7	11.7	11.1	11.0	11.3	11.5	11.7	12.0	11.8
67 NET EXPORTS	2.9	3.7	6.2	1.9	4.6	6.7	-3.8	0.7	4.8	4.0	5.1	4.7	4.5	5.3	5.7	6.1	3.2
68 NET HIDDEN GNP+	9.3	10.4	10.1	11.0	11.0	11.2	11.5	11.5	12.1	12.3	12.1	12.7	12.8	13.2	13.5	13.7	15.3

*excluding foreign trade tariffs
+illicit incomes less fictitious output

TABLE 2B: SOVIET GNP BY SECTOR OF ORIGIN IN CURRENT FACTOR COST PRICES

(billion rubles)

	1965	1970	1971	1972	1973	1974	1975	1976	1977	1978	1979	1980	1981	1982	1983	1984	1985
1 TOTAL GNP*	249.0	372.2	394.5	405.8	441.4	467.6	476.5	508.4	539.0	565.5	585.2	608.2	634.8	679.7	712.7	745.9	757.9
2 DOMESTIC ECON.	244.5	365.2	385.9	398.9	432.5	456.0	472.8	500.4	525.8	552.4	570.2	592.3	618.4	661.2	692.8	724.4	737.3
3 Factor Cost	166.7	241.4	256.8	270.8	288.9	307.2	327.5	346.4	364.5	384.0	402.7	421.6	440.4	469.2	489.3	508.8	528.2
4 Depreciation	27.3	42.2	46.2	50.3	55.6	60.7	70.5	77.0	83.0	89.1	95.5	102.2	109.2	115.9	123.9	132.1	139.8
5 Wages	132.9	189.5	200.0	209.2	221.5	233.9	243.7	255.5	267.1	279.7	291.4	303.0	314.2	327.3	338.5	348.9	359.7
6 Social Sec.	6.5	9.8	10.7	11.3	11.9	12.6	13.3	13.9	14.4	15.2	15.9	16.4	17.0	26.0	26.9	27.8	28.7
7 Net Revenues	77.8	123.8	129.2	128.1	143.6	148.8	145.3	154.0	161.3	168.4	167.4	170.8	178.0	192.0	203.6	215.7	209.1
8 NET EXPORTS	1.3	2.0	3.6	1.2	3.1	5.2	-3.3	0.7	4.7	4.1	5.2	4.9	4.7	5.3	5.9	6.5	3.6
9 NET HIDDEN GNP+	3.3	5.0	5.0	5.7	6.0	6.5	7.0	7.4	8.5	9.0	9.8	10.9	11.7	13.1	14.0	15.0	17.0
10 PROFIT SECTORS	209.4	314.0	332.2	342.6	370.9	391.5	407.1	432.2	454.7	477.2	492.5	510.7	534.0	572.7	601.7	629.8	640.4
11 Depreciation	25.7	39.6	43.4	47.3	52.1	56.9	66.3	72.2	77.9	83.7	89.8	96.1	102.9	109.4	117.1	124.9	132.3
12 Wages	111.6	159.8	168.7	175.9	185.8	196.3	204.6	215.3	225.2	235.1	244.8	253.7	263.1	274.6	284.5	293.2	301.7
13 Social Sec.	5.4	8.2	8.9	9.4	9.9	10.5	11.1	11.6	12.1	12.9	13.3	13.7	14.2	22.4	23.3	24.1	24.7
14 Net Revenues	66.6	106.4	111.2	110.0	123.1	127.8	125.1	133.1	139.5	145.5	144.6	147.2	153.7	166.4	176.8	187.6	181.7
15 TOTAL INDUSTRY	81.7	124.0	130.8	135.2	146.3	155.6	166.3	177.6	186.1	194.8	200.6	209.0	218.3	237.7	248.4	260.5	265.5
16 Depreciation	11.0	17.0	18.7	20.3	22.5	24.6	29.8	32.3	35.1	37.7	40.5	43.7	47.2	50.5	54.2	58.0	61.3
17 Wages	41.6	60.6	63.7	66.6	70.1	74.8	79.6	84.5	87.5	91.1	94.3	97.9	100.9	105.4	108.0	111.2	114.8
18 Social Sec.	3.0	4.4	4.6	4.9	5.1	5.4	5.8	6.2	6.4	6.7	6.9	7.2	7.4	12.9	13.2	13.6	14.0
19 Net Revenues	26.0	42.0	43.8	43.4	48.6	50.8	51.1	54.7	57.1	59.4	58.9	60.3	62.9	69.1	73.0	77.6	75.3
20 POWER	2.7	4.3	4.5	4.9	5.3	5.7	6.3	6.7	7.0	7.3	7.6	7.9	8.3	9.0	9.4	9.8	10.0
21 Depreciation	1.2	1.7	1.8	2.0	2.2	2.4	2.9	3.1	3.2	3.4	3.6	3.7	4.0	4.3	4.4	4.6	4.8
22 Wages	0.7	1.1	1.1	1.2	1.3	1.4	1.4	1.5	1.6	1.6	1.7	1.8	1.8	1.9	2.0	2.1	2.2
23 Social Sec.	0.0	0.1	0.1	0.1	0.1	0.1	0.1	0.1	0.1	0.1	0.1	0.1	0.1	0.2	0.2	0.2	0.2
24 Net Revenues	0.8	1.5	1.5	1.6	1.8	1.8	1.9	2.1	2.2	2.2	2.2	2.3	2.4	2.6	2.8	2.9	2.8
25 FUELS	7.6	9.8	10.0	10.0	10.6	11.0	11.9	12.9	13.6	14.3	14.9	15.9	16.7	18.9	20.0	21.4	22.4
26 Depreciation	1.6	2.4	2.5	2.6	2.8	3.0	3.5	3.8	4.1	4.5	4.8	5.4	5.8	6.3	7.0	7.6	8.3
27 Wages	3.4	3.8	3.9	3.9	4.0	4.2	4.3	4.7	4.9	5.1	5.3	5.5	5.8	6.4	6.5	6.6	6.9
28 Social Sec.	0.2	0.3	0.3	0.3	0.3	0.3	0.4	0.4	0.4	0.4	0.4	0.4	0.4	0.7	0.7	0.8	0.8
29 Net Revenues	2.4	3.3	3.3	3.2	3.5	3.6	3.7	4.0	4.2	4.4	4.4	4.6	4.8	5.5	5.9	6.4	6.4
30 METALLURGY	7.7	10.0	10.4	10.8	11.6	12.4	13.1	14.1	14.7	15.1	15.6	16.3	16.8	17.8	18.7	19.5	19.8
31 Depreciation	1.7	2.2	2.4	2.5	2.8	3.1	3.6	4.0	4.3	4.5	4.8	5.1	5.3	5.5	5.9	6.1	6.4
32 Wages	3.3	4.1	4.3	4.5	4.7	4.9	5.2	5.3	5.5	5.6	5.8	6.1	6.2	6.4	6.6	6.8	7.0
33 Social Sec.	0.2	0.3	0.3	0.3	0.3	0.3	0.3	0.4	0.4	0.4	0.4	0.4	0.4	0.7	0.7	0.7	0.7
34 Net Revenues	2.4	3.4	3.5	3.5	3.9	4.0	4.0	4.3	4.5	4.6	4.6	4.7	4.8	5.2	5.5	5.8	5.6
35 CHEMICALS	3.7	6.5	6.9	7.3	7.9	8.4	9.5	10.2	10.9	11.6	12.3	13.1	14.1	15.3	16.1	17.1	17.2
36 Depreciation	0.7	1.4	1.5	1.7	1.9	2.1	2.7	3.0	3.3	3.5	4.0	4.5	5.0	5.4	5.8	6.2	6.5
37 Wages	1.7	2.7	2.9	3.0	3.1	3.3	3.6	3.8	4.0	4.2	4.3	4.4	4.6	4.7	4.8	4.9	5.1
38 Social Sec.	0.1	0.2	0.2	0.3	0.3	0.3	0.3	0.3	0.3	0.4	0.4	0.4	0.4	0.7	0.8	0.8	0.8
39 Net Revenues	1.2	2.2	2.3	2.4	2.6	2.7	2.9	3.2	3.4	3.6	3.6	3.8	4.0	4.4	4.7	5.1	4.9
40 TOTAL MBMW	24.4	39.5	42.7	44.5	48.7	52.1	56.0	60.5	63.7	67.6	69.4	72.7	76.6	84.8	88.6	92.9	94.3
41 Depreciation	2.0	3.6	4.2	4.7	5.2	5.6	6.9	7.4	8.3	9.2	9.7	10.6	12.0	12.7	13.9	14.7	15.2
42 Wages	13.6	20.9	22.6	23.8	25.5	27.5	29.7	32.0	33.3	35.3	36.6	38.2	39.5	41.5	42.6	44.2	45.8
43 Social Sec.	1.1	1.5	1.6	1.7	1.8	2.0	2.2	2.4	2.5	2.6	2.7	2.9	3.1	6.0	6.0	6.3	6.6
44 Net Revenues	7.8	13.4	14.3	14.3	16.2	17.0	17.2	18.6	19.5	20.6	20.4	21.0	22.0	24.6	26.0	27.7	26.7

*excluding foreign trade tariffs

+illicit incomes less fictitious output

TABLE 2B: SOVIET GNP BY SECTOR OF ORIGIN IN CURRENT FACTOR COST PRICE
(billion rubles)

		1965	1970	1971	1972	1973	1974	1975	1976	1977	1978	1979	1980	1981	1982	1983	1984	1985
45	WEAPONS & SPACE	9.5	13.9	14.6	15.0	16.2	17.1	17.8	18.6	19.6	20.6	21.2	22.0	22.7	24.2	25.3	26.2	26.6
46	Depreciation	0.5	0.9	1.0	1.0	1.2	1.4	1.6	1.8	2.0	2.2	2.3	2.6	2.7	2.8	2.9	3.1	3.2
47	Wages	5.4	7.6	8.0	8.4	8.8	9.3	9.8	10.2	10.6	11.1	11.6	12.0	12.4	12.9	13.3	13.7	14.1
48	Social Sec.	0.5	0.7	0.7	0.8	0.8	0.8	0.9	0.9	1.0	1.0	1.0	1.1	1.1	1.5	1.6	1.6	1.7
49	Net Revenues	3.0	4.7	4.9	4.8	5.4	5.6	5.5	5.7	6.0	6.3	6.2	6.3	6.5	7.0	7.4	7.8	7.5
50	WOOD & PAPER	6.6	9.8	10.1	10.4	11.1	11.5	11.9	12.4	12.6	12.9	13.0	13.3	13.8	14.9	15.5	16.1	16.3
51	Depreciation	0.9	1.2	1.3	1.5	1.6	1.7	2.0	2.1	2.2	2.3	2.5	2.6	2.8	3.0	3.2	3.3	3.5
52	Wages	3.4	4.9	4.9	5.2	5.4	5.6	5.9	6.1	6.2	6.3	6.4	6.5	6.7	6.9	7.0	7.2	7.4
53	Social Sec.	0.2	0.4	0.4	0.4	0.4	0.4	0.4	0.4	0.4	0.4	0.4	0.4	0.4	0.7	0.8	0.8	0.8
54	Net Revenues	2.1	3.3	3.4	3.3	3.7	3.7	3.6	3.8	3.9	3.9	3.8	3.8	4.0	4.3	4.6	4.8	4.6
55	CONSTR. MATER.	4.5	7.1	7.5	7.8	8.4	9.0	9.6	10.0	10.3	10.8	10.9	11.3	11.7	13.0	13.5	14.0	14.2
56	Depreciation	0.7	1.1	1.2	1.3	1.4	1.6	1.9	2.0	2.2	2.3	2.4	2.6	2.7	3.0	3.2	3.4	3.6
57	Wages	2.1	3.4	3.5	3.7	3.9	4.2	4.4	4.6	4.7	4.8	4.9	5.0	5.2	5.4	5.5	5.7	5.8
58	Social Sec.	0.2	0.2	0.3	0.3	0.3	0.3	0.3	0.3	0.3	0.4	0.4	0.4	0.4	0.8	0.8	0.8	0.8
59	Net Revenues	1.4	2.4	2.5	2.5	2.8	2.9	2.9	3.1	3.2	3.3	3.2	3.3	3.4	3.8	4.0	4.2	4.0
60	LIGHT INDUSTRY	7.0	11.1	11.3	11.5	12.2	12.8	13.6	14.5	14.9	15.5	15.9	16.5	16.9	17.8	18.2	18.7	19.0
61	Depreciation	0.4	0.6	0.6	0.7	0.8	0.8	1.0	1.1	1.2	1.3	1.4	1.5	1.6	1.8	1.9	2.0	2.1
62	Wages	4.1	6.4	6.5	6.7	6.9	7.4	7.9	8.4	8.7	9.0	9.3	9.7	9.8	10.0	10.1	10.3	10.6
63	Social Sec.	0.3	0.4	0.4	0.4	0.5	0.5	0.5	0.5	0.5	0.5	0.5	0.6	0.6	0.8	0.8	0.8	0.9
64	Net Revenues	2.2	3.8	3.8	3.7	4.0	4.2	4.2	4.5	4.6	4.7	4.7	4.8	4.9	5.2	5.3	5.6	5.4
65	FOOD INDUSTRY	6.8	10.0	10.3	10.6	11.3	12.4	13.4	14.2	14.9	15.1	15.4	15.7	16.2	17.2	18.0	19.5	20.1
66	Depreciation	1.0	1.4	1.5	1.6	1.8	2.0	2.5	2.8	3.1	3.2	3.4	3.5	3.7	3.9	4.2	5.0	5.4
67	Wages	3.4	5.0	5.2	5.3	5.6	6.1	6.4	6.8	6.8	6.9	7.1	7.3	7.4	7.8	8.0	8.2	8.4
68	Social Sec.	0.2	0.2	0.2	0.2	0.2	0.3	0.3	0.3	0.4	0.4	0.4	0.4	0.4	0.6	0.6	0.6	0.6
69	Net Revenues	2.1	3.4	3.5	3.4	3.8	4.1	4.1	4.4	4.6	4.6	4.5	4.5	4.7	5.0	5.3	5.8	5.7
70	N.E.C. IND.	1.3	2.1	2.4	2.4	2.9	3.1	3.4	3.6	3.9	4.0	4.3	4.4	4.6	5.0	5.3	5.4	5.7
71	Depreciation	0.3	0.6	0.7	0.7	0.9	1.0	1.2	1.2	1.3	1.4	1.5	1.6	1.7	1.8	1.9	2.0	2.2
72	Wages	0.5	0.7	0.8	0.8	0.9	1.0	1.1	1.2	1.3	1.3	1.4	1.4	1.5	1.6	1.7	1.7	1.8
73	Social Sec.	0.1	0.1	0.1	0.1	0.1	0.1	0.1	0.1	0.1	0.1	0.1	0.1	0.1	0.1	0.1	0.2	0.2
74	Net Revenues	0.4	0.7	0.8	0.8	1.0	1.0	1.0	1.1	1.2	1.2	1.3	1.3	1.3	1.5	1.5	1.6	1.6
75	AGRICULTURE	59.2	82.2	86.1	86.0	92.9	95.8	93.4	97.9	103.5	108.0	111.5	112.8	118.5	125.1	135.5	141.6	141.9
76	Depreciation	3.6	5.4	6.0	6.4	7.2	8.0	9.2	10.2	11.3	12.2	13.1	13.9	15.1	16.2	17.4	18.5	19.9
77	Wages	35.8	47.4	49.4	49.9	52.7	54.2	53.1	55.3	58.1	60.2	62.9	63.8	66.7	69.6	75.0	77.4	78.3
78	Social Sec.	1.0	1.6	1.9	2.1	2.1	2.3	2.4	2.3	2.4	2.7	2.7	2.6	2.6	3.0	3.3	3.5	3.5
79	Net Revenues	18.8	27.9	28.8	27.6	30.8	31.3	28.7	30.2	31.8	32.9	32.8	32.5	34.1	36.3	39.8	42.2	40.3
80	Prod T and C	12.5	18.6	19.9	21.1	23.1	25.0	26.3	28.2	29.4	30.9	32.2	33.6	35.6	38.6	40.3	41.9	43.0
81	Depreciation	2.5	3.8	4.1	4.6	5.0	5.5	6.1	6.6	7.0	7.6	8.2	8.8	9.4	9.9	10.6	11.3	12.1
82	Wages	5.7	8.0	8.6	9.2	9.9	10.7	11.5	12.2	12.7	13.1	13.7	14.4	15.1	15.8	16.2	16.4	17.0
83	Social Sec.	0.4	0.5	0.6	0.6	0.6	0.6	0.6	0.7	0.7	0.8	0.8	0.8	0.9	1.6	1.7	1.7	1.7
84	Net Revenues	4.0	6.3	6.7	6.8	7.7	8.2	8.1	8.7	9.0	9.4	9.4	9.7	10.2	11.2	11.8	12.5	12.2
85	N-Prod T and C	5.8	9.3	10.1	10.7	12.1	13.0	13.8	15.3	16.2	17.5	17.7	18.9	19.8	21.2	22.1	23.2	23.5
86	Depreciation	1.4	2.1	2.4	2.6	2.9	3.1	3.5	3.9	4.1	4.4	4.7	5.1	5.5	5.7	6.2	6.6	6.9
87	Wages	2.4	3.8	4.1	4.4	4.9	5.4	5.8	6.4	6.8	7.3	7.4	7.9	8.1	8.5	8.6	8.9	9.1
88	Social Sec.	0.1	0.2	0.2	0.2	0.3	0.3	0.3	0.4	0.4	0.4	0.4	0.4	0.5	0.8	0.8	0.8	0.8
89	Net Revenues	1.8	3.1	3.4	3.4	4.0	4.3	4.3	4.7	5.0	5.3	5.2	5.4	5.7	6.1	6.5	6.9	6.7

TABLE 2B: SOVIET GNP BY SECTOR OF ORIGIN IN CURRENT FACTOR COST PRICES
(billion rubles)

	1965	1970	1971	1972	1973	1974	1975	1976	1977	1978	1979	1980	1981	1982	1983	1984	1985
90 CONSTRUCTION	25.0	40.3	43.3	45.5	48.5	51.1	53.0	55.3	58.0	60.2	62.5	64.5	67.3	72.0	74.9	78.0	80.5
91 Depreciation	1.5	2.7	3.0	3.3	3.6	4.0	4.2	4.6	5.0	5.4	5.9	6.3	6.8	7.3	8.0	8.5	9.1
92 Wages	14.9	22.9	24.7	26.4	27.6	29.1	31.1	32.1	33.7	34.9	36.5	37.8	39.2	40.9	41.9	43.2	45.3
93 Social Sec.	0.6	1.0	1.1	1.2	1.2	1.3	1.4	1.5	1.5	1.6	1.7	1.8	1.9	2.9	3.0	3.1	3.3
94 Net Revenues	7.9	13.6	14.5	14.6	16.1	16.7	16.3	17.0	17.8	18.4	18.4	18.6	19.4	20.9	22.0	23.2	22.8
95 T and D	11.1	18.1	18.9	19.8	21.5	22.5	24.0	25.2	26.8	28.8	30.0	32.1	32.9	34.4	35.2	36.7	37.0
96 Depreciation	1.0	1.6	1.7	1.8	2.0	2.2	2.7	2.9	3.1	3.3	3.6	3.7	3.8	3.9	4.1	4.4	4.6
97 Wages	6.3	9.9	10.4	11.1	11.8	12.4	13.3	13.9	14.8	16.0	16.8	18.3	18.8	19.2	19.4	20.0	20.5
98 Social Sec.	0.3	0.4	0.5	0.5	0.5	0.6	0.6	0.6	0.7	0.7	0.8	0.8	0.8	1.3	1.4	1.4	1.4
99 Net Revenues	3.5	6.1	6.3	6.4	7.1	7.3	7.4	7.7	8.2	8.8	8.8	9.2	9.5	10.0	10.3	10.9	10.5
100 OTHER PROD.	3.5	4.4	4.5	4.7	5.1	5.3	5.3	5.6	5.9	6.2	6.2	6.5	6.9	7.2	7.4	7.7	7.8
101 Depreciation	0.1	0.1	0.1	0.1	0.1	0.1	0.1	0.1	0.1	0.1	0.1	0.2	0.2	0.2	0.2	0.2	0.2
102 Wages	2.2	2.7	2.8	3.0	3.1	3.3	3.4	3.6	3.8	4.0	4.1	4.2	4.5	4.6	4.7	4.8	5.0
103 Social Sec.	0.1	0.1	0.1	0.1	0.2	0.2	0.2	0.2	0.2	0.2	0.2	0.2	0.2	0.3	0.3	0.4	0.4
104 Net Revenues	1.1	1.5	1.5	1.5	1.7	1.7	1.6	1.7	1.8	1.9	1.8	1.9	2.0	2.1	2.2	2.3	2.2
105 HOUSING	7.2	11.4	12.1	12.7	13.9	14.9	16.3	17.8	18.8	19.9	20.5	21.4	22.4	23.4	24.6	26.3	26.9
106 Depreciation	4.0	6.0	6.4	6.8	7.3	7.9	8.9	9.9	10.5	11.0	11.6	12.2	12.7	13.3	14.0	14.9	15.6
107 Wages	0.9	1.5	1.6	1.7	1.8	2.0	2.2	2.3	2.4	2.6	2.7	2.9	3.1	3.2	3.3	3.4	3.5
108 Social Sec.	0.1	0.1	0.1	0.1	0.1	0.1	0.1	0.1	0.2	0.2	0.2	0.2	0.2	0.2	0.2	0.3	0.3
109 Net Revenues	2.3	3.9	4.1	4.1	4.6	4.8	5.0	5.5	5.8	6.1	6.0	6.2	6.4	6.7	7.2	7.8	7.6
110 COMMUNAL, etc.	3.0	4.8	5.4	5.8	6.2	6.9	7.4	7.9	8.3	9.2	9.6	10.1	10.5	11.0	11.3	11.8	12.2
111 Depreciation	0.7	1.0	1.2	1.3	1.5	1.6	1.8	1.9	2.0	2.2	2.4	2.5	2.6	2.7	2.7	2.8	3.0
112 Wages	1.3	2.1	2.3	2.5	2.6	2.9	3.2	3.4	3.6	4.0	4.3	4.5	4.7	4.9	5.0	5.2	5.5
113 Social Sec.	0.1	0.1	0.1	0.1	0.1	0.1	0.1	0.1	0.2	0.2	0.2	0.2	0.2	0.2	0.2	0.3	0.3
114 Net Revenues	0.9	1.6	1.8	1.9	2.1	2.3	2.3	2.4	2.6	2.8	2.8	2.9	3.0	3.2	3.3	3.5	3.4
115 BANKING-INSUR.	0.4	0.9	1.1	1.0	1.2	1.3	1.3	1.4	1.6	1.7	1.7	1.8	1.9	2.0	2.0	2.2	2.1
116 Depreciation	0.0	0.1	0.1	0.1	0.1	0.1	0.1	0.1	0.1	0.1	0.1	0.1	0.1	0.1	0.1	0.1	0.1
117 Wages	0.3	0.5	0.6	0.6	0.7	0.8	0.8	0.9	1.0	1.1	1.1	1.2	1.3	1.4	1.4	1.5	1.5
118 Net Revenues	0.1	0.3	0.4	0.3	0.4	0.4	0.4	0.4	0.5	0.5	0.5	0.5	0.5	0.5	0.5	0.6	0.5
119 BUDGET SECTORS	35.1	51.2	53.7	56.3	61.6	64.5	65.6	68.2	71.1	75.2	77.7	81.6	84.5	88.4	91.2	94.6	96.9
120 Depreciation	1.6	2.6	2.8	3.1	3.5	3.8	4.3	4.8	5.1	5.4	5.7	6.1	6.3	6.5	6.8	7.1	7.5
121 Wages	21.2	29.7	31.3	33.3	35.7	37.6	39.0	40.2	41.9	44.5	46.6	49.3	51.1	52.7	54.0	55.7	58.0
122 Social Sec.	1.1	1.6	1.7	1.8	2.0	2.1	2.1	2.2	2.3	2.3	2.6	2.7	2.8	3.6	3.6	3.7	4.0
123 Net Revenues	11.2	17.4	18.0	18.1	20.4	21.0	20.2	21.0	21.8	22.9	22.8	23.5	24.3	25.6	26.7	28.1	27.4
124 EDUCATION	12.1	16.6	17.3	18.4	20.5	21.4	21.6	22.7	23.2	24.6	24.9	26.3	27.0	28.3	29.4	30.6	31.8
125 Depreciation	0.4	0.7	0.8	0.9	1.0	1.1	1.3	1.5	1.6	1.7	1.8	1.9	2.0	2.1	2.2	2.3	2.4
126 Wages	7.4	9.7	10.2	11.0	12.0	12.6	12.9	13.4	13.7	14.6	15.0	15.9	16.3	16.8	17.3	17.9	19.1
127 Social Sec.	0.4	0.5	0.6	0.6	0.7	0.7	0.7	0.7	0.8	0.8	0.8	0.9	0.9	1.2	1.2	1.3	1.3
128 Net Revenues	3.9	5.6	5.8	5.9	6.8	7.0	6.6	7.0	7.1	7.5	7.3	7.6	7.8	8.2	8.6	9.1	9.0
129 CULTURE & ARTS	1.7	3.0	3.2	3.2	3.3	3.4	3.6	3.9	4.1	4.2	4.4	4.7	4.9	5.1	5.1	5.3	5.5
130 Depreciation	0.1	0.2	0.2	0.2	0.2	0.2	0.3	0.3	0.3	0.3	0.4	0.4	0.4	0.4	0.4	0.4	0.4
131 Wages	1.0	1.8	1.8	1.9	1.9	2.0	2.1	2.3	2.4	2.5	2.6	2.8	2.9	3.0	3.0	3.1	3.3
132 Social Sec.	0.1	0.1	0.1	0.1	0.1	0.1	0.1	0.1	0.1	0.1	0.1	0.2	0.2	0.2	0.2	0.2	0.2
133 Net Revenues	0.5	1.0	1.1	1.0	1.1	1.1	1.1	1.2	1.3	1.3	1.3	1.4	1.4	1.5	1.5	1.6	1.6

TABLE 2B: SOVIET GNP BY SECTOR OF ORIGIN IN CURRENT FACTOR COST PRICES
(billion rubles)

		1965	1970	1971	1972	1973	1974	1975	1976	1977	1978	1979	1980	1981	1982	1983	1984	1985
134	HEALTH	7.1	10.3	10.7	11.1	12.1	12.3	12.7	13.2	14.1	14.9	15.5	16.2	16.8	17.5	18.0	18.6	18.9
135	Depreciation	0.4	0.6	0.6	0.7	0.8	0.8	0.9	0.9	1.0	1.0	1.0	1.1	1.1	1.1	1.2	1.2	1.3
136	Wages	4.2	5.9	6.2	6.5	6.9	7.1	7.5	7.8	8.3	8.9	9.4	9.9	10.3	10.6	10.9	11.2	11.5
137	Social Sec.	0.2	0.3	0.3	0.4	0.4	0.4	0.4	0.4	0.5	0.5	0.5	0.5	0.6	0.7	0.7	0.7	0.8
138	Net Revenues	2.3	3.5	3.6	3.6	4.0	4.0	3.9	4.1	4.3	4.6	4.5	4.7	4.8	5.1	5.3	5.5	5.4
134	SCIENCE	6.0	9.4	10.1	10.7	11.8	12.8	13.0	13.2	13.8	14.6	15.5	16.3	17.1	18.1	18.6	19.3	19.5
135	Depreciation	0.3	0.5	0.6	0.7	0.8	0.9	1.0	1.1	1.2	1.4	1.5	1.6	1.7	1.8	1.9	2.0	2.1
136	Wages	3.6	5.4	5.8	6.2	6.7	7.3	7.6	7.6	7.9	8.4	9.0	9.5	9.9	10.3	10.5	10.8	11.1
137	Social Sec.	0.2	0.3	0.3	0.3	0.4	0.4	0.4	0.4	0.4	0.4	0.5	0.5	0.5	0.7	0.7	0.8	0.8
138	Net Revenues	1.9	3.2	3.4	3.4	3.9	4.2	4.0	4.1	4.2	4.5	4.6	4.7	4.9	5.2	5.5	5.7	5.5
139	STATE ADMIN.	3.1	4.6	4.8	5.1	5.5	5.8	5.9	5.9	6.4	6.8	7.0	7.5	7.8	8.0	8.1	8.4	8.6
140	Depreciation	0.1	0.2	0.2	0.2	0.2	0.2	0.3	0.3	0.3	0.3	0.3	0.3	0.3	0.3	0.3	0.3	0.4
141	Wages	1.9	2.7	2.9	3.1	3.3	3.5	3.6	3.6	3.9	4.2	4.4	4.8	5.0	5.1	5.1	5.3	5.4
142	Social Sec.	0.1	0.1	0.2	0.2	0.2	0.2	0.2	0.2	0.2	0.2	0.2	0.3	0.3	0.4	0.4	0.4	0.4
143	Net Revenues	1.0	1.6	1.6	1.6	1.8	1.9	1.8	1.8	2.0	2.1	2.1	2.2	2.3	2.3	2.3	2.4	2.4
144	MILITARY ADMIN.	5.1	7.4	7.6	7.9	8.4	8.7	8.9	9.3	9.6	9.9	10.3	10.5	10.9	11.5	12.0	12.4	12.6
145	Depreciation	0.3	0.4	0.4	0.5	0.5	0.5	0.6	0.6	0.7	0.7	0.7	0.7	0.8	0.8	0.8	0.9	0.9
146	Wages	3.1	4.2	4.4	4.6	4.9	5.1	5.3	5.5	5.7	5.9	6.2	6.4	6.7	6.9	7.2	7.4	7.6
147	Social Sec.	0.1	0.2	0.2	0.3	0.3	0.3	0.3	0.3	0.3	0.3	0.3	0.3	0.4	0.4	0.5	0.5	0.5
148	Net Revenues	1.6	2.5	2.5	2.5	2.8	2.9	2.7	2.9	2.9	3.0	3.0	3.0	3.1	3.3	3.5	3.7	3.6

TABLE 2C: SOVIET GNP BY END USE IN CONSTANT FACTOR COST PRICES
(billion rubles)

	1965	1970	1971	1972	1973	1974	1975	1976	1977	1978	1979	1980	1981	1982	1983	1984	1985
1973 Prices:																	
1 TOTAL GNP*	311.0	400.9	414.6	412.7	441.5	453.1	447.1	466.2	481.2	488.5	485.4	489.0	492.4	507.7	522.8	519.4	512.9
2 growth rate	77.6	100.0	103.4	102.9	110.1	113.0	111.5	116.3	120.0	121.8	121.0	122.0	122.8	126.6	130.4	129.5	127.9
3 CONSUMPTION	130.6	163.0	168.1	169.0	177.3	182.7	183.6	187.9	191.1	194.0	194.4	192.2	190.3	191.1	197.1	199.4	199.7
4 percent of total	42.0	40.7	40.5	41.0	40.2	40.3	41.1	40.3	39.7	39.7	40.0	39.3	38.6	37.6	37.7	38.4	38.9
5 growth rate	80.1	100.0	103.1	103.7	108.8	112.1	112.7	115.3	117.2	119.0	119.3	117.9	116.8	117.3	120.9	122.4	122.5
6 per capita	84.3	100.0	102.3	101.8	105.7	108.1	107.8	109.0	110.0	110.8	110.2	107.9	105.9	105.5	107.8	108.1	107.3
7 net growth rate+	81.1	100.0	103.1	103.5	108.4	111.7	111.9	114.4	116.2	117.7	117.7	116.1	114.7	115.4	119.0	120.8	121.8
8 PURCHASED GOODS	97.8	122.3	126.0	124.4	130.6	134.3	133.1	134.7	136.5	138.1	139.0	135.1	132.3	131.9	136.0	139.1	138.9
9 Food	74.0	86.7	89.8	87.6	92.8	94.5	92.1	92.1	92.9	93.6	93.8	89.9	86.0	85.0	87.9	90.3	89.9
10 Alcohol	3.9	6.7	7.0	7.3	7.8	8.2	8.7	9.1	9.5	10.0	10.4	10.7	11.0	10.8	11.1	10.6	9.3
11 Soft Goods	17.3	24.5	24.6	24.7	24.9	26.1	26.5	27.4	27.9	28.0	28.1	28.2	28.5	29.1	29.6	30.6	31.8
12 Durables	2.6	4.4	4.6	4.7	5.1	5.5	5.8	6.0	6.3	6.5	6.6	6.4	6.9	7.0	7.4	7.6	7.9
13 PAID SERVICES	32.8	40.7	42.1	44.6	46.6	48.4	50.6	53.2	54.6	55.9	55.4	57.1	58.0	59.3	61.1	60.3	60.8
14 Total Housing	14.2	15.2	15.4	16.1	17.0	17.6	18.7	20.1	20.5	21.2	20.8	21.3	21.6	21.8	22.6	21.3	21.2
15 Personal Care	3.3	4.5	4.7	5.2	5.2	5.4	5.6	5.7	6.1	6.2	6.3	6.5	6.8	6.9	7.1	7.2	7.2
16 Transp. & Comm.	4.2	6.3	6.6	7.2	7.6	8.1	8.6	9.2	9.6	10.1	10.5	11.0	11.5	11.9	12.1	12.5	12.8
17 Other Services	4.1	6.6	6.9	7.4	7.9	8.1	8.3	8.6	8.5	8.1	7.6	7.6	7.4	7.5	8.0	7.9	7.8
18 Second Economy	7.0	8.2	8.4	8.7	9.0	9.2	9.4	9.6	9.9	10.1	10.3	10.6	10.8	11.1	11.3	11.5	11.7
19 INVESTMENT	98.6	135.2	139.4	135.2	151.8	154.3	145.6	157.0	168.1	168.7	165.1	165.1	168.8	180.4	187.2	177.8	170.3
20 percent of total	31.7	33.7	33.6	32.8	34.4	34.1	32.6	33.7	34.9	34.5	34.0	33.8	34.3	35.5	35.8	34.2	33.2
21 FIXED CAPITAL	91.1	123.6	128.4	131.6	136.2	140.9	143.8	146.8	152.0	153.9	149.1	149.9	152.2	155.8	160.8	152.2	151.6
22 Civilian Sectors	87.6	118.6	122.9	126.1	130.7	135.2	137.7	140.3	145.2	147.0	141.9	142.5	144.8	148.4	153.0	145.5	144.9
23 Non-Residential	69.6	98.2	102.4	104.7	108.1	112.9	115.5	118.2	122.0	123.8	118.6	117.9	119.2	121.5	124.4	119.9	118.6
24 Residential	18.0	20.4	20.4	21.3	22.6	22.3	22.2	22.1	23.2	23.2	23.3	24.6	25.6	26.9	28.6	25.6	26.3
25 Defense Sectors	3.5	5.0	5.5	5.6	5.6	5.7	6.1	6.5	6.8	6.9	7.2	7.3	7.4	7.4	7.8	6.7	6.6
26 INVENTORIES	6.7	13.9	10.3	8.7	14.4	11.9	9.3	12.4	12.0	12.3	11.1	14.6	11.8	17.5	19.8	14.7	13.0
27 Civilian Sectors	6.7	13.6	10.1	8.3	14.2	11.5	9.2	12.0	11.5	11.7	9.5	12.8	12.0	15.4	19.5	14.4	12.3
28 Defense Sectors	0.0	0.3	0.2	0.4	0.2	0.4	0.2	0.3	0.6	0.6	1.7	1.8	-0.1	2.1	0.3	0.3	0.7
29 FICTITIOUS OUTPUT	-1.2	-1.5	-1.7	-1.8	-2.0	-2.0	-2.1	-2.3	-2.3	-2.3	-2.4	-2.3	-2.3	-2.5	-2.5	-2.7	-2.3
30 STRATEGIC RESERVE	0.0	-3.3	-1.6	-4.6	0.1	-0.9	-2.9	-0.4	3.1	2.1	3.9	-0.2	4.1	6.0	5.4	9.4	5.9
31 NET EXPORTS	2.0	2.5	4.1	1.3	3.1	4.5	-2.5	0.5	3.2	2.6	3.4	3.1	3.0	3.5	3.8	4.1	2.1
32 GOVERNMENT	81.8	102.7	107.1	108.4	112.5	116.0	117.9	121.3	122.1	125.9	125.9	131.7	133.3	136.2	138.5	142.1	142.9
33 percent of total	26.3	25.6	25.8	26.3	25.5	25.6	26.4	26.0	25.4	25.8	25.9	26.9	27.1	26.8	26.5	27.4	27.9
34 BUDGET SERVICES	39.8	51.3	53.7	55.5	57.4	59.1	59.8	61.5	62.8	64.1	64.6	66.0	66.6	65.4	66.4	68.8	67.4
35 Civilian Services	30.2	38.1	40.0	41.4	42.8	44.4	44.9	46.2	47.4	48.5	48.8	49.5	49.8	48.5	49.1	50.7	49.5
36 Defense Services	9.5	13.1	13.7	14.1	14.6	14.7	14.9	15.2	15.4	15.6	15.7	16.5	16.8	16.9	17.3	18.1	17.8
37 DEFENSE	42.0	51.4	53.4	52.8	55.0	56.9	58.1	59.8	59.3	61.8	61.4	65.7	66.7	70.8	72.1	73.4	75.6
38 TOTAL DEFENSE	55.1	69.9	72.9	73.0	75.4	77.7	79.3	81.9	82.1	84.9	86.0	91.4	90.8	97.2	97.5	98.5	100.7
39 percent of total	17.7	17.4	17.6	17.7	17.1	17.2	17.7	17.6	17.1	17.4	17.7	18.7	18.4	19.1	18.6	19.0	19.6
40 growth rate	78.8	100.0	104.2	104.4	107.9	111.2	113.4	117.2	117.5	121.4	123.0	130.8	129.8	139.1	139.5	140.9	144.1
41 FOREIGN TARIFFS	10.2	12.5	12.5	16.2	14.9	11.8	18.1	18.2	16.4	17.9	20.9	26.5	31.2	34.3	34.5	34.4	37.3
42 POPULATION	231	243	245	248	250	252	254	257	259	261	263	266	268	270	273	275	278

*excluding foreign trade tariffs

+growth of consumption net of alcohol purchases

TABLE 2C: SOVIET GNP BY END USE IN CONSTANT FACTOR COST PRICES
       (billion rubles)

	1965	1970	1971	1972	1973	1974	1975	1976	1977	1978	1979	1980	1981	1982	1983	1984	1985
1982 Prices:																	
42 TOTAL GNP*	416.5	538.0	556.8	553.9	592.5	608.4	599.4	625.3	645.9	655.3	650.7	655.2	659.7	679.6	700.1	695.6	687.0
43 growth rate	77.4	100.0	103.5	103.0	110.1	113.1	111.4	116.2	120.1	121.8	121.0	121.8	122.6	126.3	130.1	129.3	127.7
44 CONSUMPTION	173.8	217.2	223.9	225.2	236.0	243.2	244.4	250.0	254.1	257.7	258.0	255.1	252.6	253.6	261.7	265.1	265.8
45 percent of total	41.7	40.4	40.2	40.7	39.8	40.0	40.8	40.0	39.3	39.3	39.7	38.9	38.3	37.3	37.4	38.1	38.7
46 growth rate	80.0	100.0	103.1	103.7	108.7	112.0	112.5	115.1	117.0	118.6	118.8	117.5	116.3	116.8	120.5	122.0	122.4
47 per capita	84.2	100.0	102.2	101.8	105.6	108.0	107.6	108.8	109.7	110.5	109.8	107.5	105.4	105.1	107.4	107.8	107.2
48 net growth rate+	80.9	100.0	103.1	103.5	108.4	111.6	111.9	114.4	116.1	117.6	117.5	116.0	114.6	115.2	118.9	120.8	121.8
49 PURCHASED GOODS	129.1	161.7	166.4	164.3	172.3	177.3	175.7	177.9	180.3	182.3	183.4	178.3	174.6	174.1	179.6	183.9	184.1
50 Food	96.2	112.7	116.7	113.9	120.6	122.9	119.7	119.8	120.7	121.7	121.9	116.9	111.7	110.3	114.3	117.4	116.9
51 Alcohol	4.4	7.7	8.0	8.3	8.9	9.3	9.9	10.4	10.8	11.4	11.9	12.1	12.5	12.3	12.6	12.1	10.6
52 Soft Goods	24.9	35.1	35.4	35.5	35.8	37.4	38.0	39.4	40.0	40.2	40.4	40.4	40.9	41.8	42.4	43.9	45.6
53 Durables	3.6	6.1	6.3	6.6	7.1	7.7	8.0	8.3	8.7	9.0	9.1	8.8	9.5	9.7	10.2	10.5	11.0
54 PAID SERVICES	44.7	55.6	57.5	60.9	63.6	65.9	68.7	72.1	73.8	75.4	74.6	76.8	78.0	79.6	82.1	81.1	81.7
55 Total Housing	17.9	19.1	19.4	20.3	21.4	22.2	23.5	25.3	25.8	26.7	26.2	26.9	27.2	27.4	28.5	26.8	26.6
56 Personal Care	4.1	5.5	5.8	6.4	6.4	6.7	6.9	7.0	7.5	7.7	7.7	8.0	8.4	8.5	8.7	8.9	8.9
57 Transp. & Comm.	4.7	7.1	7.4	8.1	8.6	9.1	9.7	10.4	10.8	11.5	11.9	12.4	13.0	13.5	13.7	14.1	14.5
58 Other Services	6.3	10.2	10.7	11.5	12.2	12.6	12.9	13.3	13.1	12.6	11.7	11.8	11.5	11.6	12.3	12.2	12.1
59 Second Economy	11.6	13.7	14.0	14.5	15.0	15.4	15.7	16.1	16.5	16.9	17.1	17.7	18.0	18.5	18.9	19.1	19.6
60 INVESTMENT	135.8	186.5	192.9	187.0	209.5	213.4	200.8	216.7	232.1	233.0	228.0	227.9	232.8	247.9	257.4	244.7	234.4
61 percent of total	32.6	34.7	34.6	33.8	35.4	35.1	33.5	34.7	35.9	35.6	35.0	34.8	35.3	36.5	36.8	35.2	34.1
62 FIXED CAPITAL	125.7	171.2	177.9	182.3	188.7	195.2	199.3	203.6	210.8	213.4	206.6	207.5	210.6	215.4	222.3	210.7	209.7
63 Civilian Sectors	120.8	164.2	170.2	174.6	180.9	187.3	190.8	194.5	201.3	203.8	196.5	197.3	200.4	205.1	211.5	201.4	200.4
64 Non-Residential	98.0	138.3	144.2	147.5	152.3	159.0	162.6	166.4	171.8	174.3	167.0	166.0	167.8	170.9	175.1	168.9	167.0
65 Residential	22.9	25.9	26.0	27.1	28.7	28.3	28.2	28.0	29.5	29.5	29.6	31.2	32.6	34.2	36.4	32.5	33.4
66 Defense Sectors	4.9	7.0	7.7	7.7	7.8	7.9	8.5	9.1	9.5	9.6	10.0	10.2	10.3	10.3	10.8	9.3	9.3
67 INVENTORIES	8.8	18.1	13.4	11.4	18.7	15.6	12.2	16.2	15.7	16.1	14.6	19.1	15.5	22.9	25.9	19.2	17.0
68 Civilian Sectors	8.8	17.7	13.1	10.8	18.5	15.0	12.0	15.7	15.0	15.3	12.4	16.7	15.7	20.1	25.4	18.8	16.1
69 Defense Sectors	0.0	0.4	0.3	0.6	0.3	0.6	0.2	0.5	0.8	0.8	2.2	2.4	-0.2	2.8	0.5	0.4	0.9
70 FICTITIOUS OUTPUT	-1.7	-2.2	-2.5	-2.5	-2.8	-2.9	-3.0	-3.3	-3.3	-3.3	-3.4	-3.2	-3.3	-3.6	-3.6	-3.8	-3.3
71 STRATEGIC RESERVE	0.0	-4.3	-2.1	-6.0	0.1	-1.2	-3.8	-0.5	4.1	2.7	5.1	-0.2	5.5	8.0	7.1	12.4	7.8
72 NET EXPORTS	3.0	3.7	6.2	1.9	4.7	6.8	-3.8	0.7	4.8	4.0	5.1	4.7	4.5	5.3	5.7	6.2	3.2
73 GOVERNMENT	106.9	134.2	140.0	141.8	147.1	151.7	154.2	158.6	159.7	164.6	164.7	172.2	174.3	178.0	180.9	185.8	186.7
74 percent of total	34.7	33.8	34.1	34.7	33.6	33.8	34.8	34.3	33.5	34.0	34.2	35.6	35.7	35.4	35.0	36.1	36.8
75 BUDGET SERVICES	52.6	67.8	71.0	73.5	75.9	78.2	79.1	81.3	83.0	84.7	85.4	87.3	88.1	86.6	87.8	90.9	89.1
76 Civilian Services	40.2	50.7	53.1	55.1	57.0	59.1	59.8	61.5	63.0	64.5	64.9	65.8	66.2	64.6	65.3	67.4	65.9
77 Defense Services	12.4	17.1	17.8	18.3	19.0	19.1	19.4	19.8	20.0	20.2	20.5	21.5	21.8	22.0	22.5	23.5	23.2
78 DEFENSE	54.3	66.5	69.0	68.3	71.1	73.6	75.1	77.3	76.7	79.9	79.3	84.9	86.2	91.5	93.1	94.8	97.7
79 TOTAL DEFENSE	71.6	90.9	94.9	95.0	98.1	101.2	103.2	106.7	107.0	110.5	112.0	119.1	118.1	126.6	126.9	128.1	131.0
80 percent of total	17.2	16.9	17.0	17.1	16.6	16.6	17.2	17.1	16.6	16.9	17.2	18.2	17.9	18.6	18.1	18.4	19.1
81 growth rate	78.8	100.0	104.3	104.5	108.0	111.3	113.5	117.4	117.7	121.6	123.2	131.0	130.0	139.3	139.6	140.9	144.1

*excluding foreign trade tariffs
+growth of consumption net of alcohol purchases

TABLE 2D: SOVIET GNP BY END USE IN CURRENT FACTOR COST PRICES
(billion rubles)

	1965	1970	1971	1972	1973	1974	1975	1976	1977	1978	1979	1980	1981	1982	1983	1984	1985
1 TOTAL GNP*	248.9	372.1	394.4	405.8	441.4	467.5	476.4	508.4	538.9	565.5	585.1	608.1	634.7	679.6	712.7	745.9	757.9
2 percent	100.0	100.0	100.0	100.0	100.0	100.0	100.0	100.0	100.0	100.0	100.0	100.0	100.0	100.0	100.0	100.0	100.0
3 CONSUMPTION	106.4	151.8	159.7	167.6	177.3	188.1	196.4	205.4	215.2	222.3	233.8	238.7	247.7	253.6	265.2	281.4	289.2
4 percent of total	42.7	40.8	40.5	41.3	40.2	40.2	41.2	40.4	39.9	39.3	40.0	39.2	39.0	37.3	37.2	37.7	38.2
5 PURCHASED GOODS	82.9	114.0	119.8	124.1	130.6	138.0	142.6	147.3	153.5	157.2	166.1	167.1	172.6	174.0	180.9	192.9	197.4
6 Food	64.4	82.3	86.9	89.0	92.8	96.5	97.7	98.8	102.8	104.1	109.8	108.4	110.5	110.3	114.5	122.0	123.4
7 Alcohol	3.8	6.6	6.7	7.5	7.8	8.4	9.3	10.2	10.5	10.8	11.6	12.0	12.1	12.3	12.8	14.6	15.2
8 Soft Goods	12.7	21.5	22.1	23.2	24.9	27.1	29.2	31.4	33.0	34.7	36.7	38.5	40.8	41.8	43.0	45.2	46.9
9 Durables	2.0	3.7	4.1	4.5	5.1	5.9	6.4	6.8	7.2	7.6	8.0	8.2	9.2	9.7	10.4	11.1	11.9
10 PAID SERVICES	23.5	37.8	39.9	43.5	46.7	50.2	53.8	58.1	61.7	65.0	67.6	71.6	75.1	79.6	84.3	88.5	91.8
11 Total Housing	9.3	14.2	15.0	16.0	17.0	17.9	19.3	21.2	22.3	23.3	24.1	25.0	26.0	27.4	28.9	30.6	31.5
12 Prod. Utilities	0.9	1.3	1.3	1.4	1.5	1.5	1.7	1.8	2.0	2.0	2.1	2.2	2.3	2.7	2.8	2.8	2.9
13 Housing	8.3	13.0	13.7	14.6	15.5	16.4	17.7	19.3	20.3	21.3	22.0	22.9	23.7	24.7	26.1	27.8	28.5
14 Personal Care	2.7	4.2	4.5	5.0	5.2	5.5	5.7	6.1	6.7	7.2	7.5	7.9	8.4	8.5	9.0	9.3	9.7
15 Transp. & Comm.	3.4	6.1	6.3	6.8	7.6	8.2	9.0	9.9	10.4	11.0	11.6	12.3	13.0	13.5	14.2	14.5	14.7
16 Other Services	3.2	5.9	6.5	7.2	7.9	8.7	9.3	9.8	10.0	10.3	10.1	10.7	10.9	11.6	12.5	13.2	13.5
17 Second Economy	4.9	7.4	7.7	8.5	9.0	9.8	10.4	11.2	12.4	13.2	14.4	15.6	16.8	18.5	19.7	20.9	22.3
18 INVESTMENT	75.9	125.0	132.2	132.5	151.8	159.3	155.2	172.6	189.9	199.5	206.1	212.2	221.5	247.9	263.0	269.1	265.6
19 percent of total	30.5	33.6	33.5	32.7	34.4	34.1	32.6	33.9	35.2	35.3	35.2	34.9	34.9	36.5	36.9	36.1	35.0
20 FIXED CAPITAL	69.9	114.1	121.8	129.0	136.2	144.9	154.1	161.5	171.0	181.8	184.6	191.2	200.2	215.4	227.1	232.3	237.5
21 Civilian Sectors	67.1	109.4	116.6	123.6	130.6	139.0	147.6	154.3	163.3	173.5	175.6	181.8	190.4	205.1	215.9	221.9	226.9
22 Non-Residential	55.4	90.3	96.8	102.5	108.0	116.4	124.6	131.1	138.5	148.0	149.1	153.4	159.9	170.9	179.4	185.1	187.7
23 Residential	11.8	19.1	19.8	21.2	22.6	22.7	23.0	23.3	24.8	25.6	26.6	28.4	30.4	34.2	36.6	36.8	39.1
24 Defense Sectors	2.8	4.7	5.2	5.4	5.6	5.9	6.5	7.2	7.7	8.3	9.0	9.4	9.8	10.3	11.2	10.4	10.6
25 INVENTORIES	5.6	13.4	9.9	8.6	14.4	12.3	9.9	13.5	13.4	14.1	14.5	19.2	14.3	22.9	26.3	20.8	19.3
26 Civilian Sectors	5.6	13.1	9.7	8.2	14.2	11.9	9.7	13.1	12.7	13.5	12.4	16.9	14.5	20.1	25.8	20.3	18.4
27 Defense Sectors	0.0	0.3	0.2	0.4	0.2	0.5	0.2	0.4	0.6	0.6	2.2	2.4	-0.2	2.8	0.5	0.5	1.0
28 FICTITIOUS OUTPUT	-0.9	-1.3	-1.6	-1.7	-2.0	-2.1	-2.4	-2.7	-2.7	-2.8	-3.0	-3.0	-3.2	-3.6	-3.6	-3.9	-3.4
29 Fixed Capital	-0.5	-0.6	-0.9	-1.1	-1.2	-1.4	-1.7	-2.0	-2.0	-2.0	-2.1	-2.0	-2.2	-2.5	-2.4	-2.7	-2.4
30 Inventories	-0.4	-0.7	-0.7	-0.6	-0.7	-0.7	-0.6	-0.7	-0.7	-0.8	-0.9	-1.0	-1.0	-1.1	-1.3	-1.2	-1.1
31 STRATEGIC RESERVE	0.0	-3.1	-1.6	-4.6	0.1	-1.0	-3.1	-0.4	3.5	2.3	4.7	-0.2	5.5	8.0	7.4	13.5	8.7
32 Agricultural	0.0	-2.8	-1.9	-5.1	0.5	-0.3	-3.1	-1.3	3.4	1.5	4.8	2.3	4.1	8.0	7.2	12.2	9.4
33 Other Reserves	0.0	-0.3	0.4	0.4	-0.4	-0.6	0.0	0.9	0.1	0.8	-0.1	-2.5	1.4	0.0	0.2	1.3	-0.7
34 NET EXPORTS	1.3	2.0	3.6	1.2	3.1	5.2	-3.3	0.7	4.7	4.1	5.2	4.9	4.7	5.3	5.9	6.5	3.5
35 GOVERNMENT	66.8	95.4	102.5	105.7	112.4	120.1	124.8	130.4	133.9	143.8	145.3	157.3	165.7	178.1	184.6	195.3	203.1
36 percent of total	26.8	25.6	26.0	26.0	25.5	25.7	26.2	25.7	24.8	25.4	24.8	25.9	26.1	26.2	25.9	26.2	26.8
37 BUDGET SERVICES	31.7	47.1	50.7	53.7	57.5	61.1	63.3	66.2	69.0	73.1	73.6	78.1	83.3	86.6	89.4	94.3	96.1
38 Civilian Services	23.9	34.9	37.7	40.0	42.8	45.8	47.5	49.7	52.0	55.2	55.0	58.0	62.4	64.6	66.3	69.5	70.9
39 Defense Services	7.7	12.2	13.0	13.7	14.6	15.3	15.8	16.5	17.0	17.9	18.6	20.1	20.8	22.0	23.0	24.8	25.2
40 DEFENSE	35.1	48.2	51.8	52.0	55.0	59.0	61.5	64.3	64.9	70.6	71.7	79.2	82.4	91.5	95.2	101.0	107.0
41 Weapons & Space	19.8	27.6	29.1	30.2	31.5	33.2	34.8	36.7	38.3	40.0	41.3	43.4	45.9	50.4	54.6	60.5	64.1
42 Military Admin.	10.1	14.9	15.4	16.5	17.0	19.1	20.2	20.7	21.3	24.0	24.7	29.6	30.3	30.7	31.9	33.2	34.6
43 Defense Constr.	5.2	5.7	7.4	5.2	6.4	6.8	6.6	6.8	5.3	6.7	5.7	6.2	6.2	10.4	8.6	7.2	8.3
44 TOTAL DEFENSE	45.6	65.4	70.2	71.5	75.4	80.7	84.0	88.3	90.3	97.5	101.5	111.0	112.9	126.6	129.9	136.7	143.8
45 percent of total	18.3	17.6	17.8	17.6	17.1	17.3	17.6	17.4	16.7	17.2	17.3	18.3	17.8	18.6	18.2	18.3	19.0

*excluding foreign tariffs

TABLE 2E: FACTOR COST ADJUSTMENT COEFFICIENTS BY SECTOR OF ORIGIN AND END USE
    (billion rubles)

	1965	1970	1971	1972	1973	1974	1975	1976	1977	1978	1979	1980	1981	1982	1983	1984	1985
GNP by Sector of Origin:																	
1 PROFIT SECTORS	0.97	0.97	0.97	0.97	0.97	0.97	0.97	0.98	0.98	0.98	0.98	0.98	0.98	0.98	0.98	0.98	0.98
2 TOTAL INDUSTRY	0.70	0.68	0.67	0.67	0.68	0.68	0.68	0.70	0.70	0.69	0.68	0.68	0.68	0.68	0.68	0.70	0.70
3 POWER	0.91	0.66	0.65	0.66	0.66	0.65	0.68	0.73	0.74	0.73	0.73	0.72	0.73	0.69	0.69	0.68	0.67
4 FUELS	0.89	0.61	0.57	0.54	0.54	0.53	0.55	0.55	0.54	0.51	0.50	0.51	0.54	0.46	0.46	0.48	0.49
5 METALS	1.09	0.78	0.78	0.79	0.80	0.80	0.82	0.82	0.85	0.85	0.90	0.93	0.94	0.77	0.75	0.76	0.76
6 CHEMICALS	0.86	0.82	0.78	0.77	0.73	0.70	0.72	0.73	0.73	0.74	0.75	0.77	0.77	0.95	0.96	0.94	0.91
7 TOTAL MBMW	0.92	0.92	0.95	0.89	0.94	0.91	0.91	0.93	0.92	0.91	0.87	0.87	0.86	0.96	0.94	0.93	0.89
8 WEAPONS & SPACE	1.48	1.55	1.54	1.50	1.52	1.50	1.47	1.47	1.47	1.48	1.46	1.45	1.45	1.45	1.47	1.48	1.45
9 WOOD & PAPER	1.23	1.08	1.08	1.07	1.09	1.13	1.12	1.16	1.17	1.19	1.20	1.19	1.20	1.04	1.03	1.03	1.01
10 CONSTR. MATER.	1.19	1.08	1.06	1.03	1.02	1.02	1.05	1.05	1.05	1.11	1.11	1.17	1.18	1.08	1.05	1.04	1.01
11 N.E.C. INDUSTRY	0.38	0.38	0.40	0.37	0.41	0.41	0.40	0.39	0.40	0.39	0.38	0.34	0.36	0.38	0.38	0.38	0.39
12 LIGHT INDUSTRY	0.36	0.40	0.39	0.38	0.40	0.41	0.41	0.42	0.42	0.42	0.41	0.40	0.40	0.40	0.40	0.41	0.41
13 FOOD INDUSTRY	0.24	0.26	0.25	0.26	0.26	0.26	0.27	0.29	0.29	0.27	0.27	0.27	0.26	0.26	0.27	0.30	0.34
14 AGRICULTURE	1.42	1.57	1.69	1.88	1.59	1.74	1.96	1.82	1.79	1.86	1.92	2.03	2.04	1.89	1.91	1.90	1.97
15 T & C	0.95	0.95	0.95	0.95	0.98	0.97	0.96	0.98	0.98	1.00	1.00	1.00	1.02	1.01	1.00	1.01	1.01
16 CONSTRUCTION	1.41	1.35	1.32	1.30	1.34	1.30	1.27	1.26	1.28	1.28	1.31	1.32	1.33	1.39	1.39	1.25	1.26
17 TRADE, etc.	0.91	0.95	0.94	0.92	0.94	0.94	0.94	0.93	0.94	0.96	0.95	0.95	0.94	0.98	0.97	0.99	0.98
18 OTHER PROD.	1.51	1.43	1.41	1.40	1.42	1.35	1.30	1.29	1.29	1.28	1.26	1.26	1.28	1.26	1.23	1.24	1.20
19 HOUS-COM., etc	1.96	2.08	2.09	2.04	2.12	2.14	2.22	2.26	2.28	2.34	2.28	2.28	2.30	2.29	2.27	2.29	2.23
20 BANKS & INSUR.	0.24	0.43	0.48	0.42	0.41	0.42	0.37	0.40	0.39	0.39	0.50	0.47	0.48	0.47	0.41	0.41	0.38
21 BUDGET SECTORS	1.46	1.49	1.48	1.46	1.48	1.47	1.43	1.43	1.43	1.42	1.40	1.39	1.39	1.39	1.39	1.39	1.36
22 EDUCATION	1.43	1.45	1.44	1.43	1.45	1.42	1.40	1.39	1.38	1.38	1.36	1.35	1.36	1.36	1.36	1.36	1.34
23 CULTURE & ART	1.35	1.51	1.54	1.54	1.50	1.52	1.50	1.50	1.44	1.44	1.42	1.41	1.36	1.37	1.34	1.32	1.30
24 HEALTH, etc.	1.46	1.49	1.49	1.48	1.54	1.48	1.44	1.43	1.43	1.43	1.39	1.38	1.38	1.38	1.39	1.40	1.36
25 SCIENCE	1.54	1.51	1.51	1.46	1.47	1.48	1.45	1.46	1.46	1.44	1.43	1.42	1.41	1.41	1.42	1.42	1.39
26 STATE ADMIN.	1.42	1.51	1.53	1.49	1.52	1.51	1.48	1.45	1.48	1.47	1.40	1.42	1.40	1.36	1.36	1.36	1.34
27 MILITARY ADMIN.	1.48	1.50	1.51	1.48	1.49	1.49	1.44	1.45	1.46	1.45	1.44	1.45	1.45	1.44	1.44	1.44	1.39

TABLE 2E: FACTOR COST ADJUSTMENT COEFFICIENTS BY SECTOR OF ORIGIN AND END USE
        (billion rubles)

	1965	1970	1971	1972	1973	1974	1975	1976	1977	1978	1979	1980	1981	1982	1983	1984	1985
GNP by End Use:																	
28 CONSUMPTION	0.80	0.80	0.79	0.79	0.80	0.80	0.79	0.79	0.79	0.78	0.78	0.75	0.74	0.73	0.73	0.75	0.75
29 percent of total	0.81	0.81	0.80	0.80	0.81	0.81	0.81	0.81	0.81	0.80	0.81	0.79	0.78	0.77	0.77	0.79	0.80
30 PURCHASED GOODS	0.72	0.71	0.70	0.69	0.70	0.70	0.69	0.68	0.68	0.67	0.67	0.63	0.62	0.60	0.61	0.63	0.63
31 Food	1.11	1.11	1.13	1.12	1.13	1.14	1.13	1.10	1.12	1.11	1.13	1.08	1.05	1.03	1.02	1.05	1.02
32 Alcohol	0.24	0.26	0.25	0.26	0.26	0.26	0.27	0.29	0.28	0.27	0.27	0.27	0.26	0.24	0.25	0.28	0.32
33 Soft Goods	0.36	0.40	0.39	0.38	0.40	0.41	0.41	0.42	0.42	0.42	0.41	0.40	0.40	0.40	0.40	0.41	0.41
34 Durables	0.37	0.39	0.40	0.38	0.40	0.41	0.40	0.41	0.41	0.40	0.40	0.37	0.38	0.39	0.39	0.40	0.40
35 PAID SERVICES	1.31	1.33	1.33	1.33	1.32	1.32	1.32	1.34	1.35	1.35	1.34	1.33	1.33	1.32	1.32	1.33	1.31
36 Total Housing	2.07	2.30	2.36	2.31	2.35	2.39	2.52	2.64	2.71	2.73	2.71	2.76	2.75	2.66	2.69	2.77	2.71
37 Prod. Utilities	0.82	0.73	0.71	0.72	0.73	0.72	0.75	0.80	0.81	0.80	0.80	0.80	0.80	0.76	0.76	0.75	0.73
38 Housing	1.96	2.08	2.09	2.04	2.12	2.14	2.22	2.26	2.28	2.34	2.28	2.28	2.30	2.29	2.27	2.29	2.23
39 Personal Care	1.46	1.49	1.48	1.53	1.43	1.47	1.43	1.43	1.43	1.42	1.40	1.39	1.39	1.39	1.39	1.39	1.36
40 Transp. & Comm.	0.95	0.95	0.95	0.98	0.98	0.97	0.96	0.98	0.98	1.00	1.00	1.00	1.02	1.01	1.00	1.01	1.01
41 Other Services	1.02	1.04	1.04	1.02	1.04	1.03	1.00	1.00	1.00	0.99	0.98	0.97	0.97	0.97	0.97	0.97	0.95
42 Second Economy	1.00	1.00	1.00	1.00	1.00	1.00	1.00	1.00	1.00	1.00	1.00	1.00	1.00	1.00	1.00	1.00	1.00
43 INVESTMENT	1.17	1.14	1.14	1.13	1.13	1.13	1.11	1.11	1.12	1.11	1.11	1.12	1.12	1.14	1.12	1.09	1.08
44 percent of total	1.19	1.15	1.15	1.15	1.15	1.14	1.14	1.14	1.14	1.15	1.15	1.17	1.19	1.20	1.18	1.15	1.14
45 FIXED CAPITAL	1.20	1.17	1.16	1.14	1.15	1.13	1.11	1.12	1.12	1.12	1.11	1.12	1.13	1.16	1.14	1.08	1.07
46 Civilian Sectors	1.20	1.17	1.16	1.14	1.15	1.13	1.12	1.12	1.13	1.12	1.12	1.12	1.13	1.16	1.14	1.08	1.07
47 Non-Residential	1.16	1.14	1.13	1.10	1.12	1.11	1.09	1.10	1.10	1.10	1.09	1.10	1.10	1.12	1.10	1.06	1.03
48 Residential	1.41	1.35	1.32	1.34	1.34	1.30	1.27	1.26	1.28	1.28	1.31	1.32	1.33	1.39	1.39	1.25	1.26
49 Defense Sectors	1.16	1.14	1.13	1.14	1.14	1.17	1.09	1.10	1.10	1.10	1.09	1.10	1.10	1.18	1.17	1.09	1.07
50 INVENTORIES	0.85	0.86	0.88	1.03	0.87	0.97	0.94	0.92	0.91	0.92	0.92	0.94	0.95	0.87	0.86	0.94	0.95
51 Civilian Sectors	0.85	0.86	0.88	1.03	0.86	0.97	0.94	0.92	0.91	0.92	0.93	0.95	0.95	0.86	0.86	0.94	0.95
52 Defense Sectors	0.93	0.92	0.95	0.89	0.94	0.91	0.91	0.93	0.92	0.91	0.87	0.87	0.86	0.96	0.94	0.93	0.89
53 FICTITIOUS OUTPUT	0.53	0.55	0.57	0.60	0.63	0.64	0.69	0.70	0.70	0.68	0.65	0.63	0.63	0.67	0.64	0.67	0.65
54 Fixed Capital	0.92	0.92	0.95	0.89	0.94	0.91	0.91	0.93	0.92	0.91	0.87	0.87	0.86	0.96	0.94	0.93	0.89
55 Inventories	0.36	0.40	0.39	0.38	0.40	0.41	0.41	0.42	0.42	0.42	0.41	0.40	0.40	0.40	0.40	0.41	0.41
56 STRATEGIC RESERVE	1.06	1.15	1.36	1.47	2.00	1.01	1.46	1.83	1.32	1.17	1.44	1.75	1.27	1.40	1.40	1.28	1.56
57 Agricultural	1.06	1.18	1.26	1.39	1.19	1.30	1.46	1.36	1.34	1.38	1.43	1.50	1.50	1.40	1.42	1.34	1.48
58 Other Reserves	0.94	0.92	0.95	0.89	0.94	0.91	0.92	0.93	0.92	0.91	0.87	0.87	0.86	0.90	0.94	0.93	0.89
59 NET EXPORTS	1.00	1.00	1.00	1.00	1.00	1.00	1.00	1.00	1.00	1.00	1.00	1.00	1.00	1.00	1.00	1.00	1.00
60 GOVERNMENT	1.23	1.26	1.25	1.25	1.23	1.24	1.21	1.22	1.21	1.21	1.20	1.20	1.20	1.18	1.18	1.19	1.17
61 percent of total	1.24	1.27	1.27	1.28	1.25	1.25	1.25	1.25	1.24	1.25	1.24	1.26	1.26	1.24	1.24	1.25	1.24
62 BUDGET SERVICES	1.15	1.18	1.18	1.20	1.16	1.17	1.15	1.15	1.15	1.14	1.12	1.12	1.11	1.09	1.09	1.12	1.10
63 Civilian Services	1.13	1.16	1.16	1.17	1.13	1.15	1.13	1.13	1.12	1.12	1.10	1.09	1.09	1.06	1.07	1.09	1.07
64 Defense Services	1.24	1.26	1.27	1.28	1.27	1.25	1.23	1.23	1.22	1.22	1.19	1.18	1.18	1.16	1.16	1.20	1.17
65 DEFENSE	1.30	1.33	1.33	1.32	1.31	1.31	1.29	1.29	1.29	1.29	1.30	1.30	1.30	1.27	1.28	1.26	1.24
66 Weapons & Space	1.29	1.33	1.33	1.31	1.29	1.30	1.27	1.28	1.28	1.28	1.28	1.27	1.27	1.23	1.24	1.24	1.20
67 Military Admin.	1.29	1.33	1.35	1.34	1.32	1.33	1.32	1.32	1.32	1.32	1.32	1.33	1.33	1.31	1.32	1.32	1.31
68 Defense Constr.	1.41	1.35	1.32	1.30	1.34	1.30	1.27	1.26	1.28	1.28	1.31	1.32	1.33	1.39	1.39	1.25	1.26
69 TOTAL DEFENSE	1.28	1.31	1.29	1.28	1.29	1.30	1.26	1.24	1.25	1.24	1.24	1.29	1.23	1.24	1.24	1.22	1.22
70 percent of total	1.29	1.32	1.31	1.30	1.32	1.31	1.30	1.28	1.29	1.28	1.29	1.35	1.30	1.30	1.31	1.29	1.29

Part C

Annual Growth Rates and Price Changes by Sector

TABLE 0.02: SUPPLY AND USES OF POWER
          (billion rubles)

		1965	1970	1971	1972	1973	1974	1975	1976	1977	1978	1979	1980	1981	1982	1983	1984	1985
	Current Prices:																	
1	SUPPLY	6.4	12.0	12.9	14.0	14.8	15.7	16.6	17.2	17.8	18.7	19.3	20.2	20.7	26.7	27.4	28.6	29.4
2	GVO	6.1	11.5	12.4	13.4	14.2	15.0	15.9	16.4	16.8	17.6	18.2	18.9	19.5	25.5	26.0	27.3	28.0
3	Turnover Tax	0.3	0.5	0.5	0.6	0.6	0.7	0.7	0.8	0.9	1.0	1.0	1.1	1.1	1.1	1.2	1.2	1.2
4	TC and TD	0.0	0.0	0.0	0.0	0.0	0.0	0.0	0.0	0.0	0.1	0.1	0.1	0.1	0.1	0.1	0.1	0.2
5	USES	6.4	12.0	12.9	14.0	14.8	15.7	16.6	17.2	17.8	18.7	19.3	20.2	20.7	26.7	27.4	28.6	29.4
6	Interuse	4.6	8.9	9.6	10.3	11.1	11.9	12.5	13.0	13.4	13.9	14.7	15.1	15.5	19.9	20.3	21.1	21.8
7	Domestic End Use	1.9	2.9	3.2	3.5	3.6	3.7	4.0	4.1	4.2	4.5	4.5	4.9	5.0	6.3	6.5	6.9	7.0
8	Consumption	1.5	2.4	2.5	2.8	2.9	3.1	3.1	3.3	3.3	3.5	3.6	3.8	3.9	5.0	5.2	5.4	5.6
9	Private	1.0	1.3	1.4	1.5	1.6	1.7	1.7	1.8	1.8	1.9	2.0	2.1	2.1	2.7	2.8	2.9	3.0
10	Public	0.6	1.0	1.1	1.2	1.3	1.4	1.4	1.5	1.5	1.6	1.6	1.7	1.8	2.3	2.4	2.5	2.6
11	Defense	0.4	0.6	0.6	0.7	0.7	0.7	0.8	0.8	0.9	1.0	0.9	1.0	1.0	1.3	1.3	1.4	1.4
12	Exports	0.0	0.1	0.1	0.1	0.1	0.1	0.1	0.2	0.2	0.2	0.2	0.2	0.3	0.5	0.6	0.7	0.7
	Constant 1973 Prices:																	
13	SUPPLY	8.4	11.9	12.9	14.0	14.8	15.7	16.5	17.7	18.3	19.2	19.9	20.5	21.2	22.4	22.8	23.8	24.6
14	GVO	8.0	11.4	12.4	13.4	14.2	15.0	15.8	16.9	17.3	18.1	18.8	19.4	20.1	21.3	21.8	22.7	23.4
15	Turnover Tax	0.4	0.5	0.5	0.6	0.6	0.7	0.7	0.8	0.9	1.0	1.0	1.0	1.0	1.0	1.0	1.0	1.0
16	TC and TD	0.0	0.0	0.0	0.0	0.0	0.0	0.0	0.0	0.0	0.1	0.1	0.1	0.1	0.1	0.1	0.1	0.2
17	USES	8.4	11.9	12.9	14.0	14.8	15.7	16.5	17.7	18.3	19.2	19.9	20.5	21.2	22.4	22.8	23.8	24.6
18	Interuse	6.1	8.8	9.7	10.3	11.1	11.9	12.5	13.4	13.8	14.3	15.2	15.5	16.0	16.5	17.0	17.5	18.2
19	Domestic End Use	2.5	2.9	3.2	3.5	3.6	3.7	3.9	4.2	4.3	4.6	4.6	5.0	5.1	5.4	5.5	5.7	5.8
20	Consumption	2.0	2.4	2.5	2.7	2.9	3.1	3.1	3.4	3.4	3.6	3.7	3.9	4.1	4.2	4.4	4.5	4.7
21	Private	1.2	1.3	1.4	1.5	1.6	1.7	1.7	1.9	1.9	2.0	2.1	2.2	2.2	2.3	2.3	2.4	2.5
22	Public	0.7	1.0	1.1	1.2	1.3	1.4	1.4	1.5	1.5	1.6	1.7	1.8	1.9	1.9	2.0	2.1	2.2
23	Defense	0.5	0.6	0.6	0.7	0.7	0.7	0.8	0.8	0.9	1.0	0.9	1.1	1.1	1.1	1.1	1.2	1.2
24	Exports	0.0	0.1	0.1	0.1	0.1	0.1	0.1	0.2	0.2	0.2	0.2	0.2	0.3	0.4	0.5	0.6	0.6
25	Price Index	0.76	1.01	1.00	1.00	1.00	1.00	1.00	0.97	0.97	0.97	0.97	0.97	0.97	1.20	1.20	1.20	1.20
	Supporting Data on Output in Physical Units:																	
26	Real Output	0.48	0.68	0.74	0.80	0.85	0.90	0.95	1.01	1.06	1.10	1.14	1.19	1.22	1.26	1.30	1.36	1.41
27	Physical Output	0.51	0.74	0.80	0.86	0.92	0.98	1.04	1.11	1.15	1.20	1.24	1.30	1.33	1.37	1.42	1.49	1.54
28	Prod. Losses	0.03	0.06	0.06	0.06	0.07	0.08	0.09	0.09	0.09	0.10	0.10	0.11	0.11	0.11	0.12	0.13	0.13
29	Unit Price	13.33	17.65	17.43	17.48	17.49	17.52	17.55	16.96	16.83	17.00	16.93	17.03	16.97	21.12	21.08	21.06	20.94

232

TABLE 0.03: SUPPLY AND USES OF FUELS
       (billion rubles)

	1965	1970	1971	1972	1973	1974	1975	1976	1977	1978	1979	1980	1981	1982	1983	1984	1985
Current Prices:																	
1 SUPPLY	21.8	34.9	37.4	40.0	42.7	45.3	48.5	51.8	55.0	60.1	62.5	65.3	66.7	92.6	97.3	99.3	103.1
2 GVO	12.6	22.8	23.7	24.9	26.2	27.8	29.5	31.0	32.4	33.4	33.9	35.0	35.6	58.8	60.1	60.7	61.0
3 Net Taxes	4.1	4.5	5.3	5.9	6.6	7.1	7.7	9.0	9.9	12.8	14.4	15.5	15.7	15.6	17.1	18.4	20.0
4 Taxes	4.1	4.5	5.3	5.9	6.6	7.1	7.7	9.0	9.9	14.0	15.7	16.8	17.1	17.1	17.7	18.4	20.0
5 Subsidies	0.0	0.0	0.0	0.0	0.0	0.0	0.0	0.0	0.0	1.2	1.3	1.4	1.4	1.5	0.0	0.0	0.0
6 Imports	0.2	0.3	0.5	0.8	0.9	0.7	0.8	0.8	0.8	0.9	0.7	0.5	0.5	1.0	1.2	1.4	1.3
7 TC and TD	4.9	7.3	7.9	8.4	9.0	9.7	10.5	11.0	11.9	13.0	13.5	14.3	15.0	17.2	18.3	18.8	20.8
8 USES	21.8	34.9	37.4	40.0	42.7	45.3	48.5	51.8	55.0	60.1	62.5	65.3	66.7	92.6	97.3	99.3	103.1
9 Interuse	17.6	28.0	30.1	32.1	34.0	36.3	38.9	41.4	43.9	47.5	49.8	51.7	52.7	73.8	76.8	78.1	83.2
10 Domestic End Use	3.0	4.5	4.7	5.2	5.8	5.9	6.2	6.4	6.8	8.2	8.1	8.8	9.0	10.7	12.0	12.0	11.6
11 Consumption	1.3	2.1	2.1	2.3	2.7	2.8	3.2	3.6	3.8	4.5	4.9	5.5	5.5	6.7	6.9	7.2	7.4
12 Private	0.4	0.6	0.6	0.7	0.7	0.7	0.8	0.9	1.0	1.1	1.2	1.4	1.4	1.7	1.7	1.8	1.9
13 Public	0.9	1.5	1.5	1.6	1.9	2.0	2.4	2.7	2.8	3.4	3.7	4.1	4.1	5.0	5.2	5.3	5.6
14 Inventories	0.1	0.2	0.1	0.2	0.3	0.3	0.1	0.0	0.1	0.6	0.3	0.5	0.4	0.3	1.3	0.7	0.2
15 Defense	1.6	2.2	2.5	2.7	2.8	2.8	2.9	2.9	2.9	3.1	2.9	2.9	3.1	3.8	3.8	4.1	4.0
16 Exports	1.3	2.4	2.6	2.7	2.9	3.1	3.4	3.9	4.3	4.4	4.6	4.8	5.0	8.1	8.6	9.2	8.4
Constant Prices:																	
17 SUPPLY	28.6	36.1	38.3	40.5	42.7	44.8	47.2	49.6	51.9	53.8	54.8	56.4	57.3	59.6	60.9	62.2	63.6
18 GVO	17.8	22.5	23.7	24.9	26.2	27.6	29.1	30.6	32.1	33.3	34.0	35.1	35.7	37.0	37.8	38.5	39.5
19 Net Taxes	4.5	5.7	6.0	6.3	6.6	7.0	7.3	7.7	8.1	8.4	8.6	8.9	9.0	9.3	9.5	9.7	9.9
20 Imports	0.2	0.3	0.5	0.8	0.9	0.7	0.8	0.8	0.8	0.8	0.6	0.4	0.4	0.6	0.7	0.8	0.7
21 TC and TD	6.1	7.7	8.1	8.5	9.0	9.5	10.0	10.5	11.0	11.4	11.6	12.0	12.2	12.7	12.9	13.2	13.5
22 USES	28.6	36.1	38.3	40.5	42.7	44.8	47.2	49.6	51.9	53.8	54.8	56.4	57.3	59.6	60.9	62.2	63.6
23 Interuse	22.9	29.1	31.0	32.6	34.0	35.9	37.8	39.5	41.0	42.0	43.2	44.0	44.8	46.7	47.1	48.2	50.0
24 Residual Index	0.77	0.96	0.97	0.99	1.00	1.01	1.03	1.05	1.07	1.16	1.18	1.21	1.21	1.61	1.63	1.62	1.66
25 Domestic End Use	4.0	4.6	4.8	5.2	5.8	5.9	6.1	6.3	6.6	7.4	7.2	7.9	8.0	8.1	8.8	8.8	8.6
26 Consumption	1.7	2.1	2.2	2.3	2.7	2.7	3.1	3.4	3.6	3.8	4.1	4.6	4.6	4.8	4.9	5.1	5.2
27 Private	0.5	0.6	0.7	0.7	0.7	0.7	0.8	0.9	0.9	1.0	1.0	1.2	1.2	1.3	1.3	1.3	1.3
28 Public	1.2	1.5	1.5	1.6	1.9	2.0	2.3	2.5	2.7	2.9	3.1	3.4	3.4	3.5	3.6	3.7	3.8
29 Inventories	0.1	0.2	0.1	0.2	0.3	0.3	0.1	0.0	0.1	0.5	0.3	0.4	0.3	0.2	0.8	0.5	0.1
30 Defense	2.2	2.3	2.5	2.7	2.8	2.8	2.9	2.9	2.9	3.1	2.9	2.9	3.1	3.1	3.1	3.3	3.3
31 Exports	1.7	2.4	2.5	2.7	2.9	3.0	3.4	3.8	4.2	4.4	4.4	4.5	4.5	4.8	5.0	5.2	5.0
32 General Index	0.77	0.97	0.98	0.99	1.00	1.02	1.03	1.05	1.06	1.14	1.17	1.19	1.20	1.59	1.61	1.60	1.63
33 Enterprises-1	0.71	1.02	1.00	1.00	1.00	1.01	1.01	1.01	1.01	1.00	1.00	1.00	1.00	1.59	1.59	1.58	1.55
34 Industry-2	0.75	0.97	0.98	0.99	1.00	1.01	1.02	1.05	1.06	1.14	1.17	1.18	1.18	1.64	1.65	1.64	1.64
35 Turnover Tax	0.91	0.79	0.89	0.94	1.00	1.02	1.05	1.17	1.23	1.52	1.68	1.75	1.75	1.68	1.79	1.89	2.01
36 Supply Charges	0.80	0.95	0.97	0.99	1.00	1.03	1.05	1.05	1.08	1.14	1.16	1.19	1.23	1.36	1.42	1.43	1.54
Supporting Data on Output in Physical Units:																	
37 Domestic Output	0.95	1.20	1.26	1.33	1.40	1.47	1.55	1.63	1.71	1.77	1.82	1.87	1.91	1.97	2.01	2.06	2.11
38 Domestic Use	0.83	1.03	1.08	1.14	1.19	1.26	1.31	1.36	1.41	1.46	1.50	1.55	1.59	1.63	1.66	1.69	1.76
39 Exports	0.12	0.17	0.18	0.19	0.21	0.21	0.24	0.27	0.30	0.31	0.31	0.32	0.32	0.34	0.36	0.37	0.35
40 Imports	0.009	0.014	0.023	0.035	0.044	0.031	0.035	0.034	0.034	0.035	0.026	0.017	0.017	0.025	0.030	0.035	0.032
41 Total Unit Price	22.7	28.8	29.1	29.3	29.7	30.1	30.5	31.1	31.5	33.9	34.7	35.3	35.4	47.1	47.6	47.5	48.2
42 Unit Price 1	13.3	19.0	18.8	18.7	18.8	18.9	19.0	19.0	18.9	18.8	18.7	18.7	18.7	29.8	29.8	29.5	29.0
43 Unit Price 2	17.6	22.8	22.9	23.2	23.5	23.7	23.9	24.5	24.7	26.7	27.3	27.6	27.7	38.4	38.6	38.5	38.5

TABLE 0.04: SUPPLY AND USES OF METALS
          (billion rubles)

	1965	1970	1971	1972	1973	1974	1975	1976	1977	1978	1979	1980	1981	1982	1983	1984	1985
Current Prices:																	
1 SUPPLY	24.5	40.7	42.7	45.4	48.2	51.9	55.2	58.3	60.9	63.5	64.5	65.3	65.9	79.7	82.0	84.2	87.2
2 GVO	21.6	36.4	38.0	40.4	42.7	45.2	47.8	50.7	53.2	54.9	54.9	54.8	54.9	67.6	69.9	71.9	74.2
3 Turnover Tax	0.0	0.0	0.1	0.1	0.1	0.1	0.1	0.1	0.1	0.1	0.1	0.1	0.1	0.1	0.1	0.1	0.1
4 Imports	0.7	1.1	1.1	1.2	1.4	2.3	2.7	2.6	2.4	2.8	3.6	4.1	4.3	4.8	4.4	4.4	4.4
5 TC and TD	2.2	3.2	3.5	3.7	4.0	4.3	4.6	4.9	5.2	5.7	5.9	6.3	6.6	7.2	7.6	7.8	8.5
6 USES	24.5	40.7	42.7	45.4	48.2	51.9	55.2	58.3	60.9	63.5	64.5	65.3	65.9	79.7	82.0	84.2	87.2
7 Interuse	22.5	36.2	38.3	40.9	43.3	46.7	49.7	52.2	55.0	57.6	58.6	59.0	59.9	72.0	73.9	76.2	79.3
8 Domestic End Use	0.8	1.4	1.2	1.1	1.5	1.7	2.2	2.4	2.4	2.6	2.6	2.7	2.3	3.0	3.3	3.1	3.2
9 Consumption	0.6	0.7	0.8	0.9	0.9	1.1	1.2	1.3	1.3	1.4	1.4	1.4	1.4	1.7	1.7	1.7	1.8
10 Inventories	0.3	0.6	0.3	0.2	0.4	0.5	0.8	0.8	0.8	0.7	0.8	0.8	0.4	0.7	0.9	0.7	0.7
11 Defense	0.0	0.1	0.1	0.1	0.2	0.2	0.2	0.3	0.3	0.4	0.4	0.5	0.5	0.6	0.7	0.7	0.7
12 Exports	1.3	3.1	3.2	3.3	3.4	3.5	3.4	3.7	3.5	3.3	3.3	3.6	3.7	4.7	4.8	4.9	4.9
Constant Prices:																	
13 SUPPLY	31.1	41.8	43.4	45.8	48.2	51.4	54.1	55.1	56.5	58.3	58.2	57.9	58.0	57.2	58.4	59.5	61.4
14 GVO	27.7	37.1	38.5	40.6	42.7	45.1	47.4	48.5	49.8	51.3	50.7	50.1	50.1	49.2	50.6	51.7	53.5
15 Turnover Tax	0.0	0.0	0.1	0.1	0.1	0.1	0.1	0.1	0.1	0.1	0.1	0.1	0.1	0.1	0.1	0.1	0.1
16 Imports	0.8	1.2	1.2	1.3	1.4	2.0	2.2	2.0	1.8	2.1	2.7	3.0	3.1	3.3	3.0	2.8	2.8
17 TC and TD	2.6	3.5	3.6	3.8	4.0	4.2	4.4	4.5	4.7	4.8	4.7	4.7	4.7	4.6	4.7	4.8	5.0
18 USES	31.1	41.8	43.4	45.8	48.2	51.4	54.1	55.1	56.5	58.3	58.2	57.9	58.0	57.2	58.4	59.5	61.4
19 Interuse	28.4	37.2	38.9	41.3	43.3	46.2	48.6	49.3	50.9	52.8	52.8	52.2	52.6	51.6	52.6	53.8	55.6
20 Domestic End Use	1.0	1.5	1.2	1.2	1.5	1.7	2.2	2.3	2.2	2.3	2.3	2.4	2.0	2.2	2.3	2.2	2.3
21 Consumption	0.7	0.7	0.8	0.9	0.9	1.0	1.2	1.2	1.2	1.3	1.3	1.2	1.2	1.2	1.2	1.2	1.3
22 Inventories	0.3	0.6	0.3	0.2	0.4	0.5	0.8	0.8	0.7	0.6	0.7	0.7	0.4	0.5	0.6	0.5	0.5
23 Defense	0.0	0.1	0.1	0.1	0.2	0.2	0.2	0.3	0.3	0.4	0.3	0.4	0.4	0.4	0.5	0.5	0.5
24 Exports	1.6	3.2	3.3	3.3	3.4	3.5	3.4	3.5	3.3	3.1	3.1	3.3	3.4	3.4	3.5	3.5	3.5
25 General Index	0.79	0.97	0.98	0.99	1.00	1.01	1.02	1.06	1.08	1.09	1.11	1.13	1.14	1.39	1.40	1.41	1.42
26 Enterprises	0.77	0.97	0.98	1.00	1.00	1.00	1.02	1.05	1.07	1.07	1.08	1.09	1.11	1.38	1.39	1.41	1.41
27 Ferrous Metals	0.78	0.98	0.99	1.00	1.00	1.00	1.01	1.05	1.07	1.07	1.08	1.09	1.10	1.37	1.38	1.39	1.39
28 Nonferrous	0.76	0.96	0.97	0.99	1.00	1.01	1.03	1.06	1.08	1.08	1.09	1.10	1.12	1.40	1.42	1.43	1.44
29 TC and TD	0.85	0.92	0.97	0.97	1.00	1.02	1.03	1.08	1.11	1.19	1.24	1.34	1.41	1.56	1.60	1.61	1.70
30 Imports	0.85	0.93	0.94	0.95	1.00	1.13	1.25	1.28	1.30	1.33	1.34	1.37	1.40	1.44	1.49	1.53	1.55

Supporting Data on the GVO of Ferrous Metallurgy and on the Growth Index for Nonferrous Metallurgy:

	1965	1970	1971	1972	1973	1974	1975	1976	1977	1978	1979	1980	1981	1982	1983	1984	1985
31 Ferrous Metals	6.7	11.8	12.2	13.4	14.2	15.5	16.3	17.0	18.7	19.2	19.3	19.1	19.2	23.2	23.5	24.7	25.2
32 GVO (current)	14.9	24.6	25.8	27.0	28.5	29.7	31.5	33.8	34.5	35.7	35.6	35.7	35.7	44.4	46.4	47.2	49.0
33 '70 Growth Rate	0.76	1.00	1.05	1.10	1.16	1.21	1.28	1.34	1.36	1.41	1.41	1.41	1.41	1.43	1.49	1.52	1.57
34 '73 Growth Rate	0.66	0.86	0.91	0.95	1.00	1.04	1.11	1.15	1.18	1.22	1.21	1.22	1.22	1.23	1.28	1.31	1.36
35 N-Ferrous Metals	0.56	0.80	0.87	0.93	1.00	1.07	1.13	1.16	1.21	1.23	1.24	1.25	1.28	1.30	1.30	1.33	1.38

Supporting Data on the Output of Ferrous Metals in Physical Units:

	1965	1970	1971	1972	1973	1974	1975	1976	1977	1978	1979	1980	1981	1982	1983	1984	1985
36 Domestic Output	79.9	104.9	109.5	113.5	118.9	124.0	131.0	135.0	135.3	139.4	137.3	136.5	136.5	135.0	140.6	141.9	147.7
37 Rolling	70.9	92.5	96.1	99.7	104.5	109.0	115.0	118.2	118.3	121.8	119.1	118.3	118.2	117.1	121.9	122.9	128.4
38 Pipes	9.0	12.4	13.4	13.8	14.4	15.0	16.0	16.8	17.0	17.6	18.2	18.2	18.3	17.9	18.7	18.9	19.4
39 Unit Price	0.187	0.234	0.236	0.238	0.239	0.240	0.241	0.250	0.255	0.256	0.259	0.261	0.262	0.329	0.330	0.332	0.332

Supporting Data on Growth Rate of the Total GVO:

	1965	1970	1971	1972	1973	1974	1975	1976	1977	1978	1979	1980	1981	1982	1983	1984	1985
40 '70 Growth Rate	0.65	1.00	1.04	1.11	1.17	1.24	1.31	1.34	1.39	1.43	1.43	1.44	1.45	1.46	1.50	1.54	1.59
41 '73 Growth Rate	0.56	0.85	0.89	0.95	1.00	1.06	1.12	1.15	1.19	1.22	1.22	1.23	1.24	1.25	1.28	1.32	1.36

234

TABLE 0.05: SUPPLY AND USES OF CHEMICAL PRODUCTS--Part I: UNIFIED ACCOUNTS
(billion rubles)

	1965	1970	1971	1972	1973	1974	1975	1976	1977	1978	1979	1980	1981	1982	1983	1984	1985
Current Prices																	
1 SUPPLY	15.4	27.7	30.1	32.4	34.6	38.4	41.0	43.6	46.3	48.7	51.0	54.8	58.0	59.8	62.5	65.7	68.3
2 GVO	12.0	22.5	24.4	26.6	28.8	31.5	33.9	36.2	38.6	40.5	41.9	44.1	46.5	48.8	51.6	54.2	55.4
3 Net Taxes	0.1	0.1	0.2	0.1	0.1	0.1	0.0	0.1	0.1	0.2	0.2	0.2	0.3	0.0	-0.6	-0.6	-0.6
4 Taxes	0.3	0.5	0.6	0.7	0.8	0.9	1.0	1.1	1.2	1.3	1.4	1.5	1.6	1.6	1.7	1.8	1.9
5 Subsidies	0.2	0.4	0.4	0.6	0.7	0.8	1.0	1.0	1.1	1.1	1.2	1.3	1.3	1.6	2.3	2.4	2.5
6 Imports	1.8	3.0	3.2	3.2	3.1	4.0	4.2	4.2	4.3	4.5	5.3	6.7	7.3	6.8	7.2	7.7	8.6
7 TC and TD	1.4	2.1	2.3	2.5	2.6	2.8	2.9	3.1	3.3	3.5	3.6	3.8	3.9	4.2	4.3	4.4	4.9
8 USES	15.4	27.7	30.1	32.4	34.6	38.4	41.0	43.6	46.3	48.7	51.0	54.8	58.0	59.8	62.5	65.7	68.3
9 Interuse	10.3	19.6	21.6	23.6	25.6	28.7	30.8	32.6	34.7	36.6	38.1	40.4	42.9	45.0	46.8	48.7	50.4
10 Domestic End Use	4.8	7.0	7.5	7.8	7.9	8.4	8.8	9.6	10.2	10.6	11.3	12.4	12.7	12.4	13.2	14.1	14.9
11 Consumption	2.2	3.0	3.3	3.6	3.8	4.1	4.3	4.5	4.9	5.2	5.5	5.8	6.1	6.2	6.5	6.8	7.3
12 Private	1.3	1.8	2.0	2.2	2.2	2.3	2.4	2.5	2.7	2.9	3.1	3.2	3.3	3.3	3.4	3.5	3.9
13 Public	0.9	1.2	1.3	1.4	1.6	1.8	1.9	2.0	2.2	2.3	2.4	2.6	2.8	2.9	3.1	3.3	3.4
14 Inventories	0.2	0.6	0.6	0.5	0.5	0.5	0.4	0.7	0.8	0.6	1.0	1.6	1.0	0.8	0.9	0.8	1.0
15 Defense	2.2	3.4	3.6	3.7	3.6	3.8	4.1	4.4	4.5	4.8	4.8	5.0	5.6	5.4	5.8	6.5	6.6
16 Exports	0.4	1.0	1.0	1.0	1.1	1.3	1.3	1.3	1.4	1.5	1.6	2.0	2.4	2.4	2.4	2.9	3.0
Constant Prices																	
17 SUPPLY	17.2	27.5	30.1	32.4	34.6	38.7	41.0	42.8	44.0	45.4	45.7	48.8	50.9	51.2	53.1	54.7	55.6
18 GVO	13.3	21.9	24.1	26.4	28.8	31.9	34.1	35.8	36.9	38.1	37.6	39.4	40.9	41.9	44.0	45.2	45.5
19 Net Taxes	0.3	0.3	0.3	0.2	0.1	0.1	-0.1	-0.1	-0.2	-0.2	-0.1	-0.1	0.0	-0.2	-0.8	-0.8	-0.9
20 Taxes	0.5	0.7	0.7	0.8	0.8	0.9	0.9	0.9	0.9	0.9	0.9	1.0	1.1	1.1	1.1	1.1	1.1
21 Subsidies	0.2	0.4	0.4	0.6	0.7	0.8	1.0	1.0	1.1	1.1	1.0	1.1	1.1	1.3	1.9	2.0	2.0
22 Imports	2.2	3.3	3.5	3.4	3.1	3.8	4.0	3.9	4.0	4.1	4.8	5.9	6.3	5.8	6.0	6.3	6.9
23 TC and TD	1.3	2.0	2.2	2.4	2.6	2.9	3.1	3.2	3.3	3.4	3.4	3.5	3.7	3.8	4.0	4.1	4.1
24 USES	17.2	27.5	30.1	32.4	34.6	38.7	41.0	42.8	44.0	45.4	45.7	48.8	50.9	51.2	53.1	54.7	55.6
25 Interuse	11.3	19.4	21.6	23.6	25.6	28.9	31.0	32.3	33.3	34.5	34.4	36.2	37.9	38.6	39.9	40.6	41.0
26 Domestic End Use	5.6	7.1	7.6	7.8	7.9	8.5	8.7	9.2	9.3	9.5	9.7	10.7	10.7	10.4	10.9	11.4	11.8
27 Consumption	2.8	3.2	3.4	3.7	3.8	4.1	4.1	4.1	4.2	4.4	4.6	4.7	4.9	5.0	5.2	5.3	5.6
28 Private	1.7	1.9	2.1	2.3	2.2	2.3	2.3	2.3	2.3	2.4	2.6	2.6	2.7	2.7	2.7	2.7	3.0
29 Public	1.2	1.3	1.4	1.4	1.6	1.8	1.8	1.8	1.9	1.9	2.0	2.1	2.3	2.3	2.5	2.6	2.6
30 Inventories	0.2	0.6	0.6	0.5	0.5	0.5	0.4	0.7	0.8	0.6	0.9	1.4	0.9	0.7	0.8	0.7	0.8
31 Defense	2.5	3.3	3.5	3.6	3.6	3.8	4.2	4.4	4.3	4.5	4.3	4.5	4.9	4.7	5.0	5.4	5.4
32 Exports	0.3	1.0	1.0	1.0	1.1	1.3	1.3	1.4	1.4	1.5	1.6	1.9	2.3	2.2	2.3	2.7	2.7
33 GENERAL INDEX	0.90	1.01	1.00	1.00	1.00	0.99	1.00	1.02	1.05	1.07	1.12	1.12	1.14	1.17	1.18	1.20	1.23
34 Enterprises	0.90	1.03	1.01	1.01	1.00	0.99	0.99	1.01	1.05	1.06	1.11	1.12	1.14	1.17	1.17	1.20	1.22
35 Syn. Chemicals	1.06	1.08	1.03	1.02	1.00	0.99	0.99	0.98	1.00	1.02	1.04	1.04	1.05	1.06	1.06	1.07	1.10
36 Agri. Chem.	0.97	1.05	1.02	1.02	1.00	1.00	1.01	1.02	1.05	1.07	1.18	1.19	1.20	1.21	1.21	1.22	1.22
37 Rubber	0.77	1.03	1.02	1.01	1.00	0.97	0.96	0.98	1.03	1.04	1.11	1.11	1.14	1.21	1.24	1.28	1.29
38 Paints &Consum.	0.78	0.93	0.96	0.98	1.00	1.01	1.05	1.10	1.16	1.19	1.20	1.22	1.24	1.24	1.25	1.29	1.31
39 Turnover Tax	0.58	0.71	0.81	0.90	1.00	1.05	1.13	1.24	1.35	1.44	1.45	1.47	1.52	1.52	1.54	1.58	1.66
40 TC and TD	1.08	1.07	1.06	1.05	1.00	0.99	0.96	0.97	0.99	1.02	1.06	1.07	1.06	1.11	1.09	1.08	1.20
41 Imports	0.80	0.90	0.93	0.95	1.00	1.05	1.07	1.09	1.10	1.11	1.12	1.14	1.16	1.18	1.20	1.22	1.25

TABLE 0.05: SUPPLY AND USES OF CHEMICAL PRODUCTS--Part II: OUTPUT IN ENTERPRISE PRICES BY SECTOR
      (billion rubles)

		1965	1970	1971	1972	1973	1974	1975	1976	1977	1978	1979	1980	1981	1982	1983	1984	1985
1	SYN. CHEMICALS	3.3	5.3	5.7	6.3	6.8	7.4	7.9	8.5	9.4	10.1	10.4	10.9	11.9	12.4	13.0	14.0	14.3
2	Synres&Plastics	0.9	1.6	1.7	1.9	2.1	2.2	2.5	2.6	2.8	3.0	3.0	3.1	3.5	3.5	3.8	4.2	4.4
3	Syn. Rubber	0.8	1.1	1.2	1.3	1.4	1.6	1.6	1.7	1.9	2.1	2.2	2.4	2.6	2.8	2.8	3.0	3.0
4	Syn. Fibers	1.0	1.7	1.8	1.9	2.1	2.3	2.5	2.7	3.0	3.2	3.2	3.5	3.6	3.7	4.0	4.3	4.4
5	Other Syn.	0.6	0.9	1.0	1.1	1.2	1.3	1.4	1.5	1.7	1.8	2.0	2.0	2.2	2.4	2.4	2.5	2.5
6	AGRI. CHEMICALS	1.2	2.3	2.5	2.7	2.9	3.2	3.6	3.8	4.1	4.2	4.4	4.9	5.2	5.4	5.8	6.1	6.5
7	Minerals	0.2	0.3	0.3	0.4	0.4	0.4	0.5	0.5	0.6	0.6	0.6	0.6	0.7	0.7	0.8	0.9	0.9
8	Fertilizers	0.9	1.7	1.9	2.0	2.2	2.4	2.8	2.9	3.1	3.2	3.3	3.7	3.9	4.1	4.5	4.7	5.1
9	Pesticides	0.1	0.3	0.3	0.3	0.3	0.4	0.4	0.4	0.4	0.5	0.5	0.6	0.6	0.6	0.6	0.6	0.6
10	RUBBER-PLASTICS	3.1	5.3	5.7	6.1	6.7	7.4	8.1	9.2	9.5	10.2	10.7	11.0	11.5	12.3	12.7	13.4	13.8
11	Rubber Products	2.8	4.7	5.0	5.3	5.8	6.4	7.0	7.6	8.0	8.6	9.0	9.3	9.7	10.2	10.6	11.2	11.6
12	Producer	2.2	3.7	3.9	4.1	4.6	5.0	5.5	6.0	6.3	6.8	7.0	7.2	7.5	7.9	8.1	8.5	8.7
13	Tires	1.1	1.9	2.0	2.1	2.3	2.4	2.7	3.0	3.2	3.3	3.5	3.5	3.7	3.9	4.0	4.2	4.3
14	Rubber-Asb.	1.1	1.8	1.9	2.0	2.3	2.6	2.8	2.9	3.1	3.4	3.5	3.7	3.8	4.0	4.2	4.3	4.4
15	Rub. Footware	0.5	0.7	0.7	0.7	0.8	0.8	0.8	0.8	0.8	0.8	0.9	0.9	0.9	1.0	1.1	1.2	1.3
16	Other Consum.	0.1	0.3	0.4	0.5	0.5	0.6	0.7	0.8	0.9	1.0	1.1	1.2	1.3	1.3	1.4	1.6	1.6
17	Fin. Plastics	0.3	0.6	0.7	0.8	0.9	1.0	1.1	1.3	1.5	1.6	1.7	1.7	1.8	2.1	2.1	2.2	2.2
18	PAINTS &CONSUMER	1.7	3.2	3.3	3.6	3.9	4.2	4.6	4.9	5.2	5.4	5.4	5.7	6.0	6.1	6.4	6.5	6.6
19	Paints & Lac.	0.9	1.4	1.4	1.5	1.6	1.8	1.9	2.0	2.1	2.2	2.1	2.2	2.3	2.3	2.4	2.5	2.6
20	Dyes, etc.	0.2	0.4	0.4	0.4	0.5	0.5	0.6	0.7	0.8	0.8	0.8	0.8	0.9	0.9	1.0	1.0	1.0
21	Photochemicals	0.1	0.3	0.3	0.4	0.4	0.4	0.5	0.5	0.6	0.6	0.7	0.7	0.7	0.8	0.8	0.8	0.8
22	Consumer	0.5	1.1	1.2	1.3	1.4	1.5	1.6	1.7	1.7	1.8	1.8	2.0	2.1	2.1	2.2	2.2	2.2
23	OTHER SECTORS	2.7	6.4	7.2	8.0	8.5	9.3	9.6	9.8	10.4	10.6	11.0	11.7	11.9	12.6	13.6	14.2	14.2
24	Total Group "B"	1.1	2.1	2.3	2.5	2.7	2.9	3.2	3.3	3.4	3.6	3.8	4.1	4.3	4.4	4.7	5.0	5.1

TABLE 0.05: SUPPLY AND USES OF CHEMICAL PRODUCTS--Part III: PHYSICAL OUTPUT AND PRICE INDEX FOR SELECTED SECTORS

	1965	1970	1971	1972	1973	1974	1975	1976	1977	1978	1979	1980	1981	1982	1983	1984	1985
1 Syn. Fibers	0.41	0.62	0.68	0.74	0.83	0.89	0.96	1.02	1.09	1.13	1.10	1.18	1.21	1.24	1.35	1.40	1.39
2 Unit Price	2.44	2.74	2.65	2.62	2.58	2.58	2.58	2.61	2.71	2.80	2.91	2.92	2.98	2.99	2.99	3.07	3.16
3 '73 Index	0.95	1.06	1.03	1.02	1.00	1.00	1.00	1.01	1.05	1.08	1.13	1.13	1.15	1.16	1.16	1.19	1.22
4 Synres& Plastics	0.80	1.67	1.86	2.04	2.32	2.49	2.84	3.06	3.31	3.52	3.48	3.64	4.09	4.06	4.42	4.82	5.02
5 Unit Price	1.06	0.98	0.94	0.92	0.90	0.88	0.88	0.86	0.86	0.85	0.85	0.85	0.86	0.86	0.86	0.86	0.88
6 '73 Index	1.18	1.09	1.04	1.02	1.00	0.98	0.98	0.96	0.95	0.95	0.95	0.95	0.95	0.96	0.96	0.96	0.97
7 Fertilizers	7.4	13.1	14.7	15.9	17.4	19.4	22.0	22.6	23.5	23.7	22.1	24.8	26.0	26.7	29.7	30.8	33.2
8 Unit Price	12.16	13.17	12.76	12.74	12.50	12.45	12.57	12.74	13.09	13.42	14.80	14.85	15.00	15.17	15.14	15.25	15.27
9 '73 Index	0.97	1.05	1.02	1.02	1.00	1.00	1.01	1.02	1.05	1.07	1.18	1.19	1.20	1.21	1.21	1.22	1.22
10 Tires	26.4	34.6	36.2	38.8	42.3	47.1	51.5	57.4	57.4	59.0	60.0	60.1	60.5	61.7	62.0	63.7	65.2
11 Unit Price	4.09	5.61	5.52	5.47	5.42	5.18	5.15	5.29	5.57	5.63	5.90	5.87	6.07	6.30	6.37	6.57	6.60
12 '73 Index	0.75	1.03	1.02	1.01	1.00	0.96	0.95	0.98	1.03	1.04	1.09	1.08	1.12	1.16	1.18	1.21	1.22
13 Rubber Footware	161	173	179	180	194	205	205	203	197	195	194	193	192	192	199	208	218
14 Unit Price	0.31	0.39	0.39	0.39	0.39	0.39	0.39	0.39	0.40	0.41	0.46	0.47	0.47	0.53	0.55	0.58	0.60
15 73 index	0.80	1.01	1.00	1.00	1.00	1.00	1.00	1.01	1.02	1.05	1.19	1.20	1.20	1.36	1.42	1.48	1.53
16 Paints & Lac.	1.62	2.38					3.02	3.03	3.04	3.06	2.94	2.88	3.03	2.98	3.13	3.23	3.30
17 Unit Price	0.56	0.59			0.60		0.63	0.66	0.69	0.72	0.73	0.75	0.76	0.76	0.77	0.77	0.78
18 '73 Index	0.93	0.98	0.99	1.00	1.00	1.01	1.05	1.10	1.15	1.20	1.21	1.25	1.27	1.27	1.28	1.29	1.30
19 Consumer Chem.	1.23	1.91	2.00	2.10	2.16	2.30	2.34	2.37	2.23	2.29	2.27	2.58	2.67	2.71	2.75	2.63	2.62
20 Unit Price	0.41	0.58	0.60	0.62	0.65	0.65	0.68	0.72	0.76	0.77	0.78	0.78	0.79	0.79	0.80	0.84	0.85
21 '73 Index	0.63	0.89	0.92	0.95	1.00	1.00	1.05	1.10	1.17	1.18	1.19	1.19	1.21	1.21	1.23	1.29	1.32

TABLE 0.06: SUPPLY AND USES OF MBMW PRODUCTS--PART A: UNIFIED ACCOUNTS
       (billion rubles)

	1965	1970	1971	1972	1973	1974	1975	1976	1977	1978	1979	1980	1981	1982	1983	1984	1985
Current Prices:																	
1 SUPPLY	60.6	105.6	112.0	123.5	129.2	142.1	157.9	165.1	175.9	191.2	201.3	213.7	225.7	238.0	253.5	267.1	282.0
2 Total GVO	50.9	93.0	99.0	108.3	112.3	123.0	133.9	138.5	147.4	158.7	167.8	177.6	186.2	194.1	205.5	217.3	228.8
3 a) Published	47.7	88.6	94.2	102.8	106.3	116.5	127.0	130.7	139.0	148.7	157.6	167.5	175.9	182.4	193.5	205.0	216.2
4 b) Hidden	3.2	4.4	4.8	5.5	6.0	6.5	6.9	7.8	8.4	10.0	10.2	10.1	10.3	11.7	12.0	12.3	12.6
5 Net Taxes	3.0	3.1	3.4	4.2	5.0	5.8	6.6	6.9	7.4	8.0	8.6	9.1	10.4	10.3	10.2	10.2	10.5
6 Total Taxes	3.2	3.6	3.9	4.8	5.7	6.7	7.4	7.8	8.3	8.9	9.5	10.1	11.4	11.6	12.1	12.6	13.1
7 Subsidies	0.2	0.5	0.5	0.6	0.7	0.9	0.8	0.9	0.9	0.9	0.9	1.0	1.0	1.3	1.9	2.4	2.6
8 Imports	3.1	4.9	4.7	5.7	6.4	7.4	10.7	12.6	13.7	16.9	16.9	18.5	20.1	24.0	27.7	29.2	31.3
9 TC and TD	3.6	4.6	4.9	5.3	5.5	5.9	6.7	7.1	7.4	7.6	8.0	8.5	9.0	9.6	10.1	10.4	11.4
10 USES	60.6	105.6	112.0	123.5	129.2	142.1	157.9	165.1	175.9	191.2	201.3	213.7	225.7	238.0	253.5	267.1	282.0
11 Interuse	22.9	44.9	47.9	52.4	54.3	57.6	63.9	64.3	67.5	73.0	76.1	82.2	84.9	86.7	90.7	95.2	99.2
12 Domestic End Use	35.5	56.7	60.0	66.6	69.0	77.3	87.0	92.7	98.5	107.1	113.5	119.1	127.4	136.6	147.6	155.5	166.6
13 Consumption	5.8	10.7	11.4	13.2	14.4	16.0	17.5	18.2	19.2	20.8	22.0	23.7	26.5	26.6	29.4	30.7	32.8
14 Private	4.4	8.0	8.5	10.0	10.8	12.2	13.5	14.3	15.2	16.4	17.3	18.5	20.8	21.1	22.9	23.8	25.4
15 Public	1.4	2.7	2.9	3.2	3.6	3.8	4.0	3.9	4.0	4.4	4.7	5.3	5.7	5.5	6.5	6.9	7.4
16 Capital Outlays	22.6	35.5	37.9	41.3	43.7	48.0	54.2	57.9	62.2	67.9	71.0	74.0	78.2	82.5	88.7	90.8	96.9
17 Inventories	1.2	2.9	2.6	3.3	2.7	3.9	4.5	4.6	4.9	4.9	6.5	7.3	3.4	7.4	6.3	6.8	7.4
18 Defense	5.8	7.7	8.2	8.7	8.3	9.4	10.7	12.0	12.2	13.5	13.9	14.1	19.5	20.2	23.1	27.1	29.5
19 Exports	2.1	4.1	4.3	4.5	5.9	6.9	7.1	8.1	9.9	11.1	11.7	12.4	13.4	14.8	15.3	16.4	16.3
Constant Prices:																	
20 SUPPLY	69.2	107.9	114.4	124.6	129.2	137.6	147.8	153.0	157.8	166.8	171.8	177.7	181.9	186.9	193.2	199.2	203.7
21 Total GVO	56.8	92.9	99.2	107.7	112.3	119.7	127.0	130.8	134.8	141.0	145.9	150.6	153.6	156.3	160.6	165.7	169.2
22 a) Published	53.5	88.8	94.5	102.3	106.3	113.3	120.3	123.0	126.7	131.5	136.4	141.5	144.6	146.1	150.4	155.5	158.9
23 b) Hidden	3.3	4.2	4.7	5.4	6.0	6.4	6.7	7.8	8.2	9.5	9.4	9.1	9.0	10.2	10.2	10.3	10.2
24 Net Taxes	4.0	4.2	4.5	4.9	5.0	5.2	5.7	5.8	6.1	6.4	6.7	6.9	7.0	7.0	6.8	6.7	6.9
25 Total Taxes	4.2	4.7	5.0	5.5	5.7	6.1	6.5	6.6	6.8	7.2	7.4	7.6	7.8	7.9	8.2	8.4	8.6
26 Subsidies	0.2	0.5	0.5	0.6	0.7	0.9	0.7	0.8	0.8	0.8	0.7	0.8	0.8	1.0	1.4	1.7	1.7
27 Imports	5.7	6.2	5.8	6.8	6.4	6.8	8.9	10.0	10.3	12.4	12.1	12.9	13.7	16.0	18.0	18.6	19.3
28 TC and TD	2.8	4.6	4.9	5.3	5.5	5.9	6.2	6.4	6.6	6.9	7.1	7.4	7.5	7.7	7.9	8.1	8.3
29 USES	69.2	107.9	114.4	124.6	129.2	137.6	147.8	153.0	157.8	166.8	171.8	177.7	181.9	186.9	193.2	199.2	203.7
30 Interuse	25.2	44.0	47.5	52.1	54.2	56.6	60.5	60.4	61.9	65.6	67.3	71.0	71.3	71.0	72.8	74.5	75.7
31 price index	0.91	1.02	1.01	1.01	1.00	1.02	1.06	1.06	1.09	1.11	1.13	1.16	1.19	1.22	1.25	1.28	1.31
32 Domestic End Use	41.7	59.9	62.6	67.9	69.1	74.2	80.6	84.8	86.8	91.3	94.4	96.3	99.6	104.1	108.6	112.2	116.1
33 Consumption	7.0	11.8	12.5	13.8	14.4	15.2	16.0	16.4	16.8	17.9	18.4	19.2	20.4	20.1	21.7	22.1	23.1
34 Private	5.4	9.2	9.6	10.6	10.8	11.5	12.2	12.7	13.1	13.9	14.2	14.6	15.6	15.5	16.4	16.6	17.3
35 Public	1.5	2.6	2.9	3.2	3.6	3.7	3.8	3.7	3.7	4.0	4.2	4.6	4.9	4.6	5.3	5.5	5.7
36 Capital Outlays	27.5	37.8	39.6	42.3	43.7	45.9	49.9	52.0	53.5	56.1	57.3	57.9	59.2	60.3	62.1	62.0	63.2
37 price index	0.82	0.94	0.96	0.98	1.00	1.05	1.09	1.11	1.16	1.21	1.24	1.28	1.32	1.37	1.43	1.46	1.53
38 Inventories	1.3	2.8	2.6	3.3	2.7	3.8	4.3	4.4	4.6	4.4	5.8	6.4	2.9	6.2	5.2	5.5	5.8
39 Defense	6.0	7.4	7.9	8.6	8.3	9.3	10.4	12.0	11.9	12.9	12.9	12.7	17.1	17.5	19.6	22.6	24.0
40 Exports	2.3	4.0	4.3	4.5	5.9	6.8	6.7	7.8	9.1	9.9	10.2	10.5	10.9	11.8	11.8	12.4	11.9
41 GENERAL INDEX	0.88	0.98	0.98	0.99	1.00	1.03	1.07	1.08	1.11	1.15	1.17	1.20	1.24	1.27	1.31	1.34	1.38
42 Enterprises	0.90	1.00	1.00	1.01	1.00	1.03	1.05	1.06	1.09	1.13	1.15	1.18	1.21	1.24	1.28	1.31	1.35
43 Turnover Tax	0.76	0.77	0.78	0.88	1.00	1.10	1.15	1.17	1.21	1.24	1.28	1.32	1.46	1.46	1.48	1.50	1.52
44 Imports	0.55	0.78	0.81	0.85	1.00	1.10	1.20	1.26	1.32	1.36	1.40	1.43	1.47	1.50	1.54	1.57	1.62
45 TC and TD	1.00	1.07	1.05	1.04	1.00	1.00	1.03	1.04	1.04	1.02	1.04	1.06	1.10	1.15	1.17	1.18	1.25
46 Exports	0.91	1.01	1.00	1.01	1.00	1.02	1.05	1.05	1.09	1.12	1.15	1.18	1.22	1.25	1.29	1.32	1.37

TABLE 0.06: SUPPLY AND USES OF MBMW PRODUCTS--PART B: DERIVATION OF 1973 MACHINERY PRODUCTION PRICE INDEXES

(billion rubles)   Page 1

		1965	1970	1971	1972	1973	1974	1975	1976	1977	1978	1979	1980	1981	1982	1983	1984	1985
1	TOTAL MBMW GVO	50.9	93.0	99.0	108.3	112.3	123.0	133.9	138.5	147.4	158.7	167.8	177.6	186.2	194.1	205.5	217.3	228.8
2	1973 prices	56.8	92.9	99.2	107.7	112.3	119.7	127.0	130.8	134.8	141.0	145.9	150.6	153.6	156.3	160.6	165.7	169.2
3	1973 index	0.90	1.00	1.00	1.01	1.00	1.03	1.05	1.06	1.09	1.13	1.15	1.18	1.21	1.24	1.28	1.31	1.35
4	INTERIND. USE	18.5	40.4	42.6	46.4	47.2	52.0	56.9	57.3	60.9	65.2	69.4	74.2	77.5	80.6	85.3	91.6	96.7
5	1973 prices	20.1	39.2	42.0	45.8	47.2	50.7	54.2	54.6	56.7	59.3	62.0	65.1	66.5	67.5	69.9	73.3	75.5
6	1973 index	0.92	1.03	1.01	1.01	1.00	1.02	1.05	1.05	1.07	1.10	1.12	1.14	1.17	1.19	1.22	1.25	1.28
7	Parts & Tools	9.6	18.6	19.8	21.5	22.0	24.1	26.3	26.8	28.5	30.3	31.9	33.9	35.4	37.0	39.1	41.9	44.2
8	1973 prices	11.2	18.6	19.8	21.2	22.0	23.3	25.0	25.3	26.1	26.8	27.7	28.7	29.3	29.6	30.6	32.0	32.7
9	1973 index	0.86	1.00	1.00	1.01	1.00	1.03	1.05	1.06	1.09	1.13	1.15	1.18	1.21	1.25	1.28	1.31	1.35
10	Complex Compon.	8.8	21.8	22.8	24.9	25.2	27.9	30.6	30.5	32.4	34.9	37.5	40.3	42.0	43.6	46.2	49.8	52.5
11	1973 prices	8.9	20.6	22.2	24.5	25.2	27.4	29.1	29.3	30.6	32.5	34.3	36.4	37.2	37.9	39.4	41.3	42.8
12	1973 index	1.00	1.06	1.03	1.02	1.00	1.02	1.05	1.04	1.06	1.08	1.10	1.11	1.13	1.15	1.17	1.20	1.23
13	MACHINERY*	19.6	31.8	34.1	36.9	39.0	42.4	45.9	49.2	52.6	56.0	58.9	61.8	64.5	67.1	71.1	73.5	77.5
14	1973 prices	23.1	33.4	35.2	37.3	39.0	40.7	42.9	44.7	45.7	46.6	47.9	48.7	49.1	49.4	50.0	50.3	50.7
15	1973 index	0.85	0.95	0.97	0.99	1.00	1.04	1.07	1.10	1.15	1.20	1.23	1.27	1.32	1.36	1.42	1.46	1.53
16	Power MB	0.6	0.8	0.8	0.8	0.9	0.9	0.9	1.0	1.0	1.1	1.1	1.2	1.2	1.3	1.3	1.3	1.4
17	1973 prices	0.7	0.9	0.9	0.9	0.9	0.9	0.9	0.9	0.9	0.9	0.9	0.9	0.8	0.8	0.7	0.7	0.7
18	1973 index	0.80	0.91	0.93	0.99	1.00	1.04	1.09	1.15	1.20	1.25	1.28	1.31	1.53	1.68	1.84	2.00	2.10
19	Metallurgy MB	0.3	0.4	0.4	0.5	0.5	0.6	0.6	0.7	0.7	0.7	0.7	0.7	0.7	0.8	0.8	0.9	0.9
20	1973 prices	0.4	0.4	0.4	0.5	0.5	0.6	0.5	0.6	0.6	0.6	0.5	0.5	0.5	0.6	0.6	0.6	0.6
21	1973 index	0.78	0.94	0.97	1.00	1.00	1.08	1.13	1.17	1.23	1.28	1.32	1.35	1.39	1.43	1.45	1.47	1.49
22	Hoisting-Transp.	0.6	1.1	1.1	1.2	1.2	1.3	1.4	1.5	1.6	1.6	1.7	1.8	1.8	1.9	1.9	1.9	2.0
23	1973 prices	0.7	1.1	1.1	1.2	1.2	1.3	1.3	1.4	1.4	1.4	1.4	1.4	1.3	1.3	1.3	1.3	1.3
24	1973 index	0.80	1.01	1.02	1.03	1.00	1.02	1.07	1.08	1.13	1.16	1.22	1.32	1.39	1.42	1.46	1.48	1.50
25	Railroad MB	0.8	1.1	1.2	1.2	1.2	1.3	1.4	1.5	1.6	1.7	1.7	1.7	1.8	1.8	1.9	2.0	2.0
26	1973 prices	0.8	1.1	1.1	1.2	1.2	1.3	1.3	1.4	1.4	1.4	1.4	1.4	1.3	1.3	1.3	1.3	1.4
27	1973 index	0.90	1.00	1.02	1.03	1.00	1.04	1.10	1.11	1.14	1.16	1.20	1.24	1.30	1.34	1.40	1.46	1.51
28	E-Technical MB	1.1	1.7	1.8	1.9	1.9	2.2	2.4	2.5	2.7	2.8	2.9	3.0	3.1	3.1	3.2	3.3	3.5
29	1973 prices	1.6	1.9	1.9	1.9	1.9	1.8	1.8	1.8	1.8	1.8	1.9	1.9	1.9	1.9	1.9	1.9	2.0
30	1973 index	0.72	0.90	0.94	0.98	1.00	1.18	1.33	1.38	1.45	1.50	1.53	1.56	1.62	1.64	1.67	1.71	1.78
31	Oil MB	0.5	0.7	0.7	0.8	0.8	1.0	1.0	1.1	1.2	1.3	1.4	1.5	1.5	1.6	1.8	1.9	2.0
32	1973 prices	0.6	0.8	0.8	0.8	0.8	0.8	0.8	0.8	0.9	0.9	1.0	1.0	1.1	1.1	1.2	1.2	1.3
33	1973 index	0.76	0.94	0.96	1.00	1.00	1.18	1.26	1.32	1.35	1.38	1.40	1.42	1.45	1.50	1.53	1.55	1.58
34	Trucks	1.2	1.9	2.2	2.5	2.7	3.1	3.4	3.8	4.1	4.5	4.9	5.2	5.5	5.7	5.9	6.0	6.2
35	1973 prices	1.3	2.0	2.2	2.4	2.7	2.9	3.1	2.9	3.0	3.2	3.5	3.6	3.8	3.8	3.9	3.8	3.9
36	1973 index	0.88	0.96	1.00	1.04	1.00	1.07	1.10	1.32	1.35	1.38	1.40	1.42	1.47	1.50	1.53	1.55	1.58
37	Tractors	0.7	1.0	1.1	1.2	1.3	1.3	1.4	1.6	1.8	1.8	1.9	2.1	2.2	2.3	2.3	2.4	2.5
38	1973 prices	0.8	1.1	1.1	1.2	1.3	1.4	1.4	1.5	1.5	1.5	1.5	1.5	1.5	1.5	1.6	1.6	1.6
39	1973 index	0.91	0.91	0.92	1.03	1.00	0.96	0.98	1.08	1.19	1.20	1.24	1.37	1.44	1.46	1.50	1.52	1.54

*excluding precision instruments and computers but including capital repair

TABLE 0.06: SUPPLY AND USES OF MBMW PRODUCTS--PART B: DERIVATION OF 1973 MACHINERY PRODUCTION PRICE INDEXES
         (billion rubles)

	1965	1970	1971	1972	1973	1974	1975	1976	1977	1978	1979	1980	1981	1982	1983	1984	1985
40 Construction MB	0.7	1.1	1.2	1.3	1.3	1.4	1.5	1.6	1.7	1.7	1.8	1.8	1.9	1.9	2.0	2.0	2.0
41 1973 prices	0.7	1.1	1.1	1.2	1.3	1.4	1.4	1.4	1.5	1.4	1.4	1.4	1.4	1.3	1.3	1.3	1.2
42 1973 index	1.01	1.04	1.02	1.04	1.00	1.05	1.06	1.10	1.15	1.26	1.33	1.36	1.40	1.43	1.50	1.56	1.58
43 PRECISION INSTR+	0.5	1.7	1.8	2.0	2.2	2.7	3.1	3.2	3.5	3.9	4.3	4.7	5.1	5.0	5.3	5.6	5.9
44 1973 prices	0.5	1.6	1.8	2.0	2.2	2.7	3.0	3.2	3.4	3.7	4.0	4.2	4.5	4.3	4.5	4.7	4.8
45 1973 index	1.00	1.06	1.03	1.02	1.00	1.01	1.03	1.00	1.03	1.05	1.08	1.11	1.14	1.15	1.18	1.20	1.23
46 CONSUMER PROD.	2.4	4.0	4.4	5.0	5.3	6.0	6.6	6.7	7.2	7.7	8.0	8.5	9.0	9.3	9.9	10.6	11.1
47 1973 prices	2.7	4.3	4.6	5.0	5.3	5.8	6.2	6.3	6.6	6.9	6.9	7.0	7.3	7.3	7.5	7.6	7.8
48 1973 index	0.87	0.94	0.96	1.00	1.00	1.03	1.07	1.06	1.09	1.12	1.16	1.21	1.24	1.27	1.32	1.39	1.42
49 N-PROD. USE	1.3	2.5	2.7	2.9	3.0	3.2	3.4	3.3	3.4	3.7	3.9	4.2	4.4	4.7	4.9	5.2	5.5
50 1973 prices	1.4	2.4	2.6	2.9	3.0	3.1	3.2	3.1	3.2	3.3	3.5	3.7	3.8	3.9	4.0	4.1	4.2
51 1973 index	0.93	1.03	1.01	1.01	1.00	1.02	1.05	1.05	1.08	1.10	1.12	1.14	1.17	1.20	1.23	1.26	1.29
52 DEFENSE PROD.	8.6	12.7	13.4	15.0	15.6	16.8	17.9	18.8	19.8	22.3	23.4	24.3	25.6	27.5	29.0	30.8	32.2
53 1973 prices	9.0	12.0	13.0	14.7	15.6	16.7	17.4	18.8	19.2	21.2	21.7	21.8	22.5	23.9	24.6	25.6	26.2
54 1973 index	0.96	1.06	1.03	1.02	1.00	1.01	1.03	1.00	1.03	1.05	1.08	1.11	1.14	1.15	1.18	1.20	1.23

+including computers

TABLE 0.06: SUPPLY AND USES OF MBMW PRODUCTS--PART B: DERIVATION OF 1973 MACHINERY PRODUCTION PRICE INDEXES
(billion rubles)

	1965	1970	1971	1972	1973	1974	1975	1976	1977	1978	1979	1980	1981	1982	1983	1984	1985
1 PUBLISHED GVO	47.7	88.6	94.2	102.8	106.3	116.5	127.0	130.7	139.0	148.7	157.6	167.5	175.9	182.4	193.5	205.0	216.2
2 INTERIND. USE	18.5	40.4	42.6	46.4	47.2	52.0	56.9	57.3	60.9	65.2	69.4	74.2	77.5	80.6	85.3	91.6	96.7
3 Parts & Tools	9.6	18.6	19.8	21.5	22.0	24.1	26.3	26.8	28.5	30.3	31.9	33.9	35.4	37.0	39.1	41.9	44.2
4 Power & Diesel	0.1	0.2	0.2	0.2	0.2	0.3	0.3	0.3	0.4	0.4	0.4	0.4	0.5	0.5	0.6	0.6	0.7
5 Metal&Mining MB	0.1	0.1	0.1	0.1	0.2	0.2	0.2	0.2	0.2	0.2	0.2	0.2	0.2	0.2	0.2	0.3	0.3
6 Hoisting-Transp	0.1	0.1	0.2	0.2	0.2	0.2	0.2	0.3	0.3	0.3	0.3	0.3	0.3	0.4	0.4	0.5	0.5
7 Transport MB	0.1	0.1	0.1	0.1	0.1	0.2	0.2	0.2	0.2	0.2	0.2	0.2	0.2	0.2	0.2	0.2	0.2
8 E-technical MB	0.5	0.8	0.8	0.8	0.8	0.8	0.9	0.9	0.9	1.0	1.0	1.0	1.1	1.1	1.1	1.2	1.2
9 Cable Products	1.0	1.6	1.7	1.7	1.7	1.8	1.9	2.0	2.1	2.2	2.2	2.3	2.4	2.4	2.5	2.7	2.8
10 Chemical MB	0.3	0.5	0.6	0.7	0.7	0.7	0.8	0.9	0.9	1.0	1.0	1.1	1.1	1.2	1.3	1.4	1.5
11 Metal Cutting	0.1	0.2	0.2	0.2	0.2	0.3	0.3	0.3	0.4	0.4	0.5	0.5	0.6	0.7	0.6	0.6	0.6
12 Tools & Dies	0.3	0.6	0.6	0.7	0.7	0.8	0.9	0.9	1.0	1.0	1.1	1.2	1.2	1.3	1.4	1.6	1.7
13 Abrasives	0.2	0.3	0.4	0.4	0.4	0.5	0.5	0.5	0.6	0.7	0.7	0.7	0.7	0.7	0.8	0.8	0.8
14 Precision Instr	0.2	0.5	0.5	0.5	0.5	0.6	0.6	0.7	0.7	0.8	0.9	1.0	1.0	1.1	1.2	1.3	1.4
15 Computer Parts	0.0	0.1	0.1	0.2	0.2	0.2	0.3	0.3	0.3	0.3	0.3	0.4	0.4	0.5	0.6	0.7	0.8
16 Automotive MB	0.4	0.8	0.9	1.0	1.1	1.2	1.2	1.3	1.4	1.6	1.7	1.8	1.8	1.9	1.9	2.1	2.2
17 Bearings	0.3	0.6	0.7	0.8	0.9	1.0	1.2	1.1	1.2	1.4	1.5	1.5	1.6	1.7	1.7	1.8	1.9
18 Agricultural MB	0.3	0.6	0.7	0.8	0.9	1.0	1.1	1.1	1.2	1.3	1.4	1.4	1.5	1.5	1.6	1.8	1.9
19 Construction MB	0.1	0.2	0.3	0.3	0.3	0.4	0.4	0.4	0.4	0.4	0.4	0.5	0.5	0.6	0.6	0.6	0.7
20 Const Mater. MB	0.0	0.1	0.1	0.1	0.1	0.1	0.1	0.1	0.1	0.1	0.1	0.1	0.1	0.1	0.1	0.1	0.2
21 Light Ind. MB	0.1	0.2	0.2	0.2	0.2	0.2	0.2	0.2	0.2	0.2	0.3	0.3	0.3	0.3	0.3	0.3	0.3
22 Food Ind. MB	0.0	0.1	0.1	0.1	0.1	0.1	0.1	0.1	0.1	0.1	0.1	0.1	0.1	0.1	0.1	0.2	0.2
23 Pipe Fittings	0.2	0.3	0.3	0.3	0.4	0.4	0.4	0.4	0.5	0.5	0.5	0.6	0.6	0.6	0.6	0.6	0.6
24 Electronics	0.7	2.5	2.5	2.8	2.8	3.1	3.5	3.4	3.5	3.7	4.1	4.5	4.8	4.9	5.3	5.7	6.0
25 Metal Structure	0.9	1.6	1.7	1.8	1.9	2.0	2.2	2.2	2.2	2.3	2.4	2.5	2.6	2.6	2.8	2.9	3.1
26 Metal Wares	1.3	2.5	2.6	2.8	2.8	3.0	3.3	3.3	3.3	3.5	3.6	3.7	3.9	3.9	4.2	4.4	4.7
27 Current Repair	2.2	4.0	4.2	4.5	4.7	5.1	5.5	5.8	6.3	6.7	7.0	7.6	8.0	8.4	8.9	9.5	10.0
28 Complex Componen	8.8	21.8	22.8	24.9	25.2	27.9	30.6	30.5	32.4	34.9	37.5	40.3	42.0	43.6	46.2	49.8	52.5
29 Diesel Engines	0.3	0.4	0.4	0.5	0.5	0.5	0.5	0.5	0.6	0.6	0.6	0.6	0.6	0.7	0.7	0.8	0.8
30 Transport MB	0.4	0.5	0.5	0.5	0.6	0.6	0.6	0.7	0.7	0.8	0.8	0.8	0.8	0.8	0.9	0.9	0.9
31 E-technical	2.0	3.3	3.4	3.4	3.2	3.4	3.6	3.6	3.7	3.9	4.0	4.2	4.3	4.4	4.5	4.8	4.9
32 Precision Instr	0.8	1.9	1.9	2.1	2.1	2.3	2.6	2.6	2.9	3.2	3.4	3.8	4.1	4.4	4.7	5.2	5.5
33 Automotive MB	1.6	3.1	3.6	4.0	4.3	4.7	5.0	5.3	5.8	6.4	6.9	7.2	7.4	7.6	7.8	8.2	8.8
34 Agricultural MB	1.3	2.6	2.9	3.2	3.5	3.9	4.3	4.4	4.8	5.2	5.4	5.7	5.8	6.1	6.5	7.0	7.5
35 Electronics	2.6	10.1	10.1	11.3	11.1	12.5	14.0	13.4	14.0	14.8	16.4	18.0	19.0	19.6	21.1	22.9	24.0

TABLE 0.06: SUPPLY AND USES OF MBMW PRODUCTS--PART B: DERIVATION OF 1973 MACHINERY PRODUCTION PRICE INDEXES
(billion rubles)

	1965	1970	1971	1972	1973	1974	1975	1976	1977	1978	1979	1980	1981	1982	1983	1984	1985
36 MACHINERY	14.7	24.4	26.2	28.5	30.2	33.2	36.4	38.9	41.4	44.0	46.7	48.9	50.9	52.4	55.3	57.0	59.9
37 Power	0.6	0.8	0.8	0.8	0.9	0.9	0.9	1.0	1.0	1.1	1.1	1.2	1.2	1.3	1.3	1.3	1.4
38 Metallurgy	0.3	0.4	0.4	0.5	0.5	0.6	0.6	0.7	0.7	0.7	0.7	0.7	0.7	0.8	0.8	0.9	0.9
39 Mining	0.5	0.9	1.0	1.0	1.1	1.1	1.1	1.2	1.2	1.3	1.3	1.4	1.4	1.5	1.6	1.6	1.7
40 Hoisting-Transp.	0.6	1.1	1.1	1.2	1.2	1.3	1.4	1.5	1.6	1.6	1.7	1.8	1.8	1.9	1.9	1.9	2.0
41 Railroad	0.8	1.1	1.2	1.2	1.2	1.3	1.4	1.5	1.6	1.7	1.7	1.7	1.8	1.8	1.9	2.0	2.0
42 Civil Shipbuild.	0.3	0.6	0.6	0.6	0.6	0.7	0.7	0.8	0.8	0.8	0.8	0.8	0.9	0.9	0.9	1.0	1.0
43 Civil Aircraft	1.2	2.2	2.3	2.5	2.5	2.6	2.8	3.1	3.2	3.3	3.4	3.5	3.5	3.5	3.7	3.8	4.0
44 Electrotechnical	1.1	1.7	1.8	1.9	1.9	2.2	2.4	2.5	2.7	2.8	2.9	3.0	3.1	3.1	3.2	3.3	3.5
45 Chemical MB	0.2	0.3	0.3	0.4	0.4	0.4	0.4	0.5	0.5	0.5	0.6	0.6	0.6	0.7	0.7	0.8	0.8
46 Compressors, etc	0.2	0.3	0.3	0.3	0.4	0.4	0.4	0.4	0.4	0.4	0.5	0.5	0.5	0.6	0.7	0.7	0.8
47 Pumps	0.3	0.5	0.5	0.6	0.6	0.7	0.7	0.7	0.7	0.8	0.9	0.9	1.0	1.0	1.1	1.2	1.3
48 Oil-Gas Drills	0.3	0.4	0.4	0.5	0.5	0.6	0.6	0.6	0.7	0.8	0.8	0.9	0.9	1.0	1.1	1.2	1.2
49 Oil Processing	0.2	0.3	0.3	0.3	0.3	0.4	0.4	0.5	0.5	0.5	0.6	0.6	0.6	0.6	0.7	0.7	0.8
50 Logging	0.1	0.2	0.2	0.2	0.2	0.2	0.3	0.3	0.3	0.3	0.3	0.3	0.4	0.4	0.4	0.4	0.4
51 Pulp & Paper	0.0	0.1	0.1	0.1	0.1	0.1	0.1	0.1	0.2	0.2	0.2	0.2	0.2	0.2	0.2	0.2	0.2
52 Metal-Cutting	0.5	0.9	1.0	1.0	1.1	1.2	1.4	1.5	1.7	1.8	1.9	1.9	2.0	2.1	2.2	2.4	2.5
53 Wood-Cutting	0.1	0.2	0.2	0.2	0.2	0.3	0.3	0.3	0.4	0.5	0.5	0.5	0.6	0.6	0.7	0.7	0.7
54 Forging-Press	0.2	0.3	0.3	0.3	0.3	0.4	0.4	0.5	0.5	0.5	0.6	0.6	0.6	0.7	0.8	0.8	0.9
55 Casting	0.0	0.1	0.1	0.1	0.1	0.1	0.1	0.1	0.1	0.1	0.1	0.1	0.1	0.1	0.1	0.1	0.1
56 Tools	0.1	0.2	0.3	0.3	0.3	0.3	0.4	0.4	0.4	0.4	0.5	0.5	0.5	0.6	0.6	0.7	0.7
57 Precision Instr.	0.3	1.1	1.1	1.2	1.2	1.4	1.6	1.6	1.7	1.8	2.0	2.2	2.5	2.5	2.7	2.9	3.1
58 Computers	0.1	0.6	0.7	0.9	1.0	1.2	1.6	1.7	1.8	2.1	2.3	2.5	2.7	2.5	2.6	2.7	2.9
59 Trucks and Buses	1.2	1.9	2.2	2.5	2.7	3.1	3.4	3.8	4.1	4.5	4.9	5.2	5.5	5.7	5.9	6.0	6.2
60 Tractors	0.7	1.0	1.1	1.2	1.3	1.3	1.4	1.6	1.8	1.8	1.9	2.1	2.2	2.3	2.3	2.4	2.5
61 Agricultural	1.8	2.2	2.3	2.5	2.8	3.3	3.7	4.0	4.2	4.4	4.7	4.8	5.0	5.6	5.9	6.2	6.6
62 Construction	0.7	1.1	1.2	1.3	1.3	1.4	1.5	1.6	1.7	1.7	1.8	1.8	1.9	1.9	2.0	2.0	2.0
63 Constr. Mater.	0.2	0.4	0.4	0.4	0.4	0.4	0.5	0.5	0.5	0.5	0.6	0.6	0.6	0.6	0.7	0.7	0.7
64 Light Industry	0.0	0.1	0.1	0.1	0.1	0.1	0.1	0.1	0.1	0.1	0.1	0.1	0.1	0.1	0.1	0.1	0.1
65 Food Processing	0.0	0.1	0.1	0.1	0.1	0.1	0.1	0.1	0.1	0.1	0.1	0.1	0.1	0.1	0.1	0.1	0.1
66 Processed Feed	0.1	0.2	0.2	0.2	0.2	0.2	0.3	0.3	0.3	0.3	0.3	0.3	0.3	0.3	0.4	0.4	0.4
67 Trade and Dining	0.1	0.2	0.3	0.3	0.4	0.4	0.4	0.5	0.5	0.6	0.6	0.6	0.6	0.6	0.6	0.6	0.7
68 Printing	0.0	0.1	0.1	0.1	0.1	0.1	0.1	0.1	0.1	0.1	0.1	0.1	0.1	0.1	0.1	0.1	0.1
69 N.E.C. Industry	0.1	0.2	0.2	0.2	0.2	0.2	0.3	0.3	0.3	0.3	0.3	0.3	0.3	0.3	0.4	0.4	0.4
70 Other Production	0.1	0.2	0.2	0.2	0.2	0.2	0.2	0.2	0.2	0.2	0.2	0.2	0.2	0.2	0.2	0.2	0.2
71 Sanitary-Engin.	0.4	0.7	0.8	0.9	1.0	1.1	1.2	1.2	1.2	1.4	1.4	1.5	1.6	1.6	1.6	1.6	1.6
72 Medical Equip.	0.3	0.4	0.4	0.4	0.5	0.5	0.5	0.6	0.6	0.6	0.7	0.7	0.7	0.7	0.8	0.8	0.8
73 Communal Equip.	0.1	0.1	0.1	0.1	0.1	0.2	0.1	0.2	0.2	0.2	0.2	0.2	0.2	0.2	0.2	0.2	0.2
74 Science Equip.	0.2	0.4	0.4	0.4	0.5	0.5	0.6	0.6	0.7	0.8	0.8	0.9	0.9	0.9	0.9	0.9	1.0
75 Administr Equip.	0.0	0.1	0.1	0.1	0.1	0.1	0.1	0.1	0.1	0.1	0.1	0.1	0.1	0.1	0.1	0.1	0.1
76 Defense Product.	0.3	0.5	0.6	0.7	0.8	0.8	1.0	1.1	1.1	1.3	1.4	1.6	1.7	1.7	1.8	1.8	1.8
77 CAPITAL REPAIR	5.4	9.0	9.7	10.4	11.0	11.9	12.7	13.5	14.7	15.8	16.6	17.6	18.7	19.7	21.0	22.1	23.5

TABLE 0.06: SUPPLY AND USES OF MBMW PRODUCTS--PART B: DERIVATION OF 1973 MACHINERY PRODUCTION PRICE INDEXES
(billion rubles)

	1965	1970	1971	1972	1973	1974	1975	1976	1977	1978	1979	1980	1981	1982	1983	1984	1985
78 CONSUMER PRODUCT	2.4	4.0	4.4	5.0	5.3	6.0	6.6	6.7	7.2	7.7	8.0	8.5	9.0	9.3	9.9	10.6	11.1
79 Appliances	0.3	0.7	0.7	0.8	0.8	0.8	0.9	0.9	1.0	1.0	1.0	1.1	1.2	1.2	1.2	1.3	1.3
80 Automobiles	0.2	0.3	0.6	0.8	1.0	1.3	1.6	1.6	1.7	1.8	2.0	2.1	2.3	2.5	2.6	2.7	2.8
81 M-Cycles & Bikes	0.3	0.5	0.5	0.5	0.5	0.6	0.6	0.7	0.7	0.8	0.8	0.8	0.8	0.8	0.9	0.9	0.9
82 TV & Radio	0.6	1.0	1.1	1.1	1.2	1.2	1.3	1.3	1.5	1.6	1.6	1.7	1.9	1.9	2.2	2.5	2.7
83 Refrigirators	0.2	0.4	0.4	0.4	0.4	0.5	0.5	0.5	0.5	0.6	0.6	0.6	0.6	0.7	0.7	0.8	0.8
84 Clocks & Watches	0.2	0.3	0.3	0.4	0.4	0.4	0.4	0.4	0.5	0.6	0.7	0.7	0.7	0.7	0.7	0.7	0.8
85 Sewing Machines	0.0	0.1	0.1	0.1	0.1	0.1	0.1	0.1	0.1	0.1	0.1	0.1	0.1	0.1	0.1	0.1	0.2
86 Cameras	0.0	0.1	0.0	0.1	0.1	0.2	0.2	0.2	0.1	0.1	0.1	0.1	0.1	0.1	0.0	0.1	0.1
87 Metal Wares	0.5	0.7	0.7	0.8	0.8	0.9	1.0	1.0	1.1	1.2	1.2	1.3	1.3	1.3	1.4	1.5	1.6
88 N-PROD USE MB	1.3	2.5	2.7	2.9	3.0	3.2	3.4	3.3	3.4	3.7	3.9	4.2	4.4	4.7	4.9	5.2	5.5
89 Electrotechn.	0.1	0.1	0.1	0.2	0.2	0.2	0.2	0.2	0.2	0.2	0.2	0.2	0.2	0.2	0.2	0.3	0.3
90 Cable Products	0.1	0.1	0.1	0.1	0.1	0.1	0.1	0.1	0.1	0.1	0.1	0.1	0.1	0.1	0.1	0.1	0.1
91 Precision Instr.	0.1	0.1	0.2	0.2	0.2	0.2	0.2	0.2	0.2	0.3	0.2	0.3	0.3	0.3	0.3	0.4	0.4
92 Automotive	0.1	0.2	0.2	0.2	0.2	0.3	0.3	0.3	0.4	0.4	0.4	0.4	0.5	0.5	0.5	0.5	0.5
93 Sanitary-Engin.	0.0	0.1	0.1	0.1	0.1	0.1	0.1	0.1	0.1	0.2	0.2	0.2	0.2	0.2	0.2	0.2	0.2
94 Electronics	0.7	1.4	1.5	1.6	1.6	1.8	1.9	1.9	1.9	2.1	2.3	2.5	2.6	2.7	2.9	3.2	3.3
95 Metal Wares	0.2	0.5	0.5	0.5	0.5	0.5	0.5	0.5	0.5	0.5	0.5	0.5	0.5	0.6	0.6	0.6	0.6
96 DEFENSE MB	5.4	8.3	8.6	9.5	9.6	10.3	11.0	11.0	11.4	12.3	13.2	14.2	15.3	15.8	17.0	18.5	19.6
97 HIDDEN MBMW	3.2	4.4	4.8	5.5	6.0	6.5	6.9	7.8	8.4	10.0	10.2	10.1	10.3	11.7	12.0	12.3	12.6
98 Domestic Prod.	2.4	2.9	3.2	3.9	3.4	3.4	3.8	4.3	3.7	4.8	4.8	4.4	3.6	3.8	3.8	3.8	4.6
99 Exports	0.8	1.5	1.6	1.6	2.6	3.1	3.1	3.5	4.7	5.2	5.4	5.7	6.7	7.9	8.2	8.5	8.0
100 TOTAL MBMW	50.9	93.0	99.0	108.3	112.3	123.0	133.9	138.5	147.4	158.7	167.8	177.6	186.2	194.1	205.5	217.3	228.8
101 Civilian Prod.	41.0	77.7	83.0	90.4	93.4	102.4	112.0	115.1	122.4	130.5	138.2	146.6	154.0	159.8	169.4	178.7	188.3
102 Defense Prod.	7.8	11.2	11.8	13.3	13.0	13.7	14.8	15.3	15.1	17.1	18.0	18.6	18.9	19.6	20.8	22.2	24.2
103 Exports	2.1	4.1	4.3	4.5	5.9	6.9	7.1	8.1	9.9	11.1	11.7	12.4	13.4	14.8	15.3	16.4	16.3

TABLE 0.06: SUPPLY AND USES OF MBMW PRODUCTS--PART C: GVO OF MAJOR MBMW SECTORS IN CURRENT PRODUCER PRICES
(billion rubles)

		1965	1970	1971	1972	1973	1974	1975	1976	1977	1978	1979	1980	1981	1982	1983	1984	1985
1	PUBLISHED GVO	47.7	88.6	94.2	102.8	106.3	116.5	127.0	130.7	139.0	148.7	157.6	167.5	175.9	182.4	193.5	205.0	216.2
2	INTERIND. USE	18.5	40.4	42.6	46.4	47.2	52.0	56.9	57.3	60.9	65.2	69.4	74.2	77.5	80.6	85.3	91.6	96.7
3	Parts & Tools	9.6	18.6	19.8	21.5	22.0	24.1	26.3	26.8	28.5	30.3	31.9	33.9	35.4	37.0	39.1	41.9	44.2
4	Power & Diesel	0.1	0.2	0.2	0.2	0.2	0.3	0.3	0.3	0.4	0.4	0.4	0.4	0.5	0.5	0.6	0.6	0.7
5	Metal&Mining MB	0.1	0.1	0.1	0.1	0.2	0.2	0.2	0.2	0.2	0.2	0.2	0.2	0.2	0.2	0.2	0.3	0.3
6	Hoisting-Transp	0.1	0.1	0.2	0.2	0.2	0.2	0.2	0.3	0.3	0.3	0.3	0.3	0.3	0.4	0.4	0.5	0.5
7	Transport MB	0.1	0.1	0.1	0.1	0.1	0.2	0.2	0.2	0.2	0.2	0.2	0.2	0.2	0.2	0.2	0.2	0.2
8	E-technical MB	0.5	0.8	0.8	0.8	0.8	0.8	0.9	0.9	0.9	1.0	1.0	1.0	1.1	1.1	1.1	1.2	1.2
9	Cable Products	1.0	1.6	1.7	1.7	1.7	1.8	1.9	2.0	2.1	2.2	2.2	2.3	2.4	2.4	2.5	2.7	2.8
10	Chemical MB	0.3	0.5	0.6	0.7	0.7	0.7	0.8	0.9	0.9	1.0	1.0	1.1	1.1	1.2	1.3	1.4	1.5
11	Metal Cutting	0.1	0.2	0.2	0.2	0.2	0.3	0.3	0.3	0.4	0.4	0.5	0.5	0.6	0.7	0.6	0.6	0.6
12	Tools & Dies	0.3	0.6	0.6	0.7	0.7	0.8	0.9	0.9	1.0	1.0	1.1	1.2	1.2	1.3	1.4	1.6	1.7
13	Abrasives	0.2	0.3	0.4	0.4	0.4	0.5	0.5	0.5	0.6	0.7	0.7	0.7	0.7	0.7	0.8	0.8	0.8
14	Precision Instr	0.2	0.5	0.5	0.5	0.5	0.6	0.6	0.7	0.7	0.8	0.9	1.0	1.0	1.1	1.2	1.3	1.4
15	Computer Parts	0.0	0.1	0.1	0.2	0.2	0.2	0.3	0.3	0.3	0.3	0.3	0.4	0.4	0.5	0.6	0.7	0.8
16	Automotive MB	0.4	0.8	0.9	1.0	1.1	1.2	1.2	1.3	1.4	1.6	1.7	1.8	1.8	1.9	1.9	2.1	2.2
17	Bearings	0.3	0.6	0.7	0.8	0.9	1.0	1.2	1.1	1.2	1.4	1.5	1.5	1.6	1.7	1.7	1.8	1.9
18	Agricultural MB	0.3	0.6	0.7	0.8	0.9	1.0	1.1	1.1	1.2	1.3	1.4	1.4	1.5	1.5	1.6	1.8	1.9
19	Construction MB	0.1	0.2	0.3	0.3	0.3	0.4	0.4	0.4	0.4	0.4	0.4	0.5	0.5	0.6	0.6	0.6	0.7
20	Const Mater. MB	0.0	0.1	0.1	0.1	0.1	0.1	0.1	0.1	0.1	0.1	0.1	0.1	0.1	0.1	0.1	0.1	0.2
21	Light Ind. MB	0.1	0.2	0.2	0.2	0.2	0.2	0.2	0.2	0.2	0.2	0.3	0.3	0.3	0.3	0.3	0.3	0.3
22	Food Ind. MB	0.0	0.1	0.1	0.1	0.1	0.1	0.1	0.1	0.1	0.1	0.1	0.1	0.1	0.1	0.1	0.2	0.2
23	Pipe Fittings	0.2	0.3	0.3	0.3	0.4	0.4	0.4	0.4	0.5	0.5	0.5	0.6	0.6	0.6	0.6	0.6	0.6
24	Electronics	0.7	2.5	2.5	2.8	2.8	3.1	3.5	3.4	3.5	3.7	4.1	4.5	4.8	4.9	5.3	5.7	6.0
25	Metal Structure	0.9	1.6	1.7	1.8	1.9	2.0	2.2	2.2	2.2	2.3	2.4	2.5	2.6	2.6	2.8	2.9	3.1
26	Metal Wares	1.3	2.5	2.6	2.8	2.8	3.0	3.3	3.3	3.3	3.5	3.6	3.7	3.9	3.9	4.2	4.4	4.7
27	Current Repair	2.2	4.0	4.2	4.5	4.7	5.1	5.5	5.8	6.3	6.7	7.0	7.6	8.0	8.4	8.9	9.5	10.0
28	Complex Component	8.8	21.8	22.8	24.9	25.2	27.9	30.6	30.5	32.4	34.9	37.5	40.3	42.0	43.6	46.2	49.8	52.5
29	Diesel Engines	0.3	0.4	0.4	0.5	0.5	0.5	0.5	0.5	0.6	0.6	0.6	0.6	0.6	0.7	0.7	0.8	0.8
30	Transport MB	0.4	0.5	0.5	0.5	0.6	0.6	0.6	0.7	0.7	0.8	0.8	0.8	0.8	0.8	0.9	0.9	0.9
31	E-technical	2.0	3.3	3.4	3.4	3.2	3.4	3.6	3.6	3.7	3.9	4.0	4.2	4.3	4.4	4.5	4.8	4.9
32	Precision Instr	0.8	1.9	1.9	2.1	2.1	2.3	2.6	2.6	2.9	3.2	3.4	3.8	4.1	4.4	4.7	5.2	5.5
33	Automotive MB	1.6	3.1	3.6	4.0	4.3	4.7	5.0	5.3	5.8	6.4	6.9	7.2	7.4	7.6	7.8	8.2	8.8
34	Agricultural MB	1.3	2.6	2.9	3.2	3.5	3.9	4.3	4.4	4.8	5.2	5.4	5.7	5.8	6.1	6.5	7.0	7.5
35	Electronics	2.6	10.1	10.1	11.3	11.1	12.5	14.0	13.4	14.0	14.8	16.4	18.0	19.0	19.6	21.1	22.9	24.0

TABLE 0.06: SUPPLY AND USES OF MBMW PRODUCTS--PART C: GVO OF MAJOR MBMW SECTORS IN CURRENT PRODUCER PRICES
(billion rubles)

	1965	1970	1971	1972	1973	1974	1975	1976	1977	1978	1979	1980	1981	1982	1983	1984	1985
36 MACHINERY	14.7	24.4	26.2	28.5	30.2	33.2	36.4	38.9	41.4	44.0	46.7	48.9	50.9	52.4	55.3	57.0	59.9
37 Power	0.6	0.8	0.8	0.8	0.9	0.9	0.9	1.0	1.0	1.1	1.1	1.2	1.2	1.3	1.3	1.3	1.4
38 Metallurgy	0.3	0.4	0.4	0.5	0.5	0.6	0.6	0.7	0.7	0.7	0.7	0.7	0.7	0.8	0.8	0.9	0.9
39 Mining	0.5	0.9	1.0	1.0	1.1	1.1	1.1	1.2	1.2	1.3	1.3	1.4	1.4	1.5	1.6	1.6	1.7
40 Hoisting-Transp.	0.6	1.1	1.1	1.2	1.2	1.3	1.4	1.5	1.6	1.6	1.7	1.8	1.8	1.9	1.9	1.9	2.0
41 Railroad	0.8	1.1	1.2	1.2	1.2	1.3	1.4	1.5	1.6	1.7	1.7	1.7	1.8	1.8	1.9	2.0	2.0
42 Civil Shipbuild.	0.3	0.6	0.6	0.6	0.6	0.7	0.7	0.8	0.8	0.8	0.8	0.8	0.9	0.9	0.9	1.0	1.0
43 Civil Aircraft	1.2	2.2	2.3	2.5	2.5	2.6	2.8	3.1	3.2	3.3	3.4	3.5	3.5	3.5	3.7	3.8	4.0
44 Electrotechnical	1.1	1.7	1.8	1.9	1.9	2.2	2.4	2.5	2.7	2.8	2.9	3.0	3.1	3.1	3.2	3.3	3.5
45 Chemical MB	0.2	0.3	0.3	0.4	0.4	0.4	0.4	0.5	0.5	0.5	0.6	0.6	0.6	0.7	0.7	0.8	0.8
46 Compressors, etc	0.2	0.3	0.3	0.3	0.4	0.4	0.4	0.4	0.4	0.4	0.5	0.5	0.5	0.6	0.7	0.7	0.8
47 Pumps	0.3	0.5	0.5	0.6	0.6	0.7	0.7	0.7	0.7	0.8	0.9	0.9	1.0	1.0	1.1	1.2	1.3
48 Oil-Gas Drills	0.3	0.4	0.4	0.5	0.5	0.6	0.6	0.6	0.7	0.8	0.8	0.9	0.9	1.0	1.1	1.2	1.2
49 Oil Processing	0.2	0.3	0.3	0.3	0.3	0.4	0.4	0.5	0.5	0.5	0.6	0.6	0.6	0.6	0.7	0.7	0.8
50 Logging	0.1	0.2	0.2	0.2	0.2	0.2	0.3	0.3	0.3	0.3	0.3	0.3	0.4	0.4	0.4	0.4	0.4
51 Pulp & Paper	0.0	0.1	0.1	0.1	0.1	0.1	0.1	0.1	0.2	0.2	0.2	0.2	0.2	0.2	0.2	0.2	0.2
52 Metal-Cutting	0.5	0.9	1.0	1.0	1.1	1.2	1.4	1.5	1.7	1.8	1.9	1.9	2.0	2.1	2.2	2.4	2.5
53 Wood-Cutting	0.1	0.2	0.2	0.2	0.2	0.3	0.3	0.3	0.4	0.5	0.5	0.5	0.6	0.6	0.7	0.7	0.7
54 Forging-Press	0.2	0.3	0.3	0.3	0.3	0.4	0.4	0.5	0.5	0.5	0.6	0.6	0.6	0.7	0.8	0.8	0.9
55 Casting	0.0	0.1	0.1	0.1	0.1	0.1	0.1	0.1	0.1	0.1	0.1	0.1	0.1	0.1	0.1	0.1	0.1
56 Tools	0.1	0.2	0.3	0.3	0.3	0.3	0.4	0.4	0.4	0.4	0.5	0.5	0.5	0.6	0.6	0.7	0.7
57 Precision Instr.	0.3	1.1	1.1	1.2	1.2	1.4	1.6	1.6	1.7	1.8	2.0	2.2	2.5	2.5	2.7	2.9	3.1
58 Computers	0.1	0.6	0.7	0.9	1.0	1.2	1.6	1.7	1.8	2.1	2.3	2.5	2.7	2.5	2.6	2.7	2.9
59 Trucks and Buses	1.2	1.9	2.2	2.5	2.7	3.1	3.4	3.8	4.1	4.5	4.9	5.2	5.5	5.7	5.9	6.0	6.2
60 Tractors	0.7	1.0	1.1	1.2	1.3	1.3	1.4	1.6	1.8	1.8	1.9	2.1	2.2	2.3	2.3	2.4	2.5
61 Agricultural	1.8	2.2	2.3	2.5	2.8	3.3	3.7	4.0	4.2	4.4	4.7	4.8	5.0	5.6	5.9	6.2	6.6
62 Construction	0.7	1.1	1.2	1.3	1.3	1.4	1.5	1.6	1.7	1.7	1.8	1.8	1.9	1.9	2.0	2.0	2.0
63 Constr. Mater.	0.2	0.4	0.4	0.4	0.4	0.4	0.5	0.5	0.5	0.5	0.6	0.6	0.6	0.6	0.7	0.7	0.7
64 Light Industry	0.0	0.1	0.1	0.1	0.1	0.1	0.1	0.1	0.1	0.1	0.1	0.1	0.1	0.1	0.1	0.1	0.1
65 Food Processing	0.0	0.1	0.1	0.1	0.1	0.1	0.1	0.1	0.1	0.1	0.1	0.1	0.1	0.1	0.1	0.1	0.1
66 Processed Feed	0.1	0.2	0.2	0.2	0.2	0.2	0.3	0.3	0.3	0.3	0.3	0.3	0.3	0.3	0.4	0.4	0.4
67 Trade and Dining	0.1	0.2	0.3	0.3	0.4	0.4	0.4	0.5	0.5	0.6	0.6	0.6	0.6	0.6	0.6	0.6	0.7
68 Printing	0.0	0.1	0.1	0.1	0.1	0.1	0.1	0.1	0.1	0.1	0.1	0.1	0.1	0.1	0.1	0.1	0.1
69 N.E.C. Industry	0.1	0.2	0.2	0.2	0.2	0.2	0.3	0.3	0.3	0.3	0.3	0.3	0.3	0.3	0.4	0.4	0.4
70 Other Production	0.1	0.2	0.2	0.2	0.2	0.2	0.2	0.2	0.2	0.2	0.2	0.2	0.2	0.2	0.2	0.2	0.2
71 Sanitary-Engin.	0.4	0.7	0.8	0.9	1.0	1.1	1.2	1.2	1.2	1.4	1.4	1.5	1.6	1.6	1.6	1.6	1.6
72 Medical Equip.	0.3	0.4	0.4	0.4	0.5	0.5	0.5	0.6	0.6	0.6	0.7	0.7	0.7	0.7	0.8	0.8	0.8
73 Communal Equip.	0.1	0.1	0.1	0.1	0.1	0.2	0.1	0.2	0.2	0.2	0.2	0.2	0.2	0.2	0.2	0.2	0.2
74 Science Equip.	0.2	0.4	0.4	0.4	0.5	0.5	0.6	0.6	0.7	0.8	0.8	0.9	0.9	0.9	0.9	0.9	1.0
75 Administr Equip.	0.0	0.1	0.1	0.1	0.1	0.1	0.1	0.1	0.1	0.1	0.1	0.1	0.1	0.1	0.1	0.1	0.1
76 Defense Product.	0.3	0.5	0.6	0.7	0.8	0.8	1.0	1.1	1.1	1.3	1.4	1.6	1.7	1.7	1.8	1.8	1.8
77 CAPITAL REPAIR	5.4	9.0	9.7	10.4	11.0	11.9	12.7	13.5	14.7	15.8	16.6	17.6	18.7	19.7	21.0	22.1	23.5

TABLE 0.06: SUPPLY AND USES OF MBMW PRODUCTS--PART C: GVO OF MAJOR MBMW SECTORS IN CURRENT PRODUCER PRICES
    (billion rubles)

	1965	1970	1971	1972	1973	1974	1975	1976	1977	1978	1979	1980	1981	1982	1983	1984	1985
78 CONSUMER PRODUCT	2.4	4.0	4.4	5.0	5.3	6.0	6.6	6.7	7.2	7.7	8.0	8.5	9.0	9.3	9.9	10.6	11.1
79 Appliances	0.3	0.7	0.7	0.8	0.8	0.8	0.9	0.9	1.0	1.0	1.0	1.1	1.2	1.2	1.2	1.3	1.3
80 Automobiles	0.2	0.3	0.6	0.8	1.0	1.3	1.6	1.6	1.7	1.8	2.0	2.1	2.3	2.5	2.6	2.7	2.8
81 M-Cycles & Bikes	0.3	0.5	0.5	0.5	0.5	0.6	0.6	0.7	0.7	0.8	0.8	0.8	0.8	0.8	0.9	0.9	0.9
82 TV & Radio	0.6	1.0	1.1	1.1	1.2	1.2	1.3	1.3	1.5	1.6	1.6	1.7	1.9	1.9	2.2	2.5	2.7
83 Refrigirators	0.2	0.4	0.4	0.4	0.4	0.5	0.5	0.5	0.5	0.6	0.6	0.6	0.6	0.7	0.7	0.8	0.8
84 Clocks & Watches	0.2	0.3	0.3	0.4	0.4	0.4	0.4	0.4	0.5	0.6	0.7	0.7	0.7	0.7	0.7	0.7	0.8
85 Sewing Machines	0.0	0.1	0.1	0.1	0.1	0.1	0.1	0.1	0.1	0.1	0.1	0.1	0.1	0.1	0.1	0.1	0.2
86 Cameras	0.0	0.1	0.0	0.1	0.1	0.2	0.2	0.2	0.1	0.1	0.1	0.1	0.1	0.1	0.0	0.1	0.1
87 Metal Wares	0.5	0.7	0.7	0.8	0.8	0.9	1.0	1.0	1.1	1.2	1.2	1.3	1.3	1.3	1.4	1.5	1.6
88 N-PROD USE MB	1.3	2.5	2.7	2.9	3.0	3.2	3.4	3.3	3.4	3.7	3.9	4.2	4.4	4.7	4.9	5.2	5.5
89 Electrotechn.	0.1	0.1	0.1	0.2	0.2	0.2	0.2	0.2	0.2	0.2	0.2	0.2	0.2	0.2	0.2	0.3	0.3
90 Cable Products	0.1	0.1	0.1	0.1	0.1	0.1	0.1	0.1	0.1	0.1	0.1	0.1	0.1	0.1	0.1	0.1	0.1
91 Precision Instr.	0.1	0.1	0.2	0.2	0.2	0.2	0.2	0.2	0.2	0.3	0.2	0.3	0.3	0.3	0.3	0.4	0.4
92 Automotive	0.1	0.2	0.2	0.2	0.2	0.3	0.3	0.3	0.4	0.4	0.4	0.4	0.5	0.5	0.5	0.5	0.5
93 Sanitary-Engin.	0.0	0.1	0.1	0.1	0.1	0.1	0.1	0.1	0.1	0.2	0.2	0.2	0.2	0.2	0.2	0.2	0.2
94 Electronics	0.7	1.4	1.5	1.6	1.6	1.8	1.9	1.9	1.9	2.1	2.3	2.5	2.6	2.7	2.9	3.2	3.3
95 Metal Wares	0.2	0.5	0.5	0.5	0.5	0.5	0.5	0.5	0.5	0.5	0.5	0.5	0.5	0.6	0.6	0.6	0.6
96 DEFENSE MB	5.4	8.3	8.6	9.5	9.6	10.3	11.0	11.0	11.4	12.3	13.2	14.2	15.3	15.8	17.0	18.5	19.6
97 HIDDEN MBMW	3.2	4.4	4.8	5.5	6.0	6.5	6.9	7.8	8.4	10.0	10.2	10.1	10.3	11.7	12.0	12.3	12.6
98 Domestic Prod.	2.4	2.9	3.2	3.9	3.4	3.4	3.8	4.3	3.7	4.8	4.8	4.4	3.6	3.8	3.8	3.8	4.6
99 Exports	0.8	1.5	1.6	1.6	2.6	3.1	3.1	3.5	4.7	5.2	5.4	5.7	6.7	7.9	8.2	8.5	8.0
100 TOTAL MBMW	50.9	93.0	99.0	108.3	112.3	123.0	133.9	138.5	147.4	158.7	167.8	177.6	186.2	194.1	205.5	217.3	228.8
101 Civilian Prod.	41.0	77.7	83.0	90.4	93.4	102.4	112.0	115.1	122.4	130.5	138.2	146.6	154.0	159.8	169.4	178.7	188.3
102 Defense Prod.	7.8	11.2	11.8	13.3	13.0	13.7	14.8	15.3	15.1	17.1	18.0	18.6	18.9	19.6	20.8	22.2	24.2
103 Exports	2.1	4.1	4.3	4.5	5.9	6.9	7.1	8.1	9.9	11.1	11.7	12.4	13.4	14.8	15.3	16.4	16.3

TABLE 0.06: SUPPLY AND USES OF MBMW PRODUCTS--PART D: SUPPORTING DATA ON OUTPUT IN PHYSICAL UNITS

	1965	1970	1971	1972	1973	1974	1975	1976	1977	1978	1979	1980	1981	1982	1983	1984	1985
POWER & D-ENGINES																	
1 Turbines	14.6	16.2	16.8	14.6	15.1	17.3	18.9	19.6	19.0	18.3	20.0	19.6	14.6	17.3	15.5	21.3	21.6
2 Steam Boilers	53.2	48.3	44.2	46.4	47.9	51.4	55.6	53.2	52.2	55.8	54.9	51.0	53.1	43.1	50.8	46.5	43.7
3 Diesel Engines	13.6	16.5	17.1	17.9	17.8	18.2	18.6	19.0	18.9	19.2	18.6	19.0	18.7	18.6	16.3	16.1	16.6
METALLURGY MB																	
4 Mills M&E	83.9	111.0	111.0	120.0	119.0	124.0	124.0	124.0	127.0	131.0	127.0	124.0	119.0	111.0	128.0	123.0	131.0
5 Steel Melting M&E	37.2	50.7	54.5	45.8	57.3	63.3	63.4	67.2	71.7	70.7	76.9	73.1	70.1	66.1	64.2	62.2	66.0
6 Rolling M&E	121.0	152.0	157.0	156.0	165.0	151.0	154.0	174.0	173.0	165.0	161.0	155.0	154.0	146.0	145.0	141.0	151.0
MINING MB																	
7 Coal Combines	1.05	1.19	1.29	1.21	1.17	1.27	1.27	1.36	1.32	1.36	1.37	1.46	1.44	1.37	2.9	1.22	1.23
8 Loading	2.8	2.5	2.55	2.44	2.67	2.81	2.81	3.06	3.12	3.11	3.15	3.03	3.06	3.12	3.11	3.15	3.03
9 E-Diesels	2.41	2.55	2.49	2.59	2.62	2.55	2.61	2.67	2.69	2.67	2.47	2.48	2.56	2.77	2.82	2.88	2.88
HOISTING TRANSP.																	
10 Bridge Cranes	6.8	5.7	5.8	5.8	5.9	6.4	6.5	6.9	6.7	6.7	6.4	6.3	6.1	6.2	6.2	6.3	6.6
11 Truck Cranes	11.1	15.4	16.2	16.4	17.2	19.0	19.7	20.2	21.2	21.4	21.9	21.7	n.a.	n.a.	n.a.	n.a.	n.a.
12 Tower Cranes	3.5	3.9	4.2	4.4	4.6	4.6	4.6	4.7	4.1	4.2	3.9	3.4	3.0	3.0	3.0	3.0	3.1
13 Pneumatic Cranes	1.2	2.0	2.2	2.2	2.3	2.5	2.4	2.5	2.6	2.6	2.6	2.4	2.8	2.6	2.6	2.7	2.7
14 Elevators	8.6	18.1	20.8	22.8	24.3	25.3	25.2	27.4	28.9	31.0	33.4	33.8	34.3	33.4	30.5	30.6	31.2
RAILROAD TRANSP.																	
15 Locomotives	3.3	3.8	3.8	3.9	3.8	4.0	3.9	4.1	3.7	3.9	3.7	3.8	n.a.	n.a.	n.a.	n.a.	n.a.
16 E-Diesel Trains	3.9	2.4	2.6	2.6	2.6	2.7	3.0	3.2	3.3	3.5	3.3	3.4	n.a.	n.a.	n.a.	n.a.	n.a.
17 Railroad Cars																	
18 Freight	39.6	58.3	63.7	68.9	71.8	72.4	69.9	71.9	71.2	68.3	64.6	63.0	n.a.	n.a.	n.a.	n.a.	n.a.
19 Passenger	2.0	1.8	1.9	2.0	2.0	2.1	2.1	2.1	2.1	2.1	2.1	2.0	n.a.	n.a.	n.a.	n.a.	n.a.
ELECTROTECHN. M&E																	
20 Generators for T.	14.4	10.6	12.6	13.7	16.5	16.0	17.1	16.6	17.9	18.2	17.0	16.1	13.9	12.7	12.6	13.7	13.3
21 Large E-Machines	15.3	17.0	17.6	18.9	20.4	22.2	24.0	25.9	26.7	27.9	29.6	33.4	35.1	36.0	37.8	39.9	42.2
22 E-Engines>100kvt.	5.3	5.5	5.6	6.0	6.0	6.5	7.0	7.5	7.5	7.7	7.5	7.7	7.7	7.7	7.6	8.0	8.2
23 E-Engines<100kvt.	21.6	27.8	29.6	30.5	31.7	33.5	34.7	35.4	36.0	37.6	38.2	39.9	40.6	40.7	40.4	40.6	41.2
24 Transformers	95.3	106.0	108.0	116.0	121.0	127.0	137.0	144.0	146.0	150.0	155.0	159.0	161.0	161.0	155.0	156.0	161.0
25 E-Melting M&E	198	253	274	290	287	289	303	305	303	317	305	303	309	315	314	320	327
26 E-Lamps	0.98	1.63	1.82	1.96	2.00	1.99	2.05	2.03	2.08	2.10	2.10	2.17	2.22	2.27	2.41	2.46	2.47
PUMPS & REFRIGER.																	
Pumps																	
27 Deep-Well	92.8	77.0	81.0	81.0	82.0	85.0	85.1	86.1	83.2	95.0	95.0	91.6	95.0	96.0	96.0	95.0	93.7
28 Other	0.8	1.2	1.2	1.2	1.2	1.3	1.4	1.34	1.39	1.37	1.35	1.34	1.4	1.4	1.4	1.5	1.5
29 Compressors	66.0	89.5	93.8	95.6	99.9	107.0	112.0	120.0	126.0	129.0	113.0	115.0	118.0	118.0	117.0	119.0	123.0
30 Ind. Refriger.	0.1	0.2	0.3	0.3	0.3	0.3	0.4	0.4	0.4	0.4	0.4	0.4	0.4	0.4	0.4	0.4	0.4
31 Consumer Refriger	1.7	4.1	4.6	5.0	5.4	5.4	5.6	5.8	5.8	6.1	6.0	5.9	5.9	5.8	5.7	5.7	5.9
OIL & GAS MB																	
32 Turbodrills	8.4	6.6	7.4	7.7	8.1	9.3	9.8	9.4	9.7	9.0	9.0	9.3	9.5	9.3	12.5	14.5	16.0
33 E-drills	0.22	0.12	0.11	0.09	0.11	0.10	0.10	0.11	0.10	0.08	0.02	n.a.	n.a.	n.a.	n.a.	n.a.	n.a.
34 Oil Expl. M&E	0.52	0.48	n.a.	n.a.	n.a.	n.a.	0.54	0.51	0.50	0.51	0.47	0.52	0.54	0.56	0.58	0.56	0.57
35 Oil Process M&E	140	127	139	157	159	172	171	164	171	180	188	184	n.a.	n.a.	n.a.	n.a.	n.a.

TABLE 0.06: SUPPLY AND USES OF MBMW PRODUCTS--PART D: SUPPORTING DATA ON OUTPUT IN PHYSICAL UNITS

	1965	1970	1971	1972	1973	1974	1975	1976	1977	1978	1979	1980	1981	1982	1983	1984	1985
MACHINE TOOLS																	
36 Metal Cutting	186	202	207	211	214	226	231	233	238	238	230	216	205	195	190	188	182
37 Manual	185.9	200.4	204.7	208.1	210.4	221.5	225.4	226.8	231.4	230.6	222.0	207.1	194.9	184.4	178.6	174.7	164.2
38 Automated	0.05	1.59	2.3	2.9	3.6	4.5	5.55	6.13	6.55	7.37	7.97	8.9	10.1	10.6	11.4	13.3	17.8
39 Forging-Press	34.6	41.3	42.3	44.0	46.5	48.9	50.5	51.9	54.4	55.5	56.3	57.2	57.1	57.3	57.4	55.3	52.7
40 Automatic Lines	0.2	0.6	0.6	0.6	0.8	0.7	0.7	0.7	0.8	0.8	0.8	0.8	0.8	0.8	1.0	1.1	1.1
41 Robots	0.0	0.0	0.0	0.0	0.0	0.0	0.1	n.a.	n.a.	n.a.	n.a.	1.4	2.5	4.5	8.7	11.1	13.2
AUTOMOTIVE MB																	
42 Trucks (units)	0.38	0.52	0.56	0.60	0.63	0.67	0.70	0.72	0.73	0.76	0.78	0.79	n.a.	n.a.	n.a.	n.a.	n.a.
43 Trucks (capacity)	1.38	2.27	2.43	2.55	2.64	2.75	2.90	3.02	3.32	3.43	3.88	4.00	n.a.	n.a.	n.a.	n.a.	n.a.
44 Trailers	92.9	124	133	137	142	148	158	158	164	167	171	178	n.a.	n.a.	n.a.	n.a.	n.a.
45 Buses	35.5	47.4	49.3	51.9	56.0	61.0	67.0	70.1	74.6	77.4	79.2	85.3	n.a.	n.a.	n.a.	n.a.	n.a.
46 Autos	0.20	0.34	0.53	0.73	0.92	1.12	1.20	1.24	1.28	1.31	1.31	1.33	1.32	1.31	1.32	1.33	1.33
47 M-Bikes	0.71	0.83	0.87	0.90	0.93	0.96	1.03	1.06	1.09	1.10	1.08	1.09	1.10	1.11	1.13	1.15	1.15
48 Bikes	3.87	4.44	4.55	4.63	4.78	4.83	5.01	5.07	5.23	5.41	5.36	5.45	5.56	5.72	5.82	6.07	6.16
49 Bearings	0.53	0.67	0.71	0.76	0.80	0.85	0.91	0.95	0.99	1.02	1.03	1.05	1.06	1.08	1.08	1.09	1.09
AGRICULTURAL MB																	
50 Tractors (units)	0.36	0.46	0.47	0.48	0.50	0.53	0.55	0.56	0.57	0.58	0.56	0.56	0.56	0.56	0.56	0.57	0.59
51 Tracklaying	0.16	0.22	0.23	0.23	0.23	0.24	0.25	n.a.	n.a.	n.a.	n.a.	n.a.	n.a.	n.a.	n.a.	n.a.	n.a.
52 Wheeled	0.2	0.24	0.25	0.25	0.27	0.29	0.3	n.a.	n.a.	n.a.	n.a.	n.a.	n.a.	n.a.	n.a.	n.a.	n.a.
53 Tractors (power)	21.0	29.4	31.3	33.4	36.3	39.8	41.4	44.0	45.4	47.0	46.6	47.0	47.9	47.9	49.3	50.7	52.8
54 Trailers	132	296	339	319	277	283	290	299	299	305	309	312	321	315	317	323	346
CONSTRUCTION MB																	
55 Excavators	21.6	30.8	33.2	34.9	35.8	37.1	39.0	40.4	41.5	41.1	41.7	42.0	42.3	42.7	41.8	41.9	42.6
56 Buldozers	20.1	33.5	38.0	40.2	45.6	47.1	51.1	49.1	52.3	45.3	44.9	45.5	46.1	43.0	42.6	41.6	41.3
57 Scrapers	7.3	9.8	10.8	11.7	12.6	13.5	14.5	14.1	12.9	11.8	12.0	11.4	11.3	9.9	11.2	11.2	10.2
58 Motor Graders	4.2	4.6	5.6	6.0	5.9	6.3	6.5	6.8	6.9	6.6	5.8	6.2	6.0	6.3	6.1	5.1	5.0
SANITARY-ENGIN.																	
59 Pipe Fittings	54	68	71	68	74	79	81	80	83	83	80	79	78	80	83	84	84
60 Heating Boilers	2.8	3.7	3.7	3.8	3.9	4.1	4.3	n.a.	n.a.	n.a.	n.a.	n.a.	n.a.	n.a.	n.a.	n.a.	n.a.
61 Radiators	25.5	29.7	32.2	34.2	36.6	40	42.1	n.a.	n.a.	n.a.	n.a.	n.a.	n.a.	n.a.	n.a.	n.a.	n.a.
62 Air Stoves	8	10.6	11.3	12	12.3	11.9	12.2	n.a.	n.a.	n.a.	n.a.	n.a.	n.a.	n.a.	n.a.	n.a.	n.a.
63 Sewage Pipes	0.32	0.37	0.38	0.39	0.41	0.43	0.44	n.a.	n.a.	n.a.	n.a.	n.a.	n.a.	n.a.	n.a.	n.a.	n.a.
64 Trailing Pipes	3.1	3.1	3.0	2.9	2.9	2.7	2.8	n.a.	n.a.	n.a.	n.a.	n.a.	n.a.	n.a.	n.a.	n.a.	n.a.
65 Enameled Bathtabs	1.61	1.89	1.89	1.94	2.02	2.23	2.34	n.a.	n.a.	n.a.	n.a.	n.a.	n.a.	n.a.	n.a.	n.a.	n.a.
66 Water Heaters	0.66	0.89	0.96	1.00	0.97	0.76	0.77	n.a.	n.a.	n.a.	n.a.	n.a.	n.a.	n.a.	n.a.	n.a.	n.a.

TABLE 0.06: SUPPLY AND USES OF MBMW PRODUCTS--PART D: SUPPORTING DATA ON OUTPUT IN PHYSICAL UNITS

	1965	1970	1971	1972	1973	1974	1975	1976	1977	1978	1979	1980	1981	1982	1983	1984	1985
APPLIANCES, ETC.																	
67 Clocks & Watches	30.6	40.2	42.1	44.1	47.5	50.6	55.1	57.9	60.8	63.3	64.9	66.7	68.6	69.1	69.1	67.1	67.2
68 Radio (Stereo)	5.16	7.82	8.79	8.84	8.62	8.75	8.38	8.46	8.65	8.73	8.45	8.48	8.70	8.91	9.30	9.39	8.85
69 TV Sets	3.66	6.68	5.82	5.98	6.27	6.57	6.96	7.06	7.07	7.17	7.27	7.53	8.19	8.35	8.58	9.00	9.37
70 -/- color sets	0.00	0.05	0.06	0.08	0.17	0.41	0.59	0.81	1.12	1.43	1.81	2.26	2.71	3.12	3.41	3.60	4.02
71 Tape Players	0.45	1.19	1.39	1.64	1.80	2.11	2.53	2.60	2.60	2.62	2.74	3.05	3.22	3.45	3.73	4.08	4.67
72 Washing Machines	3.43	5.24	4.05	2.99	3.01	3.32	2.92	3.51	3.65	3.70	3.66	3.83	3.39	4.00	4.25	4.53	5.07
73 Vacuum Cleaners	0.80	1.51	1.74	2.17	2.66	3.32	2.92	2.66	2.75	2.93	3.10	3.22	3.36	3.49	3.59	3.81	4.07
74 Sewing Machines	0.80	1.51	1.74	2.17	2.66	3.32	2.92	1.36	1.36	1.36	1.32	1.32	1.35	1.36	1.39	1.42	1.50
75 Cameras	1.05	2.05	2.22	2.38	2.57	2.60	3.03	3.25	3.57	3.85	4.06	4.26	4.38	4.06	2.99	2.21	2.09
76 E-Kettles	0.50	1.52	1.87	2.20	2.68	2.98	3.44	n.a.	n.a.	n.a.	n.a.	3.91	3.95	3.92	3.90	4.22	4.29
77 E-Irons	3.60	8.60	10.10	10.80	12.20	13.50	14.50	n.a.	n.a.	n.a.	n.a.	12.56	12.99	13.33	14.08	14.74	15.44
METAL WORKS																	
78 Metal Structures	3.2	4.1	4.5	4.7	4.9	5.2	5.4	5.4	5.0	5.1	4.9	4.6	4.6	4.9	5.2	5.3	5.2
Consumer Wares																	
79 Meat Grinders	0.8	3.2	5.2	6.6	7.6	8.4	8.6	n.a.	n.a.	n.a.	n.a.	n.a.	n.a.	n.a.	n.a.	n.a.	n.a.
80 Baby Carriages	0.8	1.47	1.59	1.75	1.94	2.17	2.44	n.a.	n.a.	n.a.	n.a.	n.a.	n.a.	n.a.	n.a.	n.a.	n.a.
81 Dishware	330	400	430	460	490	505	525	n.a.	n.a.	n.a.	n.a.	n.a.	n.a.	n.a.	n.a.	n.a.	n.a.

TABLE 0.07: SUPPLY AND USES OF WOOD AND PAPER PRODUCTS
(billion rubles)

	1965	1970	1971	1972	1973	1974	1975	1976	1977	1978	1979	1980	1981	1982	1983	1984	1985
Current Prices																	
1 SUPPLY	14.9	24.2	25.3	26.8	27.8	29.0	31.3	32.0	32.8	33.8	34.2	36.1	37.5	44.9	46.2	47.4	49.7
2 GVO	12.0	19.5	20.3	21.2	22.0	22.8	23.7	24.3	24.6	25.2	25.2	25.9	26.7	33.9	35.2	36.3	37.6
3 Turnover Tax	0.1	0.2	0.2	0.3	0.3	0.3	0.3	0.4	0.4	0.4	0.4	0.4	0.4	0.4	0.5	0.5	0.5
4 Imports	0.6	1.1	1.2	1.4	1.4	1.5	2.5	2.2	2.3	2.3	2.6	3.5	3.8	3.7	3.4	3.4	3.8
5 TC and TD	2.2	3.4	3.6	3.9	4.1	4.4	4.8	5.1	5.5	5.9	6.0	6.3	6.6	6.9	7.1	7.2	7.8
6 USES	14.9	24.2	25.3	26.8	27.8	29.0	31.3	32.0	32.8	33.8	34.2	36.1	37.5	44.9	46.2	47.4	49.7
7 Interuse	10.2	17.1	18.0	19.1	19.4	20.1	21.3	21.7	21.8	22.5	22.4	23.0	23.8	30.0	30.7	31.6	32.7
8 Domestic End Use	4.2	5.8	6.1	6.5	7.0	7.2	8.2	8.6	9.1	9.6	9.9	11.0	11.8	12.8	13.3	13.4	14.5
9 Consumption	3.1	4.2	4.7	4.9	5.2	5.5	6.1	6.4	6.7	7.2	7.6	8.5	9.1	9.3	9.7	10.2	11.0
10 Private	2.6	3.5	3.9	4.1	4.3	4.6	5.0	5.2	5.4	5.8	6.2	6.9	7.4	7.4	7.8	8.1	8.7
11 Public	0.5	0.7	0.8	0.8	0.9	1.0	1.1	1.2	1.3	1.4	1.4	1.6	1.7	1.9	1.9	2.1	2.2
12 Capital Outlays	0.3	0.5	0.5	0.5	0.6	0.6	0.7	0.7	0.8	0.9	0.9	1.0	1.1	1.1	1.2	1.2	1.2
13 Inventories	0.3	0.6	0.3	0.4	0.6	0.4	0.7	0.7	0.8	0.6	0.6	0.7	0.7	1.1	1.2	0.7	0.8
14 Defense	0.4	0.5	0.6	0.6	0.6	0.6	0.7	0.8	0.8	0.9	0.8	0.8	0.9	1.3	1.3	1.4	1.4
15 Exports	0.5	1.3	1.2	1.2	1.4	1.8	1.8	1.7	1.9	1.7	1.9	2.0	1.9	2.1	2.2	2.4	2.5
Constant Prices																	
16 SUPPLY	23.5	26.9	27.5	27.9	27.8	27.9	28.5	28.0	27.5	27.2	26.3	27.0	27.3	28.0	27.9	28.3	28.8
17 GVO	19.1	21.5	21.8	22.0	22.0	22.1	22.2	22.0	21.6	21.3	20.4	20.6	20.7	21.6	21.7	22.0	22.3
18 Turnover Tax	0.1	0.2	0.2	0.3	0.3	0.3	0.3	0.3	0.3	0.3	0.3	0.3	0.3	0.3	0.3	0.3	0.3
19 Imports	0.8	1.3	1.4	1.5	1.4	1.3	1.9	1.7	1.6	1.6	1.8	2.4	2.5	2.1	1.9	1.9	2.0
20 TC and TD	3.5	4.0	4.1	4.1	4.1	4.1	4.1	4.1	4.0	4.0	3.8	3.8	3.8	4.0	4.0	4.1	4.2
21 USES	23.5	26.9	27.5	27.9	27.8	27.9	28.5	28.0	27.5	27.2	26.3	27.0	27.3	28.0	27.9	28.3	28.8
22 Interuse	16.5	18.8	19.5	20.0	19.4	19.4	19.5	18.7	17.9	17.6	16.7	16.7	16.6	17.4	17.4	17.8	18.0
23 Residual Index	0.62	0.91	0.92	0.96	1.00	1.04	1.10	1.16	1.22	1.28	1.34	1.38	1.43	1.72	1.76	1.77	1.82
24 Domestic End Use	6.2	6.6	6.6	6.6	7.0	6.9	7.6	7.9	8.1	8.2	8.2	8.8	9.2	9.1	9.1	9.1	9.4
25 Consumption	4.5	4.8	5.0	5.1	5.2	5.3	5.6	5.9	6.0	6.2	6.4	6.9	7.3	7.0	7.0	7.1	7.4
26 Private	3.7	4.1	4.2	4.3	4.3	4.4	4.6	4.8	4.9	5.0	5.2	5.6	5.9	5.6	5.6	5.7	6.0
27 Public	0.8	0.8	0.9	0.8	0.9	0.9	1.0	1.1	1.2	1.2	1.2	1.3	1.4	1.4	1.4	1.4	1.5
28 Capital Outlays	0.5	0.6	0.6	0.6	0.6	0.6	0.6	0.6	0.7	0.7	0.7	0.7	0.8	0.7	0.7	0.7	0.7
29 Inventories	0.5	0.7	0.3	0.4	0.6	0.4	0.7	0.7	0.7	0.5	0.5	0.6	0.6	0.6	0.6	0.4	0.4
30 Defense	0.7	0.6	0.6	0.6	0.6	0.6	0.6	0.7	0.6	0.7	0.6	0.6	0.6	0.8	0.8	0.8	0.8
31 Exports	0.8	1.5	1.4	1.3	1.4	1.6	1.5	1.5	1.5	1.3	1.4	1.5	1.4	1.5	1.4	1.4	1.4
32 GENERAL INDEX	0.63	0.90	0.92	0.96	1.00	1.04	1.10	1.14	1.19	1.24	1.30	1.34	1.38	1.60	1.66	1.67	1.72
33 Domestic	0.63	0.91	0.93	0.96	1.00	1.03	1.07	1.11	1.14	1.18	1.23	1.26	1.29	1.57	1.62	1.65	1.68
34 Logging	0.65	0.90	0.94	0.98	1.00	1.03	1.04	1.07	1.10	1.17	1.19	1.22	1.23	1.82	1.87	1.88	1.89
35 Furniture	0.63	0.89	0.91	0.95	1.00	1.06	1.10	1.13	1.17	1.22	1.26	1.29	1.32	1.38	1.48	1.50	1.55
36 Woodworking	0.59	0.88	0.91	0.95	1.00	1.02	1.09	1.15	1.20	1.23	1.31	1.35	1.42	1.56	1.60	1.64	1.68
37 Pulp & Paper	0.71	1.02	1.00	1.00	1.00	1.00	0.98	0.98	0.98	0.98	1.02	1.02	1.02	1.30	1.30	1.30	1.35
38 Parts, etc.	0.63	0.88	0.86	0.90	1.00	1.13	1.18	1.18	1.21	1.29	1.31	1.32	1.35	1.46	1.66	1.69	1.74
39 Imports	0.70	0.84	0.87	0.90	1.00	1.10	1.30	1.34	1.38	1.42	1.47	1.50	1.54	1.74	1.80	1.83	1.87
40 TC and TD	0.62	0.85	0.89	0.95	1.00	1.07	1.16	1.25	1.37	1.49	1.58	1.65	1.72	1.72	1.76	1.76	1.88

TABLE 0.07: SUPPLY AND USES OF WOOD AND PAPER PRODUCTS
   (billion rubles)

	1965	1970	1971	1972	1973	1974	1975	1976	1977	1978	1979	1980	1981	1982	1983	1984	1985
Supporting Data on GVO of Major Wood and Paper Product Sectors:																	
41 Logging	4.1	5.9	6.1	6.3	6.5	6.7	6.9	6.9	7.0	7.1	7.1	7.3	7.4	10.9	11.2	11.6	11.7
42 Production	2.7	4.2	4.5	4.7	4.9	5.1	5.3	5.3	5.4	5.5	5.5	5.6	5.7	8.7	9.0	9.3	9.4
43 Firewood	1.4	1.6	1.6	1.6	1.6	1.6	1.6	1.6	1.6	1.6	1.6	1.7	1.7	2.2	2.2	2.3	2.3
44 Wood-Working	6.1	10.1	10.5	11.1	11.5	12.0	12.5	12.9	13.0	13.4	13.5	14.0	14.6	17.0	17.7	18.1	18.9
45 Sawmills	1.7	2.7	2.8	3.0	3.0	3.1	3.3	3.4	3.4	3.4	3.4	3.5	3.6	4.1	4.2	4.3	4.6
46 Plywood	0.3	0.5	0.5	0.5	0.5	0.6	0.6	0.6	0.6	0.6	0.6	0.6	0.7	0.8	0.8	0.8	0.9
47 Prefab. Housing	0.2	0.3	0.3	0.3	0.3	0.3	0.3	0.3	0.3	0.3	0.3	0.3	0.3	0.4	0.5	0.5	0.6
48 Containers	0.6	0.9	0.9	1.0	1.0	1.1	1.2	1.2	1.2	1.2	1.2	1.2	1.2	1.5	1.5	1.5	1.6
49 Constr. Parts	0.8	1.3	1.4	1.4	1.5	1.5	1.5	1.6	1.6	1.6	1.6	1.7	1.7	2.0	2.1	2.1	1.9
50 Furniture	1.8	3.3	3.4	3.6	4.0	4.2	4.3	4.4	4.6	5.0	5.1	5.3	5.7	6.4	6.8	7.1	7.5
51 Matches	0.1	0.1	0.1	0.1	0.1	0.1	0.1	0.1	0.1	0.1	0.1	0.1	0.1	0.1	0.1	0.1	0.2
52 Other Woodworks	0.6	1.0	1.1	1.1	1.1	1.2	1.2	1.2	1.2	1.2	1.2	1.2	1.3	1.6	1.7	1.7	1.6
53 Structures	0.1	0.2	0.3	0.3	0.3	0.4	0.4	0.4	0.4	0.4	0.4	0.4	0.4	0.5	0.6	0.6	0.5
54 Consumer	0.5	0.8	0.8	0.8	0.8	0.8	0.8	0.8	0.8	0.8	0.8	0.8	0.8	1.1	1.1	1.1	1.1
55 Pulp & Paper	1.5	3.1	3.2	3.4	3.6	3.8	3.9	4.1	4.2	4.3	4.2	4.2	4.3	5.5	5.9	6.1	6.4
56 Pulp	0.6	1.4	1.5	1.5	1.7	1.7	1.8	1.8	1.9	2.0	1.9	1.9	2.0	2.7	2.8	2.9	3.1
57 Paper	0.8	1.7	1.8	1.9	1.9	2.0	2.2	2.3	2.3	2.3	2.3	2.3	2.3	2.9	3.1	3.2	3.3
58 Producer	0.7	1.6	1.7	1.7	1.7	1.8	2.0	2.0	2.0	2.1	2.0	2.0	2.0	2.6	2.7	2.8	2.9
59 Consumer	0.1	0.1	0.1	0.2	0.2	0.2	0.2	0.2	0.3	0.3	0.3	0.3	0.3	0.3	0.4	0.4	0.4
60 Wood Chem.	0.3	0.4	0.4	0.4	0.4	0.4	0.4	0.4	0.4	0.4	0.4	0.4	0.4	0.5	0.5	0.5	0.6
Supporting Data on Output in Physical Units:																	
61 Logging	378.9	385.0	384.7	382.9	387.8	388.5	395.1	384.7	376.8	361.8	354.0	356.6	358.2	355.9	355.7	367.9	368.0
62 Production	273.8	298.5	298.3	297.6	304.3	303.7	312.9	301.9	296.1	283.9	273.0	277.7	277.3	272.6	275.4	282.8	281.2
63 Firewood	105.1	86.5	86.4	85.3	83.5	84.8	82.2	82.8	80.7	77.9	81.0	78.9	80.9	83.3	80.3	85.1	86.8
64 -/- as fuels	33.5	26.6	26.6	25.7	25.1	24.0	23.8	24.6	24.6	24.1	24.2	22.8	22.9	23.3	23.1	22.6	23.5
65 -/- quality	3.14	3.25	3.25	3.32	3.33	3.53	3.45	3.37	3.28	3.23	3.35	3.46	3.53	3.58	3.48	3.77	3.69
66 Unit Price	1.09	1.52	1.59	1.65	1.68	1.72	1.75	1.80	1.85	1.96	2.01	2.05	2.07	3.06	3.14	3.15	3.18
67 Production	1.00	1.41	1.51	1.58	1.62	1.67	1.70	1.77	1.83	1.94	2.00	2.02	2.05	3.20	3.26	3.28	3.33
68 Firewood	0.13	0.19	0.19	0.19	0.19	0.19	0.19	0.19	0.19	0.20	0.20	0.21	0.21	0.26	0.27	0.27	0.27
69 Sawmills	111.0	116.4	118.8	118.7	116.2	114.7	116.2	112.8	109.2	106.0	99.6	98.2	98.1	97.5	97.0	97.3	98.2
70 Unit Price	1.55	2.31	2.39	2.50	2.62	2.68	2.86	3.02	3.15	3.23	3.44	3.55	3.71	4.25	4.36	4.40	4.65
71 Furniture	1.8	2.8	3.1	3.3	3.6	3.9	4.3	4.4	4.6	4.9	5.1	5.4	5.7	6.4	6.8	7.1	7.5
72 67,75 & 82 index	1.00	1.15	1.10	1.10	1.10	1.09	1.00	0.99	1.00	1.03	1.00	0.99	1.00	1.00	1.00	1.00	1.00
73 Pulp	3.23	5.11	5.41	5.68	6.07	6.34	6.84	7.20	7.45	7.58	7.05	7.12	7.32	7.44	7.91	8.15	8.37
74 Unit Price	0.20	0.27	0.27	0.27	0.27	0.27	0.26	0.26	0.26	0.26	0.27	0.27	0.27	0.36	0.36	0.36	0.37
75 Paper & Carton	4.89	6.71	7.09	7.42	7.89	8.20	8.59	8.92	9.07	9.24	8.73	8.74	8.96	8.98	9.56	9.85	10.02
76 Paper	3.23	4.19	4.41	4.61	4.91	5.04	5.22	5.39	5.46	5.55	5.25	5.29	5.40	5.44	5.67	5.86	5.99
77 Carton	1.66	2.52	2.68	2.81	2.98	3.16	3.37	3.53	3.61	3.69	3.48	3.45	3.56	3.54	3.89	3.99	4.03
78 Unit Price	0.17	0.26	0.25	0.25	0.25	0.25	0.25	0.25	0.25	0.25	0.26	0.26	0.26	0.32	0.32	0.32	0.33
79 Parts&Structures	8.2	10.1	11.3	11.2	10.7	10.0	9.7	10.2	9.9	9.5	9.4	9.6	9.6	10.0	9.7	9.5	8.2
80 Unit Price	10.68	14.81	14.49	15.30	16.91	19.04	20.02	19.83	20.48	21.83	22.11	22.30	22.79	24.70	28.17	28.62	29.49

TABLE O.08: SUPPLY AND USES OF CONSTRUCTION MATERIALS
(billion rubles)

	1965	1970	1971	1972	1973	1974	1975	1976	1977	1978	1979	1980	1981	1982	1983	1984	1985
Current Prices																	
1 SUPPLY	12.6	20.9	22.1	23.4	25.4	27.5	29.6	30.9	31.8	32.7	33.2	34.3	35.1	41.9	44.0	45.4	47.5
2 GVO	8.6	14.8	15.6	16.5	17.8	19.3	20.7	21.6	21.9	22.3	22.5	22.9	23.2	28.8	30.1	31.0	32.1
3 Turnover Tax	0.1	0.2	0.2	0.2	0.2	0.2	0.2	0.2	0.2	0.2	0.2	0.3	0.3	0.3	0.3	0.4	0.4
4 Imports	0.1	0.3	0.3	0.3	0.4	0.4	0.4	0.5	0.5	0.6	0.6	0.6	0.6	0.8	0.9	1.0	1.0
5 TC and TD	3.8	5.6	6.0	6.4	7.0	7.6	8.3	8.6	9.2	9.6	9.9	10.5	11.0	12.0	12.7	13.0	14.0
6 USES	12.6	20.9	22.1	23.4	25.4	27.5	29.7	30.9	31.7	32.6	33.2	34.2	35.1	41.9	44.1	45.4	47.6
7 Interuse	11.7	19.6	21.1	22.2	23.9	26.1	28.2	29.2	29.8	30.6	31.2	32.0	33.0	38.8	40.6	42.9	44.1
8 Domestic End Use	0.8	1.2	0.9	1.1	1.4	1.3	1.3	1.6	2.0	2.0	1.9	2.2	2.0	3.0	3.3	2.4	3.2
9 Consumption	0.5	0.6	0.6	0.6	0.7	0.8	0.8	0.8	0.8	0.8	0.8	0.8	0.8	1.2	1.2	1.2	1.3
10 Private	0.1	0.1	0.1	0.1	0.1	0.1	0.1	0.1	0.1	0.1	0.1	0.1	0.1	0.2	0.2	0.2	0.2
11 Public	0.4	0.5	0.5	0.5	0.6	0.7	0.7	0.7	0.7	0.7	0.7	0.7	0.7	1.0	1.0	1.0	1.1
12 Inventories	0.2	0.4	0.1	0.2	0.4	0.3	0.3	0.5	0.8	0.7	0.7	1.0	0.8	1.2	1.5	0.5	1.1
13 Defense	0.1	0.2	0.2	0.3	0.3	0.3	0.3	0.3	0.4	0.5	0.4	0.4	0.4	0.6	0.6	0.7	0.8
14 Exports	0.1	0.1	0.1	0.1	0.1	0.1	0.1	0.1	0.1	0.1	0.1	0.1	0.1	0.1	0.1	0.1	0.2
Constant Prices																	
15 SUPPLY	14.9	21.3	23.0	24.0	25.4	26.4	27.9	28.9	29.2	29.7	28.9	29.1	29.4	30.2	31.0	31.8	32.6
16 GVO	10.5	14.9	16.1	16.8	17.8	18.6	19.6	20.3	20.6	21.0	20.4	20.5	20.8	21.2	21.8	22.3	22.8
17 Turnover Tax	0.1	0.2	0.2	0.2	0.2	0.2	0.2	0.2	0.2	0.2	0.2	0.3	0.3	0.3	0.3	0.4	0.4
18 Imports	0.1	0.4	0.4	0.3	0.4	0.4	0.3	0.4	0.4	0.4	0.4	0.4	0.4	0.5	0.5	0.5	0.5
19 TC and TD	4.1	5.9	6.4	6.6	7.0	7.3	7.7	8.0	8.1	8.3	8.0	8.1	8.2	8.4	8.6	8.8	9.0
20 USES	14.9	21.3	23.0	24.0	25.4	26.4	27.9	28.9	29.2	29.7	28.9	29.1	29.4	30.2	31.0	31.8	32.6
21 Interuse	13.8	20.0	22.0	22.7	23.9	25.0	26.5	27.2	27.3	27.9	27.1	27.2	27.7	28.0	28.6	30.1	30.2
22 Domestic End Use	0.9	1.3	0.9	1.1	1.4	1.3	1.2	1.5	1.8	1.8	1.7	1.8	1.6	2.2	2.3	1.7	2.2
23 Consumption	0.6	0.6	0.6	0.6	0.7	0.7	0.7	0.7	0.7	0.7	0.7	0.7	0.7	0.9	0.8	0.8	0.9
24 Private	0.1	0.1	0.1	0.1	0.1	0.1	0.1	0.1	0.1	0.1	0.1	0.1	0.1	0.1	0.1	0.1	0.1
25 Public	0.5	0.5	0.5	0.5	0.6	0.6	0.6	0.6	0.6	0.6	0.6	0.6	0.6	0.8	0.7	0.7	0.8
26 Inventories	0.2	0.4	0.1	0.2	0.4	0.3	0.2	0.5	0.7	0.6	0.6	0.8	0.7	0.9	1.1	0.4	0.8
27 Defense	0.1	0.2	0.2	0.3	0.3	0.3	0.3	0.3	0.3	0.4	0.4	0.3	0.3	0.4	0.4	0.5	0.6
28 Exports	0.1	0.1	0.1	0.1	0.1	0.1	0.1	0.1	0.1	0.1	0.1	0.1	0.1	0.1	0.1	0.1	0.1
29 General Index	0.85	0.98	0.96	0.98	1.00	1.04	1.06	1.07	1.09	1.10	1.15	1.18	1.19	1.39	1.42	1.43	1.46
30 Domestic Prices	0.82	0.99	0.97	0.98	1.00	1.04	1.06	1.06	1.06	1.06	1.11	1.11	1.12	1.36	1.38	1.39	1.41
31 Cement	0.76	0.95	0.96	0.97	1.00	1.02	1.05	1.05	1.05	1.05	1.12	1.12	1.12	1.44	1.45	1.46	1.48
32 Concrete	0.84	1.01	0.99	0.99	1.00	1.02	1.04	1.04	1.04	1.04	1.07	1.07	1.08	1.34	1.36	1.36	1.37
33 Wall Materials	0.73	0.94	0.95	0.96	1.00	1.08	1.13	1.13	1.13	1.13	1.22	1.22	1.22	1.55	1.58	1.60	1.63
34 Asb.-Cement	0.84	1.00	1.00	1.00	1.00	1.00	1.00	1.00	1.00	1.00	1.00	1.00	1.00	1.20	1.20	1.20	1.20
35 Roofing	0.77	1.00	1.00	1.00	1.00	1.00	1.00	1.00	1.00	1.00	1.00	1.00	1.00	1.21	1.21	1.21	1.21
36 Ceramics	0.91	1.00	1.00	1.00	1.00	1.00	1.00	1.00	1.00	1.00	1.00	1.00	1.00	1.11	1.11	1.11	1.11
37 Lenoleum, etc.	0.97	1.00	1.00	1.00	1.00	1.00	1.00	1.00	1.00	1.00	1.00	1.00	1.00	1.03	1.03	1.03	1.03
38 Other	0.84	1.01	0.94	0.97	1.00	1.06	1.09	1.09	1.09	1.09	1.12	1.12	1.12	1.36	1.38	1.39	1.41
39 T & C	0.92	0.95	0.94	0.96	1.00	1.04	1.07	1.07	1.13	1.16	1.23	1.29	1.34	1.43	1.48	1.48	1.55
40 Imports	0.70	0.84	0.87	0.90	1.00	1.10	1.30	1.34	1.38	1.42	1.47	1.50	1.54	1.74	1.80	1.83	1.87

TABLE 0.08: SUPPLY AND USES OF CONSTRUCTION MATERIALS
(billion rubles)

	1965	1970	1971	1972	1973	1974	1975	1976	1977	1978	1979	1980	1981	1982	1983	1984	1985
Supporting Data on GVO of Major Construction Materials Sectors:																	
41 Cement	0.9	1.4	1.5	1.6	1.7	1.9	2.0	2.1	2.1	2.1	2.2	2.2	2.3	2.8	2.9	3.0	3.1
42 Concrete	2.9	5.3	5.6	5.9	6.4	6.9	7.3	7.6	7.8	7.9	8.0	8.1	8.2	10.3	10.8	11.1	11.6
43 Wall Materials	1.4	2.3	2.4	2.5	2.6	2.8	3.0	3.1	3.1	3.1	3.0	3.0	3.0	3.8	3.9	4.0	4.1
44 Asb.-Cement	0.2	0.3	0.4	0.4	0.4	0.4	0.5	0.5	0.4	0.4	0.4	0.4	0.5	0.5	0.6	0.6	0.6
45 Roofing	0.2	0.3	0.3	0.3	0.4	0.4	0.4	0.4	0.4	0.4	0.4	0.4	0.4	0.5	0.5	0.5	0.6
46 Ceramics	0.2	0.3	0.3	0.3	0.3	0.3	0.3	0.4	0.4	0.4	0.4	0.4	0.4	0.5	0.6	0.6	0.6
47 Lenoleum, etc.	0.6	1.1	1.1	1.1	1.2	1.2	1.4	1.5	1.6	1.6	1.7	1.8	1.8	1.9	2.0	2.1	2.2
48 Other	2.2	3.8	4.0	4.4	4.8	5.3	5.7	6.0	6.1	6.3	6.4	6.5	6.6	8.4	8.8	9.0	9.4
49 Mining	0.9	1.6	1.7	1.8	2.0	2.2	2.3	2.4	2.4	2.5	2.5	2.6	2.6	3.2	3.3	3.3	3.4
50 Manufacturing	1.3	2.2	2.3	2.5	2.8	3.1	3.4	3.6	3.7	3.8	3.9	3.9	4.0	5.2	5.5	5.7	6.0
Supporting Data on Output in Physical Units:																	
51 Cement	72.4	95.2	100.3	104.3	109.5	115.1	122.1	124.2	127.1	127.0	123.0	125.0	127.0	124.0	128.0	130.0	131.0
52 Unit Price	1.20	1.51	1.52	1.53	1.58	1.61	1.66	1.66	1.66	1.66	1.77	1.77	1.77	2.28	2.30	2.31	2.34
53 Concrete	56.1	84.6	90.8	96.1	102.9	108.5	114.2	118.7	121.3	123.2	120.8	122.2	124.0	124.0	128.0	132.0	137.0
54 Unit Price	5.17	6.24	6.13	6.15	6.19	6.33	6.43	6.43	6.43	6.43	6.65	6.65	6.66	8.31	8.40	8.44	8.46
55 Wall Materials	46.6	56.9	58.7	60.1	61.6	61.8	63.0					58.0					59.1
56 Unit Price	3.09	4.00	4.05	4.10	4.25	4.60	4.80					5.17					6.94
57 Asb.-Cement	4.2	5.8	6.2	6.6	7.0	7.4	7.8	8.1	7.3	7.3	7.3	7.3	7.5	7.6	7.9	8.1	8.3
58 Unit Price	0.50	0.60	0.60	0.60	0.60	0.60	0.60	0.60	0.60	0.60	0.60	0.60	0.60	0.72	0.72	0.72	0.72
59 Roofing	1.1	1.3	1.4	1.4	1.6	1.7	1.8	1.9	1.8	1.9	1.8	1.7	1.7	1.7	1.8	1.9	1.9
60 Unit Price	0.19	0.24	0.24	0.24	0.24	0.24	0.24	0.24	0.24	0.24	0.24	0.24	0.24	0.29	0.29	0.29	0.29
61 Ceramics	27.6	36.7	38.9	40.6	43.2	45.0	47.8	48.8	51.4	54.0	52.4	55.8	61.4	65.1	68.8	70.7	72.7
62 Unit Price	0.65	0.72	0.72	0.72	0.72	0.72	0.72	0.72	0.72	0.72	0.72	0.72	0.72	0.80	0.80	0.80	0.80
63 Lenoleum	31.2	57.4	59.8	59.9	61.1	64.2	71.9	78.7	83.0	85.2	87.8	93.1	95.1	96.2	99.9	106.0	113.0
64 Unit Price	1.85	1.90	1.90	1.90	1.90	1.90	1.90	1.90	1.90	1.90	1.90	1.90	1.90	1.96	1.96	1.96	1.96
65 Gravel, etc.	211.0	295.0	330.0	345.0	375.0	394.0	415.0										
	0.63	0.76	0.71	0.73	0.75	0.80	0.82										

TABLE 0.09: SUPPLY AND USES OF LIGHT INDUSTRIAL PRODUCTS--Part I: UNIFIED ACCOUNTS
       (billion rubles)

	1965	1970	1971	1972	1973	1974	1975	1976	1977	1978	1979	1980	1981	1982	1983	1984	1985
Current Prices:																	
1 SUPPLY	58.3	89.4	95.4	99.0	107.3	112.5	119.7	124.8	129.9	134.2	138.4	147.6	153.2	165.7	168.1	170.2	177.2
2 GVO	40.6	62.7	66.4	68.1	76.4	80.2	84.4	87.9	91.3	94.1	96.5	100.2	102.7	113.3	114.3	115.1	118.6
3 Turnover Tax	11.3	13.8	15.0	15.7	15.5	15.8	16.7	17.5	18.5	19.0	20.1	21.9	22.6	22.3	22.6	22.8	22.9
4 Imports	4.3	9.5	10.3	11.3	11.3	12.2	14.1	14.6	15.1	15.6	15.8	18.8	21.1	23.1	24.1	25.0	28.0
5 TC and TD	2.1	3.4	3.7	3.9	4.1	4.3	4.5	4.8	5.1	5.5	6.0	6.7	6.8	7.0	7.1	7.3	7.7
6 USES	58.3	89.4	95.4	99.0	107.3	112.5	119.7	124.8	129.9	134.2	138.4	147.6	153.2	165.7	168.1	170.2	177.2
7 Interuse	28.9	43.6	46.6	47.6	54.4	56.9	59.9	62.2	64.6	65.9	67.2	69.6	70.8	79.4	79.6	80.1	82.5
8 Domestic End Use	28.5	44.3	47.2	49.5	50.9	53.5	57.5	59.9	62.7	65.4	68.3	75.0	79.2	82.9	85.3	86.8	91.6
9 Consumption	26.7	41.2	43.5	45.7	47.4	49.8	53.6	56.4	59.9	62.0	65.8	71.1	76.3	77.4	78.9	81.3	84.3
10 Private	25.2	39.2	41.4	43.5	45.1	47.4	51.0	53.6	56.8	58.7	62.4	67.6	72.7	73.7	75.0	77.3	80.3
11 Public	1.5	2.0	2.1	2.2	2.3	2.4	2.6	2.8	3.1	3.3	3.4	3.5	3.6	3.7	3.9	4.0	4.0
12 Fixed Invest.	0.1	0.2	0.2	0.2	0.2	0.2	0.2	0.2	0.2	0.2	0.2	0.2	0.2	0.2	0.2	0.3	0.3
13 Inventories	1.1	2.1	2.7	2.7	2.4	2.6	2.8	2.3	1.6	2.1	1.2	2.5	1.5	3.7	4.6	3.5	5.3
14 Defense	0.6	0.8	0.8	0.9	0.9	0.9	0.9	1.0	1.0	1.1	1.1	1.1	1.2	1.6	1.6	1.7	1.7
15 Exports	0.9	1.5	1.6	1.8	1.9	2.1	2.3	2.6	2.7	2.9	2.9	3.0	3.2	3.3	3.2	3.4	3.2
Constant Prices:																	
16 SUPPLY	76.4	102.5	107.0	107.5	107.3	108.9	111.3	113.9	114.9	116.2	114.6	117.4	120.3	118.6	118.6	119.0	121.9
17 GVO	55.5	73.6	76.8	76.8	76.4	77.5	79.4	81.6	82.8	84.1	82.9	84.5	85.6	82.5	82.4	82.3	83.6
18 Turnover Tax	12.2	14.5	15.0	14.9	15.5	15.7	15.8	16.3	16.4	16.5	16.3	16.1	16.5	17.1	17.1	17.2	17.3
19 Imports	5.7	10.5	11.0	11.6	11.3	11.5	11.8	11.6	11.3	11.1	11.0	12.4	13.6	14.6	14.7	15.0	16.5
20 TC and TD	2.9	3.9	4.1	4.1	4.1	4.2	4.3	4.4	4.5	4.5	4.5	4.5	4.6	4.5	4.5	4.5	4.5
21 USES	76.4	102.5	107.0	107.5	107.3	108.9	111.3	113.9	114.9	116.2	114.6	117.4	120.3	118.6	118.6	119.0	121.9
22 Interuse	35.7	49.8	52.2	51.4	54.4	55.8	57.6	59.1	59.8	60.2	58.2	57.8	58.2	59.9	59.6	59.6	60.5
23 Domestic End Use	39.5	50.9	52.8	53.9	50.9	51.1	51.6	52.4	52.8	53.5	54.0	57.1	59.5	56.4	56.8	57.1	59.3
24 Consumption	37.0	47.4	48.6	49.7	47.4	47.6	48.1	49.2	50.3	50.6	51.9	54.1	57.0	52.6	52.5	53.4	54.4
25 Private	35.0	45.1	46.3	47.3	45.1	45.3	45.7	46.8	47.7	47.9	49.2	51.4	54.3	50.1	49.9	50.7	51.8
26 Public	2.1	2.3	2.3	2.4	2.3	2.3	2.3	2.4	2.6	2.7	2.7	2.7	2.7	2.5	2.6	2.6	2.6
27 Fixed Invest.	0.1	0.2	0.2	0.2	0.2	0.2	0.2	0.2	0.2	0.2	0.2	0.2	0.2	0.2	0.2	0.3	0.3
28 Inventories	1.6	2.4	3.1	3.0	2.4	2.4	2.5	2.0	1.3	1.7	0.9	1.9	1.1	2.5	3.0	2.3	3.4
29 Defense	0.9	0.9	0.9	1.0	0.9	0.9	0.8	0.9	0.9	1.0	0.9	0.9	1.0	1.1	1.1	1.1	1.3
30 Exports	1.3	1.8	1.9	2.1	1.9	2.0	2.1	2.4	2.4	2.5	2.4	2.5	2.6	2.3	2.2	2.3	2.1

TABLE 0.09: SUPPLY AND USES OF LIGHT INDUSTRIAL PRODUCTS--Part II: DERIVATION OF THE GENERAL PRICE INDEX

		1965	1970	1971	1972	1973	1974	1975	1976	1977	1978	1979	1980	1981	1982	1983	1984	1985
1	General Index	0.76	0.87	0.89	0.92	1.00	1.03	1.08	1.10	1.13	1.15	1.21	1.26	1.27	1.40	1.42	1.43	1.45
2	Interuse	0.81	0.88	0.89	0.93	1.00	1.02	1.04	1.05	1.08	1.09	1.15	1.20	1.22	1.33	1.34	1.34	1.36
3	End Use	0.72	0.87	0.89	0.92	1.00	1.05	1.12	1.15	1.19	1.23	1.27	1.31	1.34	1.47	1.50	1.52	1.55
4	Enterprises	0.73	0.85	0.86	0.89	1.00	1.04	1.06	1.08	1.10	1.12	1.16	1.19	1.20	1.37	1.39	1.40	1.42
5	Turnover Tax	0.92	0.95	1.00	1.05	1.00	1.00	1.05	1.07	1.13	1.15	1.24	1.36	1.37	1.31	1.32	1.32	1.32
6	Imports	0.75	0.90	0.94	0.97	1.00	1.06	1.20	1.26	1.34	1.40	1.44	1.52	1.55	1.59	1.64	1.67	1.70
7	Enterprises	0.73	0.85	0.86	0.89	1.00	1.04	1.06	1.08	1.10	1.12	1.16	1.19	1.20	1.37	1.39	1.40	1.42
8	Fibers	0.77	0.85	0.86	0.88	1.00	1.01	1.03	1.05	1.06	1.08	1.14	1.15	1.17	1.34	1.34	1.34	1.35
9	Cotton	0.87	0.89	0.89	0.89	1.00	1.00	1.00	1.00	1.00	1.00	1.07	1.07	1.07	1.27	1.27	1.27	1.27
10	Silk	0.55	0.77	0.81	0.85	1.00	1.04	1.13	1.20	1.23	1.27	1.41	1.42	1.47	1.68	1.67	1.68	1.69
11	Wool	0.73	0.84	0.85	0.87	1.00	1.01	1.02	1.03	1.07	1.09	1.10	1.12	1.16	1.25	1.25	1.25	1.25
12	Flax	0.54	0.81	0.81	0.81	1.00	1.00	1.00	1.00	1.00	1.00	1.00	1.00	1.00	1.15	1.15	1.15	1.15
13	Yarns	0.75	0.89	0.89	0.91	1.00	1.01	1.03	1.05	1.07	1.08	1.11	1.13	1.15	1.22	1.22	1.24	1.25
14	Cotton	0.82	0.86	0.87	0.89	1.00	1.03	1.05	1.07	1.08	1.09	1.13	1.16	1.18	1.48	1.49	1.50	1.51
15	Wool	0.66	0.88	0.89	0.89	1.00	1.01	1.03	1.05	1.08	1.09	1.12	1.13	1.15	1.09	1.10	1.10	1.10
16	Flax	0.83	0.95	0.95	0.95	1.00	1.00	1.00	1.00	1.00	1.00	1.00	1.00	1.00	1.13	1.13	1.13	1.13
17	Fabrics	0.71	0.83	0.84	0.86	1.00	1.06	1.09	1.10	1.14	1.15	1.19	1.22	1.21	1.44	1.46	1.46	1.49
18	Cotton	0.69	0.71	0.72	0.74	1.00	1.09	1.12	1.13	1.16	1.14	1.18	1.19	1.19	1.51	1.53	1.54	1.59
19	Silk	0.77	0.87	0.90	0.92	1.00	1.07	1.16	1.17	1.25	1.30	1.40	1.45	1.46	1.72	1.75	1.75	1.77
20	Wool													1.21	1.37	1.37	1.37	1.37
21	Quantity	0.64	0.92	0.93	0.94	1.00	1.03	1.07	1.08	1.12	1.15	1.20	1.23	1.28	1.63	1.66	1.69	1.70
22	Quality												1.00	1.01	1.02	1.02	1.03	1.04
23	Flax	0.89	0.93	0.93	0.93	1.00	1.00	1.00	1.00	1.00	1.00	1.00	1.00	1.00	1.18	1.18	1.18	1.18
24	Other	0.77	0.77	0.77	0.77	1.00	1.00	1.00	1.00	1.00	1.00	1.00	1.00	1.00	1.00	1.00	1.00	1.00
25	Finished Goods	0.71	0.85	0.87	0.89	1.00	1.04	1.07	1.08	1.11	1.13	1.17	1.20	1.22	1.41	1.43	1.44	1.47
26	Sewn Goods													1.29	1.44	1.47	1.52	1.55
27	Quantity	0.71	0.83	0.84	0.86	1.00	1.06	1.09	1.10	1.14	1.15	1.19	1.22	1.21	1.44	1.46	1.46	1.49
28	Quality												1.00	0.99	0.98	0.97	0.95	0.94
29	Knitted Goods	0.54	0.88	0.91	0.94	1.00	1.00	1.00	1.00	1.00	1.01	1.04	1.06	1.12	1.30	1.34	1.36	1.38
30	Knitwear													1.18	1.41	1.45	1.48	1.52
31	Quantity	0.60	0.89	0.91	0.93	1.00	1.00	1.00	1.00	1.00	1.00	1.04	1.09	1.11	1.31	1.34	1.34	1.34
32	Quality												1.00	0.98	0.97	0.96	0.94	0.92
33	Hosiery	0.43	0.87	0.91	0.96	1.00	1.00	1.00	1.00	1.00	1.03	1.06	1.01	1.01	1.09	1.11	1.12	1.11
34	Footware	0.93	0.93	0.95	0.96	1.00	1.02	1.06	1.10	1.12	1.19	1.23	1.27	1.33	1.37	1.39	1.43	1.45
35	Leather	0.74	0.84	0.86	0.93	1.00	1.07	1.10	1.12	1.14	1.16	1.18	1.25	1.26	1.34	1.34	1.34	1.37
36	Furs	0.74	0.81	0.84	0.89	1.00	1.03	1.09	1.08	1.11	1.17	1.15	1.15	1.19	1.25	1.25	1.32	1.37
37	Rugs	0.66	0.86	0.90	0.91	1.00	1.02	1.09	1.12	1.14	1.20	1.26	1.32	1.37	1.48	1.53	1.57	1.61

TABLE 0.09: SUPPLY AND USES OF LIGHT INDUSTRIAL PRODUCTS--Part III: OUTPUT IN ENTERPRISE PRICES AND TURNOVER TAX BY SECTOR
(billion rubles)

	1965	1970	1971	1972	1973	1974	1975	1976	1977	1978	1979	1980	1981	1982	1983	1984	1985
1 TOTAL TEXTILES	25.8	38.1	40.5	41.7	48.9	51.7	54.2	56.0	58.2	59.9	61.2	63.3	64.4	74.3	75.2	75.0	77.0
2 Fiber	4.7	6.1	6.6	6.7	7.9	8.2	8.9	9.1	9.5	9.6	9.7	10.3	10.6	12.2	12.2	11.4	12.0
3 Yarn	7.3	11.4	12.2	12.5	14.2	14.8	15.3	15.9	16.5	16.9	17.2	17.4	17.4	18.1	18.4	18.1	18.1
4 Fabrics	10.6	14.5	15.3	15.6	18.8	20.1	20.9	21.7	22.5	22.8	23.4	23.6	24.1	29.7	29.9	30.4	31.1
5 Other	3.3	6.0	6.5	7.0	7.9	8.5	9.1	9.3	9.7	10.5	10.9	12.0	12.3	14.3	14.7	15.0	15.9
6 COTTON INDUSTRY	10.3	11.9	12.7	12.9	16.3	17.2	17.8	18.0	18.6	18.8	19.1	20.0	20.3	25.0	25.2	25.5	27.1
7 Cotton Fiber	2.4	2.8	3.1	3.1	3.7	3.7	4.0	3.9	4.1	4.1	4.0	4.5	4.6	5.3	4.9	4.4	5.0
8 Cotton Textiles	7.5	8.4	8.8	9.0	11.7	12.6	12.8	13.1	13.5	13.6	14.0	14.3	14.5	18.2	18.7	19.4	20.4
9 Cotton Yarn	2.5	2.9	3.1	3.2	3.7	3.8	3.9	4.0	4.1	4.2	4.3	4.5	4.6	5.7	5.9	6.0	6.2
10 Cotton Fabrics	4.9	5.3	5.6	5.7	7.8	8.6	8.7	8.9	9.2	9.2	9.5	9.6	9.7	12.2	12.5	13.1	13.7
11 Other	0.1	0.1	0.1	0.1	0.2	0.2	0.2	0.2	0.2	0.2	0.2	0.2	0.2	0.3	0.3	0.3	0.4
12 Other Ind. Use	0.5	0.7	0.8	0.8	0.9	0.9	1.0	1.0	1.1	1.1	1.1	1.2	1.2	1.5	1.6	1.7	1.7
13 SILK INDUSTRY	1.5	2.4	2.6	2.7	3.2	3.6	4.0	4.2	4.5	4.7	5.0	5.2	5.5	6.4	6.7	6.7	6.8
14 Silk Weaving	0.6	0.9	0.9	1.0	1.1	1.4	1.5	1.6	1.7	1.7	1.9	1.9	2.0	2.4	2.5	2.5	2.5
15 Silk Fabrics	0.9	1.4	1.5	1.6	1.8	2.0	2.3	2.4	2.6	2.8	3.0	3.1	3.2	3.7	3.9	3.9	4.0
16 Other Ind. Use	0.0	0.1	0.1	0.1	0.2	0.2	0.2	0.2	0.2	0.2	0.2	0.2	0.3	0.3	0.3	0.3	0.3
17 WOOL INDUSTRY	7.9	14.5	15.4	15.8	18.0	19.1	20.2	21.2	22.4	23.3	24.3	25.3	26.0	28.1	28.0	27.4	27.1
18 Wool Fiber	1.4	1.9	2.0	2.1	2.5	2.6	2.8	3.0	3.1	3.3	3.4	3.5	3.6	3.9	4.1	3.9	3.9
19 Wool Textiles	6.3	12.1	12.9	13.2	14.8	15.7	16.5	17.2	18.0	18.6	19.2	19.6	19.9	21.4	21.1	20.5	20.1
20 Wool Yarn	3.4	6.6	7.1	7.3	8.4	8.9	9.3	9.8	10.2	10.5	10.9	11.2	11.2	10.6	10.6	10.2	10.0
21 Wool Fabrics	2.8	5.5	5.8	5.9	6.4	6.8	7.1	7.4	7.8	8.1	8.3	8.4	8.8	10.9	10.5	10.3	10.1
22 Consumer Goods	0.2	0.5	0.5	0.6	0.7	0.8	0.9	1.0	1.3	1.4	1.7	2.2	2.4	2.7	2.9	3.0	3.1
23 Wool Rugs	0.2	0.4	0.5	0.5	0.6	0.7	0.8	0.9	1.1	1.3	1.5	1.9	2.1	2.4	2.5	2.6	2.7
24 Other	0.0	0.1	0.1	0.1	0.1	0.1	0.1	0.1	0.2	0.2	0.2	0.3	0.3	0.4	0.4	0.4	0.4
25 FLAX INDUSTRY	2.7	3.6	3.8	3.8	4.1	4.1	4.1	4.2	4.2	4.1	3.7	3.2	3.1	3.8	4.1	4.0	4.0
26 Flax Fiber	0.3	0.5	0.5	0.5	0.6	0.5	0.6	0.7	0.6	0.5	0.4	0.4	0.3	0.6	0.7	0.6	0.5
27 Flax Textiles	2.3	3.1	3.3	3.3	3.6	3.6	3.5	3.6	3.6	3.6	3.3	2.9	2.7	3.2	3.4	3.4	3.5
28 Flax Yarn	1.4	1.9	2.0	2.0	2.1	2.1	2.1	2.2	2.2	2.2	2.0	1.7	1.6	1.8	1.9	1.9	1.9
29 Flax Fabrics	0.9	1.2	1.3	1.3	1.4	1.4	1.4	1.4	1.4	1.4	1.3	1.2	1.1	1.4	1.5	1.6	1.6
30 KNITTED GOODS	2.2	4.2	4.5	4.8	5.2	5.5	5.7	5.9	6.1	6.3	6.7	6.9	7.2	8.2	8.5	8.7	9.1
31 Production	2.2	4.1	4.4	4.6	5.1	5.3	5.5	5.7	5.9	6.0	6.4	6.6	6.8	7.8	8.1	8.3	8.6
32 Knitted Wear	1.3	2.5	2.7	2.8	3.2	3.2	3.3	3.4	3.5	3.6	3.8	4.1	4.3	4.9	5.1	5.2	5.4
33 Hosiery	0.5	1.1	1.1	1.2	1.3	1.4	1.4	1.5	1.5	1.5	1.6	1.6	1.6	1.8	1.9	2.0	2.0
34 Industrial	0.4	0.5	0.6	0.6	0.6	0.7	0.8	0.8	0.9	0.9	0.9	0.9	0.9	1.1	1.1	1.1	1.2
35 Services	0.0	0.1	0.1	0.2	0.2	0.2	0.2	0.2	0.3	0.3	0.3	0.3	0.4	0.4	0.4	0.4	0.5
36 OTHER TEXTILES	1.3	1.6	1.6	1.7	2.0	2.1	2.2	2.5	2.5	2.5	2.5	2.5	2.5	2.7	2.7	2.8	2.9
37 Haberdashery	0.4	0.5	0.5	0.6	0.7	0.8	0.9	1.0	1.0	1.1	1.1	1.2	1.2	1.2	1.2	1.2	1.3
38 Jute & Hemp	0.5	0.5	0.5	0.5	0.6	0.6	0.6	0.7	0.7	0.6	0.6	0.5	0.5	0.5	0.5	0.5	0.5
39 Heavy Felt	0.2	0.3	0.3	0.3	0.4	0.4	0.4	0.4	0.4	0.4	0.4	0.4	0.4	0.4	0.4	0.4	0.4
40 Other	0.2	0.3	0.3	0.3	0.3	0.3	0.3	0.3	0.3	0.4	0.4	0.4	0.4	0.6	0.6	0.7	0.7
41 Woven	0.2	0.2	0.2	0.2	0.2	0.2	0.2	0.2	0.2	0.2	0.2	0.2	0.2	0.3	0.3	0.3	0.3
42 Ropes and Nets	0.0	0.1	0.1	0.1	0.1	0.1	0.1	0.1	0.1	0.2	0.2	0.2	0.2	0.3	0.3	0.4	0.4

TABLE 0.09: SUPPLY AND USES OF LIGHT INDUSTRIAL PRODUCTS--Part III: OUTPUT IN ENTERPRISE PRICES AND TURNOVER TAX BY SECTOR
       (billion rubles)

	1965	1970	1971	1972	1973	1974	1975	1976	1977	1978	1979	1980	1981	1982	1983	1984	1985
43 SEWN GOODS	9.3	16.7	17.7	18.1	18.4	19.0	20.1	21.3	22.0	23.0	23.9	25.0	26.0	26.0	26.0	26.7	27.7
44 Production	8.6	15.7	16.7	16.9	17.2	17.7	18.7	19.8	20.6	21.4	22.2	23.2	24.2	24.2	24.2	24.9	25.9
45 Commodity	6.5	11.8	12.5	12.7	12.9	13.3	14.0	14.9	15.5	16.1	16.7	17.4	18.3	18.2	18.0	18.7	19.4
46 Supplies	2.1	3.9	4.2	4.2	4.3	4.4	4.7	5.0	5.2	5.3	5.6	5.8	6.0	6.1	6.2	6.2	6.5
47 Services	0.7	1.0	1.0	1.1	1.2	1.3	1.3	1.4	1.4	1.6	1.7	1.8	1.8	1.8	1.8	1.8	1.8
48 LEATHER AND FURS	5.4	7.7	8.0	8.1	8.8	9.3	9.8	10.3	10.8	11.0	11.2	11.7	12.0	12.7	12.8	13.1	13.5
49 Leather Prod.	2.0	2.9	3.0	3.1	3.3	3.5	3.6	3.7	3.9	3.9	3.7	3.8	3.8	4.0	4.0	4.0	4.1
50 Leather Prod.	1.0	1.3	1.3	1.2	1.3	1.4	1.4	1.4	1.5	1.5	1.5	1.5	1.5	1.5	1.5	1.5	1.5
51 Artificial L.	0.6	1.1	1.2	1.3	1.4	1.5	1.6	1.7	1.7	1.7	1.6	1.7	1.7	1.8	1.8	1.8	1.9
52 Haberdashery	0.4	0.5	0.5	0.6	0.6	0.6	0.6	0.6	0.7	0.7	0.6	0.6	0.6	0.7	0.7	0.7	0.7
53 Furs	1.0	1.4	1.5	1.6	1.7	1.9	2.0	2.2	2.3	2.2	2.4	2.6	2.8	3.0	2.9	2.9	3.0
54 Natural	0.9	1.2	1.3	1.4	1.5	1.6	1.8	1.9	1.9	1.8	1.9	1.9	1.8	1.9	1.8	1.9	1.9
55 Artificial	0.1	0.2	0.2	0.2	0.2	0.2	0.3	0.3	0.4	0.4	0.5	0.8	1.0	1.1	1.1	1.1	1.0
56 Footware	2.3	3.3	3.4	3.3	3.6	3.7	3.9	4.2	4.4	4.7	4.8	5.0	5.2	5.5	5.7	5.9	6.2
57 Production	2.1	2.9	3.0	2.9	3.1	3.3	3.4	3.7	3.8	4.1	4.2	4.4	4.6	4.7	4.8	5.1	5.3
58 Services	0.2	0.4	0.4	0.4	0.5	0.5	0.5	0.5	0.6	0.6	0.6	0.7	0.7	0.8	0.8	0.9	0.9
59 Other	0.1	0.1	0.1	0.1	0.2	0.2	0.2	0.2	0.2	0.2	0.2	0.2	0.2	0.2	0.2	0.2	0.2
60 OTHER INDUSTRIES	0.1	0.2	0.2	0.2	0.3	0.3	0.3	0.3	0.3	0.3	0.3	0.3	0.3	0.3	0.3	0.3	0.4

Turnover Tax by Sector:

	1965	1970	1971	1972	1973	1974	1975	1976	1977	1978	1979	1980	1981	1982	1983	1984	1985
61 Textiles	6.1	7.4	8.1	8.5	8.3	8.5	9.0	9.4	9.9	10.2	10.8	11.8	12.1	12.0	12.1	12.2	12.3
62 -/- in Apparel	4.6	5.6	6.1	6.3	6.3	6.4	6.7	7.1	7.4	7.7	8.1	8.8	9.1	9.0	9.1	9.2	9.2
63 Cotton	2.0	2.5	2.7	2.8	2.8	2.8	3.0	3.2	3.3	3.4	3.6	3.9	4.1	4.0	4.1	4.1	4.1
64 Silk	1.7	2.1	2.3	2.4	2.3	2.4	2.5	2.6	2.8	2.9	3.0	3.3	3.4	3.3	3.4	3.4	3.4
65 Wool	1.9	2.3	2.6	2.7	2.6	2.7	2.8	3.0	3.1	3.2	3.4	3.7	3.8	3.8	3.8	3.9	3.9
66 Flax	0.1	0.1	0.1	0.1	0.1	0.1	0.1	0.1	0.1	0.1	0.1	0.1	0.1	0.1	0.1	0.1	0.1
67 Other	0.4	0.4	0.5	0.5	0.5	0.5	0.5	0.6	0.6	0.6	0.6	0.7	0.7	0.7	0.7	0.7	0.7
68 Knitted Goods	2.5	3.0	3.3	3.5	3.4	3.5	3.7	3.9	4.1	4.2	4.4	4.8	5.0	4.9	5.0	5.0	5.0
69 Other	2.7	3.3	3.6	3.8	3.7	3.8	4.0	4.2	4.5	4.6	4.9	5.3	5.5	5.4	5.5	5.6	5.6

TABLE 0.09: SUPPLY AND USES OF LIGHT INDUSTRIAL PRODUCTS--Part IV: PHYSICAL OUTPUT AND UNIT PRICE BY SECTOR

		1965	1970	1971	1972	1973	1974	1975	1976	1977	1978	1979	1980	1981	1982	1983	1984	1985
1	Cotton Fiber	1.84	2.13	2.36	2.36	2.47	2.48	2.65	2.59	2.70	2.73	2.50	2.80	2.89	2.79	2.59	2.34	2.64
2	Unit Price	1.30	1.33	1.33	1.33	1.50	1.50	1.50	1.50	1.50	1.50	1.60	1.60	1.60	1.90	1.90	1.90	1.90
3	Cotton Yarn	1.29	1.44	1.50	1.51	1.54	1.56	1.57	1.58	1.60	1.63	1.62	1.64	1.65	1.63	1.66	1.69	1.74
4	Unit Price	1.94	2.03	2.07	2.12	2.37	2.45	2.50	2.53	2.56	2.59	2.68	2.74	2.79	3.50	3.52	3.55	3.59
5	Cotton Fabrics	70.77	74.82	77.16	76.08	78.39	78.57	78.10	79.00	79.02	80.49	80.27	80.63	81.40	80.67	82.09	84.63	86.41
6	Unit Price	6.92	7.14	7.21	7.44	10.00	10.89	11.16	11.28	11.58	11.41	11.79	11.93	11.89	15.12	15.26	15.42	15.91
7	Raw Silk	2.65	3.02	3.08	3.12	3.00	3.44	3.46	3.41	3.56	3.59	3.49	3.60	3.65	3.72	3.90	3.88	3.91
8	Unit Price	2.09	2.94	3.06	3.24	3.80	3.95	4.28	4.56	4.68	4.84	5.34	5.39	5.58	6.37	6.36	6.39	6.43
9	Silk Fabrics	9.37	12.41	12.73	13.48	14.01	14.47	15.17	15.88	16.09	16.19	16.15	16.32	16.51	16.54	17.04	17.10	17.18
10	Unit Price	10.09	11.38	11.83	12.02	13.13	14.10	15.29	15.40	16.38	17.05	18.42	19.00	19.17	22.56	23.01	22.93	23.19
11	Wool Fiber	218	256	270	275	284	299	315	326	333	339	350	360	357	357	369	358	353
12	Unit Price	0.64	0.74	0.75	0.76	0.88	0.89	0.90	0.91	0.94	0.96	0.97	0.98	1.02	1.10	1.10	1.10	1.10
13	Wool Yarn	0.24	0.35	0.37	0.38	0.39	0.41	0.42	0.43	0.44	0.45	0.45	0.46	0.45	0.45	0.45	0.43	0.42
14	Unit Price	14.26	18.94	19.07	19.18	21.52	21.65	22.21	22.68	23.19	23.44	24.16	24.34	24.82	23.50	23.61	23.73	23.74
15	Wool Fabrics	0.37	0.50	0.52	0.52	0.53	0.54	0.55	0.57	0.57	0.58	0.57	0.56	0.57	0.55	0.52	0.50	0.49
16	Unit Price	7.80	11.11	11.25	11.35	12.11	12.53	12.91	13.13	13.58	13.91	14.51	14.85	15.48	19.76	20.10	20.46	20.63
17	Rugs	19.6	30.3	33.1	35.9	39.0	44.7	47.7	52.3	62.9	67.8	77.5	93.4	100.0	104.0	106.0	108.0	109.0
18	Unit Price	10.20	13.20	13.90	13.93	15.38	15.66	16.77	17.21	17.49	18.44	19.35	20.34	21.00	22.79	23.58	24.07	24.77
19	Flax Fiber	0.48	0.46	0.49	0.46	0.44	0.4	0.49	0.51	0.48	0.38	0.31	0.29	0.26	0.41	0.47	0.39	0.35
20	Unit Price	0.71	1.05	1.05	1.05	1.30	1.30	1.30	1.30	1.30	1.30	1.30	1.30	1.30	1.50	1.50	1.50	1.50
21	Flax Yarn	0.21	0.25	0.26	0.26	0.27	0.27	0.26	0.27	0.27	0.27	0.25	0.21	0.20	0.20	0.21	0.21	0.21
22	Unit Price	6.67	7.60	7.60	7.60	8.00	8.00	8.00	8.00	8.00	8.00	8.00	8.00	8.00	9.00	9.00	9.00	9.00
23	Flax Fabrics	0.59	0.73	0.77	0.78	0.80	0.80	0.77	0.78	0.79	0.80	0.73	0.65	0.62	0.64	0.69	0.73	0.75
24	Unit Price	1.60	1.68	1.68	1.68	1.80	1.80	1.80	1.80	1.80	1.80	1.80	1.80	1.80	2.13	2.13	2.13	2.13
25	Knitwear	0.90	1.23	1.27	1.29	1.36	1.39	1.42	1.46	1.51	1.55	1.59	1.62	1.65	1.61	1.64	1.68	1.73
26	Unit Price	1.40	2.07	2.12	2.18	2.33	2.33	2.33	2.33	2.33	2.33	2.41	2.54	2.58	3.06	3.12	3.12	3.13
27	Hosiery	1.35	1.34	1.31	1.34	1.41	1.47	1.50	1.54	1.56	1.60	1.64	1.67	1.73	1.78	1.82	1.87	1.91
28	Unit Price	0.40	0.81	0.86	0.90	0.94	0.94	0.94	0.95	0.94	0.97	0.99	0.95	0.95	1.03	1.04	1.05	1.04
29	Ropes and Nets	0.16	0.28					0.47	0.53	0.57	0.62	0.67	0.76	0.94	1.19	1.26	1.38	1.57
30	Unit Price	0.20	0.20	0.20	0.20	0.26	0.26	0.26	0.26	0.26	0.26	0.26	0.26	0.26	0.26	0.26	0.26	0.26
31	Soft Leather	8.78	10.92	11.26	10.73	10.93	10.98	11.16	11.49	11.79	11.63	11.38	11.52	11.31	11.17	11.40	11.40	11.38
32	Unit Price	9.11	10.99	11.10	11.65	12.35	13.21	13.44	13.49	13.57	13.67	13.80	13.89	14.15	14.77	14.65	14.65	14.94
33	Coarse Leather	0.12	0.17	0.17	0.15	0.15	0.15	0.15	0.15	0.15	0.15	0.14	0.13	0.13	0.12	0.13	0.13	0.12
34	Unit Price	6.67	7.06	7.35	8.33	9.00	9.67	10.00	10.33	10.67	10.97	11.21	12.31	12.31	13.31	13.36	13.36	13.71
35	Footwear	486	679	682	647	666	684	698	724	736	740	740	742	738	734	745	764	788
36	Unit Price	4.33	4.33	4.40	4.48	4.65	4.75	4.93	5.10	5.22	5.54	5.73	5.92	6.17	6.39	6.48	6.65	6.73
37	Natural Furs*	2.57	3.05	3.23	3.22	3.23	3.34	3.43	3.76	3.74	3.26	3.57	3.47	3.30	3.27	3.09	2.99	2.96
38	Coats - Adults	0.39	0.34	0.29	0.35	0.31	0.30	0.34	0.43	0.46	0.48	0.66	0.67	0.67	0.70	0.77	0.76	0.78
39	Coats - Child.	1.98	1.59	1.54	1.44	1.2	1.2	1.14	1.31	1.09	0.94	0.91	0.78	0.73	0.72	0.72	0.72	0.76
40	Hats & Collars	36.3	48.6	53.0	52.6	54.4	56.7	58.3	62.9	63.1	53.8	57.7	56.2	53.0	52.1	47.4	45.6	44.5
41	Unit Price	0.35	0.38	0.40	0.42	0.47	0.48	0.51	0.51	0.52	0.55	0.54	0.54	0.56	0.59	0.59	0.62	0.64
40	Artificial Furs*	0.65	1.42	1.31	1.47	1.40	1.43	1.64	1.96	2.08	2.39	2.98	4.50	5.81	6.36	6.36	6.40	6.14
41	Coats - Adults	0.34	1.14	1.00	1.14	1.14	1.10	1.49	1.89	2.14	2.43	3.24	5.22	6.90	7.40	7.30	7.24	6.68
42	Coats- Child.	0.36	0.65	0.64	0.74	0.65	0.54	0.50	0.61	0.60	0.61	0.91	1.13	1.36	1.73	2.02	2.48	3.53
43	Hats	6.5	9.0	8.9	9.6	8.7	10.3	9.2	9.4	8.4	10.3	9.3	10.5	11.7	13.8	13.8	13.3	10.9
44	Unit Price	0.17	0.17	0.17	0.17	0.17	0.17	0.17	0.17	0.17	0.17	0.17	0.17	0.17	0.17	0.17	0.17	0.17

*weighted average

TABLE 0.09: SUPPLY AND USES OF LIGHT INDUSTRIAL PRODUCTS--Part V: DERIVATION OF IMPORTS IN DOMESTIC PRICES
(billion rubles)

	1965	1970	1971	1972	1973	1974	1975	1976	1977	1978	1979	1980	1981	1982	1983	1984	1985
1 TOTAL	4.32	9.48	10.35	11.25	11.28	12.23	14.11	14.56	15.09	15.56	15.80	18.78	21.08	23.15	24.05	25.02	28.05
2 MATERIALS		1.78	1.77	1.64	1.65	1.63	1.74	1.74	1.72	1.83	2.10	2.14	1.90	2.16	3.54	2.56	2.83
3 Fiber (Crops)		0.63	0.54	0.45	0.41	0.41	0.42	0.41	0.33	0.28	0.35	0.36	0.29	0.39	0.88	0.72	0.82
4 mil. tons		0.31	0.26	0.21	0.17	0.17	0.17	0.16	0.13	0.11	0.13	0.13	0.10	0.11	0.25	0.20	0.23
5 Fiber (Wool)		0.58	0.67	0.62	0.72	0.77	0.84	0.87	0.92	1.06	1.29	1.20	1.30	1.34	2.19	1.40	1.58
6 mil. tons		0.08	0.09	0.08	0.10	0.10	0.11	0.11	0.11	0.13	0.13	0.12	0.13	0.13	0.15	0.09	0.11
7 Yarn		0.46	0.46	0.49	0.40	0.34	0.37	0.38	0.46	0.42	0.40	0.53	0.29	0.39	0.39	0.37	0.39
8 mil. tons		0.06	0.06	0.07	0.05	0.06	0.07	0.07	0.10	0.08	0.07	0.11	0.10	0.11	0.11	0.10	0.11
9 a) Crops		0.04	0.04	0.05	0.04	0.05	0.06	0.06	0.09	0.07	0.06	0.10	0.10	0.11	0.11	0.10	0.11
10 b) Wool		0.02	0.02	0.02	0.01	0.01	0.01	0.01	0.01	0.01	0.01	0.01	0.00	0.00	0.00	0.00	0.00
11 Leather		0.12	0.10	0.09	0.13	0.11	0.10	0.09	0.00	0.07	0.06	0.06	0.03	0.05	0.08	0.06	0.04
12 g.r.		0.06	0.05	0.04	0.06	0.06	0.05	0.04	0.00	0.04	0.03	0.03	0.01	0.03	0.04	0.03	0.02
13 FABRICS		1.09	1.26	1.52	1.43	1.49	1.99	1.86	1.99	1.91	1.95	2.43	2.74	3.91	3.93	3.98	4.45
14 Cotton		0.45	0.47	0.53	0.48	0.53	0.67	0.68	0.74	0.73	0.67	0.92	1.20	1.37	1.69	1.76	2.24
15 bln. m.		0.15	0.15	0.17	0.14	0.15	0.18	0.18	0.19	0.19	0.17	0.23	0.30	0.27	0.33	0.34	0.42
16 Wool		0.24	0.31	0.34	0.32	0.38	0.46	0.35	0.34	0.31	0.32	0.33	0.41	0.44	0.22	0.18	0.16
17 mil. m.		12.00	15.10	16.60	14.50	16.20	19.10	14.50	13.60	12.50	12.20	12.50	15.40	13.70	6.90	5.60	4.90
18 Silk		0.23	0.28	0.40	0.35	0.35	0.61	0.54	0.58	0.51	0.62	0.76	0.72	0.74	0.87	0.86	0.82
19 bln. m.		0.08	0.10	0.13	0.10	0.10	0.16	0.14	0.14	0.12	0.13	0.16	0.15	0.13	0.15	0.15	0.14
20 Jute		0.10	0.11	0.13	0.14	0.12	0.13	0.17	0.15	0.14	0.16	0.18	0.19	0.22	0.25	0.25	0.27
21 mil. m.		0.12	0.13	0.15	0.14	0.11	0.12	0.15	0.13	0.12	0.13	0.15	0.15	0.16	0.17	0.17	0.18
22 Rugs		0.07	0.09	0.12	0.13	0.11	0.11	0.12	0.19	0.22	0.19	0.23	0.22	1.14	0.89	0.92	0.96
23 mil. sq. m.		5.20	6.30	8.70	8.60	7.20	6.60	6.90	10.60	11.90	9.60	11.30	10.40	39.04	30.62	31.60	33.12
24 APPAREL, etc.		3.82	4.11	4.42	4.55	5.32	6.08	6.27	6.70	6.86	6.78	8.56	10.36	10.50	10.42	11.61	13.00
25 Sewn Goods		2.11	2.26	2.32	2.32	2.72	3.06	3.12	3.43	3.61	3.45	4.13	4.53	4.73	4.90	5.43	6.26
26 g.r.		0.44	0.48	0.50	0.50	0.61	0.68	0.71	0.78	0.82	0.78	0.96	1.05	1.10	1.14	1.29	1.59
27 Leather & Fur		0.05	0.05	0.05	0.10	0.10	0.15	0.18	0.15	0.17	0.22	0.38	0.55	0.62	0.57	0.57	0.58
28 g.r.		0.01	0.01	0.01	0.02	0.02	0.03	0.04	0.03	0.03	0.04	0.08	0.11	0.12	0.11	0.11	0.12
29 Knitwear		1.40	1.49	1.71	1.78	2.10	2.34	2.38	2.52	2.52	2.48	3.15	4.06	3.90	3.80	4.38	4.83
30 g.r.		0.20	0.21	0.24	0.25	0.30	0.33	0.34	0.36	0.36	0.35	0.45	0.62	0.60	0.58	0.67	0.81
31 Linen & Towels		0.03	0.05	0.06	0.06	0.06	0.08	0.09	0.07	0.07	0.07	0.09	0.19	0.26	0.27	0.30	0.38
32 g.r.		0.01	0.02	0.02	0.02	0.02	0.03	0.03	0.02	0.02	0.02	0.03	0.06	0.09	0.09	0.10	0.13
33 Hosiery		0.16	0.18	0.19	0.20	0.23	0.31	0.36	0.39	0.36	0.40	0.58	0.72	0.68	0.58	0.63	0.68
34 g.r.		0.02	0.02	0.02	0.02	0.03	0.03	0.04	0.04	0.04	0.04	0.06	0.08	0.08	0.06	0.07	0.08
35 Hats		0.07	0.08	0.10	0.10	0.11	0.14	0.14	0.14	0.13	0.16	0.24	0.31	0.32	0.30	0.29	0.28
36 g.r.		0.01	0.01	0.01	0.01	0.02	0.02	0.02	0.02	0.02	0.02	0.03	0.04	0.05	0.04	0.04	0.04
37 FOOTWARE		2.36	2.63	2.94	2.85	3.05	3.46	3.78	3.75	4.01	3.97	4.43	4.79	5.15	4.60	5.24	6.05
38 mil. pairs		77.9	84.1	91.2	85.0	94.3	104.9	110.6	107.3	109.6	104.9	113.4	116.0	122.2	105.9	117.6	139.4
39 a) Leather		60.7	63.2	60.9	60.9	65.9	69.7	69.8	70.5	68.2	62.8	66.1	71.3	71.9	59.6	68.9	85.9
40 b) Art. Leather		17.2	20.9	30.3	24.1	28.4	35.2	40.8	36.8	41.4	42.1	47.3	44.7	50.3	46.3	48.7	53.5
41 HABERDASHERY		0.42	0.59	0.73	0.80	0.74	0.84	0.91	0.93	0.95	1.01	1.22	1.29	1.44	1.57	1.64	1.72
42 g.r.		0.06	0.08	0.10	0.11	0.11	0.12	0.13	0.13	0.14	0.14	0.17	0.18	0.21	0.22	0.23	0.25

TABLE 0.10: SUPPLY AND USES OF FOOD INDUSTRIAL PRODUCTS--Part I: UNIFIED ACCOUNTS
(billion rubles)

	1965	1970	1971	1972	1973	1974	1975	1976	1977	1978	1979	1980	1981	1982	1983	1984	1985
Current Prices:																	
1 SUPPLY	88.2	120.1	127.0	130.6	138.1	148.1	156.4	157.2	165.6	171.2	177.0	183.6	192.6	203.3	210.4	214.8	209.9
2 GVO	57.6	79.0	83.1	86.9	91.4	98.2	103.6	102.5	107.5	109.9	112.5	113.9	116.0	123.9	131.0	136.4	137.9
3 Turnover Tax	19.3	25.3	27.8	26.2	28.0	30.3	31.0	31.9	33.1	36.8	37.0	38.4	42.6	43.1	43.8	42.3	34.9
4 Imports	3.1	4.4	4.1	4.3	5.1	5.2	6.4	6.8	8.1	7.1	9.4	12.1	14.1	16.1	14.9	14.8	14.6
5 TC and TD	8.2	11.4	12.0	13.2	13.6	14.4	15.4	16.0	16.9	17.4	18.2	19.2	19.9	20.2	20.7	21.3	22.5
6 USES	88.2	120.1	127.0	130.6	138.1	148.1	156.4	157.2	165.6	171.2	177.0	183.6	192.6	203.3	210.4	214.8	209.9
7 Interuse	26.1	34.1	36.4	38.0	39.8	42.6	45.1	44.9	47.2	48.0	49.3	50.0	51.2	53.6	56.5	58.8	59.1
8 Domestic End Use	60.8	83.9	88.6	90.9	96.6	102.8	109.1	110.5	116.6	121.2	125.8	132.0	139.4	147.6	151.6	153.5	148.3
9 Consumption	56.8	79.7	84.9	89.3	93.6	98.1	103.8	108.3	111.8	116.5	120.8	126.6	131.7	138.0	142.2	146.9	146.4
10 Private	53.8	75.2	80.2	84.3	88.3	92.6	98.1	102.3	105.6	110.1	114.1	119.7	124.5	130.6	134.5	139.0	138.3
11 a) Food, etc.	37.9	50.2	53.4	55.8	58.2	60.7	64.1	66.6	67.8	70.1	72.0	75.3	78.0	80.3	83.0	86.3	90.6
12 b) Alcohol	16.0	24.9	26.9	28.5	30.1	31.9	34.0	35.7	37.8	39.9	42.1	44.4	46.5	50.3	51.5	52.8	47.7
13 Public	3.0	4.5	4.7	5.0	5.3	5.5	5.7	6.0	6.2	6.4	6.7	6.9	7.2	7.4	7.7	7.9	8.1
14 Inventories	2.3	2.0	1.3	-1.1	0.3	1.8	2.2	-0.9	1.6	1.4	1.4	1.4	3.6	5.3	5.0	2.0	-2.7
15 Defense	1.7	2.3	2.4	2.6	2.7	2.9	3.1	3.2	3.3	3.4	3.6	3.9	4.1	4.3	4.4	4.5	4.6
16 Exports	1.3	2.1	2.0	1.7	1.7	2.7	2.2	1.8	1.8	2.0	2.0	1.6	2.0	2.1	2.3	2.6	2.5
Constant Prices:																	
17 SUPPLY	101.3	125.8	131.8	131.6	138.1	146.5	152.3	150.6	157.0	157.5	160.3	162.4	165.1	167.5	173.1	174.8	168.0
18 GVO	66.5	82.8	86.2	87.8	91.4	96.6	100.3	97.9	101.5	102.1	102.9	102.0	101.0	102.5	107.8	110.9	110.4
19 Turnover Tax	21.4	26.0	28.4	26.2	28.0	30.2	30.7	31.5	32.6	33.5	33.3	34.3	36.6	36.3	36.8	35.4	29.4
20 Food	8.3	7.9	8.0	7.7	8.3	8.2	8.5	8.4	8.7	9.0	8.5	8.3	8.4	8.7	9.0	9.1	6.2
21 Alcohol	13.1	18.1	20.4	18.5	19.7	22.0	22.2	23.0	23.9	24.5	24.8	26.0	28.2	27.7	27.8	26.2	23.2
22 Imports	3.5	4.6	4.2	4.3	5.1	5.1	6.2	6.4	7.7	6.6	8.6	10.8	12.3	13.3	12.3	12.0	11.7
23 TC and TD	10.0	12.4	12.9	13.2	13.6	14.5	15.0	14.7	15.2	15.3	15.4	15.3	15.2	15.4	16.2	16.6	16.6
24 USES	101.3	125.8	131.8	131.6	138.1	146.5	152.3	150.6	157.0	157.5	160.3	162.4	165.1	167.5	173.1	174.8	168.0
25 Interuse	30.3	35.8	37.8	38.4	39.8	41.9	43.7	42.9	44.6	44.3	44.8	44.5	44.4	44.3	46.6	47.8	47.5
26 Domestic End Use	69.5	87.9	91.9	91.4	96.6	101.9	106.5	106.0	110.7	111.3	113.7	116.4	118.9	121.4	124.6	125.0	118.5
27 Consumption	65.0	83.4	88.1	89.9	93.6	97.3	101.3	103.8	106.1	106.8	109.1	111.6	112.2	113.5	116.9	119.7	116.9
28 Private	61.5	78.7	83.2	84.9	88.3	91.8	95.8	98.1	100.2	101.0	103.0	105.5	106.0	107.4	110.5	113.2	110.5
29 a) Food, etc.	45.1	53.3	56.4	56.4	58.2	60.0	61.7	62.4	62.4	61.4	61.4	61.5	62.1	64.8	67.8	69.6	73.5
30 1973 index	0.84	0.94	0.95	0.99	1.00	1.01	1.04	1.07	1.09	1.14	1.17	1.22	1.25	1.24	1.23	1.24	1.23
31 b) Alcohol	16.5	25.5	26.9	28.5	30.1	31.9	34.0	35.7	37.8	39.5	41.6	44.0	43.9	42.6	42.8	43.6	37.0
32 1973 index	0.97	0.98	1.00	1.00	1.00	1.00	1.00	1.00	1.00	1.01	1.01	1.01	1.06	1.18	1.21	1.21	1.29
33 Public	3.4	4.7	4.9	5.0	5.3	5.4	5.5	5.7	5.9	5.9	6.1	6.1	6.2	6.1	6.3	6.4	6.5
34 Inventories	2.6	2.0	1.3	-1.1	0.3	1.8	2.1	-0.9	1.5	1.3	1.3	1.3	3.2	4.4	4.1	1.6	-2.1
35 Defense	2.0	2.4	2.5	2.6	2.7	2.9	3.0	3.0	3.1	3.2	3.3	3.5	3.6	3.6	3.6	3.7	3.7
36 Exports	1.5	2.2	2.1	1.7	1.7	2.7	2.1	1.7	1.7	1.9	1.8	1.5	1.7	1.8	1.9	2.1	2.0

260

	1965	1970	1971	1972	1973	1974	1975	1976	1977	1978	1979	1980	1981	1982	1983	1984	1985
1 General Index	0.87	0.95	0.96	0.99	1.00	1.01	1.03	1.04	1.06	1.09	1.10	1.13	1.17	1.21	1.22	1.23	1.25
2 Enterprises	0.87	0.95	0.96	0.99	1.00	1.02	1.03	1.05	1.06	1.08	1.09	1.12	1.15	1.21	1.21	1.23	1.25
3 Turnover Tax	0.90	0.97	0.98	1.00	1.00	1.00	1.01	1.01	1.02	1.10	1.11	1.12	1.16	1.19	1.19	1.19	1.19
4 Food	0.75	0.91	0.92	1.00	1.00	1.01	1.04	1.05	1.06	1.07	1.11	1.15	1.24	1.34	1.34	1.36	1.36
5 Alcohol	1.00	1.00	1.00	1.00	1.00	1.00	1.00	1.00	1.00	1.11	1.11	1.11	1.14	1.14	1.14	1.14	1.14
6 TC & TD	0.82	0.92	0.93	1.00	1.00	1.00	1.02	1.08	1.11	1.14	1.18	1.26	1.30	1.30	1.28	1.28	1.36
7 Imports	0.87	0.95	0.96	0.99	1.00	1.02	1.03	1.05	1.06	1.08	1.09	1.12	1.15	1.21	1.21	1.23	1.25
8 Sugar-Bread-Con.	0.76	0.93	0.95	0.99	1.00	1.01	1.01	1.02	1.03	1.03	1.04	1.06	1.08	1.16	1.17	1.19	1.19
9 Sugar	0.72	0.90	0.92	1.00	1.00	1.00	1.00	1.00	1.00	1.00	1.02	1.05	1.10	1.29	1.28	1.28	1.28
10 Unit Price	0.30	0.38	0.39	0.42	0.42	0.42	0.42	0.42	0.42	0.42	0.43	0.44	0.46	0.54	0.54	0.54	0.54
11 Bread, etc.	0.84	0.96	0.97	0.98	1.00	1.02	1.03	1.05	1.06	1.07	1.06	1.07	1.08	1.10	1.11	1.12	1.11
12 Unit Price	17.99	20.77	20.89	21.03	21.54	21.95	22.15	22.54	22.84	23.06	22.79	22.99	23.34	23.67	23.85	24.06	23.85
13 Confectioneries	0.68	0.92	0.95	1.00	1.00	1.00	1.01	1.01	1.01	1.02	1.05	1.05	1.06	1.10	1.13	1.17	1.19
14 Unit Price	1.16	1.59	1.63	1.72	1.72	1.72	1.74	1.74	1.53	1.54	1.59	1.61	1.62	1.69	1.73	1.81	1.84
15 Fuit/Veg & Tea	0.84	0.97	0.97	1.00	1.00	1.00	1.02	1.02	1.02	1.03	1.04	1.05	1.07	1.10	1.12	1.13	1.13
16 Veg Oils	0.85	0.96	0.96	1.00	1.00	1.00	1.02	1.03	1.04	1.04	1.04	1.04	1.06	1.08	1.10	1.12	1.12
17 Unit Price	0.93	1.05	1.05	1.09	1.09	1.09	1.12	1.13	1.13	1.13	1.13	1.14	1.16	1.18	1.20	1.22	1.22
18 Fruit/Veg	0.78	0.98	0.98	1.00	1.00	1.00	1.00	1.01	1.01	1.03	1.07	1.08	1.11	1.13	1.16	1.17	1.17
19 Dry Fruit	0.51	0.88	0.88	1.00	1.00	1.00	1.00	1.06	1.06	1.06	1.14	1.14	1.16	1.19	1.19	1.19	1.19
20 Unit Price	0.58	1.00	1.00	1.14	1.14	1.14	1.14	1.21	1.21	1.21	1.30	1.30	1.32	1.36	1.36	1.36	1.36
21 Canned Foods	0.85	1.00	1.00	1.00	1.00	1.00	1.00	1.00	1.00	1.02	1.05	1.06	1.09	1.11	1.16	1.16	1.16
22 Unit Price	0.17	0.20	0.20	0.20	0.20	0.20	0.20	0.20	0.20	0.20	0.21	0.21	0.22	0.22	0.23	0.23	0.23
23 Tea	0.73	0.90	0.90	1.00	1.00	1.00	1.03	1.03	1.03	1.03	1.03	1.05	1.06	1.13	1.13	1.13	1.10
24 Unit Price	1.54	1.91	1.91	2.12	2.12	2.12	2.18	2.18	2.19	2.18	2.18	2.23	2.25	2.39	2.39	2.39	2.34
25 Alc & Tobacco	0.88	0.96	0.98	1.00	1.00	1.00	1.02	1.02	1.02	1.04	1.07	1.12	1.21	1.34	1.35	1.36	1.42
26 Alcohol	0.97	0.98	1.00	1.00	1.00	1.00	1.00	1.00	1.00	1.02	1.02	1.04	1.07	1.20	1.22	1.22	1.23
27 Unit Price	0.45	0.46	0.47	0.47	0.47	0.47	0.47	0.47	0.47	0.48	0.48	0.49	0.51	0.57	0.58	0.58	0.58
28 Tobacco	0.58	0.89	0.90	1.00	1.00	1.00	1.10	1.10	1.11	1.11	1.24	1.37	1.62	1.75	1.75	1.75	1.86
29 Unit Price	0.23	0.36	0.36	0.40	0.40	0.40	0.44	0.44	0.44	0.44	0.50	0.55	0.65	0.70	0.70	0.70	0.74
30 Other Food	0.85	0.96	0.96	1.00	1.00	1.00	1.02	1.03	1.04	1.04	1.04	1.04	1.06	1.08	1.10	1.12	1.12
31 Total Meats	0.97	0.98	0.99	1.01	1.00	1.02	1.04	1.06	1.06	1.08	1.12	1.14	1.14	1.15	1.14	1.16	1.16
32 Physical Index	0.96	0.99	0.99	0.99	1.00	1.01	1.01	1.02	1.03	1.04	1.04	1.05	1.05	1.05	1.06	1.06	1.07
33 Quality Index	0.99	1.01	1.00	0.99	1.00	0.99	0.97	0.96	0.97	0.96	0.93	0.92	0.91	0.91	0.92	0.92	0.92
34 Meats	1.00	1.00	1.00	1.00	1.00	1.00	1.00	1.00	1.00	1.00	1.00	1.00	1.00	1.00	1.00	1.00	1.00
35 Unit Price	2.01	2.00	2.00	2.00	2.00	2.00	2.00	2.00	2.00	2.00	2.00	2.00	2.00	2.00	2.00	2.00	2.01
36 Deli Meats	0.79	0.94	0.94	0.95	1.00	1.06	1.08	1.12	1.17	1.20	1.24	1.27	1.27	1.28	1.32	1.35	1.39
37 Unit Price	1.43	1.70	1.70	1.70	1.80	1.90	1.95	2.01	2.11	2.16	2.23	2.28	2.29	2.30	2.37	2.44	2.50
38 Canned Foods	0.80	1.00	1.00	1.00	1.00	1.00	1.00	1.14	1.16	1.32	1.31	1.31	1.32	1.40	1.40	1.40	1.38
39 Unit Price	0.40	0.50	0.50	0.50	0.50	0.50	0.50	0.57	0.58	0.66	0.66	0.66	0.66	0.70	0.70	0.70	0.69
40 Total Dairy	0.93	0.98	0.98	1.00	1.00	1.02	1.02	1.05	1.03	1.04	1.06	1.09	1.12	1.12	1.10	1.09	1.09
41 Quality Index	1.07	1.02	1.02	1.00	1.00	0.98	0.98	0.96	0.97	0.96	0.94	0.91	0.89	0.89	0.91	0.92	0.91
42 Butter	1.00	1.00	1.00	1.00	1.00	1.00	1.00	1.00	1.00	1.00	1.00	1.00	1.00	1.00	1.00	1.00	1.00
43 Unit Price	2.90	2.90	2.90	2.90	2.90	2.90	2.90	2.90	2.90	2.90	2.90	2.90	2.90	2.90	2.90	2.90	2.86
44 Milk Prod.	1.00	1.00	1.00	1.00	1.00	1.00	1.00	1.00	1.00	1.00	1.00	1.00	1.00	1.00	1.00	1.00	1.00
45 Unit Price	0.22	0.22	0.22	0.22	0.22	0.22	0.22	0.22	0.22	0.22	0.22	0.22	0.22	0.22	0.22	0.22	0.22
46 Cheese	1.00	1.00	1.00	1.00	1.00	1.00	1.00	1.00	1.00	1.00	1.00	1.00	1.00	1.00	1.00	1.00	1.00
47 Unit Price	1.46	1.46	1.46	1.46	1.46	1.46	1.46	1.46	1.46	1.46	1.46	1.46	1.46	1.46	1.46	1.46	1.46
48 Canned Foods	0.88	0.92	0.96	1.00	1.00	1.00	1.00	1.00	1.00	1.03	1.22	1.23	1.34	1.46	1.48	1.52	1.55
49 Unit Price	0.13	0.14	0.14	0.15	0.15	0.15	0.15	0.15	0.15	0.15	0.18	0.18	0.20	0.22	0.22	0.23	0.23

TABLE 0.10: SUPPLY AND USES OF FOOD INDUSTRIAL PRODUCTS--Part II: DERIVATION OF THE GENERAL INDEX

	1965	1970	1971	1972	1973	1974	1975	1976	1977	1978	1979	1980	1981	1982	1983	1984	1985
50 Fish Industry	0.95	0.98	0.98	0.99	1.00	1.04	1.05	1.05	1.06	1.09	1.09	1.11	1.11	1.31	1.35	1.37	1.39
51 Physical Index	1.00	1.00	1.00	1.00	1.00	1.03	1.03	1.03	1.03	1.04	1.05	1.06	1.07	1.25	1.28	1.29	1.30
52 Quality Index	1.05	1.02	1.02	1.01	1.00	0.99	0.98	0.98	0.97	0.96	0.96	0.96	0.96	0.95	0.95	0.94	0.93
53 Fishing	1.00	1.00	1.00	1.00	1.00	1.04	1.04	1.04	1.04	1.05	1.06	1.07	1.08	1.26	1.31	1.32	1.33
54 Unit Price	0.70	0.70	0.70	0.70	0.70	0.73	0.73	0.73	0.73	0.74	0.74	0.75	0.76	0.88	0.92	0.92	0.93
55 Canned Foods	1.00	1.00	1.00	1.00	1.00	1.00	1.00	1.00	1.00	1.01	1.01	1.01	1.01	1.17	1.17	1.17	1.17
56 Unit Price	0.70	0.70	0.70	0.70	0.70	0.70	0.70	0.70	0.70	0.71	0.71	0.71	0.71	0.82	0.82	0.82	0.82
57 -/- per ton	2.00	2.00	2.00	2.00	2.01	2.00	2.01	2.02	2.02	2.03	2.03	2.03	2.03	2.36	2.36	2.36	2.36
58 Flour	0.75	0.90	0.91	0.93	1.00	1.08	1.12	1.17	1.27	1.32	1.32	1.38	1.48	1.57	1.62	1.67	1.76
59 Unit Price	0.16	0.19	0.19	0.20	0.21	0.23	0.24	0.25	0.27	0.28	0.28	0.29	0.31	0.33	0.34	0.35	0.37

TABLE 0.10: SUPPLY AND USES OF FOOD INDUSTRIAL PRODUCTS--Part III: OUTPUT IN ENTERPRISE PRICES AND TURNOVER TAX BY SECTOR
(billion rubles)

		1965	1970	1971	1972	1973	1974	1975	1976	1977	1978	1979	1980	1981	1982	1983	1984	1985
1	FOOD PRODUCTS	25.8	36.2	37.3	38.5	40.9	42.9	45.8	46.5	48.2	49.2	50.6	52.5	54.3	59.2	62.5	64.1	63.2
2	Sugar,Bread,Conf	12.7	16.5	16.5	17.4	18.8	18.7	19.5	19.6	20.3	21.0	20.7	21.1	21.5	24.5	25.1	25.8	25.6
3	Salt	0.5	0.6	0.6	0.6	0.6	0.6	0.6	0.6	0.6	0.6	0.6	0.6	0.6	0.6	0.6	0.6	0.6
4	Producer	0.4	0.5	0.5	0.5	0.5	0.5	0.5	0.5	0.5	0.5	0.5	0.5	0.5	0.5	0.5	0.5	0.5
5	Consumer	0.1	0.1	0.1	0.1	0.1	0.1	0.1	0.1	0.1	0.1	0.1	0.1	0.1	0.1	0.1	0.1	0.1
6	Sugar	4.0	4.6	4.3	4.5	5.4	5.0	5.4	5.0	6.1	6.2	5.7	5.6	5.6	8.0	8.2	8.3	7.9
7	Producer	1.7	1.8	1.7	1.9	2.3	2.1	2.4	2.0	2.3	2.3	2.3	2.3	2.4	3.4	3.3	3.3	2.8
8	Consumer	2.3	2.8	2.6	2.6	3.1	2.9	3.1	3.0	3.8	3.9	3.4	3.3	3.2	4.6	4.9	5.0	5.1
9	Bread Prod.	5.5	6.7	7.0	7.2	7.3	7.5	7.8	8.1	8.2	8.5	8.4	8.7	8.9	9.0	9.2	9.4	9.2
10	Bread, etc.	5.2	6.3	6.6	6.8	6.9	7.0	7.3	7.6	7.6	7.9	7.9	8.1	8.3	8.4	8.5	8.6	8.5
11	Producer	0.3	0.4	0.4	0.4	0.4	0.4	0.4	0.5	0.5	0.5	0.5	0.5	0.5	0.5	0.5	0.5	0.5
12	Consumer	4.8	5.9	6.1	6.3	6.5	6.6	6.9	7.1	7.2	7.4	7.4	7.6	7.8	7.9	8.0	8.0	8.0
13	Macaroni	0.3	0.4	0.4	0.4	0.5	0.5	0.5	0.6	0.6	0.6	0.6	0.6	0.6	0.6	0.7	0.7	0.7
14	Confectioneries	2.7	4.6	4.7	5.1	5.4	5.6	5.6	5.9	5.4	5.7	6.0	6.2	6.4	6.8	7.1	7.5	7.9
15	Producer	0.6	0.5	0.5	0.6	0.6	0.6	0.5	0.6	0.5	0.6	0.6	0.7	0.8	0.8	0.8	0.9	1.0
16	Consumer	2.1	4.1	4.2	4.5	4.8	5.0	5.2	5.3	4.9	5.1	5.4	5.5	5.6	6.0	6.3	6.6	6.9
17	Sugar-Base	1.5	3.0	3.1	3.3	3.5	3.7	3.8	3.8	3.6	3.7	3.9	4.0	4.0	4.3	4.5	4.8	4.9
18	Other	0.6	1.1	1.1	1.2	1.3	1.3	1.4	1.4	1.4	1.4	1.5	1.5	1.6	1.7	1.8	1.9	1.9
19	Veg/Fruit & Tea	5.1	6.6	7.0	7.2	7.4	8.4	8.8	8.0	8.9	8.8	9.1	8.9	9.3	9.9	10.7	10.7	10.7
20	Veg Oils	3.5	4.0	4.2	4.3	4.2	5.0	5.1	4.6	4.9	5.0	4.9	4.7	4.9	5.1	5.4	5.3	5.1
21	Producer	2.4	2.6	2.7	2.8	2.7	3.5	3.5	3.0	3.3	3.3	3.2	3.1	3.2	3.3	3.5	3.4	3.2
22	Consumer	1.1	1.4	1.5	1.5	1.5	1.6	1.6	1.6	1.7	1.7	1.7	1.7	1.7	1.8	1.9	1.9	2.0
23	Oil&Margerine	0.8	1.1	1.2	1.2	1.2	1.3	1.3	1.3	1.4	1.4	1.4	1.4	1.4	1.5	1.6	1.6	1.7
24	Non-Food	0.3	0.3	0.3	0.3	0.3	0.3	0.3	0.3	0.3	0.3	0.3	0.3	0.3	0.3	0.3	0.3	0.3
25	Fruit/Veg Prod.	1.2	2.1	2.2	2.3	2.5	2.6	2.9	2.6	3.0	2.8	3.1	3.0	3.1	3.4	3.9	3.9	4.1
26	Producer	0.1	0.2	0.2	0.2	0.2	0.2	0.3	0.3	0.3	0.3	0.3	0.3	0.3	0.3	0.4	0.4	0.4
27	Consumer	1.1	1.9	2.0	2.1	2.3	2.4	2.6	2.3	2.7	2.5	2.8	2.7	2.9	3.1	3.5	3.6	3.7
28	Fruit/Veg	0.3	0.4	0.4	0.5	0.5	0.5	0.5	0.4	0.6	0.5	0.6	0.5	0.5	0.6	0.8	0.8	0.9
29	Canned Foods	0.8	1.5	1.6	1.6	1.7	1.9	2.1	1.8	2.1	2.0	2.2	2.2	2.3	2.6	2.7	2.8	2.9
30	Tea	0.3	0.5	0.5	0.6	0.7	0.7	0.8	0.8	0.9	1.0	1.1	1.2	1.3	1.4	1.4	1.5	1.4
31	Producer	0.1	0.1	0.1	0.1	0.1	0.1	0.1	0.1	0.2	0.2	0.2	0.2	0.2	0.2	0.2	0.2	0.2
32	Consumer	0.3	0.4	0.4	0.5	0.5	0.6	0.6	0.7	0.8	0.8	0.9	1.0	1.1	1.1	1.2	1.2	1.2

TABLE 0.10: SUPPLY AND USES OF FOOD INDUSTRIAL PRODUCTS--Part III: OUTPUT IN ENTERPRISE PRICES AND TURNOVER TAX BY SECTOR
(billion rubles)

		1965	1970	1971	1972	1973	1974	1975	1976	1977	1978	1979	1980	1981	1982	1983	1984	1985
33	Alcohol& Tobacco	3.9	6.2	6.6	6.7	7.3	7.9	8.6	8.9	9.4	9.5	10.0	10.7	11.2	11.9	12.3	12.1	10.8
34	Alcohol	3.1	5.0	5.2	5.2	5.6	6.1	6.7	7.0	7.4	7.5	7.9	8.2	8.4	8.9	9.2	8.9	7.5
35	Producer	0.4	0.5	0.5	0.5	0.6	0.7	0.7	0.8	0.8	0.8	0.9	0.9	0.9	1.0	1.0	1.1	1.0
36	Consumer	2.7	4.5	4.7	4.7	5.0	5.4	6.0	6.2	6.6	6.7	7.0	7.3	7.5	7.9	8.2	7.8	6.6
37	N-Alc Beverages	0.2	0.3	0.3	0.4	0.4	0.4	0.4	0.4	0.5	0.5	0.5	0.5	0.5	0.5	0.6	0.6	0.6
38	Tobacco	0.8	1.2	1.5	1.5	1.7	1.7	1.9	2.0	2.0	2.0	2.1	2.4	2.8	3.0	3.1	3.2	3.3
39	Producer	0.1	0.1	0.2	0.2	0.3	0.3	0.3	0.3	0.3	0.3	0.4	0.4	0.5	0.5	0.5	0.5	0.5
40	Consumer	0.7	1.1	1.2	1.3	1.4	1.5	1.6	1.6	1.7	1.7	1.8	2.0	2.4	2.5	2.6	2.7	2.8
41	Other Foods	3.9	6.6	6.8	6.9	7.1	7.5	8.5	9.4	9.2	9.4	10.3	11.3	11.7	12.4	13.8	14.8	15.5
42	Producer	1.4	2.6	2.8	2.7	2.9	3.0	3.6	4.5	4.1	4.2	4.9	5.6	5.8	6.2	7.0	7.4	7.9
43	Consumer	2.5	3.9	4.0	4.2	4.2	4.5	4.8	4.9	5.1	5.2	5.4	5.7	5.9	6.2	6.8	7.4	7.6
44	Perfume	0.6	1.0	1.0	1.0	1.0	1.1	1.2	1.2	1.3	1.3	1.4	1.4	1.5	1.6	1.7	1.8	1.9
45	Hygiene Items	0.2	0.3	0.3	0.3	0.3	0.4	0.4	0.4	0.4	0.4	0.4	0.5	0.5	0.5	0.5	0.6	0.6
46	Other	1.7	2.6	2.7	2.8	2.9	3.0	3.2	3.3	3.4	3.5	3.6	3.8	4.0	4.2	4.6	4.9	5.1
47	MEATS & DAIRY	21.3	28.5	31.1	32.9	33.9	37.5	38.8	36.5	39.3	40.5	40.9	40.1	40.4	41.0	44.3	46.5	48.1
48	Meats	12.6	17.7	19.9	21.2	21.1	23.7	24.9	22.3	24.2	25.3	25.8	25.0	25.4	25.3	27.5	29.1	30.1
49	Producer	5.8	6.5	7.3	7.5	7.6	8.6	8.8	7.3	7.9	8.2	8.4	7.7	7.9	7.4	7.8	8.0	8.4
50	Meats	5.4	6.0	6.8	7.0	7.1	8.0	8.2	6.8	7.3	7.6	7.8	7.2	7.4	6.8	7.3	7.5	7.8
51	Gelatin	0.4	0.5	0.5	0.5	0.5	0.6	0.6	0.5	0.6	0.6	0.6	0.5	0.6	0.5	0.5	0.6	0.6
52	Consumer	6.8	11.2	12.7	13.7	13.5	15.1	16.1	15.0	16.3	17.1	17.4	17.3	17.5	17.9	19.7	21.1	21.6
53	Meats	6.6	10.8	12.2	13.1	12.9	14.5	15.5	14.5	15.8	16.5	16.8	16.7	16.8	17.3	19.0	20.3	20.8
54	Meats	4.3	7.0	8.1	8.7	8.1	9.2	9.8	8.5	9.5	9.9	9.9	9.7	9.8	10.2	11.4	12.3	12.3
55	Deli Meats	2.3	3.8	4.1	4.4	4.8	5.4	5.7	5.9	6.3	6.6	6.9	7.0	7.0	7.1	7.6	8.0	8.5
56	Canned Foods	0.3	0.4	0.5	0.6	0.6	0.5	0.6	0.5	0.6	0.6	0.6	0.6	0.7	0.6	0.7	0.7	0.8
57	Dairy	8.7	10.8	11.2	11.7	12.8	13.8	13.9	14.2	15.1	15.2	15.1	15.1	15.0	15.7	16.8	17.4	18.0
58	Producer	2.3	2.7	2.9	3.0	3.2	3.8	3.7	3.9	4.0	4.0	4.1	4.2	4.3	4.5	4.7	4.9	5.3
59	Consumer	6.4	8.1	8.3	8.7	9.6	10.0	10.2	10.3	11.1	11.2	11.0	10.9	10.7	11.2	12.1	12.6	12.7
60	Butter	3.1	2.8	3.0	3.1	3.6	3.7	3.6	3.7	4.1	4.0	3.9	3.7	3.5	3.6	4.2	4.4	4.4
61	Milk Prod.	2.6	4.3	4.2	4.5	4.7	5.0	5.3	5.3	5.4	5.6	5.5	5.6	5.6	5.8	6.1	6.3	6.5
62	Cheese	0.5	0.7	0.7	0.7	0.8	0.8	0.8	0.9	1.0	1.0	1.0	0.9	1.0	1.0	1.1	1.1	1.2
63	Canned Foods	0.1	0.2	0.2	0.2	0.2	0.2	0.2	0.2	0.2	0.2	0.3	0.3	0.3	0.3	0.3	0.3	0.3
64	Other	0.2	0.3	0.3	0.3	0.3	0.3	0.3	0.3	0.3	0.3	0.4	0.4	0.4	0.4	0.4	0.4	0.4
65	FISH INDUSTRY	4.7	6.4	6.5	6.9	7.6	8.4	9.1	9.2	8.8	8.7	9.1	9.2	9.4	11.2	11.5	12.1	12.5
66	Producer	2.1	2.9	3.1	3.3	3.8	4.3	4.7	4.5	4.2	4.0	4.2	4.2	4.2	4.8	4.8	5.1	5.2
67	Consumer	2.6	3.5	3.4	3.6	3.8	4.1	4.4	4.7	4.6	4.7	4.9	5.0	5.1	6.4	6.7	7.0	7.3
68	Fish	1.9	2.5	2.3	2.5	2.5	2.7	2.9	3.0	2.9	2.8	2.8	3.0	3.0	4.0	4.3	4.6	4.7
69	Canned Foods	0.7	1.0	1.1	1.2	1.2	1.4	1.5	1.7	1.7	1.9	2.1	2.0	2.1	2.4	2.4	2.4	2.6
70	FLOUR & CEREALS	5.8	7.9	8.2	8.6	9.0	9.5	9.9	10.3	11.2	11.6	11.9	12.2	12.0	12.5	12.7	13.6	14.1
71	Producer	4.5	5.8	6.0	6.3	6.8	7.3	7.6	8.1	8.9	9.2	9.6	10.0	9.7	10.2	10.3	11.3	11.7
72	Consumer	1.3	2.1	2.2	2.3	2.2	2.2	2.3	2.2	2.3	2.4	2.3	2.2	2.3	2.3	2.4	2.3	2.4
73	All Canned Foods	1.9	3.0	3.3	3.6	3.7	4.0	4.4	4.2	4.6	4.6	5.1	5.0	5.3	5.8	6.1	6.2	6.6

TABLE 0.10: SUPPLY AND USES OF FOOD INDUSTRIAL PRODUCTS--Part III: OUTPUT IN ENTERPRISE PRICES AND TURNOVER TAX BY SECTOR
   (billion rubles)

	1965	1970	1971	1972	1973	1974	1975	1976	1977	1978	1979	1980	1981	1982	1983	1984	1985
Supporting Data on GVO of Group "B" Industries:																	
74 Total Group "B"	64.3	105.3	111.5	117.1	124.3	132.1	141.3	144.1	152.2	158.1	162.6	169.6	175.1	188.1	197.0	204.2	208.3
Percent of Total Group "B"																	
75 Total Food	54.6	49.0	48.4	48.2	47.2	47.1	46.8	45.6	45.5	44.9	44.2	43.0	42.4	42.3	43.1	42.9	42.5
76 Sugar	3.6	2.7	2.3	2.2	2.5	2.2	2.2	2.1	2.5	2.4	2.1	2.0	1.8	2.5	2.5	2.4	2.5
77 Flour & Cereals	2.0	2.0	2.0	1.9	1.8	1.7	1.6	1.5	1.5	1.5	1.4	1.3	1.3	1.2	1.2	1.1	1.1
78 Bread	7.5	5.6	5.5	5.4	5.2	5.0	4.9	4.9	4.7	4.7	4.5	4.5	4.5	4.2	4.0	3.9	3.8
79 Confectioneries	3.3	3.9	3.8	3.8	3.8	3.8	3.7	3.7	3.2	3.2	3.3	3.2	3.2	3.2	3.2	3.2	3.3
80 Veg Oils	1.3	1.1	1.1	1.0	0.9	1.0	0.9	0.9	0.9	0.9	0.9	0.8	0.8	0.8	0.8	0.8	0.8
81 Beverages	4.5	4.6	4.5	4.3	4.3	4.4	4.5	4.6	4.7	4.5	4.6	4.6	4.6	4.5	4.5	4.2	3.5
82 Alcoholic	4.2	4.3	4.2	4.0	4.0	4.1	4.2	4.3	4.3	4.2	4.3	4.3	4.3	4.2	4.2	3.8	3.2
83 Non-Alcoholic	0.3	0.3	0.3	0.3	0.3	0.3	0.3	0.3	0.3	0.3	0.3	0.3	0.3	0.3	0.3	0.3	0.3
84 Canned Foods	2.9	2.9	3.0	3.0	3.0	3.0	3.1	2.9	3.0	2.9	3.1	3.0	3.0	3.1	3.1	3.0	3.2
85 Tea	0.4	0.4	0.4	0.4	0.4	0.4	0.5	0.5	0.5	0.5	0.5	0.6	0.6	0.6	0.6	0.6	0.6
86 Tobacco	1.1	1.1	1.1	1.1	1.1	1.1	1.1	1.1	1.1	1.1	1.1	1.2	1.4	1.3	1.3	1.3	1.4
87 Meats	10.2	10.3	10.9	11.2	10.4	11.0	11.0	10.0	10.4	10.5	10.3	9.8	9.6	9.2	9.6	10.0	10.0
88 Butter	4.8	2.7	2.7	2.7	2.9	2.8	2.5	2.5	2.7	2.5	2.4	2.2	2.0	1.9	2.1	2.1	2.1
89 Dairy Prod.	4.0	4.0	3.8	3.8	3.8	3.8	3.7	3.7	3.6	3.6	3.4	3.3	3.2	3.1	3.1	3.1	3.1
90 Fish Prod.	3.0	2.4	2.1	2.1	2.1	2.0	2.0	2.1	1.9	1.8	1.7	1.8	1.7	2.2	2.2	2.3	2.3
91 Other	6.0	5.4	5.3	5.1	5.0	4.9	5.0	5.0	4.8	4.8	4.7	4.8	4.7	4.6	4.8	4.9	5.0
Turnover Tax by Sector:																	
92 Total	19.3	25.3	27.8	26.2	28.0	30.3	31.0	31.9	33.1	36.8	37.0	38.4	42.6	43.1	43.8	42.3	34.9
93 Sugar	2.1	2.3	2.1	2.3	2.7	2.5	2.7	2.6	3.0	3.1	2.9	2.8	2.9	4.0	4.1	4.2	0.0
94 Confections	0.7	0.9	0.9	1.0	1.1	1.1	1.1	1.2	1.2	1.2	1.2	1.2	1.3	1.4	1.4	1.5	1.6
95 Veg. Oils	0.9	0.9	1.0	1.0	1.0	1.1	1.1	1.1	1.1	1.1	1.1	1.1	1.2	1.2	1.2	1.2	1.2
96 Alcohol	13.1	18.1	20.4	18.5	19.7	22.0	22.2	23.0	23.9	27.2	27.6	28.8	32.1	31.5	31.7	29.9	26.5
97 Tobacco	0.6	0.9	1.1	1.1	1.3	1.3	1.4	1.5	1.5	1.5	1.6	1.8	2.2	2.2	2.3	2.4	2.5
98 Flour & Cereals	1.1	1.2	1.2	1.2	1.2	1.2	1.2	1.2	1.2	1.3	1.3	1.2	1.3	1.2	1.3	1.3	1.3
99 Other Food	0.7	1.0	1.0	1.0	1.1	1.1	1.2	1.3	1.3	1.4	1.4	1.4	1.6	1.6	1.7	1.8	1.9

TABLE 0.10: SUPPLY AND USES OF FOOD INDUSTRIAL PRODUCTS--Part IV: OUTPUT IN PHYSICAL UNITS

	1965	1970	1971	1972	1973	1974	1975	1976	1977	1978	1979	1980	1981	1982	1983	1984	1985
1 Sugar	13.24	12.23	11.03	10.86	13.07	11.72	12.86	11.78	14.63	14.90	13.31	12.74	12.16	14.89	15.22	15.39	14.70
2 Refined	11.04	10.22	9.03	8.90	10.71	9.45	10.38	9.25	12.04	12.21	10.65	10.13	9.50	12.10	12.40	12.50	11.80
3 Granulated	2.20	2.01	2.00	1.96	2.36	2.27	2.48	2.53	2.59	2.69	2.66	2.61	2.66	2.79	2.82	2.89	2.90
4 Bread, etc.	30.40	32.20	33.30	34.10	34.10	34.20	35.20	36.10	35.80	36.80	37.00	37.90	38.20	38.10	38.50	39.00	38.70
5 Flour	37	42	43	44	43	42	42	42	42	42	43	42					
6 Macaroni	1.25	1.18	1.19	1.33	1.34	1.22	1.34	1.48	1.49	1.45	1.50	1.55	1.62	1.63	1.60	1.61	1.65
7 Confectioneries	2.32	2.90	2.89	2.96	3.14	3.27	3.25	3.39	3.53	3.70	3.77	3.86	3.95	4.02	4.10	4.15	4.29
8 Sugar-Base	1.35	1.75	1.74	1.76	1.86	1.96	1.93	1.99	2.03	2.14	2.18	2.22	2.26	2.29	2.33	2.36	2.42
9 Other	0.97	1.15	1.15	1.2	1.28	1.31	1.316	1.4	1.5	1.56	1.59	1.64	1.69	1.73	1.77	1.79	1.87
10 Veg Oil	2.77	2.78	2.92	2.83	2.68	3.41	3.34	2.78	2.94	2.97	2.82	2.65	2.61	2.63	2.78	2.68	2.55
11 Margerine	0.67	0.76	0.85	0.85	0.88	1.00	1.00	1.04	1.17	1.23	1.27	1.26	1.36	1.43	1.48	1.43	1.41
12 Veg Fats		0.28				0.37											
13 Canned Foods	7.1	10.7	11.3	12.1	13.0	14.2	14.6	14.5	15.1	15.0	16.1	15.3	15.9	16.6	17.1	17.2	18.0
14 Fruit & Veg	4.69	7.39	7.68	7.98	8.83	9.74	9.78	9.74	10.21	10.08	10.86	10.18	10.65	11.50	11.81	11.86	12.30
15 Meat	0.72	0.82	0.97	1.29	1.15	1.14	1.14	0.92	0.97	0.84	0.93	0.93	1.00	0.92	0.99	1.06	1.17
16 Milk	0.71	1.10	1.15	1.17	1.28	1.38	1.47	1.46	1.45	1.41	1.40	1.36	1.32	1.33	1.38	1.36	1.41
17 Fish	0.98	1.39	1.50	1.66	1.74	1.94	2.21	2.38	2.47	2.67	2.91	2.83	2.93	2.85	2.92	2.92	3.12
18 Dry Fruit	57.30	34.70	36.20	40.10	43.90	37.80	45.20	33.20	46.80	41.10	46.80	39.60	36.50	38.80	56.50	57.00	61.20
19 Frozen Veg	1.10	4.00	3.90	3.40	2.80	3.60	2.60	3.10	3.90	2.30	2.00	2.40	2.90	3.90	3.30	2.00	3.40
20 Tea	0.20	0.27	0.28	0.29	0.31	0.33	0.35	0.38	0.43	0.45	0.48	0.53	0.56	0.57	0.59	0.62	0.62
21 Alcohol		980					1271					1380	1401	1397	1412	1388	1230
22 Vodka		243					260					295	292	277	277	281	238
23 Vine		318					440					472	479	473	474	453	335
24 Beer	317	419	441	469	508	540	571	592	619	641	634	613	630	647	661	654	657
25 Sigarettes	304	323	334	304	323	343	364	375	378	377	360	364	365	359	369	373	381
26 Soap	1.19	0.96	0.84	0.82	0.79	1.00	1.04	1.03	0.93	0.94	0.95	1.04	1.07	1.08	1.09	1.02	0.98
27 per capita	5.1	4.0	3.4	3.3	3.2	3.9	4.1	4.0	3.6	3.6	3.6	3.9	4.0	4.0	4.0	3.7	3.5
28 Meats	5.25	7.14	8.18	8.72	8.35	9.37	9.86	8.37	9.12	9.58	9.58	9.15	9.28	9.27	10.11	10.66	10.81
29 High Grade	4.79	6.49	7.43	7.91	7.57	8.53	8.97	7.62	8.36	8.78	8.80	8.42	8.55	8.55	9.33	9.84	9.95
30 Low Grade	0.46	0.65	0.75	0.81	0.78	0.84	0.89	0.75	0.76	0.80	0.78	0.73	0.73	0.72	0.78	0.82	0.86
31 Meat Materials	0.40	0.60	0.70	0.80	0.80	0.85	0.91	0.98	1.05	1.13	1.20	1.24	1.28	1.34	1.43	1.49	1.57
32 Deli Meats	1.61	2.29	2.44	2.57	2.66	2.81	2.95	2.95	3.01	3.08	3.10	3.07	3.06	3.08	3.19	3.30	3.41
33 Total Milk Units	32.34	40.15	41.64	43.50	47.59	51.30	51.68	52.79	56.14	56.36	56.14	56.14	55.77	58.37	62.46	64.84	66.92
34 Butter	1.07	0.96	1.02	1.08	1.24	1.26	1.23	1.26	1.41	1.38	1.33	1.28	1.21	1.26	1.46	1.50	1.52
35 in milk units	16.53	14.47	16.15	17.12	19.17	20.65	20.50	21.28	22.85	22.27	21.73	21.97	21.07	22.47	24.48	25.55	26.02
36 Milk Prod.	11.70	19.70	19.70	19.90	21.20	23.00	23.60	23.40	24.30	24.80	25.00	25.30	25.70	26.40	27.80	28.60	29.80
37 Cheese	0.31	0.48	0.46	0.48	0.54	0.57	0.56	0.61	0.66	0.69	0.70	0.65	0.66	0.70	0.74	0.78	0.81
38 in milk units	3.41	4.80	4.60	5.28	5.94	6.27	6.16	6.71	7.52	7.80	7.91	7.35	7.46	7.91	8.51	8.97	9.32
39 Other Milk Prod.	0.70	1.18	1.18	1.19	1.27	1.38	1.42	1.40	1.46	1.49	1.50	1.52	1.54	1.58	1.67	1.72	1.79
40 Fish Industry	5.77	7.83	7.79	8.21	9.01	9.62	10.36	10.50	9.70	9.20	9.40	9.50	9.60	10.00	9.90	10.60	10.70
41 Producer	2.90	4.13	4.20	4.51	5.03	5.50	6.12	5.80	5.31	4.82	5.20	5.01	4.86	5.08	5.42	6.11	6.21
42 Consumer	2.87	3.70	3.59	3.70	3.99	4.12	4.24	4.70	4.39	4.38	4.20	4.49	4.74	4.92	4.48	4.49	4.49
43 Fish	2.53	3.21	3.06	3.12	3.38	3.44	3.47	3.87	3.53	3.45	3.18	3.50	3.72	3.92	3.46	3.47	3.40
44 Canned Foods	0.35	0.49	0.53	0.58	0.61	0.68	0.77	0.83	0.86	0.93	1.02	0.99	1.02	1.00	1.02	1.02	1.09
45 Fish Consumption	2.91	3.74	3.63	3.74	4.03	4.16	4.27	4.73	4.43	4.46	4.31	4.67	4.80	4.97	4.80	4.84	4.91
46 per capita	12.60	15.40	14.80	15.10	16.10	16.50	16.80	18.40	17.10	17.10	16.40	17.60	17.90	18.40	17.60	17.60	17.70
47 Population	231	243	245	247.5	250	252	254	257	259	261	263	265.5	268	270	272.5	275	277.5
48 Imports	0.04	0.04	0.04	0.04	0.04	0.04	0.03	0.03	0.04	0.08	0.11	0.18	0.06	0.044	0.32	0.35	0.42

TABLE 0.10: SUPPLY AND USES OF FOOD INDUSTRIAL PRODUCTS--Part V: DERIVATION OF IMPORTS IN DOMESTIC PRICES
(billion rubles)

	1965	1970	1971	1972	1973	1974	1975	1976	1977	1978	1979	1980	1981	1982	1983	1984	1985
1 TOTAL	3.06	4.39	4.06	4.28	5.14	5.17	6.43	6.75	8.14	7.08	9.43	12.05	14.10	16.14	14.90	14.78	14.57
2 Rice, milled		0.10	0.07	0.08	0.05	0.08	0.11	0.13	0.21	0.21	0.33	0.42	0.70	0.47	0.18	0.21	0.08
3 physical		0.32	0.23	0.28	0.15	0.22	0.28	0.32	0.46	0.41	0.63	0.69	1.28	0.86	0.32	0.38	0.13
4 Total Sugar		1.53	0.84	0.97	1.42	1.08	1.86	2.18	2.62	2.24	2.37	2.96	3.30	5.39	4.40	4.21	2.74
5 physical		3.01	1.61	1.71	2.53	1.89	3.26	3.72	4.75	3.99	4.06	4.87	5.14	7.27	5.93	5.7	4.51
6 a) Raw Sugar		3.00	1.59	1.66	2.49	1.86	3.24	3.34	4.29	3.99	3.77	3.80	4.20	6.16	4.80	4.97	4.31
7 b) Refined		0.01	0.02	0.05	0.04	0.03	0.02	0.38	0.46	0.00	0.29	1.06	0.94	1.11	1.13	0.73	0.20
8 Fruits		0.36	0.34	0.34	0.27	0.32	0.30	0.39	0.39	0.39	0.42	0.47	0.45	0.49	0.30	0.38	0.30
9 physical		0.18	0.17	0.15	0.12	0.14	0.13	0.16	0.16	0.16	0.16	0.18	0.17	0.18	0.11	0.14	0.11
10 a) Sulfured		0.05	0.04	0.05	0.04	0.04	0.03	0.04	0.06	0.05	0.05	0.05	0.05	0.05	0.03	0.04	0.03
11 b) Dried		0.13	0.13	0.10	0.08	0.10	0.10	0.12	0.10	0.11	0.11	0.13	0.12	0.13	0.08	0.10	0.08
12 Meats		0.32	0.46	0.26	0.26	1.04	1.04	0.72	1.19	0.36	1.22	1.65	1.96	1.88	1.98	1.62	1.72
13 physical		0.16	0.23	0.13	0.13	0.52	0.52	0.36	0.62	0.18	0.61	0.82	0.98	0.94	0.99	0.81	0.86
14 a) Frozen Meats		0.14	0.17	0.09	0.09	0.05	0.45	0.29	0.53	0.11	0.53	0.72	0.88	0.85	0.88	0.80	0.75
15 b) Canned Meats		0.02	0.06	0.04	0.04	0.07	0.07	0.07	0.09	0.07	0.08	0.10	0.10	0.09	0.11	0.11	0.11
16 c) Other		0.16	0.23	0.13	0.13	0.93	0.52	0.36	0.57	0.18	0.61	0.83	0.98	0.94	0.99	0.71	0.86
17 Butter		0.00	0.00	0.00	0.69	0.03	0.03	0.03	0.23	0.12	0.57	0.76	0.69	0.45	0.60	0.60	0.84
18 physical		0.00	0.00	0.00	0.23	0.01	0.01	0.01	0.08	0.04	0.19	0.25	0.23	0.15	0.20	0.20	0.28
19 Fish		0.06	0.03	0.03	0.02	0.03	0.04	0.04	0.05	0.12	0.17	0.28	0.09	0.07	0.59	0.66	0.78
20 physical		0.04	0.02	0.02	0.02	0.02	0.03	0.03	0.04	0.08	0.11	0.18	0.06	0.04	0.32	0.36	0.42
21 Flour & Cereals		0.11	0.11	0.12	0.15	0.16	0.18	0.21	0.27	0.24	0.48	0.61	0.94	0.55	0.23	0.35	0.16
22 physical		0.26	0.26	0.27	0.31	0.32	0.34	0.38	0.46	0.39	0.79	0.96	1.57	0.91	0.39	0.58	0.26
23 Canned Veg/Fruit		0.46	0.52	0.53	0.51	0.51	0.50	0.51	0.55	0.59	0.64	0.69	0.70	0.85	0.80	0.55	0.57
24 physical		0.46	0.51	0.53	0.51	0.50	0.50	0.51	0.55	0.57	0.61	0.65	0.64	0.77	0.69	0.47	0.49
25 a) Vegetables		0.25	0.30	0.32	0.33	0.32	0.32	0.32	0.37	0.38	0.42	0.42	0.39	0.46	0.41	0.18	0.19
26 b) Fruits		0.21	0.21	0.21	0.18	0.18	0.18	0.19	0.18	0.19	0.19	0.23	0.25	0.31	0.28	0.29	0.30
27 Veg Oils		0.12	0.12	0.12	0.12	0.12	0.12	0.27	0.27	0.35	0.42	0.76	1.15	1.75	1.44	1.59	1.68
28 physical		0.06	0.06	0.06	0.06	0.06	0.06	0.13	0.13	0.17	0.20	0.36	0.60	0.87	0.71	0.77	0.81
29 Tea		0.07	0.08	0.10	0.08	0.12	0.14	0.12	0.11	0.10	0.10	0.15	0.15	0.16	0.18	0.20	0.24
30 physical		0.04	0.04	0.05	0.04	0.06	0.07	0.06	0.06	0.05	0.05	0.07	0.07	0.07	0.08	0.09	0.11
31 Alcohol		0.43	0.48	0.50	0.46	0.54	0.76	0.76	0.80	0.88	0.96	1.09	1.16	1.37	1.44	1.54	2.24
32 g.r.		0.20	0.22	0.24	0.23	0.27	0.38	0.38	0.40	0.43	0.47	0.52	0.54	0.57	0.59	0.63	0.65
33 Sigarettes		0.62	0.76	0.96	0.87	0.87	0.99	1.02	1.04	1.03	1.19	1.36	1.84	1.79	1.95	1.94	2.07
34 physical		41.60	45.80	52.60	47.70	48.90	53.40	55.60	55.90	54.30	57.20	58.10	73.50	66.50	72.50	70.10	68.40
35 Soap & Cosmetics		0.12	0.15	0.16	0.14	0.18	0.24	0.26	0.31	0.36	0.46	0.71	0.83	0.78	0.65	0.70	0.80
36 g.r.		0.06	0.07	0.08	0.07	0.09	0.12	0.13	0.16	0.18	0.23	0.35	0.46	0.41	0.34	0.37	0.42
37 Other Imports		0.10	0.10	0.10	0.10	0.10	0.10	0.10	0.10	0.10	0.10	0.15	0.15	0.15	0.15	0.20	0.35

TABLE 0.11: SUPPLY AND USES OF N.E.C. INDUSTRY PRODUCTS
(billion rubles)   Page 1

	1965	1970	1971	1972	1973	1974	1975	1976	1977	1978	1979	1980	1981	1982	1983	1984	1985

Current Prices:

	1965	1970	1971	1972	1973	1974	1975	1976	1977	1978	1979	1980	1981	1982	1983	1984	1985
1 SUPPLY	8.6	13.8	14.9	16.0	18.1	19.4	21.5	23.1	24.5	26.1	27.6	29.6	31.1	33.9	35.2	36.5	38.1
2 GVO	7.4	12.3	13.0	13.8	15.6	16.6	17.8	18.7	20.0	21.0	21.8	23.1	24.1	26.7	27.7	28.9	30.2
3 Producer	3.8	6.7	7.0	7.4	8.6	9.2	9.8	10.4	11.0	11.8	12.1	12.5	13.0	14.7	15.2	15.7	16.5
4 Consumer	3.6	5.6	6.0	6.4	7.0	7.4	8.0	8.3	9.0	9.2	9.7	10.6	11.1	12.0	12.5	13.2	13.7
4 Net Taxes	0.5	0.2	0.5	0.6	0.8	0.9	1.6	2.1	2.2	2.6	3.2	3.8	3.7	3.5	3.5	3.3	3.4
5 Taxes	0.5	0.9	1.1	1.3	1.5	1.7	1.9	2.2	2.4	2.7	3.3	3.9	3.8	3.5	3.5	3.3	3.4
6 Subsidies	0.0	0.7	0.6	0.8	0.7	0.8	0.3	0.0	0.2	0.1	0.1	0.1	0.1	0.0	0.0	0.0	0.0
7 Imports	0.2	0.6	0.7	0.8	0.8	0.9	1.1	1.2	1.2	1.3	1.4	1.5	2.0	2.4	2.6	2.9	3.0
8 TC and TD	0.5	0.7	0.7	0.8	0.9	1.0	1.0	1.1	1.1	1.2	1.2	1.2	1.3	1.3	1.4	1.4	1.5
9 USES	8.6	13.8	14.9	16.0	18.1	19.4	21.5	23.1	24.5	26.1	27.6	29.6	31.1	33.9	35.2	36.5	38.1
10 Interuse	4.1	6.1	6.7	7.0	8.0	8.7	9.6	10.2	10.8	11.0	11.2	11.8	12.4	14.8	15.9	17.0	17.6
11 Domestic End Use	4.4	7.4	8.0	8.8	9.8	10.3	11.5	12.3	13.2	14.5	15.8	17.2	18.1	18.6	18.8	18.8	19.8
12 Consumption	4.3	7.2	7.9	8.6	9.3	10.1	11.0	12.1	12.8	13.9	15.0	16.4	17.5	17.0	17.0	17.7	18.7
13 Private	3.4	5.8	6.4	6.9	7.5	8.2	8.8	9.6	10.3	11.2	12.2	13.5	14.4	13.7	13.5	14.0	14.9
14 Public	0.9	1.4	1.5	1.7	1.8	1.9	2.2	2.4	2.5	2.7	2.8	2.9	3.1	3.4	3.5	3.7	3.8
15 Inventories	0.1	0.2	0.1	0.2	0.3	0.2	0.3	0.3	0.3	0.3	0.5	0.6	0.4	1.2	1.5	0.7	0.7
16 Defense	0.0	0.0	0.0	0.0	0.1	0.1	0.1	0.1	0.1	0.3	0.2	0.2	0.2	0.3	0.3	0.4	0.4
17 Exports	0.1	0.2	0.3	0.3	0.4	0.4	0.4	0.5	0.5	0.6	0.6	0.6	0.6	0.6	0.6	0.7	0.7

Constant Prices:

	1965	1970	1971	1972	1973	1974	1975	1976	1977	1978	1979	1980	1981	1982	1983	1984	1985
18 SUPPLY	9.8	13.6	14.9	16.0	18.1	19.2	21.3	22.7	24.1	25.2	25.4	26.5	27.0	27.7	28.3	28.6	28.7
19 GVO	8.4	12.0	12.9	13.7	15.6	16.7	17.8	18.7	20.0	21.0	21.2	21.9	22.3	22.6	23.1	23.3	23.5
20 Producer	4.9	6.6	7.0	7.4	8.6	9.3	9.8	10.4	11.0	11.8	12.0	12.2	12.7	13.1	13.3	13.3	13.3
21 Consumer	3.6	5.3	5.9	6.3	7.0	7.4	8.0	8.3	9.0	9.2	9.2	9.7	9.7	9.6	9.8	10.1	10.2
22 Net Taxes	0.6	0.3	0.5	0.6	0.8	0.8	1.5	1.9	2.0	2.0	2.0	2.2	2.1	2.3	2.2	2.1	2.1
23 Taxes	0.6	1.0	1.1	1.3	1.5	1.6	1.8	1.9	2.2	2.1	2.1	2.3	2.3	2.2	2.2	2.1	2.1
24 Subsidies	0.0	0.7	0.5	0.8	0.7	0.8	0.3	0.1	0.2	0.1	0.2	0.1	0.2	0.0	0.0	0.0	0.0
25 Imports	0.3	0.7	0.8	0.9	0.8	0.8	1.0	1.0	1.0	1.0	1.0	1.1	1.4	1.5	1.7	1.8	1.8
26 TC and TD	0.5	0.7	0.7	0.8	0.9	0.9	1.0	1.1	1.1	1.2	1.2	1.2	1.3	1.3	1.3	1.3	1.3
27 USES	9.8	13.6	14.9	16.0	18.1	19.2	21.3	22.7	24.1	25.2	25.4	26.5	27.0	27.7	28.3	28.6	28.7
28 Interuse	5.3	6.0	6.6	6.9	8.0	8.7	9.6	10.3	10.8	11.0	11.1	11.5	12.1	13.1	13.9	14.4	14.2
29 Domestic End Use	4.3	7.4	8.0	8.8	9.7	10.1	11.2	11.9	12.7	13.6	13.7	14.4	14.4	14.1	14.0	13.6	14.0
30 Consumption	4.2	7.2	7.9	8.5	9.3	9.8	10.8	11.5	12.3	13.0	13.1	13.6	13.8	13.0	12.5	12.8	13.1
31 Private	3.4	5.8	6.4	6.9	7.5	8.0	8.6	9.2	9.9	10.5	10.6	11.2	11.4	10.4	10.0	10.1	10.4
32 Public	0.9	1.4	1.5	1.7	1.8	1.9	2.2	2.3	2.4	2.5	2.4	2.4	2.5	2.6	2.6	2.6	2.7
33 Inventories	0.1	0.2	0.1	0.2	0.3	0.2	0.3	0.3	0.3	0.3	0.5	0.5	0.3	0.9	1.2	0.5	0.5
34 Defense	0.0	0.0	0.0	0.0	0.1	0.1	0.1	0.1	0.1	0.3	0.2	0.2	0.2	0.2	0.2	0.3	0.4
35 Exports	0.1	0.2	0.3	0.3	0.4	0.4	0.4	0.5	0.5	0.6	0.6	0.6	0.5	0.5	0.5	0.5	0.5

General Price Index and Price Indexes for Selected N.E.C. Industrial Sectors:

	1965	1970	1971	1972	1973	1974	1975	1976	1977	1978	1979	1980	1981	1982	1983	1984	1985
36 General Index	0.88	1.01	1.00	1.00	1.00	1.01	1.01	1.02	1.02	1.04	1.09	1.12	1.15	1.22	1.25	1.28	1.33
37 Enterprises	0.88	1.03	1.01	1.01	1.00	1.00	1.00	1.00	1.00	1.00	1.00	1.03	1.05	1.08	1.18	1.20	1.24	1.29
38 Producer	0.77	1.01	1.01	1.00	1.00	1.00	1.00	1.00	1.00	1.00	1.01	1.02	1.02	1.13	1.15	1.18	1.24
39 Consumer	1.00	1.05	1.02	1.01	1.00	1.00	1.00	1.00	1.00	1.00	1.05	1.09	1.15	1.25	1.27	1.31	1.34
40 Turnover Tax	0.82	0.94	0.99	1.00	1.00	1.06	1.08	1.11	1.12	1.27	1.54	1.65	1.69	1.55	1.56	1.58	1.63
41 Imports	0.80	0.88	0.89	0.90	1.00	1.10	1.15	1.20	1.25	1.30	1.37	1.42	1.45	1.55	1.57	1.60	1.65
42 TC and TD	0.94	0.97	0.99	1.00	1.00	1.00	1.00	1.00	1.00	1.01	1.01	1.01	1.03	1.04	1.07	1.09	1.13

TABLE 0.11: SUPPLY AND USES OF N.E.C. INDUSTRY PRODUCTS

(billion rubles)

	1965	1970	1971	1972	1973	1974	1975	1976	1977	1978	1979	1980	1981	1982	1983	1984	1985
43 Feed	0.77	1.04	1.03	1.02	1.00	0.99	0.98	0.94	0.93	0.91	0.88	0.85	0.84	0.94	0.95	0.97	1.00
44 Glass	0.76	0.93	0.94	0.95	1.00	1.06	1.09	1.18	1.24	1.33	1.46	1.60	1.63	1.78	1.83	1.91	2.10
45 China	0.90	1.06	1.03	1.02	1.00	1.00	1.00	1.00	1.00	1.00	1.04	1.09	1.15	1.25	1.27	1.31	1.32

Supporting Data on GVO of Major N.E.C. Industrial Sectors:

	1965	1970	1971	1972	1973	1974	1975	1976	1977	1978	1979	1980	1981	1982	1983	1984	1985
47 Feed	1.4	3.0	3.2	3.4	4.2	4.6	4.9	5.2	5.7	6.1	6.3	6.5	6.8	7.7	8.0	8.4	8.7
48 Glass and China	1.0	1.6	1.6	1.8	1.9	2.1	2.3	2.4	2.6	2.7	2.9	3.0	3.1	3.3	3.5	3.6	3.8
49 Producer	0.7	1.1	1.1	1.2	1.3	1.4	1.5	1.6	1.7	1.8	1.9	2.0	2.0	2.2	2.3	2.4	2.6
50 Consumer	0.3	0.5	0.5	0.6	0.6	0.7	0.8	0.8	0.9	0.9	1.0	1.0	1.1	1.1	1.2	1.2	1.2
51 a) China	0.2	0.3	0.3	0.4	0.4	0.5	0.5	0.5	0.5	0.5	0.6	0.6	0.7	0.7	0.8	0.8	0.8
52 b) Chrystal etc	0.1	0.1	0.1	0.2	0.2	0.3	0.3	0.3	0.4	0.4	0.4	0.4	0.4	0.4	0.5	0.5	0.5
53 Pharm. & Micro.	0.9	1.6	1.7	1.8	2.0	2.1	2.1	2.3	2.3	2.5	2.6	2.7	2.8	3.0	3.0	3.1	3.3
54 Pharmaceuticals	0.5	0.8	0.9	1.0	1.0	1.1	1.1	1.2	1.2	1.3	1.4	1.5	1.5	1.6	1.6	1.6	1.8
55 Microbiology	0.4	0.7	0.8	0.8	0.9	1.0	1.0	1.1	1.1	1.2	1.2	1.2	1.3	1.4	1.4	1.4	1.5
56 Printing & Copy	0.7	1.0	1.1	1.1	1.2	1.2	1.3	1.3	1.4	1.5	1.6	1.7	1.8	2.1	2.3	2.5	2.7
57 Producer	0.3	0.4	0.4	0.4	0.5	0.5	0.5	0.5	0.6	0.6	0.7	0.7	0.7	0.8	0.8	0.8	0.9
58 Consumer	0.4	0.6	0.7	0.7	0.7	0.7	0.8	0.8	0.8	0.9	0.9	1.0	1.1	1.3	1.5	1.7	1.8
59 Water Supply	1.2	1.7	1.8	1.8	2.0	2.0	2.1	2.1	2.2	2.3	2.3	2.4	2.5	2.9	2.9	3.0	3.0
60 Producer	1.1	1.5	1.6	1.6	1.8	1.8	1.9	1.9	2.0	2.1	2.1	2.2	2.3	2.6	2.6	2.7	2.7
61 Consumer	0.1	0.2	0.2	0.2	0.2	0.2	0.2	0.2	0.2	0.2	0.2	0.2	0.2	0.3	0.3	0.3	0.3
62 Other Consumer	2.1	3.5	3.6	4.0	4.3	4.6	5.0	5.4	5.7	5.8	6.1	6.7	7.1	7.7	8.0	8.3	8.7
63 Toys	0.3	0.6	0.6	0.7	0.7	0.7	0.8	0.8	0.9	0.9	0.9	1.0	1.1	1.2	1.3	1.4	1.5
64 Music Instrum.	0.1	0.2	0.2	0.2	0.2	0.2	0.2	0.2	0.2	0.2	0.2	0.3	0.3	0.3	0.3	0.4	0.4
65 Everyday Serv.	0.3	0.5	0.5	0.5	0.6	0.6	0.7	0.7	0.7	0.7	0.8	0.8	0.9	1.1	1.2	1.3	1.4
66 Jewelry	0.2	0.5	0.6	0.9	1.1	1.2	1.3	1.5	1.7	1.7	1.9	2.2	2.3	2.5	2.5	2.4	2.4
67 a) Domestic	0.1	0.4	0.4	0.7	0.8	0.8	1.0	1.1	1.3	1.3	1.5	1.8	1.9	2.0	2.0	1.8	1.8
68 b) Exports	0.1	0.2	0.2	0.2	0.3	0.3	0.3	0.4	0.4	0.5	0.5	0.5	0.5	0.6	0.6	0.6	0.6
69 Crafts	0.5	0.7	0.7	0.7	0.7	0.8	0.8	0.8	0.9	0.9	0.9	0.9	0.9	1.0	1.0	1.0	1.0
70 Other Sectors	0.7	1.0	1.0	1.0	1.0	1.1	1.2	1.3	1.3	1.4	1.4	1.5	1.6	1.6	1.7	1.8	1.9

Supporting Data on Turnover Tax by Sector:

	1965	1970	1971	1972	1973	1974	1975	1976	1977	1978	1979	1980	1981	1982	1983	1984	1985
70 China	0.2	0.3	0.3	0.4	0.4	0.4	0.4	0.5	0.5	0.5	0.5	0.5	0.5	0.5	0.5	0.5	0.5
71 Jewelry	0.1	0.2	0.3	0.5	0.7	0.7	0.9	1.0	1.3	1.6	2.1	2.7	2.6	2.3	2.3	2.1	2.2
72 Crafts and Other	0.2	0.4	0.4	0.4	0.4	0.5	0.6	0.6	0.6	0.6	0.7	0.7	0.7	0.7	0.7	0.7	0.7

Supporting Data on Output in Physical Units:

	1965	1970	1971	1972	1973	1974	1975	1976	1977	1978	1979	1980	1981	1982	1983	1984	1985
73 Feed	15.5	23.7	n.a.	n.a.	n.a.	n.a.	41.8	46.3	51.3	56.4	60.3	64.4	n.a.	n.a.	n.a.	n.a.	n.a.
74 Unit Price	9.29	12.49	n.a.	n.a.	12.00	n.a.	11.82	11.32	11.11	10.89	10.51	10.16	n.a.	n.a.	n.a.	n.a.	n.a.
75 Glass	190	231	237	248	254	258	269	267	269	266	255	245	245	243	247	247	243
76 Unit Price	0.39	0.48	0.48	0.48	0.51	0.54	0.56	0.60	0.63	0.68	0.75	0.82	0.83	0.91	0.93	0.97	1.07
77 China	0.44	0.59	0.65	0.74	0.83	0.93	0.99	1.05	1.06	1.08	1.08	1.11	1.11	1.14	1.16	1.16	1.17
78 Unit Price	0.45	0.54	0.52	0.52	0.51	0.51	0.51	0.51	0.51	0.51	0.53	0.56	0.59	0.64	0.65	0.67	0.68
79 Water Supply	n.a.	n.a.	n.a.	n.a.	n.a.	n.a.	n.a.	n.a.	n.a.	n.a.	n.a.	288	285	275	278	280	282
80 Agriculture	n.a.	n.a.	n.a.	n.a.	n.a.	n.a.	n.a.	n.a.	n.a.	n.a.	n.a.	161	163	151	153	152	150
81 Industry	n.a.	n.a.	n.a.	n.a.	n.a.	n.a.	n.a.	n.a.	n.a.	n.a.	n.a.	105	99	101	101	104	107
82 Consumption	n.a.	n.a.	n.a.	n.a.	n.a.	n.a.	n.a.	n.a.	n.a.	n.a.	n.a.	22	23	23	24	24	25

TABLE 0.12: SUPPLY AND USES OF AGRICULTURAL PRODUCTS--Part A: UNIFIED ACCOUNTS
(billion rubles)

	1965	1970	1971	1972	1973	1974	1975	1976	1977	1978	1979	1980	1981	1982	1983	1984	1985
Current Prices:																	
1 SUPPLY	73.3	97.5	101.0	102.5	117.8	115.6	114.7	125.0	130.9	136.5	141.7	144.7	152.9	161.0	174.1	184.2	186.0
2 GVO	71.0	103.8	108.1	108.8	121.9	122.1	122.3	132.4	141.7	147.0	151.9	152.6	160.0	170.3	207.9	217.0	219.5
3 Net Subsidies	2.2	13.2	14.6	15.5	14.4	15.2	17.8	18.5	21.0	22.4	23.2	21.6	23.4	25.5	50.6	50.9	52.5
4 Imports	0.9	1.4	1.7	3.2	3.8	2.1	3.5	4.0	3.0	4.1	5.1	5.6	8.1	7.8	7.9	9.0	9.4
5 TC and TD	3.6	5.5	5.8	5.9	6.5	6.7	6.7	7.1	7.2	7.8	7.8	8.1	8.2	8.4	8.9	9.1	9.6
6 USES	73.3	97.5	101.0	102.5	117.8	115.6	114.7	125.0	130.9	136.5	141.7	144.7	152.9	161.0	174.1	184.2	186.0
7 Interuse	45.5	61.9	66.7	70.7	75.3	79.3	83.6	86.3	90.3	93.0	98.2	101.3	103.2	107.2	115.0	122.3	124.2
8 a) Agriculture	15.8	20.0	22.1	24.0	25.5	26.3	27.7	30.2	31.3	32.6	35.6	38.1	38.8	38.7	42.6	46.4	47.1
9 b) Industry	29.7	41.9	44.6	46.6	49.8	53.0	55.9	56.1	59.0	60.4	62.6	63.2	64.4	68.4	72.4	75.9	77.1
10 Domestic End Use	27.4	34.8	33.5	31.1	41.7	35.0	30.2	38.2	39.9	43.2	42.8	42.8	49.0	53.3	58.7	61.5	61.3
11 Consumption	23.7	28.6	28.9	29.1	30.1	30.1	29.8	30.4	31.5	32.3	34.5	35.0	37.5	38.8	41.4	42.5	43.2
12 Private	23.1	27.8	28.0	28.3	29.1	29.0	28.6	29.1	30.1	30.8	32.9	33.4	35.8	37.1	39.5	40.5	41.2
13 Public	0.6	0.8	0.9	0.9	1.0	1.1	1.2	1.3	1.4	1.5	1.6	1.6	1.7	1.7	1.9	2.0	2.0
14 Capital Outlays	1.0	1.7	2.3	1.3	1.9	2.0	1.4	2.1	2.4	2.5	2.3	2.0	2.4	3.0	3.0	2.7	2.6
15 Inventories	0.8	5.6	2.5	1.6	8.3	1.9	-1.9	5.2	2.3	5.1	0.5	1.7	2.8	4.3	7.1	5.5	5.5
16 Losses	2.0	1.3	1.4	2.8	0.7	1.2	3.1	1.5	1.1	2.2	2.1	2.5	3.5	1.5	2.2	1.7	3.7
17 Reserves	0.0	-2.4	-1.5	-3.6	0.6	-0.2	-2.1	-1.0	2.6	1.1	3.4	1.6	2.7	5.7	5.1	9.1	6.3
18 Exports	0.3	0.8	0.8	0.7	0.8	1.3	0.9	0.5	0.7	0.3	0.7	0.6	0.7	0.5	0.4	0.4	0.4
Constant Prices:																	
19 SUPPLY*	90.8	115.7	117.3	113.3	132.3	126.7	120.2	129.7	130.7	137.9	132.6	130.3	129.9	137.1	144.4	145.2	144.1
20 GVO	86.1	108.9	110.0	104.7	122.0	118.1	110.9	119.4	121.4	127.3	121.8	119.6	117.6	124.8	131.7	132.0	130.8
21 Imports	0.9	1.3	1.6	3.1	3.8	2.1	3.2	3.8	2.8	3.7	4.4	4.6	6.2	5.9	5.3	5.9	6.2
22 TC and TD	3.8	5.6	5.7	5.6	6.5	6.5	6.2	6.5	6.5	7.0	6.5	6.2	6.1	6.5	7.4	7.2	7.2
23 USES*	90.8	115.7	117.3	113.3	132.3	126.7	120.2	129.7	130.7	137.9	132.6	130.3	129.9	137.1	144.4	145.2	144.1
24 Interuse	57.5	78.5	82.7	82.6	89.7	91.3	91.8	94.6	95.6	100.2	97.5	96.2	93.0	97.3	105.6	106.4	106.3
25 a) Agriculture	19.2	21.0	22.5	23.1	25.5	25.5	25.2	27.3	26.8	28.2	28.5	29.9	28.5	28.4	27.0	28.2	28.1
26 b) Industry*	38.3	57.5	60.2	59.5	64.2	65.9	66.7	67.4	68.9	71.9	69.0	66.3	64.5	68.9	78.6	78.1	78.2
27 -/- Subsidized	29.7	44.7	46.7	46.2	49.8	51.1	51.8	52.3	53.4	55.8	53.5	51.4	50.0	53.5	61.0	60.6	60.7
28 -/-Price Index	1.00	0.94	0.95	1.01	1.00	1.04	1.08	1.07	1.10	1.08	1.17	1.23	1.29	1.28	1.19	1.25	1.27
29 Domestic End Use	32.9	36.4	33.8	30.1	41.7	34.1	27.6	34.6	34.4	37.5	34.6	33.7	36.4	39.5	38.5	38.5	37.6
30 Consumption	28.3	29.9	29.3	28.0	30.1	29.1	27.0	27.4	27.1	28.0	27.7	27.4	27.6	28.6	26.7	26.3	26.2
31 Private	27.6	29.1	28.4	27.0	29.1	28.1	25.9	26.3	25.9	26.7	26.4	26.1	26.3	27.2	25.4	25.0	24.9
32 Public	0.7	0.8	0.9	0.9	1.0	1.1	1.1	1.2	1.2	1.3	1.3	1.3	1.3	1.3	1.3	1.3	1.3
33 Capital Outlays	1.2	1.7	2.1	1.4	1.9	2.2	1.4	2.0	2.2	2.2	2.1	1.8	2.1	2.5	2.7	2.3	2.2
34 Inventories	1.0	5.9	2.5	1.5	8.3	1.8	-1.7	4.7	2.0	4.4	0.4	1.3	2.1	3.2	4.5	3.3	3.3
35 Losses	2.4	1.4	1.4	2.7	0.7	1.2	2.8	1.4	0.9	1.9	1.7	2.0	2.6	1.1	1.4	1.0	2.2
36 Reserves	0.0	-2.5	-1.6	-3.5	0.6	-0.2	-1.9	-0.9	2.2	0.9	2.7	1.3	2.0	4.2	3.2	5.6	3.8
37 Exports	0.4	0.8	0.8	0.7	0.8	1.2	0.8	0.4	0.6	0.2	0.6	0.4	0.5	0.4	0.3	0.3	0.3
38 General Index	0.83	0.96	0.99	1.04	1.00	1.03	1.10	1.11	1.16	1.15	1.24	1.28	1.36	1.36	1.56	1.62	1.65
39 Total GVO	0.82	0.95	0.98	1.04	1.00	1.03	1.10	1.11	1.17	1.15	1.25	1.28	1.36	1.36	1.58	1.64	1.68
40 TC & TD	0.95	0.99	1.01	1.06	1.00	1.03	1.10	1.10	1.10	1.12	1.21	1.31	1.33	1.30	1.20	1.26	1.33

*in procurement prices that exclude subsidies

TABLE 0.12: SUPPLY AND USES OF AGRICULTURAL PRODUCTS--Part A: UNIFIED ACCOUNTS
       (billion rubles)

	1965	1970	1971	1972	1973	1974	1975	1976	1977	1978	1979	1980	1981	1982	1983	1984	1985
41 Official '73 GVO	88.3	108.4	109.9	105.1	121.9	119.2	112.9	120.1	125.0	128.3	123.5	121.3	120.8	127.4	135.2	134.4	132.9
42 Official Index	0.80	0.96	0.98	1.04	1.00	1.02	1.08	1.10	1.13	1.15	1.23	1.26	1.32	1.34	1.54	1.61	1.65

Crops in 1973 Prices:

	1965	1970	1971	1972	1973	1974	1975	1976	1977	1978	1979	1980	1981	1982	1983	1984	1985
43 Official '73 GVO	40.7	51.8	51.2	46.9	60.1	54.1	48.7	57.6	56.6	59.4	55.4	54.3	53.3	58.2	61.6	60.4	58.5
44 Derived '73 GVO	40.5	51.8	50.9	46.3	60.1	53.2	47.5	57.5	55.0	58.8	53.9	53.1	50.1	56.4	59.0	58.3	56.5
45 Grain	10.9	16.3	15.9	14.7	19.2	17.0	12.2	19.0	16.9	20.3	15.5	16.6	13.9	16.5	16.9	15.3	16.9
46 Wheat	5.2	8.7	8.6	7.5	9.6	7.3	5.8	8.5	8.1	10.6	7.9	8.6	7.1	7.4	6.8	6.0	6.8
47 Rye	1.7	1.3	1.3	1.0	1.1	1.5	0.9	1.4	0.9	1.4	0.8	1.0	1.0	1.5	1.8	1.4	1.6
48 Corn	0.9	1.1	1.0	1.1	1.5	1.4	0.8	1.1	1.3	1.0	1.0	1.1	1.1	1.7	1.5	1.5	1.6
49 Barley	1.4	2.6	2.4	2.5	3.8	3.7	2.5	4.8	3.6	4.3	3.3	3.0	2.5	3.0	3.5	2.9	3.2
50 Oats	0.4	1.0	1.0	1.0	1.2	1.0	0.9	1.2	1.3	1.3	1.0	1.1	0.9	1.2	1.3	1.3	1.4
51 Millet	0.2	0.1	0.1	0.1	0.3	0.2	0.1	0.2	0.1	0.2	0.1	0.1	0.1	0.2	0.2	0.1	0.2
52 Buckweat	0.2	0.3	0.3	0.2	0.3	0.2	0.1	0.2	0.2	0.2	0.2	0.2	0.1	0.3	0.2	0.3	0.3
53 Rice	0.2	0.4	0.4	0.4	0.5	0.5	0.6	0.6	0.6	0.6	0.7	0.8	0.7	0.7	0.7	0.8	0.7
54 Other Grain	0.7	0.8	0.7	0.8	0.9	1.0	0.6	0.9	0.8	0.8	0.5	0.7	0.5	0.7	1.0	1.0	1.0
55 Cotton	3.0	3.7	3.8	3.9	4.1	4.5	4.2	4.4	4.7	4.3	4.5	4.8	4.6	4.2	4.2	4.5	4.7
56 Sugarbeets	2.5	2.7	2.5	2.6	3.0	2.7	2.3	3.4	3.2	3.2	2.6	2.8	2.1	2.5	2.8	2.9	2.8
57 Sunflower	1.1	1.1	1.1	0.9	1.5	1.5	1.1	0.9	1.1	1.0	1.2	1.0	1.0	1.2	1.4	1.3	1.4
58 Flax	0.8	0.7	0.8	0.7	0.7	0.6	0.8	0.8	0.8	0.6	0.5	0.4	0.4	0.7	0.7	0.6	0.5
59 Potatoes	10.1	11.0	10.6	8.9	12.3	9.2	10.1	9.7	9.5	9.8	10.4	7.6	8.2	8.9	9.4	9.7	8.3
60 Vegetables	2.9	3.4	3.4	3.2	4.2	4.0	3.8	4.1	3.9	4.5	4.4	4.4	4.4	4.9	4.8	5.1	4.6
61 Fruits and Nuts	2.3	3.3	3.5	2.7	3.8	3.6	4.1	4.3	4.3	4.1	4.6	4.2	4.9	5.2	5.2	5.3	4.7
62 Oil Crops	0.2	0.3	0.3	0.2	0.3	0.3	0.4	0.3	0.3	0.3	0.3	0.3	0.3	0.3	0.3	0.3	0.3
63 Tea	0.2	0.3	0.3	0.3	0.3	0.3	0.3	0.4	0.4	0.4	0.5	0.5	0.5	0.6	0.6	0.6	0.6
64 Tobacco	0.4	0.5	0.5	0.6	0.6	0.6	0.6	0.6	0.6	0.6	0.6	0.6	0.5	0.6	0.8	0.8	0.8
65 Other Commod.	0.3	0.3	0.3	0.3	0.2	0.2	0.2	0.1	0.1	0.1	0.1	0.1	0.1	0.1	0.2	0.2	0.2
66 Gardening	0.4	0.6	0.7	0.6	0.6	0.7	0.7	0.8	0.8	0.8	0.9	0.8	0.9	1.0	1.1	0.9	0.9
67 Unfin. Prod.	0.1	0.8	0.7	-0.1	0.6	0.5	0.4	0.1	0.4	0.3	0.4	0.4	0.5	0.5	0.8	0.8	0.7
68 Feed Crops	5.4	6.7	6.7	6.7	8.5	7.5	6.4	8.5	8.0	8.4	7.5	8.5	7.8	9.3	9.8	10.0	9.2

Husbandry Products in 1973 Prices:

	1965	1970	1971	1972	1973	1974	1975	1976	1977	1978	1979	1980	1981	1982	1983	1984	1985
69 Official '73 GVO	47.6	56.6	58.7	58.2	61.8	65.1	64.2	62.5	68.4	68.9	68.1	67.0	67.5	69.2	73.6	74.0	74.4
70 Derived '73 GVO	45.6	57.1	59.0	58.4	61.8	64.9	63.4	62.0	66.4	68.5	67.8	66.5	67.5	68.4	72.7	73.7	74.3
71 Meats	23.3	28.7	31.0	31.7	31.5	34.1	35.0	31.7	34.3	36.2	35.7	35.2	35.5	35.9	38.3	39.6	39.9
72 Milk	14.9	17.0	17.1	17.1	18.1	18.8	18.6	18.4	19.5	19.4	19.1	18.6	18.2	18.7	19.8	20.1	20.2
73 Eggs	2.5	3.5	3.9	4.1	4.4	4.8	4.9	4.8	5.3	5.5	5.7	5.8	6.1	6.2	6.5	6.6	6.6
74 Wool	1.8	2.1	2.2	2.1	2.2	2.4	2.4	2.3	2.4	2.4	2.4	2.3	2.4	2.3	2.4	2.4	2.3
75 Livestock	2.3	4.8	3.9	2.3	4.5	3.8	1.4	3.6	3.8	3.8	3.6	3.3	4.0	3.8	4.4	3.6	3.7
76 Productive	0.8	1.1	1.4	0.8	1.4	1.5	0.7	1.2	1.4	1.4	1.2	1.0	1.2	1.5	1.6	1.4	1.3
77 Young (Pub.)	0.5	2.9	1.9	0.8	2.3	1.7	-0.2	1.7	1.7	1.5	1.6	1.2	1.5	1.6	2.2	1.7	1.6
78 Young (Pri.)	0.3	0.3	0.2	0.0	0.2	0.2	0.0	0.2	0.2	0.2	0.2	0.1	0.1	0.2	0.1	0.0	-0.1
79 Losses	0.7	0.5	0.4	0.7	0.6	0.5	0.9	0.5	0.5	0.8	0.6	1.0	1.2	0.6	0.5	0.5	0.9
80 Other	0.8	0.9	1.0	1.0	1.1	1.1	1.1	1.2	1.2	1.2	1.4	1.3	1.4	1.4	1.5	1.5	1.5
81 Honey	0.3	0.4	0.4	0.3	0.4	0.4	0.3	0.3	0.4	0.3	0.3	0.3	0.3	0.3	0.4	0.3	0.4
82 Silk Cocoons	0.2	0.2	0.2	0.3	0.3	0.2	0.2	0.3	0.3	0.3	0.3	0.3	0.3	0.3	0.3	0.3	0.3
83 Other	0.3	0.3	0.4	0.4	0.4	0.5	0.5	0.5	0.5	0.6	0.7	0.6	0.7	0.8	0.8	0.8	0.8

TABLE O.12: SUPPLY AND USES OF AGRICULTURAL PRODUCTS--Part II: STATE AND OTHER PURCHASES
    (billion rubles)

Page 1

	1965	1970	1971	1972	1973	1974	1975	1976	1977	1978	1979	1980	1981	1982	1983	1984	1985
State Purchases:																	
1 TOTAL CROPS	13.28	20.51	19.85	19.97	25.07	23.79	22.7?	28.12	27.25	30.15	28.38	28.35	30.04	33.82	42.12	39.34	40.00
2 Grain	3.26	7.13	6.34	6.25	8.79	7.34	5.61	9.75	7.29	10.78	7.30	8.29	7.96	9.81	12.21	9.09	12.00
3 Cotton	2.50	3.82	3.93	3.88	4.12	4.83	4.59	4.91	5.33	5.00	5.44	5.92	5.86	5.75	6.53	6.69	6.80
4 Sugarbeets	1.93	2.03	1.83	2.32	2.73	2.26	2.15	3.05	3.11	2.86	2.45	2.34	2.40	2.89	3.58	3.46	3.31
5 Sunflower	0.84	0.89	0.86	0.74	1.18	1.15	0.76	0.70	0.89	0.80	0.87	0.72	0.80	0.93	1.10	0.98	1.11
6 Flax	0.62	0.79	0.75	0.73	0.72	0.54	0.88	0.84	0.81	0.56	0.48	0.36	0.47	0.51	0.63	0.59	0.60
7 All Vegetables	1.64	2.49	2.64	2.82	3.32	3.23	3.78	3.99	4.64	4.78	5.58	4.79	5.47	6.43	8.00	8.26	7.48
8 Fruits and Nuts	1.33	1.87	1.93	1.54	2.28	2.43	2.89	2.89	2.93	3.06	3.76	3.37	4.06	4.40	5.27	5.51	4.28
9 Other Crops	1.16	1.49	1.57	1.69	1.93	2.01	2.06	1.99	2.25	2.31	2.50	2.56	3.03	3.11	4.80	4.76	4.42
10 TOTAL HUSBANDRY	17.67	31.67	35.95	37.40	38.77	42.65	44.27	42.36	47.53	49.09	50.79	48.85	51.12	52.29	73.11	78.49	80.20
11 Meats	9.22	18.45	21.54	22.38	22.19	24.86	25.52	22.37	25.56	26.92	26.22	24.96	26.07	26.30	37.43	40.34	41.22
12 Milk	5.75	8.77	9.42	9.71	10.82	11.41	12.11	13.26	14.65	14.52	16.09	15.38	15.90	16.52	23.03	24.74	25.34
13 Eggs	0.83	1.63	1.97	2.22	2.53	2.79	2.95	3.15	3.49	3.70	3.78	4.04	4.24	4.37	6.07	6.48	6.66
14 Wool	1.21	1.97	2.07	2.04	2.14	2.36	2.38	2.23	2.44	2.52	3.05	2.92	3.07	3.16	4.39	4.71	4.66
15 Other	0.66	0.85	0.95	1.05	1.09	1.23	1.31	1.35	1.39	1.43	1.65	1.55	1.83	1.94	2.19	2.22	2.32
16 Honey	0.26	0.34	0.38	0.39	0.44	0.49	0.52	0.52	0.58	0.57	0.62	0.60	0.73	0.77	0.88	0.89	0.93
17 Silk Cocoons	0.13	0.17	0.19	0.22	0.22	0.25	0.26	0.29	0.28	0.30	0.31	0.33	0.38	0.39	0.44	0.44	0.46
18 Other	0.26	0.34	0.38	0.44	0.44	0.49	0.52	0.53	0.54	0.56	0.72	0.62	0.72	0.77	0.88	0.89	0.93
Other Purchases:																	
19 TRADE	1.36	2.34	2.12	2.06	2.30	2.52	2.56	2.92	3.31	3.60	3.97	4.51	4.91	5.47	5.95	5.95	5.99
20 State	0.19	0.75	0.45	0.34	0.67	0.78	0.89	1.17	1.47	1.51	1.77	1.96	2.08	2.18	1.61	1.12	0.81
21 from Farms	0.16	0.66	0.38	0.30	0.60	0.66	0.70	0.59	0.59	0.60	0.71	0.97	0.88	0.91	0.94	0.95	0.70
22 from Private	0.03	0.09	0.07	0.04	0.07	0.12	0.19	0.59	0.88	0.91	1.06	0.99	1.20	1.27	0.67	0.17	0.11
23 Cooperative	1.17	1.59	1.67	1.72	1.63	1.74	1.67	1.75	1.84	2.09	2.2	2.55	2.83	3.29	4.34	4.83	5.18
24 EX-VILLAGE	3.1	3.6	3.5	3.9	3.9	4.1	4.3	5.0	5.0	5.7	5.8	6.6	7.5	7.5	7.3	7.6	7.5
25 ALL PURCHASES	35.6	58.4	61.8	63.7	70.4	73.4	74.2	78.8	83.5	88.9	89.3	88.7	93.9	99.6	129.0	132.1	134.4
26 All State	31.1	52.9	56.3	57.7	64.5	67.2	67.9	71.7	76.3	80.8	80.9	79.2	83.2	88.3	116.8	119.0	121.0
27 in 1973 Prices	37.8	55.7	57.4	55.6	64.5	65.1	61.6	64.7	65.3	70.0	64.9	62.0	61.1	64.9	74.2	72.5	72.2
28 All Coop. Trade	0.9	1.3	1.4	1.4	1.3	1.4	1.3	1.3	1.4	1.6	1.7	2.0	2.2	2.7	3.9	4.4	4.8
29 All Ex-Village	3.6	4.2	4.1	4.6	4.6	4.8	5.0	5.8	5.8	6.5	6.7	7.5	8.5	8.6	8.3	8.7	8.6

272

TABLE 0.12: SUPPLY AND USES OF AGRICULTURAL PRODUCTS--Part II: STATE AND OTHER PURCHASES

	1965	1970	1971	1972	1973	1974	1975	1976	1977	1978	1979	1980	1981	1982	1983	1984	1985
Procurement Price Index by Commodity:																	
30 Grains	0.92	1.00	1.02	1.07	1.00	1.03	1.15	1.09	1.10	1.16	1.20	1.23	1.41	1.45	1.66	1.67	1.68
31 Unit Price	89.8	97.3	98.9	104.2	97.1	100.1	111.8	105.9	107.2	112.4	116.2	119.5	137.0	140.7	161.6	161.7	163.3
32 Cotton	0.82	1.03	1.03	0.99	1.00	1.07	1.09	1.10	1.13	1.15	1.19	1.21	1.27	1.35	1.55	1.47	1.44
33 Unit Price	44.2	55.4	55.4	53.2	53.8	57.4	58.4	59.3	60.8	61.7	64.1	65.1	68.5	72.4	83.2	79.0	77.7
34 Sugarbeets	0.81	0.81	0.81	0.97	1.00	1.00	1.00	1.02	1.04	1.02	1.01	1.02	1.28	1.29	1.37	1.30	1.30
35 Unit Price	2.86	2.84	2.85	3.41	3.51	3.50	3.47	3.58	3.66	3.57	3.54	3.59	4.49	4.51	4.81	4.55	4.55
36 Sunflower	1.02	0.91	0.93	0.93	1.00	1.03	0.93	0.87	0.94	0.93	0.97	1.00	1.03	1.04	1.28	1.29	1.25
37 Unit Price	21.59	19.31	19.72	19.73	21.26	21.99	19.79	18.57	20.00	19.85	20.62	21.36	21.81	22.20	27.11	27.32	26.56
38 All Vegetables	0.83	1.00	1.02	1.12	1.00	1.11	1.18	1.21	1.24	1.28	1.44	1.49	1.59	1.62	1.88	1.87	1.87
39 Unit Price	9.32	11.27	11.48	12.65	11.25	12.52	13.31	13.57	13.93	14.35	16.22	16.75	17.87	18.26	21.12	21.02	21.07
40 Fruits and Nuts	1.01	1.03	1.04	0.99	1.00	1.05	1.16	1.02	1.06	1.13	1.19	1.15	1.18	1.21	1.48	1.45	1.47
41 Unit Price	29.69	30.26	30.39	28.89	29.27	30.64	33.84	29.86	31.04	33.01	34.72	33.70	34.54	35.29	43.19	42.37	43.06
42 Meats	0.68	0.97	1.00	0.98	1.00	1.01	1.00	1.03	1.07	1.09	1.07	1.08	1.11	1.12	1.45	1.50	1.50
43 Unit Price	1.59	2.27	2.35	2.30	2.34	2.37	2.35	2.40	2.51	2.54	2.51	2.52	2.61	2.62	3.39	3.50	3.51
44 Milk	0.73	0.94	0.98	0.98	1.00	1.00	1.05	1.16	1.18	1.18	1.34	1.32	1.40	1.39	1.78	1.83	1.82
45 Unit Price	14.86	19.19	20.00	20.06	20.42	20.45	21.51	23.59	24.10	24.04	27.27	26.89	28.65	28.49	36.32	37.38	37.21
46 Eggs	0.86	0.98	0.99	0.99	1.00	0.98	0.97	1.04	1.03	1.02	1.00	1.02	1.02	1.02	1.37	1.42	1.43
47 Unit Price	7.90	9.01	9.12	9.14	9.20	9.03	8.91	9.57	9.48	9.41	9.20	9.37	9.39	9.41	12.59	13.08	13.13
48 Wool	0.70	0.96	0.99	0.99	1.00	1.03	1.02	1.03	1.08	1.09	1.29	1.29	1.33	1.40	1.90	1.99	2.06
49 Unit Price	3.67	4.93	5.10	5.10	5.14	5.28	5.24	5.31	5.55	5.60	6.63	6.64	6.82	7.19	9.75	10.24	10.60

TABLE 0.12: SUPPLY AND USES OF AGRICULTURAL PRODUCTS--Part III: OUTPUT IN PHYSICAL UNITS

	1965	1970	1971	1972	1973	1974	1975	1976	1977	1978	1979	1980	1981	1982	1983	1984	1985
1 TOTAL GRAINS	121.1	186.8	181.2	168.2	222.5	195.7	140.1	223.8	195.7	237.4	179.3	189.1	158.2	186.8	192.2	172.6	191.7
2 Wheat	59.7	99.7	98.8	86.0	109.8	83.9	66.2	96.9	92.2	120.8	90.2	98.2	81.1	84.3	77.5	68.6	78.1
3 Rye	16.2	13.0	12.8	9.6	10.8	15.2	9.1	14.0	8.5	13.6	8.1	10.2	9.6	14.8	17.3	14.0	15.7
4 Corn	8.0	9.4	8.6	9.8	13.2	12.1	7.3	10.1	11.0	9.0	8.4	9.5	9.4	14.7	13.3	13.6	14.4
5 Barley	20.3	38.2	34.6	36.8	55.0	54.2	35.8	69.5	52.7	62.1	48.0	43.5	36.1	43.0	50.0	41.8	46.5
6 Oats	6.2	14.2	14.6	14.1	17.5	15.3	12.5	18.1	18.4	18.6	15.2	15.5	12.4	16.8	18.8	19.2	20.5
7 Millet	2.2	2.1	2.0	2.1	4.4	2.9	1.1	3.2	2.0	2.2	1.6	1.9	1.9	2.5	2.5	1.9	2.9
8 Buckweat	1.0	1.1	1.2	0.8	1.3	1.0	0.5	0.9	1.0	1.0	0.8	1.0	0.6	1.1	1.0	1.1	1.2
9 Rice	0.6	1.3	1.4	1.6	1.8	1.9	2.0	2.0	2.2	2.1	2.4	2.8	2.5	2.5	2.6	2.7	2.6
10 Other Grain	6.9	7.8	7.2	7.4	8.7	9.2	5.6	9.1	7.7	8.0	4.6	6.5	4.6	7.1	9.2	9.7	9.8
11 TOTAL GRAINS	121.1	186.8	181.2	168.2	222.5	195.7	140.1	223.8	195.7	237.4	179.3	189.1	158.2	186.8	192.2	172.6	191.7
12 Commodity	41.1	80.8	70.5	67.2	100.0	80.8	54.8	101.3	79.1	109.4	74.9	85.1	69.7	83.1	90.2	70.1	87.8
13 Procurement	36.3	73.3	64.1	60.0	90.5	73.3	50.2	92.1	68.0	95.9	62.8	69.4	58.1	69.7	75.6	56.2	73.5
14 Trade	0.0	0.2	0.1	0.1	0.1	0.1	0.1	0.2	0.2	0.2	0.2	0.1	0.1	0.1	0.0	0.0	0.0
15 Coop Trade	0.2	0.4	0.3	0.3	0.2	0.2	0.2	0.2	0.2	0.2	0.2	0.2	0.1	0.1	0.1	0.1	0.1
16 Ex-Village	4.6	7.1	6.1	6.9	9.3	7.3	4.4	9.0	10.9	13.3	11.9	15.5	11.5	13.3	14.5	13.8	14.2
17 Non-Commodity	80.0	106.0	110.7	101.0	122.5	114.9	85.3	122.5	116.6	128.0	104.4	104.0	88.5	103.7	102.0	102.5	103.9
18 COTTON	5.7	6.9	7.1	7.3	7.7	8.4	7.9	8.3	8.8	8.1	8.5	9.1	8.6	7.9	7.9	8.5	8.8
19 SUGARBEETS	72.3	78.9	72.2	76.4	87.0	77.9	66.3	99.9	93.1	92.5	76.2	81.0	60.8	71.4	81.8	85.4	82.4
20 sugar content	0.18	0.167					0.166					0.153					0.154
21 Commodity	67.5	71.4	64.3	68.0	77.8	64.5	61.9	85.1	84.9	80.1	69.3	65.2	53.5	64.1	74.5	76.0	72.8
22 Non-Commodity	4.8	7.5	7.9	8.4	9.2	13.4	4.4	14.8	8.2	12.4	6.9	15.8	7.3	7.3	7.3	9.4	9.6
23 SUNFLOWER	5.45	6.14	5.66	5.05	7.39	6.78	4.99	5.28	5.90	5.33	5.41	4.62	4.68	5.34	5.06	4.53	5.26
24 oil content	0.46	0.45					0.47					0.476					0.456
25 Commodity	4.27	5.16	4.71	4.02	6.14	5.64	4.14	4.31	4.89	4.41	4.50	3.80	3.90	4.50	4.30	3.80	4.40
26 Procurement	3.89	4.61	4.36	3.75	5.55	5.23	3.84	3.77	4.45	4.03	4.22	3.37	3.65	4.19	4.04	3.60	4.18
27 Ex-Village	0.38	0.55	0.35	0.27	0.59	0.41	0.30	0.54	0.44	0.38	0.28	0.43	0.25	0.31	0.26	0.20	0.22
28 Non-Commodity	1.18	0.98	0.95	1.03	1.25	1.14	0.85	0.97	1.01	0.92	0.91	0.82	0.78	0.84	0.76	0.73	0.86
29 FLAX	0.48	0.46	0.49	0.46	0.44	0.40	0.49	0.51	0.48	0.38	0.31	0.28	0.26	0.41	0.47	0.39	0.34
30 Commodity	0.43	0.43	0.46	0.44	0.42	0.36	0.48	0.48	0.44	0.33	0.30	0.26	0.25	0.40	0.46	0.37	0.33
31 Non-Commodity	0.05	0.03	0.03	0.02	0.02	0.04	0.01	0.03	0.04	0.05	0.01	0.02	0.01	0.01	0.01	0.02	0.01
32 POTATOES	88.7	96.8	92.7	78.3	108.2	81.0	88.7	85.1	83.7	86.1	91.0	67.0	72.1	78.2	82.9	85.5	73.0
33 starch content	0.155	0.15					0.145										
34 Commodity	15.8	18.1	18.0	16.4	23.1	17.3	20.5	19.6	23.1	20.9	22.8	16.8	18.9	21.3	24.1	24.5	21.5
35 Procurement	9.9	11.2	11.5	11.1	15.4	11.1	14.5	13.4	17.1	14.9	16.4	11.1	13.5	15.7	18.2	18.6	15.7
36 from Public	7.23	9.41	10.01	8.88	12.32	9.10	12.04	10.18	12.65	11.18	12.14	7.55	9.86	11.46	13.65	14.32	11.78
37 % of total	0.73	0.84	0.87	0.80	0.80	0.82	0.83	0.76	0.74	0.75	0.74	0.68	0.73	0.73	0.75	0.77	0.75
38 from Private	2.67	1.79	1.49	2.22	3.08	2.00	2.47	3.22	4.45	3.73	4.26	3.55	3.65	4.24	4.55	4.28	3.92
39 State Trade	0.0	0.1	0.1	0.1	0.1	0.1	0.2	0.2	0.2	0.2	0.2	0.2	0.2	0.2	0.1	0.1	0.0
40 Coop Trade	0.2	0.2	0.2	0.3	0.2	0.2	0.2	0.2	0.2	0.2	0.2	0.2	0.2	0.2	0.2	0.2	0.2
41 Ex-Village	5.7	6.5	6.2	4.9	7.4	5.8	5.6	5.8	5.5	5.6	6.0	5.3	5.0	5.2	5.6	5.6	5.6
42 Public Farms	1.22	1.08	0.84	0.73	1.20	0.56	0.49	1.34	0.99	0.57	0.92	0.47	0.54	0.32	0.52	0.61	0.67
43 Private Farms	4.44	5.45	5.35	4.18	6.16	5.27	5.12	4.43	4.56	5.05	5.08	4.85	4.48	4.92	5.09	5.03	4.89
44 Non-Commodity	72.9	78.7	74.7	61.9	85.1	63.7	68.2	65.5	60.6	65.2	68.2	50.2	53.2	56.9	58.8	61.0	51.5
45 Public Farms	24.12	23.02	23.13	19.75	28.33	19.12	23.45	22.93	20.22	21.45	23.85	15.72	16.62	16.79	18.7	20.72	16.51
46 Private Farms	48.77	55.68	51.56	42.15	56.76	44.57	44.75	42.57	40.37	43.74	44.34	34.48	36.58	40.11	40.10	40.28	34.99

TABLE 0.12: SUPPLY AND USES OF AGRICULTURAL PRODUCTS--Part III: OUTPUT IN PHYSICAL UNITS

		1965	1970	1971	1972	1973	1974	1975	1976	1977	1978	1979	1980	1981	1982	1983	1984	1985
47	VEGETABLES	17.6	21.2	20.8	19.9	25.9	24.8	23.4	25.0	24.1	27.9	27.2	27.3	27.1	30.0	29.5	31.5	28.1
48	Commodity	9.9	13.8	14.0	13.7	17.3	17.0	16.3	18.5	18.7	20.9	20.5	20.1	19.4	21.8	22.0	23.0	21.9
49	Procurement	7.7	10.9	11.5	11.2	14.1	14.7	13.9	16.0	16.2	18.4	18.0	17.5	17.1	19.5	19.7	20.7	19.8
50	from Public	7.16	10.25	10.81	10.53	13.25	13.82	13.21	15.04	15.07	17.11	16.74	16.10	15.73	17.94	17.93	19.25	18.41
51	% of total	0.93	0.94	0.94	0.94	0.94	0.94	0.95	0.94	0.93	0.93	0.93	0.92	0.92	0.92	0.91	0.93	0.93
52	from Private	0.54	0.65	0.69	0.67	0.85	0.88	0.70	0.96	1.13	1.29	1.26	1.40	1.37	1.56	1.77	1.45	1.39
53	State Trade	0.1	0.2	0.1	0.1	0.2	0.2	0.3	0.3	0.4	0.3	0.3	0.3	0.3	0.3	0.2	0.1	0.1
54	Coop Trade	0.3	0.4	0.4	0.4	0.4	0.4	0.4	0.4	0.4	0.4	0.3	0.3	0.3	0.4	0.4	0.4	0.3
55	Ex-Village	1.8	2.3	2.0	2.0	2.6	1.7	1.7	1.8	1.7	1.8	1.8	2.0	1.7	1.6	1.7	1.8	1.7
56	Non-Commodity	7.7	7.4	6.8	6.2	8.6	7.8	7.1	6.5	5.4	7.0	6.7	7.2	7.7	8.2	7.5	8.5	6.2
57	Public Farms	2.98	2.44	1.97	1.91	3.38	2.41	1.81	1.96	1.49	2.17	1.52	1.71	1.44	1.98	1.78	2.51	1.31
58	Private Farms	4.72	4.96	4.83	4.29	5.22	5.39	5.29	4.54	3.91	4.83	5.18	5.49	6.26	6.22	5.72	5.99	4.89
59	FRUITS & NUTS	8.10	11.69	12.31	9.57	13.35	12.44	14.24	15.26	15.28	14.37	16.30	14.67	17.29	18.37	18.39	18.55	16.43
60	Commodity	5.99	8.13	8.16	7.16	9.60	9.87	10.74	11.48	11.47	11.03	12.74	11.72	13.50	14.31	14.26	15.22	12.17
61	Procurement	4.48	6.18	6.35	5.33	7.79	7.93	8.54	9.68	9.44	9.27	10.83	10.00	11.74	12.46	12.19	13.00	9.94
62	from Public	4.26	5.87	6.03	4.90	7.09	7.02	7.17	7.70	7.47	7.12	8.43	7.71	9.11	10.04	10.18	10.59	8.51
63	from Private	0.22	0.31	0.32	0.43	0.70	0.91	1.37	1.98	1.97	2.15	2.40	2.29	2.63	2.42	2.01	2.41	1.43
64	State Trade	0.06	0.20	0.11	0.08	0.16	0.19	0.25	0.25	0.33	0.26	0.31	0.27	0.26	0.25	0.17	0.12	0.08
65	Coop Trade	0.29	0.35	0.34	0.35	0.33	0.35	0.39	0.31	0.34	0.30	0.32	0.29	0.30	0.32	0.38	0.42	0.43
66	Ex-Village	1.16	1.40	1.36	1.40	1.32	1.40	1.56	1.24	1.36	1.20	1.28	1.16	1.20	1.28	1.52	1.68	1.72
67	Non-Commodity	2.11	3.56	4.15	2.41	3.75	2.57	3.50	3.78	3.81	3.34	3.56	2.95	3.79	4.06	4.13	3.33	4.26
68	Public Farms	0.75	0.64	0.74	0.57	0.53	0.46	0.63	0.68	0.69	0.60	0.64	0.53	0.68	0.73	0.74	0.60	0.77
69	Private Farms	1.42	2.92	3.41	1.83	3.22	2.11	2.87	3.10	3.13	2.74	2.92	2.42	3.10	3.33	3.39	2.73	3.49

OTHER COMMODITY OUTPUT OF CROPS

		1965	1970	1971	1972	1973	1974	1975	1976	1977	1978	1979	1980	1981	1982	1983	1984	1985
70	Other Oil Crops	0.62	0.83	0.80	0.47	0.76	0.64	0.93	0.71	0.72	0.88	0.67	0.71	0.64	0.80	0.83	0.65	0.70
71	Tea	0.20	0.27	0.28	0.29	0.31	0.33	0.35	0.38	0.43	0.45	0.48	0.53	0.56	0.57	0.59	0.62	0.62
72	Tobacco	0.21	0.26	0.25	0.29	0.30	0.31	0.29	0.31	0.31	0.28	0.29	0.28	0.27	0.30	0.38	0.38	0.38
73	Total Oil Crops	6.07	6.97	6.46	5.52	8.15	7.42	5.92	5.99	6.62	6.21	6.08	5.33	5.32	6.14	5.89	5.18	5.96

OTHER NON-COMMODITY OUTPUT OF CROPS (FEED)

		1965	1970	1971	1972	1973	1974	1975	1976	1977	1978	1979	1980	1981	1982	1983	1984	1985
74	Total	287.3	358.0	360.3	360.3	454.9	401.1	343.2	458.2	427.6	448.4	402.8	455.6	417.6	499.5	523.5	538.9	491.5
75	Corn	181.0	212.0	210.9	206.1	281.7	226.5	193.0	277.0	247.0	251.0	230.0	266.0	232.0	294.0	298.0	312.0	331.0
76	Beets, etc	23.8	35.7	36.7	39.7	47.1	43.9	34.4	50.0	45.3	45.7	38.4	41.6	36.6	45.5	48.1	58.3	59.0
77	Straw, etc.	82.5	110.3	112.7	114.5	126.1	130.7	115.8	131.2	135.3	151.7	134.4	148.0	149.0	160.0	177.4	168.6	101.5

		1965	1970	1971	1972	1973	1974	1975	1976	1977	1978	1979	1980	1981	1982	1983	1984	1985
78	MEATS	10.00	12.30	13.30	13.60	13.50	14.60	15.00	13.60	14.70	15.50	15.30	15.10	15.20	15.40	16.40	17.00	17.10
79	Commodity	7.00	9.40	10.40	10.90	10.70	11.80	12.20	10.80	11.60	12.30	12.10	11.70	11.90	12.00	13.20	13.70	13.90
80	Procurement	5.81	8.11	9.18	9.71	9.47	10.47	10.86	9.31	10.19	10.59	10.43	9.91	10.00	10.02	11.04	11.53	11.75
81	from Public	5.23	7.22	7.99	8.35	8.24	9.11	9.45	8.38	9.48	9.85	9.70	9.32	9.60	9.72	10.71	11.18	11.40
82	% of total	0.90	0.89	0.87	0.86	0.87	0.87	0.87	0.90	0.93	0.93	0.93	0.94	0.96	0.97	0.97	0.97	0.97
83	from Private	0.58	0.89	1.19	1.36	1.23	1.36	1.41	0.93	0.71	0.74	0.73	0.59	0.40	0.30	0.33	0.35	0.35
84	Coop Trade	0.17	0.20	0.17	0.17	0.16	0.17	0.17	0.19	0.19	0.20	0.22	0.25	0.29	0.34	0.36	0.35	0.37
85	Ex-Village	1.02	1.09	1.05	1.02	1.07	1.16	1.17	1.30	1.22	1.51	1.45	1.54	1.61	1.64	1.80	1.82	1.78
86	Noncommodity	3.00	2.90	2.90	2.70	2.80	2.80	2.80	2.80	3.10	3.20	3.20	3.40	3.30	3.40	3.20	3.30	3.20
87	Losses	0.77	0.78	0.66	0.63	0.81	0.82	0.90	1.14	0.96	1.00	1.01	1.10	1.04	1.06	0.94	1.06	0.91
88	Consumption	2.23	2.12	2.24	2.07	1.99	1.98	1.90	1.66	2.14	2.20	2.19	2.30	2.26	2.34	2.26	2.24	2.29

TABLE 0.12: SUPPLY AND USES OF AGRICULTURAL PRODUCTS--Part III: OUTPUT IN PHYSICAL UNITS

		1965	1970	1971	1972	1973	1974	1975	1976	1977	1978	1979	1980	1981	1982	1983	1984	1985
89	MILK	72.60	83.00	83.20	83.20	88.30	91.80	90.80	89.70	94.90	94.70	93.20	90.90	88.90	91.00	96.50	97.90	98.60
90	Commodity	40.90	48.00	49.40	50.80	55.60	58.30	58.70	58.30	62.80	62.40	61.20	59.40	57.80	60.10	65.20	66.90	68.20
91	Procurement	38.70	45.70	47.10	48.40	53.00	55.80	56.30	56.20	60.80	60.40	59.00	57.20	55.50	58.00	63.40	66.20	68.10
92	from Public	37.15	44.33	45.22	45.98	50.35	53.01	53.49	53.39	57.76	57.38	56.05	53.77	53.28	56.84	62.77	65.54	67.42
93	% of total	0.96	0.97	0.96	0.95	0.95	0.95	0.95	0.95	0.95	0.95	0.95	0.94	0.96	0.98	0.99	0.99	0.99
94	from Private	1.55	1.37	1.88	2.42	2.65	2.79	2.82	2.81	3.04	3.02	2.95	3.43	2.22	1.16	0.63	0.66	0.68
95	Ex-Village	2.20	2.30	2.30	2.40	2.60	2.50	2.40	2.10	2.00	2.00	2.20	2.20	2.30	2.10	1.80	0.70	0.10
96	Non-Commodity	31.70	35.00	33.80	32.40	32.70	33.50	32.10	31.40	32.10	32.30	32.00	31.50	31.10	30.90	31.30	31.00	30.40
97	Production	9.31	8.79	8.86	8.93	8.81	9.41	9.17	9.40	9.62	9.86	10.12	9.86	8.95	6.86	5.75	3.97	2.59
98	Consumption	22.39	26.21	24.94	23.47	23.89	24.09	22.93	22.00	22.48	22.44	21.88	21.64	22.15	24.04	25.55	27.03	27.81
99	EGGS	29.10	40.70	45.10	47.90	51.20	55.50	57.40	56.20	61.20	64.50	65.80	67.90	70.90	72.40	75.10	76.50	77.30
100	Commodity	13.90	22.10	25.60	28.20	31.50	34.60	36.70	36.20	40.40	43.10	44.70	46.60	49.10	50.50	52.70	53.90	55.00
101	Procurement	10.50	18.10	21.60	24.30	27.50	30.90	33.10	32.90	36.80	39.30	41.10	43.10	45.20	46.40	48.20	49.50	50.70
102	from Public	7.77	16.11	19.44	22.36	25.58	28.74	30.78	30.93	34.96	38.12	39.87	41.81	43.84	45.01	46.75	48.02	49.18
103	% of total	0.74	0.89	0.90	0.92	0.93	0.93	0.93	0.94	0.95	0.97	0.97	0.97	0.97	0.97	0.97	0.97	0.97
104	from Private	2.73	1.99	2.16	1.94	1.92	2.16	2.32	1.97	1.84	1.18	1.23	1.29	1.36	1.39	1.45	1.48	1.52
105	Coop Trade	0.30	0.50	0.60	0.60	0.60	0.60	0.50	0.40	0.40	0.40	0.34	0.30	0.40	0.46	0.54	0.54	0.50
106	Ex-Village	3.10	3.50	3.40	3.30	3.40	3.10	3.10	2.90	3.20	3.40	3.26	3.20	3.50	3.64	3.96	3.86	3.80
107	Non-Commodity	15.20	18.60	19.50	19.70	19.70	20.90	20.70	20.00	20.80	21.40	21.10	21.30	21.80	21.90	22.40	22.60	22.30
108	Production	1.83	3.02	3.11	3.03	3.61	4.01	4.23	4.48	4.82	3.80	4.22	4.37	5.08	4.95	5.82	6.30	6.48
109	Consumption	13.37	15.58	16.39	16.67	16.09	16.89	16.47	15.52	15.98	17.60	16.88	16.94	16.72	16.95	16.58	16.30	15.82
110	WOOL	0.36	0.42	0.43	0.42	0.43	0.46	0.47	0.44	0.46	0.47	0.47	0.45	0.46	0.45	0.46	0.47	0.45
111	Procurement	0.33	0.40	0.41	0.40	0.42	0.45	0.45	0.42	0.44	0.45	0.46	0.44	0.45	0.44	0.45	0.46	0.44
112	Ex-Village	0.03	0.02	0.02	0.02	0.01	0.01	0.02	0.02	0.02	0.02	0.01	0.01	0.01	0.01	0.01	0.01	0.01
113	HONEY	0.19	0.21	0.20	0.18	0.22	0.20	0.17	0.19	0.21	0.18	0.19	0.18	0.19	0.19	0.21	0.19	0.20
114	SILK COCOONS	0.04	0.03	0.04	0.04	0.04	0.04	0.04	0.05	0.04	0.05	0.05	0.05	0.05	0.05	0.05	0.05	0.05

TABLE 0.12: SUPPLY AND USES OF AGRICULTURAL PRODUCTS--Part IV: HOUSEHOLD CONSUMPTION AND PRODUCTION

	1965	1970	1971	1972	1973	1974	1975	1976	1977	1978	1979	1980	1981	1982	1983	1984	1985
Household Consumption:																	
1 Potatoes	32.80	31.59	31.36	29.95	30.50	30.49	30.48	30.58	31.08	30.54	30.25	28.94	27.87	29.70	29.70	29.70	28.86
2 per capita	142	130	128	121	122	121	120	119	120	117	115	109	104	110	109	108	104
3 Vegetables	16.63	19.93	20.83	19.80	21.25	21.92	22.61	22.10	22.79	24.01	25.77	25.75	26.53	27.27	27.80	28.05	28.31
4 per capita	72	82	85	80	85	87	89	86	88	92	98	97	99	101	102	102	102
5 Fruits and Nuts	6.47	8.51	9.56	8.91	10.25	9.32	9.91	10.02	10.62	10.70	9.99	10.09	10.72	11.34	11.99	12.38	12.77
6 per capita	28	35	39	36	41	37	39	39	41	41	38	38	40	42	44	45	46
7 Eggs	28.64	38.64	42.63	45.79	48.75	51.66	54.86	53.71	57.50	60.55	61.81	63.45	66.20	67.23	69.76	70.95	72.15
8 per capita	124	159	174	185	195	205	216	209	222	232	235	239	247	249	256	258	260
9 Population	231	243	245	248	250	252	254	257	259	261	263	266	268	270	273	275	278
Private Production:																	
10 Potatoes	55.88	62.92	58.40	48.55	66.00	51.84	52.33	50.21	49.38	52.52	53.69	42.88	44.70	49.27	49.74	49.59	43.80
11 % of total	0.63	0.65	0.63	0.62	0.61	0.64	0.59	0.59	0.59	0.61	0.59	0.64	0.62	0.63	0.60	0.58	0.60
12 Vegetables	7.22	8.06	7.70	7.16	8.81	8.18	7.96	7.50	6.99	8.09	8.43	9.01	9.49	9.60	9.44	9.45	8.15
13 % of total	0.41	0.38	0.37	0.36	0.34	0.33	0.34	0.30	0.29	0.29	0.31	0.33	0.35	0.32	0.32	0.30	0.29
14 Fruits	3.09	4.98	5.43	4.01	5.57	4.77	6.19	6.63	6.80	6.39	6.92	6.16	7.23	7.35	7.30	7.24	7.07
15 % of total	0.38	0.43	0.44	0.42	0.42	0.38	0.43	0.43	0.45	0.44	0.42	0.42	0.42	0.40	0.40	0.39	0.43
16 Meats	4.00	4.31	4.66	4.62	4.46	4.67	4.65	4.08	4.26	4.65	4.59	4.68	4.56	4.62	4.76	4.76	4.79
17 % of total	0.40	0.35	0.35	0.34	0.33	0.32	0.31	0.30	0.29	0.30	0.30	0.31	0.30	0.30	0.29	0.28	0.28
18 Milk	26.14	29.88	29.12	28.29	29.14	29.38	28.15	26.91	27.52	27.46	27.03	27.27	26.67	27.30	27.99	28.39	28.59
19 % of total	0.36	0.36	0.35	0.34	0.33	0.32	0.31	0.30	0.29	0.29	0.29	0.30	0.30	0.30	0.29	0.29	0.29
20 Eggs	19.50	21.57	22.55	22.51	22.02	22.76	22.39	20.79	21.42	22.58	21.71	21.73	21.98	22.44	22.53	22.19	21.64
21 % of total	0.67	0.53	0.50	0.47	0.43	0.41	0.39	0.37	0.35	0.35	0.33	0.32	0.31	0.31	0.30	0.29	0.28
22 Wool	0.07	0.08	0.08	0.08	0.09	0.09	0.09	0.08	0.08	0.09	0.10	0.10	0.10	0.11	0.11	0.11	0.11
23 % of total	0.20	0.19	0.20	0.21	0.21	0.20	0.20	0.19	0.19	0.20	0.22	0.22	0.22	0.24	0.24	0.24	0.26
24 Honey	0.11	0.11	0.11	0.10	0.12	0.12	0.10	0.12	0.13	0.11	0.12	0.11	0.11	0.11	0.13	0.12	0.13
25 % of total	0.58	0.54	0.55	0.56	0.55	0.58	0.59	0.61	0.60	0.61	0.61	0.61	0.58	0.59	0.62	0.63	0.65

TABLE 0.12: SUPPLY AND USES OF AGRICULTURAL PRODUCTS--Part V: DERIVATION OF IMPORTS IN DOMESTIC PRICES
(billion rubles)

	1965	1970	1971	1972	1973	1974	1975	1976	1977	1978	1979	1980	1981	1982	1983	1984	1985
1 TOTAL	0.88	1.35	1.65	3.22	3.76	2.14	3.45	3.96	2.99	4.12	5.07	5.45	8.12	7.80	7.91	9.04	9.44
2 constant prices	0.93	1.28	1.59	3.08	3.76	2.09	3.18	3.75	2.79	3.65	4.41	4.57	6.20	5.88	5.27	5.95	6.16
3 Grains	0.11	0.21	0.34	1.61	2.32	0.71	1.78	2.19	1.12	2.55	3.10	3.44	5.78	5.37	5.37	6.67	7.10
4 physical	1.21	2.14	3.48	15.50	23.90	7.13	15.91	20.64	10.47	22.68	26.71	27.92	42.19	38.16	33.26	41.97	44.21
5 constant prices	0.12	0.21	0.34	1.50	2.32	0.69	1.54	2.00	1.02	2.20	2.59	2.71	4.10	3.70	3.23	4.07	4.29
6 Oil Crops	0.00	0.01	0.01	0.07	0.16	0.02	0.08	0.34	0.29	0.19	0.37	0.25	0.32	0.35	0.38	0.27	0.24
7 physical	0.02	0.04	0.05	0.38	0.77	0.07	0.42	1.83	1.46	0.97	1.81	1.15	1.46	1.58	1.42	0.99	0.92
8 constant prices	0.00	0.01	0.01	0.08	0.16	0.01	0.09	0.38	0.30	0.20	0.38	0.24	0.30	0.33	0.30	0.21	0.19
9 Vegetables	0.16	0.18	0.25	0.34	0.18	0.21	0.19	0.26	0.26	0.26	0.24	0.22	0.38	0.31	0.38	0.40	0.40
10 physical	0.17	0.16	0.21	0.27	0.16	0.17	0.14	0.19	0.19	0.18	0.15	0.13	0.21	0.17	0.18	0.19	0.19
11 constant prices	0.19	0.18	0.24	0.30	0.18	0.19	0.16	0.21	0.21	0.20	0.17	0.15	0.24	0.19	0.20	0.21	0.21
12 Fruits & Nuts	0.16	0.22	0.24	0.25	0.26	0.27	0.31	0.27	0.28	0.30	0.33	0.36	0.37	0.42	0.50	0.50	0.50
13 physical	0.53	0.72	0.79	0.86	0.88	0.89	0.92	0.92	0.90	0.90	0.96	1.05	1.07	1.20	1.16	1.17	1.17
14 constant prices	0.16	0.21	0.23	0.25	0.26	0.26	0.27	0.27	0.27	0.27	0.28	0.31	0.32	0.35	0.34	0.35	0.35
15 Tobacco	0.20	0.29	0.33	0.36	0.36	0.36	0.36	0.28	0.32	0.24	0.28	0.34	0.40	0.48	0.40	0.40	0.38
16 physical	0.05	0.07	0.08	0.09	0.09	0.09	0.09	0.07	0.08	0.06	0.07	0.08	0.10	0.12	0.10	0.10	0.10
17 constant prices	0.20	0.28	0.32	0.36	0.36	0.36	0.36	0.28	0.32	0.24	0.28	0.34	0.40	0.48	0.40	0.40	0.38
18 Eggs	0.06	0.05	0.06	0.09	0.07	0.07	0.07	0.07	0.07	0.07	0.07	0.07	0.06	0.05	0.06	0.07	0.05
19 physical	0.70	0.60	0.70	1.00	0.80	0.80	0.80	0.70	0.70	0.70	0.80	0.70	0.60	0.50	0.50	0.50	0.40
20 constant prices	0.06	0.05	0.06	0.09	0.07	0.07	0.07	0.06	0.06	0.06	0.07	0.06	0.05	0.04	0.04	0.04	0.04
21 Coffee Beans	0.12	0.24	0.25	0.26	0.24	0.28	0.32	0.28	0.18	0.24	0.30	0.31	0.28	0.30	0.31	0.34	0.35
22 physical	0.06	0.10	0.11	0.13	0.12	0.14	0.16	0.13	0.07	0.10	0.13	0.13	0.12	0.12	0.12	0.13	0.15
23 constant prices	0.12	0.21	0.22	0.26	0.24	0.28	0.32	0.26	0.14	0.20	0.26	0.27	0.24	0.24	0.24	0.26	0.30
24 Husbandry	0.04	0.06	0.06	0.07	0.07	0.10	0.22	0.12	0.18	0.10	0.20	0.23	0.26	0.26	0.26	0.16	0.14
25 g.r.	0.02	0.03	0.03	0.03	0.03	0.05	0.11	0.06	0.09	0.05	0.10	0.11	0.13	0.13	0.13	0.08	0.07
26 constant prices	0.05	0.07	0.07	0.08	0.08	0.12	0.26	0.14	0.21	0.12	0.23	0.27	0.30	0.30	0.30	0.19	0.16
27 Seeds	0.04	0.07	0.11	0.17	0.09	0.11	0.12	0.16	0.29	0.18	0.16	0.25	0.28	0.26	0.24	0.24	0.27
28 g.r.	0.02	0.03	0.05	0.08	0.05	0.06	0.06	0.08	0.14	0.09	0.08	0.12	0.14	0.13	0.12	0.12	0.14
29 constant prices	0.04	0.06	0.10	0.15	0.08	0.10	0.11	0.14	0.26	0.16	0.14	0.22	0.25	0.23	0.22	0.22	0.24

TABLE 0.13: CAPITAL CONSTRUCTION--Part A: OUTPUT BY SECTOR IN CURRENT PRICES
      (billion rubles)

	1965	1970	1971	1972	1973	1974	1975	1976	1977	1978	1979	1980	1981	1982	1983	1984	1985
1 TOTAL GVO	40.0	67.6	74.7	77.4	80.9	86.4	91.7	94.2	96.2	99.6	101.1	103.4	106.4	115.1	119.3	132.3	136.3
2 CAPITAL OUTLAYS	36.3	63.3	68.7	73.4	75.9	81.2	86.3	88.6	91.7	96.3	96.7	98.7	101.8	107.6	113.1	126.6	129.7
3 New Investment	30.1	53.9	58.4	62.4	64.0	68.4	72.4	73.7	75.9	79.7	79.2	80.1	82.3	86.9	91.3	103.6	105.4
4 Capital Repair	6.2	9.4	10.3	11.0	11.9	12.8	13.9	14.9	15.8	16.6	17.5	18.6	19.5	20.7	21.8	23.0	24.3
5 DEFENSE CONST.	3.7	4.2	5.6	4.0	4.8	5.2	5.2	5.4	4.1	5.2	4.4	4.7	4.7	7.5	6.2	5.8	6.6
6 INVENTORIES	0.0	0.1	0.4	0.0	0.2	0.1	0.2	0.2	0.4	-1.9	0.0	0.0	-0.1	0.0	0.0	0.0	0.0
7 TOTAL INDUSTRY	11.6	19.7	21.0	22.6	23.6	25.6	26.9	27.3	28.0	28.8	28.8	29.8	30.9	32.7	34.2	38.8	40.2
8 New Investment	9.9	16.9	18.0	19.4	20.2	21.8	22.9	23.0	23.5	24.2	23.8	24.6	25.3	26.4	27.5	31.7	32.6
9 Capital Repair	1.7	2.8	3.0	3.2	3.4	3.8	4.0	4.3	4.5	4.6	5.0	5.2	5.6	6.3	6.7	7.1	7.6
10 Unfinished	1.2	1.7	2.3	3.6	1.2	2.9	2.3	2.6	4.2	2.6	3.0	-2.0	0.9	-0.9	-0.6	1.0	1.7
11 POWER	1.8	2.7	2.8	2.8	2.8	3.0	3.1	3.0	3.2	3.2	3.2	3.6	3.7	3.9	3.9	4.6	5.0
12 New Investment	1.6	2.4	2.5	2.5	2.5	2.6	2.7	2.6	2.7	2.7	2.7	3.1	3.1	3.2	3.2	3.8	4.2
13 Capital Repair	0.2	0.3	0.3	0.3	0.3	0.4	0.4	0.4	0.5	0.5	0.5	0.5	0.6	0.7	0.7	0.8	0.8
14 FUELS	2.6	4.1	4.4	4.6	4.9	5.5	5.8	6.1	6.6	7.1	7.6	8.4	8.9	9.9	10.6	11.4	13.0
15 New Investment	2.4	3.7	4.0	4.2	4.5	5.0	5.3	5.5	6.0	6.5	6.9	7.7	8.2	9.1	9.7	10.5	12.0
16 Oil-Gas Explr.	0.8	1.1	1.2	1.3	1.4	1.5	1.6	1.6	1.7	2.0	2.1	2.3	2.5	2.6	2.8	3.1	3.3
17 Coal Mining	0.8	1.1	1.2	1.2	1.2	1.2	1.2	1.2	1.3	1.4	1.4	1.5	1.4	1.5	1.6	1.8	1.9
18 Manufacturing	0.8	1.5	1.6	1.7	1.9	2.3	2.5	2.7	3.0	3.1	3.4	3.9	4.3	5.0	5.3	5.6	6.8
19 Capital Repair	0.2	0.4	0.4	0.4	0.4	0.5	0.5	0.6	0.6	0.6	0.7	0.7	0.7	0.8	0.9	0.9	1.0
20 TOTAL METALS	1.4	2.0	2.2	2.4	2.5	2.8	3.1	2.9	2.8	2.8	3.0	3.0	3.1	3.3	3.3	3.5	3.7
21 New Investment	1.2	1.6	1.8	2.0	2.0	2.3	2.5	2.3	2.2	2.2	2.3	2.3	2.3	2.4	2.4	2.5	2.6
22 Mining	0.3	0.4	0.4	0.4	0.4	0.4	0.4	0.4	0.4	0.4	0.4	0.5	0.5	0.5	0.5	0.5	0.5
23 Manufacturing	0.9	1.2	1.4	1.6	1.6	1.9	2.1	1.9	1.8	1.8	1.9	1.8	1.8	1.9	1.9	2.0	2.1
24 Capital Repair	0.2	0.4	0.4	0.4	0.5	0.5	0.6	0.6	0.6	0.6	0.7	0.7	0.8	0.9	0.9	1.0	1.1
25 CHEMICALS	1.2	1.7	1.7	2.0	2.2	2.5	2.8	2.8	3.0	3.2	2.8	2.5	2.5	2.8	2.8	3.5	3.0
26 New Investment	1.1	1.5	1.5	1.7	1.9	2.2	2.5	2.5	2.6	2.8	2.4	2.1	2.1	2.3	2.3	2.9	2.4
27 Capital Repair	0.1	0.2	0.2	0.3	0.3	0.3	0.3	0.3	0.4	0.4	0.4	0.4	0.4	0.5	0.5	0.6	0.6
28 TOTAL MBMW	1.6	4.4	4.7	5.0	5.3	5.8	6.1	6.6	6.5	6.5	6.4	6.2	6.4	6.3	6.7	8.1	7.8
29 New Investment	1.3	3.8	4.1	4.4	4.7	5.0	5.3	5.7	5.6	5.6	5.4	5.2	5.2	5.0	5.4	6.7	6.3
30 Capital Repair	0.3	0.6	0.6	0.6	0.7	0.8	0.8	0.9	0.9	0.9	1.0	1.0	1.2	1.3	1.3	1.4	1.5
31 WOOD AND PAPER	0.8	1.1	1.2	1.3	1.4	1.3	1.5	1.5	1.6	1.5	1.6	1.6	1.7	1.8	1.9	1.8	2.0
32 New Investment	0.6	0.8	0.9	1.0	1.1	1.0	1.1	1.1	1.2	1.1	1.1	1.1	1.2	1.2	1.3	1.2	1.3
33 Capital Repair	0.2	0.3	0.3	0.3	0.3	0.3	0.4	0.4	0.4	0.4	0.5	0.5	0.5	0.6	0.6	0.6	0.7
34 CONSTR MATER	0.6	1.2	1.3	1.5	1.5	1.5	1.5	1.3	1.4	1.4	1.4	1.4	1.5	1.5	1.5	1.7	1.6
35 New Investment	0.5	1.0	1.1	1.2	1.2	1.2	1.2	1.0	1.0	1.0	1.0	1.0	1.1	1.0	1.0	1.1	1.0
36 Capital Repair	0.1	0.2	0.2	0.3	0.3	0.3	0.3	0.3	0.4	0.4	0.4	0.4	0.4	0.5	0.5	0.6	0.6
37 LIGHT INDUSTRY	0.3	0.6	0.7	0.8	0.8	0.8	0.8	0.8	0.8	0.9	0.9	1.0	1.0	1.0	1.1	1.4	1.2
38 New Investment	0.2	0.5	0.5	0.6	0.6	0.6	0.6	0.6	0.6	0.7	0.6	0.7	0.7	0.7	0.8	1.0	0.8
39 Capital Repair	0.1	0.1	0.2	0.2	0.2	0.2	0.2	0.2	0.2	0.2	0.3	0.3	0.3	0.3	0.3	0.4	0.4
40 FOOD INDUSTRY	0.8	1.4	1.4	1.5	1.5	1.7	1.5	1.5	1.4	1.4	1.3	1.4	1.4	1.4	1.4	1.8	1.8
41 New Investment	0.7	1.2	1.2	1.3	1.3	1.4	1.2	1.2	1.1	1.1	0.9	1.0	1.0	1.0	0.9	1.3	1.3
42 Capital Repair	0.1	0.2	0.2	0.2	0.2	0.3	0.3	0.3	0.3	0.3	0.4	0.4	0.4	0.4	0.5	0.5	0.5

TABLE 0.13: CAPITAL CONSTRUCTION--Part A: OUTPUT BY SECTOR IN CURRENT PRICES
(billion rubles)

	1965	1970	1971	1972	1973	1974	1975	1976	1977	1978	1979	1980	1981	1982	1983	1984	1985
43 NEC INDUSTRY	0.4	0.6	0.6	0.7	0.7	0.7	0.7	0.8	0.8	0.8	0.8	0.7	0.7	0.8	0.9	1.1	1.1
44 New Investment	0.3	0.4	0.4	0.5	0.5	0.5	0.5	0.5	0.5	0.5	0.5	0.4	0.4	0.5	0.5	0.7	0.7
45 Capital Repair	0.1	0.2	0.2	0.2	0.2	0.2	0.2	0.3	0.3	0.3	0.3	0.3	0.3	0.3	0.4	0.4	0.4
46 AGRICULTURE	4.8	9.7	11.6	12.6	13.6	14.8	15.7	15.9	16.1	16.5	16.7	17.1	17.3	18.0	18.9	19.9	20.1
47 New Investment	4.2	8.6	10.4	11.4	12.3	13.4	14.2	14.3	14.5	14.8	14.9	15.1	15.2	15.8	16.5	17.3	17.3
48 Installations	2.8	6.1	7.6	8.0	8.5	9.4	9.3	9.7	9.8	10.1	10.2	10.3	10.5	10.9	11.1	11.6	11.3
49 Husbandry	1.5	3.4	4.0	4.2	4.4	4.9	4.8	5.0	5.1	5.1	5.0	4.6	4.3	3.9	3.9	4.1	4.1
50 Crops & Other	1.3	2.7	3.6	3.8	4.1	4.5	4.5	4.7	4.7	5.1	5.2	5.7	6.2	7.0	7.2	7.5	7.2
51 Land & Water	1.4	2.5	2.8	3.4	3.8	4.0	4.9	4.6	4.7	4.7	4.7	4.8	4.7	4.9	5.4	5.7	6.0
52 Capital Repair	0.6	1.1	1.2	1.2	1.3	1.4	1.5	1.6	1.6	1.7	1.8	2.0	2.1	2.2	2.4	2.6	2.8
53 Unfinished	0.7	1.0	1.5	1.6	0.8	0.6	1.3	1.5	1.4	1.0	1.2	0.6	0.5	0.6	0.6	0.4	0.8
54 TOTAL T & C	3.0	5.1	5.6	6.2	6.4	7.0	8.0	8.6	9.0	10.8	10.9	10.4	11.1	11.9	12.5	13.4	13.2
55 New Investment	2.2	4.0	4.4	4.9	5.1	5.5	6.4	6.9	7.2	8.9	8.9	8.3	8.9	9.6	10.1	10.9	10.6
56 Oil-Gas & Roads	1.5	2.7	3.1	3.4	3.8	4.0	4.0	3.4	3.8	3.7	4.1	3.9	5.0	5.4	6.0	6.1	6.7
57 Installations	0.7	1.3	1.3	1.5	1.3	1.5	2.4	3.5	3.4	5.2	4.8	4.4	3.9	4.2	4.1	4.8	3.9
58 Capital Repair	0.8	1.1	1.2	1.3	1.4	1.5	1.6	1.7	1.8	1.9	2.0	2.1	2.2	2.3	2.4	2.5	2.6
59 Unfinished	0.2	0.2	0.1	0.3	0.1	0.1	0.2	1.3	1.3	1.0	1.2	0.5	0.6	0.3	0.2	-0.7	-0.9
60 CONSTRUCTION	0.3	1.1	1.3	1.3	1.2	1.3	1.3	1.4	1.4	1.4	1.4	1.6	1.4	1.3	1.2	1.3	1.2
61 New Investment	0.3	1.1	1.2	1.2	1.1	1.2	1.2	1.3	1.3	1.3	1.3	1.4	1.2	1.1	1.0	1.1	1.0
62 Capital Repair	0.0	0.0	0.1	0.1	0.1	0.1	0.1	0.1	0.1	0.1	0.1	0.2	0.2	0.2	0.2	0.2	0.2
63 TRADE, etc.	1.8	2.6	2.7	2.9	2.9	3.2	3.1	3.3	3.4	3.6	3.9	4.0	4.1	4.4	4.7	4.8	4.5
64 New Investment	1.4	2.2	2.2	2.3	2.3	2.5	2.4	2.5	2.6	2.8	3.0	3.0	3.1	3.3	3.6	3.7	3.4
65 Trade	0.7	1.3	1.3	1.4	1.2	1.3	1.2	1.3	1.3	1.4	1.6	1.4	1.5	1.7	1.9	1.8	1.6
66 Supply	0.2	0.3	0.3	0.3	0.3	0.4	0.4	0.4	0.4	0.4	0.5	0.5	0.5	0.5	0.5	0.6	0.6
67 Procurement	0.4	0.5	0.5	0.5	0.5	0.6	0.6	0.6	0.6	0.6	0.6	0.6	0.6	0.6	0.6	0.7	0.6
68 Forestry	0.1	0.1	0.1	0.1	0.1	0.2	0.2	0.2	0.2	0.2	0.2	0.2	0.2	0.2	0.2	0.3	0.3
69 Other	0.0	0.0	0.0	0.0	0.1	0.1	0.1	0.1	0.1	0.2	0.2	0.3	0.3	0.3	0.3	0.3	0.3
70 Capital Repair	0.4	0.4	0.5	0.6	0.6	0.7	0.7	0.8	0.8	0.8	0.9	1.0	1.0	1.1	1.1	1.1	1.1
71 Unfinished	0.1	0.2	0.1	0.3	0.0	0.2	0.1	0.1	0.0	0.1	0.1	-0.1	0.1	0.1	0.2	0.0	0.0
72 HOUSING	9.1	15.2	16.2	16.8	17.4	18.1	19.1	19.7	20.6	21.4	21.7	22.6	23.6	25.6	27.5	31.4	32.8
73 New Investment	7.4	12.8	13.5	13.9	14.3	14.8	15.5	15.7	16.3	16.8	16.8	17.5	18.2	20.0	21.6	25.3	26.4
74 Capital Repair	1.7	2.4	2.7	2.9	3.1	3.3	3.6	4.0	4.3	4.6	4.9	5.1	5.4	5.6	5.9	6.1	6.4
75 Unfinished	0.3	0.4	0.4	0.3	-0.2	0.0	0.3	0.5	0.5	0.6	0.4	-0.2	-0.2	0.1	0.1	0.7	0.4
76 COMMUNAL, etc.	1.4	3.0	3.1	3.2	3.3	3.4	3.7	3.9	4.0	4.2	4.2	4.2	4.5	4.9	5.2	6.2	6.5
77 New Investment	1.3	2.8	2.9	3.0	3.1	3.1	3.4	3.5	3.6	3.8	3.8	3.8	4.1	4.5	4.7	5.7	6.0
78 Capital Repair	0.1	0.2	0.2	0.2	0.2	0.3	0.3	0.4	0.4	0.4	0.4	0.4	0.4	0.4	0.5	0.5	0.5
79 EDUCATION	2.0	3.1	3.2	3.4	3.3	3.7	3.9	3.8	4.0	3.8	3.8	3.6	3.6	3.8	3.8	4.4	4.7
80 New Investment	1.6	2.4	2.4	2.5	2.3	2.7	2.8	2.7	2.8	2.6	2.5	2.2	2.2	2.3	2.3	2.8	3.1
81 Capital Repair	0.4	0.7	0.8	0.9	1.0	1.0	1.1	1.1	1.2	1.2	1.3	1.4	1.4	1.5	1.5	1.6	1.6
82 CULTURE	0.6	0.7	0.8	0.8	0.8	0.8	0.9	0.8	0.9	0.9	0.9	0.9	0.8	0.8	0.8	1.1	1.3
83 New Investment	0.5	0.6	0.6	0.6	0.6	0.6	0.7	0.6	0.7	0.7	0.7	0.7	0.6	0.6	0.6	0.9	1.0
84 Capital Repair	0.1	0.1	0.2	0.2	0.2	0.2	0.2	0.2	0.2	0.2	0.2	0.2	0.2	0.2	0.2	0.2	0.3

TABLE 0.13: CAPITAL CONSTRUCTION--Part A: OUTPUT BY SECTOR IN CURRENT PRICES
(billion rubles)

	1965	1970	1971	1972	1973	1974	1975	1976	1977	1978	1979	1980	1981	1982	1983	1984	1985
85 HEALTH, etc.	0.7	1.3	1.4	1.6	1.2	1.2	1.2	1.0	1.1	1.3	1.2	1.2	1.2	1.2	1.2	1.5	1.6
86 New Investment	0.5	1.0	1.1	1.3	0.9	0.8	0.8	0.6	0.6	0.7	0.6	0.6	0.6	0.6	0.6	0.8	0.9
87 Capital Repair	0.2	0.3	0.3	0.3	0.3	0.4	0.4	0.4	0.5	0.6	0.6	0.6	0.6	0.6	0.6	0.7	0.7
88 SCIENCE	0.4	0.8	1.0	1.2	1.2	1.3	1.4	1.7	1.9	2.1	2.1	2.1	2.0	2.0	2.0	2.3	2.1
89 New Investment	0.3	0.6	0.8	1.0	1.0	1.1	1.2	1.5	1.7	1.9	1.9	1.9	1.8	1.7	1.7	2.0	1.8
90 Capital Repair	0.1	0.2	0.2	0.2	0.2	0.2	0.2	0.2	0.2	0.2	0.2	0.2	0.2	0.3	0.3	0.3	0.3
91 ADMINISTRATION	0.2	0.3	0.3	0.3	0.3	0.3	0.3	0.4	0.4	0.4	0.4	0.4	0.4	0.4	0.5	0.6	0.6
92 New Investment	0.1	0.2	0.2	0.2	0.2	0.2	0.2	0.3	0.3	0.3	0.3	0.3	0.3	0.3	0.3	0.4	0.4
93 Capital Repair	0.1	0.1	0.1	0.1	0.1	0.1	0.1	0.1	0.1	0.1	0.1	0.1	0.1	0.1	0.2	0.2	0.2
94 DEFENSE INDUSTRY	0.7	1.0	1.0	1.0	1.0	1.0	1.0	1.2	1.2	1.3	1.1	1.1	1.2	1.1	1.2	1.4	1.4
95 New Investment	0.5	0.7	0.7	0.7	0.7	0.7	0.7	0.8	0.8	0.9	0.7	0.7	0.8	0.7	0.8	0.9	0.9
96 Capital Repair	0.2	0.3	0.3	0.3	0.3	0.3	0.3	0.4	0.4	0.4	0.4	0.4	0.4	0.4	0.4	0.5	0.5

TABLE O.13: CAPITAL CONSTRUCTION--Part B: OUTPUT BY SECTOR IN CONSTANT 1973 PRICES
  (billion rubles)

Page 1

	1965	1970	1971	1972	1973	1974	1975	1976	1977	1978	1979	1980	1981	1982	1983	1984	1985
1 TOTAL	57.7	74.8	79.6	80.1	81.1	84.8	87.5	86.7	86.5	84.7	81.8	81.8	81.4	83.6	84.9	83.5	83.7
2 General Index	0.69	0.90	0.94	0.97	1.00	1.02	1.05	1.09	1.11	1.18	1.24	1.26	1.31	1.38	1.40	1.59	1.63
3 INDUSTRY	15.5	22.1	22.7	23.4	23.6	25.0	25.6	25.2	25.1	24.3	22.7	23.0	23.1	23.1	23.6	23.7	24.1
4 General Index	0.75	0.89	0.92	0.97	1.00	1.02	1.05	1.09	1.11	1.19	1.27	1.29	1.34	1.42	1.45	1.63	1.67
5 Power	2.5	3.1	3.2	2.9	2.8	2.9	3.0	2.8	2.8	2.6	2.5	2.8	2.7	2.7	2.6	2.9	3.1
6 Fuels	3.9	4.6	4.7	4.8	4.9	5.3	5.3	5.4	5.7	5.7	5.7	6.1	6.3	6.6	6.9	6.6	7.4
7 Metals	1.8	2.2	2.4	2.5	2.5	2.8	2.9	2.7	2.6	2.5	2.4	2.4	2.3	2.4	2.4	2.2	2.2
8 Chemicals	1.6	1.9	1.9	2.0	2.2	2.5	2.7	2.7	2.7	2.7	2.3	2.0	2.0	2.0	2.0	2.2	1.8
9 MBMW	2.1	4.8	5.0	5.2	5.3	5.7	5.9	6.1	5.9	5.6	5.2	4.9	4.9	4.5	4.8	5.1	4.8
10 Wood & Paper	0.9	1.2	1.3	1.3	1.4	1.3	1.4	1.4	1.5	1.3	1.2	1.2	1.3	1.3	1.3	1.1	1.2
11 Constr Mater	0.8	1.4	1.5	1.5	1.5	1.5	1.5	1.3	1.2	1.2	1.1	1.1	1.2	1.1	1.1	1.0	1.0
12 Light Industry	0.4	0.7	0.7	0.8	0.8	0.8	0.8	0.8	0.8	0.8	0.7	0.8	0.8	0.7	0.8	0.8	0.7
13 Food Industry	1.0	1.5	1.5	1.6	1.5	1.6	1.4	1.4	1.3	1.2	1.0	1.1	1.1	1.0	1.0	1.1	1.1
14 NEC Industry	0.5	0.6	0.6	0.7	0.7	0.7	0.7	0.7	0.7	0.7	0.6	0.6	0.5	0.6	0.6	0.7	0.7
15 AGRICULTURE	8.0	11.2	12.7	13.3	13.6	14.0	14.5	13.5	13.4	12.3	11.8	11.6	11.1	10.9	11.1	9.9	9.6
16 T & C	3.8	5.5	5.8	6.4	6.5	6.9	7.6	7.9	8.0	9.2	8.7	8.1	8.4	8.7	9.0	8.7	8.5
17 CONSTRUCTION	0.4	1.2	1.4	1.3	1.2	1.3	1.2	1.3	1.3	1.2	1.1	1.3	1.1	0.9	0.9	0.8	0.7
18 T & D, etc.	2.4	2.8	2.8	2.9	2.9	3.1	3.0	3.0	2.9	2.9	2.9	2.8	2.8	2.9	2.9	2.7	2.5
19 HOUSING	14.0	16.2	16.7	16.9	17.4	17.8	18.5	18.7	19.3	19.6	19.0	19.6	19.9	20.4	21.5	21.8	22.1
20 COMMUNAL	2.1	3.2	3.2	3.2	3.3	3.3	3.6	3.7	3.8	3.9	3.7	3.6	3.8	3.9	4.1	4.3	4.4
21 OTHER SERVICES	5.5	6.0	6.0	6.2	5.6	5.9	6.1	5.7	6.0	5.9	5.7	5.4	5.2	5.1	5.0	5.2	5.4
22 DEFENSE, etc.	5.9	6.6	7.9	6.4	6.9	7.3	7.3	7.5	6.4	7.4	6.2	6.3	6.1	7.7	6.8	6.2	6.5
23 Science	0.6	0.9	1.1	1.2	1.2	1.3	1.3	1.5	1.7	1.9	1.8	1.7	1.6	1.5	1.5	1.5	1.3
24 Industry	0.8	1.1	1.0	1.0	1.0	1.0	0.9	1.1	1.1	1.1	0.9	0.9	0.9	0.8	0.9	0.9	0.8
25 Construction	4.5	4.7	5.8	4.1	4.8	5.1	5.0	4.9	3.7	4.4	3.5	3.7	3.6	5.4	4.5	3.8	4.3

Derivation of Construction Price Indexes for Selected Sectors:

	1965	1970	1971	1972	1973	1974	1975	1976	1977	1978	1979	1980	1981	1982	1983	1984	1985
26 POWER	0.70	0.86	0.89	0.96	1.00	1.02	1.04	1.09	1.14	1.23	1.27	1.27	1.33	1.43	1.46	1.59	1.62
27 Unit Cost	n.a.	18.0	18.5	20.0	20.9	21.4	21.7	22.7	23.8	25.6	26.6	26.6	27.9	29.8	30.5	33.3	33.9
28 in value terms	1.4	2.2	2.3	2.3	2.3	2.2	2.8	2.7	2.4	2.1	2.9	3.4	2.9	2.8	2.9	3.4	4.0
29 in phys. terms	n.a.	12.2	12.4	11.5	11.0	10.3	12.9	11.9	10.1	8.2	10.9	12.8	10.4	9.3	9.5	10.2	11.8
30 OIL-GAS EXPLR.	0.60	0.85	0.90	0.95	1.00	1.06	1.15	1.20	1.25	1.35	1.45	1.55	1.62	1.69	1.73	1.92	1.96
31 Unit Cost		0.21					0.29					0.39	0.41	0.42	0.43	0.48	0.49
32 in value terms	0.8	1.1	1.2	1.3	1.4	1.5	1.6	1.6	1.7	2.0	2.1	2.3	2.5	2.6	2.8	3.1	3.3
33 in phys. terms		5.15					5.42					5.93	6.17	6.17	6.48	6.46	6.81
34 MANUFACTURING	0.80	0.90	0.92	0.97	1.00	1.02	1.04	1.07	1.10	1.16	1.24	1.26	1.30	1.38	1.41	1.60	1.63
35 Unit Cost	10.77	12.24	12.49	n.a.	13.53	13.78	14.10	14.51	14.82	15.68	16.80	17.08	n.a.	n.a.	n.a.	n.a.	n.a.
36 in value terms	7.1	13.0	14.3	15.7	16.4	17.9	18.9	19.5	19.8	20.4	19.7	19.8	20.4	21.0	21.9	25.4	25.3
37 in phys. terms	66.2	106.5	114.5	n.a.	120.9	130.1	134.2	134.6	133.7	130.2	117.1	116.2	n.a.	n.a.	n.a.	n.a.	n.a.

282

TABLE 0.13: CAPITAL CONSTRUCTION--Part B: OUTPUT BY SECTOR IN CONSTANT 1973 PRICES
   (billion rubles)

Page 2

	1965	1970	1971	1972	1973	1974	1975	1976	1977	1978	1979	1980	1981	1982	1983	1984	1985
38 AGRICULTURE	0.60	0.87	0.91	0.95	1.00	1.06	1.08	1.17	1.20	1.34	1.42	1.47	1.55	1.65	1.70	2.00	2.09
39 Husbandry	n.a.	0.85	0.88	0.93	1.00	1.06	1.07	1.16	1.18	1.30	1.35	1.38	1.39	1.43	1.48	1.83	1.93
40 Unit Cost	n.a.	0.62	0.64	0.68	0.73	0.78	0.78	0.84	0.86	0.95	0.99	1.01	1.02	1.05	1.08	1.34	1.41
41 in value terms	1.2	3.1	3.5	3.8	4.2	4.6	4.5	4.6	4.7	4.7	4.5	4.5	4.3	3.8	3.8	4.0	3.9
42 in phys. terms	n.a.	5.1	5.4	5.6	5.7	5.9	5.8	5.4	5.4	5.0	4.6	4.5	4.2	3.6	3.5	3.0	2.8
43 large livestock	n.a.	3.9	4.1	4.5	4.6	4.9	4.8	4.6	4.6	4.2	4.0	3.9	3.6	3.1	3.0	2.6	2.4
44 other livestock	n.a.	1.2	1.3	1.1	1.1	1.0	1.0	0.8	0.8	0.8	0.6	0.6	0.6	0.5	0.5	0.4	0.4
45 Land & Water	0.62	0.90	0.95	0.96	1.00	1.05	1.09	1.19	1.23	1.38	1.49	1.55	1.70	1.81	1.86	2.12	2.21
46 Unit Cost	1.04	1.49	1.57	1.60	1.66	1.74	1.81	1.98	2.04	2.30	2.47	2.58	2.81	3.01	3.08	3.52	3.66
47 in value terms	1.1	1.8	2.1	2.6	3.1	3.3	3.9	3.0	3.5	3.4	3.6	3.8	3.8	4.0	4.4	4.8	4.9
48 in phys. terms	1.10	1.21	1.35	1.62	1.87	1.91	2.16	1.51	1.71	1.46	1.46	1.48	1.34	1.33	1.44	1.36	1.33
49 OIL-GAS LINES	0.77	0.91	0.96	0.97	1.00	1.04	1.09	1.13	1.18	1.23	1.29	1.33	1.38	1.40	1.41	1.53	1.56
50 Unit Cost	n.a.	9.78	10.33	10.43	10.72	11.13	11.71	12.11	12.67	13.21	13.81	14.29	14.77	15.05	15.10	16.37	16.74
51 in value terms	0.5	0.6	0.9	0.9	1.3	1.7	1.4	0.9	1.2	1.1	1.4	1.4	1.3	1.8	2.3	2.0	2.8
52 in phys. terms	n.a.	5.8	8.4	8.8	11.7	15.1	12.3	7.3	9.3	8.4	9.8	10.1	8.8	12.2	15.1	12.3	16.6
53 FREEWAYS	0.80	0.94	0.98	0.97	1.00	1.01	1.06	1.11	1.15	1.20	1.26	1.30	1.33	1.34	1.35	1.45	1.47
52 Unit Cost	n.a.	11.8	12.3	12.2	12.5	12.7	13.2	13.9	14.4	15.0	15.8	16.3	16.6	16.8	16.9	18.2	18.4
53 in value terms	1.1	2.1	2.2	2.5	2.5	2.3	2.6	2.5	2.6	2.6	2.7	2.5	3.7	3.6	3.7	4.1	4.4
54 in phys. terms	n.a.	18.1	18.2	20.4	20.4	18.3	19.4	18.1	18.2	17.3	17.4	15.1	22.3	21.2	22.0	22.5	24.0
55 TRADE	0.74	0.93	0.97	0.99	1.00	1.02	1.04	1.11	1.18	1.26	1.36	1.41	1.47	1.54	1.61	1.77	1.81
56 Unit Cost	0.45	0.57	0.59	0.61	0.61	0.63	0.64	0.67	0.72	0.77	0.83	0.86	0.90	0.94	0.98	1.08	1.10
57 in value terms	0.6	1.0	1.2	1.2	1.2	1.1	1.2	1.3	1.3	1.4	1.4	1.4	1.5	1.5	1.6	1.6	1.6
58 in phys. terms	1.31	1.79	2.03	1.99	1.99	1.80	1.84	1.86	1.86	1.77	1.63	1.66	1.63	1.58	1.62	1.52	1.45
59 HOUSING	0.65	0.94	0.97	0.99	1.00	1.02	1.03	1.05	1.07	1.09	1.14	1.16	1.19	1.26	1.28	1.44	1.49
60 Quality Index	0.87	0.95	0.96	0.98	1.00	1.01	1.02	1.03	1.05	1.06	1.08	1.09	1.11	1.12	1.14	1.16	1.18
61 Unit Cost	7.46	11.70	12.21	12.75	13.12	13.48	13.83	14.24	14.69	15.17	16.16	16.53	17.29	18.44	19.11	21.89	23.01
62 in value terms	7.3	12.4	13.1	13.6	14.5	14.8	15.2	15.1	15.8	16.2	16.4	17.8	18.4	19.9	21.5	24.6	26.0
63 in phys. terms	97.6	106.0	107.6	106.7	110.5	110.1	109.9	106.2	107.8	106.8	101.5	107.7	106.4	107.9	112.5	112.4	113.0
64 EDUCATION	0.63	0.91	0.95	0.98	1.00	1.01	1.03	1.06	1.07	1.09	1.11	1.13	1.16	1.22	1.25	1.46	1.52
65 Unit Cost	0.66	0.94	0.99	1.02	1.04	1.05	1.08	1.10	1.12	1.14	1.15	1.17	1.20	1.27	1.30	1.52	1.58
66 in value terms	1.5	1.9	2.0	2.0	2.0	2.2	2.4	2.3	2.3	2.2	2.1	2.0	2.0	2.0	2.1	2.4	2.7
67 in phys. terms	2.27	2.06	2.06	1.99	1.88	2.09	2.26	2.08	2.04	1.91	1.84	1.70	1.67	1.54	1.58	1.57	1.73
68 general	1.72	1.58	1.60	1.56	1.43	1.66	1.74	1.56	1.45	1.34	1.23	1.14	1.11	0.99	0.99	1.00	1.14
69 pre-school	0.55	0.48	0.46	0.43	0.45	0.43	0.52	0.52	0.59	0.57	0.61	0.56	0.56	0.55	0.59	0.57	0.59

Supporting Data on the Use of Reinforced Concrete Per 1000 Square Meters of Floor Space:

	1965	1970	1971	1972	1973	1974	1975	1976	1977	1978	1979	1980	1981	1982	1983	1984	1985
70 TOTAL	56.1	84.6	90.8	96.1	102.9	108.5	114.2	118.7	121.3	123.2	120.8	122.2	116.3	113.4	118.6	121.4	120.2
71 Industry	20.7	33.3	36.2	n.a.	42.7	42.8	46.6	48.7	50.4	50.6	44.5	45.5	n.a.	n.a.	n.a.	n.a.	n.a.
72 per 1000 sq. m.	312	313	316	n.a.	353	329	347	362	377	389	380	392	n.a.	n.a.	n.a.	n.a.	n.a.
73 Housing	29.4	39.5	42.2	n.a.	46.6	48.4	49.9	50.3	51.2	50.1	50.9	54.9	n.a.	n.a.	n.a.	n.a.	n.a.
74 per 1000 sq. m.	430	466	461	n.a.	469	478	504	509	528	521	557	566	n.a.	n.a.	n.a.	n.a.	n.a.
75 Services	6.1	11.7	12.4	n.a.	13.6	17.3	17.8	19.7	19.7	22.5	25.4	21.8	n.a.	n.a.	n.a.	n.a.	n.a.
76 per 1000 sq. m.	298	311	313	n.a.	336	352	344	350	364	372	403	397	n.a.	n.a.	n.a.	n.a.	n.a.

TABLE 0.14: SUPPLY AND USES OF OTHER SECTORS' PRODUCTS
(billion rubles)

	1965	1970	1971	1972	1973	1974	1975	1976	1977	1978	1979	1980	1981	1982	1983	1984	1985
Current Prices																	
1 SUPPLY	3.6	4.7	5.1	5.3	5.5	5.9	6.2	6.5	6.8	7.1	7.2	7.5	7.8	8.4	8.7	8.9	9.7
2 GVO	2.9	3.9	4.2	4.4	4.6	5.0	5.2	5.5	5.7	6.0	6.1	6.3	6.6	7.1	7.4	7.6	8.1
3 Imports	0.1	0.1	0.1	0.1	0.1	0.1	0.1	0.1	0.2	0.2	0.2	0.2	0.2	0.2	0.2	0.2	0.3
4 TC and TD	0.6	0.7	0.8	0.8	0.8	0.8	0.9	0.9	0.9	0.9	0.9	1.0	1.0	1.1	1.1	1.1	1.3
5 USES	3.6	4.7	5.1	5.3	5.5	5.9	6.2	6.5	6.8	7.1	7.2	7.5	7.8	8.4	8.7	8.9	9.7
6 Interuse	1.7	2.1	2.3	2.4	2.5	2.6	2.7	2.7	2.7	2.7	2.8	2.8	2.9	3.0	3.0	3.0	3.1
7 Domestic End Use	1.9	2.5	2.7	2.8	2.9	3.2	3.4	3.6	3.9	4.2	4.2	4.5	4.7	5.2	5.5	5.7	6.4
8 Consumption	1.9	2.4	2.5	2.7	2.8	2.9	3.1	3.3	3.5	3.7	3.9	4.1	4.4	4.8	5.2	5.4	5.8
9 Private	1.7	2.1	2.2	2.4	2.5	2.6	2.7	2.9	3.1	3.3	3.5	3.7	3.9	4.3	4.7	4.9	5.3
10 Public	0.2	0.3	0.3	0.3	0.3	0.3	0.4	0.4	0.4	0.4	0.4	0.4	0.5	0.5	0.5	0.5	0.5
11 Inventories	0.0	0.2	0.2	0.1	0.1	0.2	0.2	0.2	0.3	0.2	0.2	0.3	0.2	0.3	0.3	0.2	0.3
12 Defense	0.0	0.0	0.0	0.0	0.0	0.1	0.1	0.1	0.1	0.2	0.1	0.1	0.1	0.1	0.1	0.1	0.2
13 Exports	0.0	0.1	0.1	0.1	0.1	0.1	0.1	0.2	0.2	0.2	0.2	0.2	0.2	0.2	0.2	0.2	0.2
Constant Prices																	
14 SUPPLY	4.1	5.1	5.2	5.3	5.5	5.8	5.9	6.2	6.3	6.5	6.3	6.4	6.5	6.5	6.7	6.8	7.1
15 GVO	3.4	4.3	4.4	4.5	4.6	4.9	4.9	5.1	5.2	5.4	5.2	5.2	5.3	5.3	5.4	5.5	5.7
16 Imports	0.1	0.1	0.1	0.1	0.1	0.1	0.1	0.1	0.2	0.2	0.2	0.2	0.2	0.2	0.2	0.2	0.3
17 TC and TD	0.6	0.7	0.8	0.8	0.8	0.8	0.9	0.9	0.9	0.9	0.9	0.9	0.9	1.0	1.0	1.1	1.1
18 USES	4.1	5.1	5.2	5.3	5.5	5.8	5.9	6.2	6.3	6.5	6.3	6.4	6.5	6.5	6.7	6.8	7.1
19 Interuse	1.7	2.1	2.2	2.3	2.5	2.6	2.6	2.7	2.7	2.7	2.7	2.7	2.8	2.8	2.8	2.8	2.8
20 Domestic End Use	2.4	2.9	3.0	2.9	2.9	3.1	3.2	3.3	3.5	3.6	3.4	3.5	3.5	3.6	3.7	3.9	4.2
21 Consumption	2.4	2.8	2.8	2.8	2.8	2.8	2.8	3.0	3.1	3.1	3.1	3.1	3.2	3.3	3.5	3.6	3.7
22 Private	2.1	2.4	2.5	2.5	2.5	2.5	2.5	2.6	2.7	2.8	2.7	2.8	2.8	3.0	3.2	3.3	3.4
23 Public	0.3	0.3	0.3	0.3	0.3	0.3	0.3	0.3	0.3	0.3	0.3	0.3	0.3	0.3	0.3	0.3	0.3
24 Inventories	0.0	0.2	0.2	0.1	0.1	0.2	0.2	0.2	0.3	0.2	0.2	0.3	0.2	0.2	0.2	0.2	0.2
25 Defense	0.0	0.0	0.0	0.0	0.0	0.1	0.1	0.1	0.1	0.2	0.1	0.1	0.1	0.1	0.1	0.1	0.2
26 Exports	0.0	0.1	0.1	0.1	0.1	0.1	0.1	0.2	0.2	0.2	0.2	0.2	0.2	0.2	0.2	0.2	0.2
27 General Index	0.87	0.92	0.98	0.99	1.00	1.02	1.06	1.06	1.08	1.10	1.14	1.18	1.21	1.29	1.30	1.31	1.36
28 Enterprises	0.85	0.91	0.95	0.99	1.00	1.03	1.06	1.07	1.09	1.12	1.17	1.20	1.24	1.37	1.39	1.40	1.44
29 Forestry	1.00	1.00	1.00	1.00	1.00	1.00	1.00	1.00	1.00	1.00	1.00	1.00	1.00	1.40	1.40	1.40	1.40
30 Publishing	0.80	0.85	0.91	0.98	1.00	1.05	1.09	1.11	1.15	1.19	1.28	1.33	1.39	1.45	1.48	1.50	1.56
31 unit price	0.043	0.046	0.049	0.053	0.055	0.057	0.060	0.061	0.063	0.065	0.070	0.073	0.076	0.079	0.081	0.082	0.086
32 Other	0.80	0.95	0.98	1.00	1.00	1.02	1.05	1.07	1.08	1.10	1.13	1.15	1.17	1.20	1.23	1.24	1.25

TABLE 0.14: SUPPLY AND USES OF OTHER SECTORS' PRODUCTS
(billion rubles)

	1965	1970	1971	1972	1973	1974	1975	1976	1977	1978	1979	1980	1981	1982	1983	1984	1985
Supporting Data on GVO of Major Other Production Sectors:																	
33 Forestry	0.8	0.9	0.9	0.9	0.9	0.9	1.0	1.0	1.0	1.0	1.0	1.0	1.0	1.4	1.4	1.4	1.4
34 Recycling	0.3	0.8	0.9	1.0	1.0	1.1	1.1	1.2	1.3	1.4	1.3	1.4	1.5	1.5	1.6	1.6	1.6
35 Metals	0.3	0.6	0.6	0.7	0.7	0.8	0.8	0.9	0.9	1.1	1.0	1.0	1.1	1.1	1.2	1.2	1.2
36 Wood and Paper	0.1	0.2	0.2	0.2	0.2	0.2	0.2	0.2	0.2	0.3	0.2	0.3	0.3	0.3	0.3	0.3	0.3
37 Chemicals	0.0	0.1	0.1	0.1	0.1	0.1	0.1	0.1	0.1	0.1	0.1	0.1	0.1	0.1	0.1	0.1	0.1
38 Glass	0.0	0.0	0.1	0.1	0.1	0.1	0.1	0.1	0.1	0.1	0.1	0.1	0.1	0.1	0.1	0.1	0.1
39 Publishing	0.9	1.3	1.5	1.6	1.8	1.9	2.0	2.1	2.2	2.3	2.5	2.6	2.8	3.0	3.2	3.4	3.7
40 M & R Studios	0.2	0.3	0.3	0.3	0.4	0.4	0.4	0.4	0.4	0.5	0.5	0.5	0.5	0.5	0.5	0.5	0.5
41 Other Product	0.7	0.6	0.6	0.6	0.6	0.7	0.7	0.8	0.8	0.8	0.8	0.8	0.8	0.8	0.8	0.8	0.9
42 Public	0.1	0.1	0.1	0.1	0.1	0.2	0.2	0.2	0.2	0.2	0.2	0.2	0.1	0.1	0.1	0.1	0.1
43 Private	0.6	0.5	0.5	0.5	0.5	0.5	0.5	0.6	0.6	0.6	0.6	0.6	0.7	0.7	0.7	0.7	0.8
Supporting Data on Output in Physical Units:																	
Recycling																	
45 Ferrous Metals	n.a.	n.a.	n.a.	n.a.	n.a.	n.a.	n.a.	n.a.	n.a.	n.a.	n.a.	n.a.	88.3	88.0	90.3	91.5	93.0
46 Wood	n.a.	n.a.	n.a.	n.a.	n.a.	n.a.	n.a.	n.a.	n.a.	n.a.	n.a.	n.a.	30.3	38.7	45.7	48.8	53.7
47 Paper	n.a.	n.a.	n.a.	n.a.	n.a.	n.a.	n.a.	n.a.	n.a.	n.a.	n.a.	n.a.	1.86	2.01	2.56	2.71	2.87
48 Rubber	n.a.	n.a.	n.a.	n.a.	n.a.	n.a.	n.a.	n.a.	n.a.	n.a.	n.a.	n.a.	0.30	0.32	0.38	0.39	0.30
49 Chemicals	n.a.	n.a.	n.a.	n.a.	n.a.	n.a.	n.a.	n.a.	n.a.	n.a.	n.a.	n.a.	0.00	0.01	0.05	0.16	0.28
50 Glass	n.a.	n.a.	n.a.	n.a.	n.a.	n.a.	n.a.	n.a.	n.a.	n.a.	n.a.	n.a.	0.43	0.54	0.58	0.65	1.03
51 Publishing	20.5	27.7	30.1	29.8	31.8	32.9	33.3	34.3	34.9	35.1	35.6	35.6	36.6	37.6	39.4	41.3	43
52 Books	13.0	14.5	17.2	16.0	16.5	17.9	18.0	18.6	18.8	18.8	19.1	19.2	20.5	21.3	22.3	24.0	25.3
53 Periodicals	7.5	13.2	12.9	13.8	15.3	15.0	15.3	15.7	16.1	16.3	16.5	16.4	16.1	16.3	17.1	17.3	17.7
54 Total Movies*	259	335	335	360	373	393	420	426	420	438	450	448	444	449	435	460	459
55 Movies	167	218	214	234	245	261	282	285	280	297	310	315	310	311	299	321	313
56 Features	920	1170	1210	1260	1280	1320	1380	1410	1400	1410	1400	1330	1340	1380	1360	1390	1460
57 Records	0.14	0.17	0.18	0.18	0.19	0.20	0.20	0.20	0.20	0.20	0.21	0.21	0.21	0.21	0.19	0.18	0.17

* ten feature films are assumed to be equivalent to one movie

TABLE 0.15: TRANSPORTATION AND COMMUNICATIONS SERVICES
       (billion rubles)

	1965	1970	1971	1972	1973	1974	1975	1976	1977	1978	1979	1980	1981	1982	1983	1984	1985
Current Prices:																	
1 TOTAL SERVICES	24.9	37.6	40.4	42.8	46.5	50.2	54.0	57.3	60.5	64.2	67.1	70.6	73.5	80.3	84.6	87.1	94.0
2 TOTAL PROD	18.0	25.7	27.7	29.5	31.7	34.1	36.7	38.6	41.1	43.8	45.2	47.6	49.8	55.2	58.0	59.5	66.0
3 Transportation	17.3	24.6	26.6	28.3	30.4	32.6	35.1	36.8	39.1	41.5	42.8	45.0	47.0	52.3	55.0	56.3	62.3
4 Communication	0.7	1.1	1.1	1.2	1.3	1.5	1.6	1.8	2.0	2.3	2.4	2.6	2.8	2.9	3.0	3.2	3.7
5 Total Industry	17.0	24.2	26.0	27.8	29.8	32.2	34.7	36.5	38.9	41.5	42.9	45.2	47.4	52.6	55.3	56.7	62.7
6 Power	0.0	0.0	0.0	0.0	0.0	0.0	0.0	0.0	0.0	0.1	0.1	0.1	0.1	0.1	0.1	0.1	0.2
7 Fuels	4.3	6.2	6.7	7.1	7.6	8.2	8.8	9.2	10.0	11.0	11.4	12.1	12.7	14.9	15.9	16.3	18.2
8 Metallurgy	1.8	2.6	2.8	3.0	3.2	3.5	3.8	4.1	4.4	4.8	5.0	5.3	5.6	6.2	6.5	6.7	7.3
9 Chemicals	1.0	1.2	1.3	1.5	1.6	1.7	1.8	1.9	2.1	2.2	2.3	2.4	2.5	2.8	2.9	3.0	3.4
10 Total MBMW	2.5	3.2	3.4	3.7	3.9	4.2	4.7	5.0	5.2	5.3	5.5	5.8	6.1	6.6	7.0	7.2	8.1
11 Wood and Paper	1.8	2.7	2.9	3.1	3.3	3.5	3.8	4.0	4.3	4.6	4.7	4.9	5.1	5.4	5.5	5.6	6.1
12 Constr. Mater.	3.6	5.3	5.6	6.0	6.6	7.2	7.8	8.1	8.6	9.0	9.3	9.8	10.3	11.3	12.0	12.3	13.2
13 N.E.C. Ind.	0.2	0.2	0.2	0.2	0.3	0.3	0.3	0.3	0.3	0.3	0.3	0.3	0.4	0.4	0.4	0.4	0.5
14 Light Ind.	0.5	0.8	0.9	0.9	1.0	1.1	1.1	1.2	1.2	1.3	1.3	1.4	1.4	1.5	1.5	1.5	1.7
15 Food Ind.	1.3	2.0	2.2	2.3	2.4	2.5	2.6	2.7	2.8	2.9	3.0	3.1	3.2	3.4	3.5	3.6	4.0
16 Agriculture	0.8	1.3	1.4	1.4	1.6	1.6	1.7	1.8	1.9	2.0	2.0	2.1	2.1	2.2	2.3	2.4	2.8
17 Other Prod.	0.2	0.2	0.3	0.3	0.3	0.3	0.3	0.3	0.3	0.3	0.3	0.3	0.3	0.4	0.4	0.4	0.5
18 HOUSEHOLD SERV.	4.8	8.3	8.7	9.1	10.2	11.0	12.0	12.9	13.4	14.0	14.8	15.5	16.2	16.8	17.9	18.2	18.5
19 Transportation	4.0	7.1	7.5	7.8	8.8	9.5	10.3	11.1	11.6	12.1	12.7	13.3	13.9	14.3	15.1	15.3	15.5
20 Communication	0.8	1.2	1.2	1.3	1.4	1.5	1.7	1.8	1.8	1.9	2.1	2.2	2.3	2.5	2.8	2.9	3.0
21 N-PROD SERV.	2.1	3.6	4.0	4.2	4.6	5.1	5.3	5.8	6.0	6.4	7.1	7.5	7.5	8.3	8.7	9.4	9.5
Constant Prices:																	
22 TOTAL SERVICES	28.7	39.8	42.2	44.1	46.5	49.0	51.3	53.2	54.4	56.2	57.1	58.3	59.2	61.0	62.9	64.5	65.8
23 TOTAL PROD	20.4	27.4	28.9	30.2	31.8	33.2	34.7	35.7	36.7	37.7	37.5	38.2	38.5	39.9	41.1	41.9	42.9
24 Total Industry	19.3	25.8	27.2	28.5	29.9	31.4	32.9	33.8	34.8	35.7	35.6	36.3	36.7	37.9	38.9	39.7	40.7
25 Power	0.0	0.0	0.0	0.0	0.0	0.0	0.0	0.0	0.0	0.1	0.1	0.1	0.1	0.1	0.1	0.1	0.2
26 Fuels	5.2	6.5	6.9	7.2	7.7	8.0	8.5	8.9	9.3	9.7	9.9	10.2	10.4	10.8	11.0	11.2	11.5
27 Metallurgy	2.1	2.8	2.9	3.0	3.2	3.4	3.6	3.6	3.7	3.8	3.8	3.8	3.8	3.8	3.9	4.0	4.1
28 Chemicals	0.8	1.2	1.3	1.4	1.6	1.7	1.8	1.9	2.0	2.1	2.0	2.1	2.2	2.3	2.5	2.5	2.5
29 Total MBMW	2.0	3.2	3.4	3.7	3.9	4.1	4.3	4.5	4.6	4.8	5.0	5.2	5.3	5.5	5.7	5.8	6.0
30 Wood and Paper	2.7	3.2	3.2	3.3	3.3	3.3	3.3	3.3	3.3	3.2	3.1	3.1	3.1	3.2	3.3	3.3	3.4
31 Constr. Mater.	3.9	5.5	6.0	6.2	6.6	6.9	7.3	7.5	7.6	7.8	7.6	7.6	7.7	7.9	8.1	8.3	8.5
32 N.E.C. Ind.	0.2	0.2	0.2	0.3	0.3	0.3	0.3	0.4	0.4	0.4	0.4	0.4	0.4	0.4	0.4	0.4	0.4
33 Light Ind.	0.7	0.9	0.9	0.9	1.0	1.0	1.0	1.0	1.0	1.0	1.0	1.0	1.0	1.0	1.0	1.0	1.0
34 Food Ind.	1.8	2.2	2.3	2.4	2.4	2.6	2.7	2.6	2.7	2.7	2.8	2.7	2.7	2.8	2.9	3.0	3.0
35 Agriculture	0.9	1.4	1.4	1.4	1.6	1.6	1.5	1.6	1.6	1.7	1.6	1.6	1.5	1.6	1.9	1.8	1.8
36 Other Prod.	0.2	0.2	0.3	0.3	0.3	0.3	0.3	0.3	0.3	0.3	0.3	0.3	0.3	0.4	0.4	0.4	0.4
37 HOUSEHOLD SERV.	5.9	8.6	9.1	9.6	10.2	10.8	11.6	12.1	12.3	12.9	13.5	14.0	14.6	14.8	15.3	15.7	16.2
38 N-PROD SERV.	2.4	3.8	4.2	4.3	4.6	4.9	5.0	5.4	5.4	5.6	6.0	6.2	6.1	6.2	6.4	6.9	6.7
39 '73 TRANS INDEX	0.87	0.95	0.96	0.97	1.00	1.02	1.05	1.08	1.11	1.14	1.18	1.22	1.25	1.33	1.36	1.36	1.46
40 Prod Sectors	0.88	0.94	0.96	0.98	1.00	1.03	1.06	1.08	1.12	1.16	1.20	1.25	1.29	1.38	1.41	1.42	1.54
41 Household Serv.	0.81	0.97	0.96	0.95	1.00	1.02	1.04	1.07	1.09	1.09	1.09	1.11	1.11	1.13	1.17	1.16	1.15
42 OFFICIAL INDEX*	0.97	0.98	0.99	1.00	1.00	1.00	1.01	1.01	1.02	1.02	1.04	1.05	1.06	1.16	1.16	1.16	1.26

* official 1973 price index derived using published data on tonns/kilometers

TABLE 0.15: TRANSPORTATION AND COMMUNICATION SERVICES
         (billion rubles)

	1965	1970	1971	1972	1973	1974	1975	1976	1977	1978	1979	1980	1981	1982	1983	1984	1985

Supporting Data on Total Transportation Services in Physical Units (tonns of freight and passengers per kilometer):

	1965	1970	1971	1972	1973	1974	1975	1976	1977	1978	1979	1980	1981	1982	1983	1984	1985
37 Prod Sectors	2.83	3.99	4.28	4.49	4.85	5.20	5.53	5.82	6.08	6.47	6.57	6.83	7.08	7.19	7.53	7.74	7.88
38 Published	2.76	3.83	4.09	4.28	4.62	4.94	5.20	5.43	5.63	5.95	5.98	6.18	6.34	6.36	6.61	6.68	6.68
39 Gas Pipe Lines	0.05	0.13	0.15	0.17	0.19	0.22	0.28	0.34	0.40	0.47	0.54	0.60	0.68	0.77	0.86	1.00	1.13
40 Fishing	0.02	0.03	0.04	0.04	0.04	0.04	0.05	0.05	0.05	0.05	0.05	0.05	0.06	0.06	0.06	0.07	0.07
41 Household Serv.	0.37	0.55	0.59	0.62	0.66	0.70	0.75	0.78	0.80	0.84	0.87	0.90	0.94	0.95	0.97	0.99	1.02

TABLE 0.16: TRADE AND DISTRIBUTION SERVICES
         (billion rubles)

	1965	1970	1971	1972	1973	1974	1975	1976	1977	1978	1979	1980	1981	1982	1983	1984	1985
Current Prices:																	
1 TOTAL SERVICES	15.2	22.3	23.7	25.5	26.6	28.0	29.9	31.5	33.0	34.7	36.2	38.8	40.0	40.6	42.0	43.0	45.0
2 TOTAL PROD.	15.2	22.2	23.6	25.3	26.4	27.8	29.6	31.1	32.6	34.3	35.8	38.3	39.5	40.0	41.4	42.4	44.3
3 Total Industry	12.0	17.5	18.7	20.3	21.0	22.2	24.0	25.2	26.7	27.9	29.3	31.6	32.7	33.1	34.1	35.0	36.7
4 Fuels	0.6	1.1	1.2	1.3	1.4	1.5	1.7	1.8	1.9	2.0	2.1	2.2	2.3	2.3	2.4	2.5	2.6
5 Metallurgy	0.4	0.6	0.7	0.7	0.8	0.8	0.8	0.8	0.8	0.9	0.9	1.0	1.0	1.0	1.1	1.1	1.2
6 Chemicals	0.5	0.9	1.0	1.0	1.0	1.1	1.1	1.2	1.2	1.3	1.3	1.4	1.4	1.4	1.4	1.4	1.5
7 Total MBMW	1.1	1.4	1.5	1.6	1.7	1.7	2.0	2.1	2.2	2.3	2.5	2.7	2.9	3.0	3.1	3.2	3.3
8 Wood and Paper	0.4	0.7	0.7	0.8	0.8	0.9	1.0	1.1	1.2	1.3	1.3	1.4	1.5	1.5	1.6	1.6	1.7
9 Constr. Mater.	0.2	0.3	0.4	0.4	0.4	0.4	0.5	0.5	0.6	0.6	0.6	0.7	0.7	0.7	0.7	0.7	0.8
10 N.E.C. Ind.	0.3	0.5	0.6	0.6	0.6	0.7	0.7	0.8	0.8	0.8	0.8	0.9	0.9	0.9	1.0	1.0	1.1
11 Light Ind.	1.6	2.6	2.8	3.0	3.1	3.2	3.4	3.6	3.9	4.2	4.7	5.3	5.4	5.5	5.6	5.8	6.0
12 Food Ind.	6.9	9.4	9.8	10.9	11.2	11.9	12.8	13.3	14.1	14.5	15.1	16.0	16.6	16.8	17.2	17.7	18.5
13 Agriculture	2.8	4.2	4.4	4.5	4.9	5.1	5.0	5.3	5.3	5.8	5.9	6.0	6.1	6.2	6.6	6.7	6.8
14 Other Prod.	0.4	0.5	0.5	0.5	0.5	0.5	0.6	0.6	0.6	0.6	0.6	0.7	0.7	0.7	0.7	0.7	0.8
15 N-PROD SERVICE*	0.0	0.1	0.1	0.2	0.2	0.2	0.3	0.4	0.4	0.4	0.4	0.5	0.5	0.6	0.6	0.6	0.7
Constant Prices:																	
16 TOTAL SERVICES	17.7	23.5	24.6	25.1	26.6	27.8	28.5	28.9	29.7	30.3	30.0	30.1	30.1	30.5	32.0	32.5	32.9
17 TOTAL PROD.	17.7	23.4	24.5	24.9	26.4	27.6	28.3	28.5	29.3	30.0	29.6	29.7	29.7	30.0	31.5	32.0	32.4
18 Total Industry	14.5	18.8	19.7	20.2	21.0	22.1	23.0	23.1	23.8	24.1	24.2	24.3	24.5	24.4	25.3	25.8	26.1
19 Fuels	0.9	1.2	1.2	1.3	1.4	1.4	1.5	1.6	1.6	1.7	1.7	1.8	1.8	1.9	1.9	2.0	2.0
20 Metallurgy	0.5	0.7	0.7	0.8	0.8	0.8	0.9	0.9	0.9	1.0	0.9	0.9	0.9	0.8	0.9	0.9	0.9
21 Chemicals	0.5	0.8	0.9	1.0	1.0	1.1	1.2	1.3	1.3	1.4	1.4	1.4	1.5	1.4	1.5	1.5	1.6
22 Total MBMW	0.8	1.4	1.5	1.6	1.7	1.8	1.9	1.9	2.0	2.1	2.1	2.2	2.3	2.2	2.2	2.3	2.3
23 Wood and Paper	0.7	0.8	0.8	0.8	0.8	0.8	0.8	0.8	0.8	0.8	0.8	0.8	0.8	0.8	0.8	0.8	0.8
24 Constr. Mater.	0.2	0.4	0.4	0.4	0.4	0.4	0.5	0.5	0.5	0.5	0.5	0.5	0.5	0.5	0.5	0.5	0.5
25 N.E.C. Ind.	0.3	0.5	0.5	0.5	0.6	0.6	0.7	0.7	0.8	0.8	0.8	0.8	0.8	0.9	0.9	0.9	0.9
26 Light Ind.	2.2	3.0	3.1	3.1	3.1	3.2	3.3	3.3	3.4	3.4	3.4	3.4	3.5	3.4	3.4	3.4	3.5
27 Food Ind.	8.2	10.2	10.6	10.8	11.2	11.9	12.3	12.0	12.5	12.5	12.6	12.5	12.4	12.6	13.2	13.6	13.6
28 Agriculture	2.8	4.2	4.3	4.2	4.9	4.9	4.6	4.8	4.9	5.2	4.9	4.7	4.6	4.9	5.5	5.4	5.4
29 Other Prod.	0.4	0.5	0.5	0.5	0.5	0.5	0.6	0.6	0.6	0.6	0.6	0.7	0.7	0.7	0.7	0.7	0.9
30 N-PROD SERVICE*	0.0	0.1	0.1	0.2	0.2	0.2	0.3	0.4	0.4	0.3	0.3	0.4	0.4	0.5	0.5	0.5	0.5
31 GENERAL INDEX	0.86	0.95	0.96	1.01	1.00	1.01	1.05	1.09	1.11	1.14	1.21	1.29	1.33	1.33	1.31	1.32	1.37

*consignment trade surcharge

TABLE 0.17: SOCIO-CULTURAL SERVICES, SCIENCE AND STATE ADMINISTRATION
(billion rubles)

	1965	1970	1971	1972	1973	1974	1975	1976	1977	1978	1979	1980	1981	1982	1983	1984	1985
Current Prices:																	
1 TOTAL SERVICES	40.3	58.5	62.1	66.7	72.1	76.7	80.7	84.6	89.1	94.8	98.3	104.3	108.9	115.0	120.0	125.3	130.9
2 Prod. Services	2.5	3.6	3.8	4.2	4.5	4.7	5.0	5.3	5.7	6.0	6.3	6.7	7.2	8.0	8.3	8.5	8.9
3 Utilities	1.2	1.6	1.7	1.9	2.0	2.1	2.2	2.3	2.4	2.5	2.6	2.7	2.9	3.6	3.7	3.8	4.0
4 Personal Care	1.4	2.0	2.1	2.3	2.5	2.6	2.8	3.0	3.3	3.5	3.7	4.0	4.3	4.4	4.6	4.7	4.9
5 HC & E*	6.7	10.0	10.7	11.5	12.3	13.2	13.9	14.8	15.6	16.6	17.6	18.4	19.1	20.1	21.2	22.3	23.3
6 Public Services	5.4	8.5	9.2	9.9	10.7	11.5	12.2	13.0	13.8	14.8	15.7	16.5	17.1	18.1	19.1	20.1	21.0
7 Imputed Rent	1.3	1.5	1.5	1.6	1.6	1.7	1.7	1.8	1.8	1.8	1.9	1.9	2.0	2.0	2.1	2.2	2.3
8 Banking & Insur.	1.8	2.1	2.2	2.4	2.9	3.2	3.5	3.6	4.1	4.5	3.5	4.0	4.0	4.4	5.0	5.4	5.8
9 Education	11.2	15.4	16.3	17.5	18.9	20.0	20.7	21.8	22.8	24.1	24.8	26.1	26.9	28.1	29.1	30.2	31.6
10 Culture & Arts	1.5	2.3	2.5	2.6	2.7	2.9	3.1	3.4	3.6	3.8	4.0	4.3	4.5	4.7	4.8	5.0	5.2
11 Health, etc.**	7.2	10.4	10.8	11.3	12.0	12.7	13.6	14.4	15.3	16.1	16.9	17.8	18.6	19.4	19.9	20.4	21.1
12 Science	6.7	10.8	11.7	12.8	14.1	15.1	15.7	15.7	16.2	17.3	18.2	19.7	20.9	22.2	23.4	24.9	26.1
13 State Admin.	2.7	3.7	4.1	4.3	4.5	4.8	5.0	5.2	5.5	6.0	6.5	6.9	7.2	7.5	7.7	8.0	8.2
14 Trade Consign.	0.0	0.1	0.1	0.2	0.2	0.2	0.3	0.4	0.4	0.4	0.4	0.5	0.5	0.6	0.6	0.6	0.7
Purchases by:																	
15 Households	9.5	14.6	15.7	17.2	18.4	19.8	20.9	22.1	22.9	23.9	24.6	25.7	26.7	28.5	30.1	31.4	33.0
16 State Budget	26.1	36.7	39.0	40.9	44.8	47.3	49.6	51.9	54.4	57.9	60.9	65.1	68.1	71.3	73.6	76.0	78.7
17 Enterprises	4.7	7.1	7.4	8.7	8.9	9.6	10.1	10.6	11.8	13.0	12.9	13.5	14.1	15.2	16.3	18.0	19.2
Constant Prices:																	
18 TOTAL SERVICES	49.6	62.8	65.6	68.5	72.1	73.7	76.4	78.0	80.8	82.8	83.3	86.3	87.7	88.4	90.3	91.4	92.6
19 Prod. Services	3.0	3.8	4.0	4.2	4.5	4.6	4.8	5.1	5.5	5.7	5.9	6.2	6.4	6.4	6.7	6.8	7.0
20 Utilities	1.4	1.7	1.8	1.9	2.0	2.0	2.1	2.3	2.4	2.5	2.6	2.7	2.8	2.9	2.9	3.0	3.1
21 Personal Care	1.6	2.1	2.2	2.3	2.5	2.5	2.7	2.9	3.1	3.2	3.4	3.5	3.7	3.6	3.8	3.9	3.9
22 HC & E*	7.7	10.3	10.9	11.5	12.3	12.8	13.5	14.0	14.7	15.3	16.1	16.3	16.6	16.8	17.4	17.9	18.1
23 Banking & Insur.	2.1	2.2	2.3	2.4	2.9	3.1	3.3	3.4	3.8	4.0	2.8	3.1	3.1	3.3	3.7	3.9	4.1
24 Education	14.6	17.5	17.9	18.2	18.9	19.1	19.5	19.8	20.2	20.3	20.5	21.0	20.9	21.1	21.1	21.1	21.5
25 Culture & Arts	1.8	2.4	2.6	2.6	2.7	2.9	3.0	3.3	3.4	3.5	3.5	3.6	3.5	3.6	3.6	3.7	3.8
26 Health, etc.**	8.7	11.3	11.7	12.2	12.0	12.1	12.2	12.2	12.4	12.6	12.7	12.8	12.8	12.9	13.0	13.1	13.2
27 Science	8.7	11.2	12.0	12.9	14.1	14.2	15.0	14.9	15.5	15.7	16.0	17.4	18.2	18.1	18.7	18.5	18.4
28 State Admin.	3.1	3.9	4.2	4.3	4.5	4.6	4.8	4.9	5.0	5.3	5.3	5.4	5.6	5.6	5.7	5.8	5.8
29 Trade Consign.	0.0	0.1	0.1	0.2	0.2	0.2	0.3	0.4	0.4	0.4	0.4	0.5	0.5	0.5	0.5	0.5	0.6
Purchases by:																	
29 Households	11.7	15.7	16.5	17.6	18.4	19.0	19.8	20.4	20.8	20.9	20.8	21.3	21.5	21.9	22.7	22.9	23.4
30 State Budget	32.1	39.5	41.2	42.1	44.8	45.4	47.0	47.9	49.3	50.5	51.5	53.9	54.8	54.8	55.4	55.4	55.7
31 Enterprises	5.8	7.6	7.9	8.8	8.9	9.3	9.6	9.8	10.7	11.4	10.9	11.1	11.3	11.7	12.3	13.1	13.6
32 GENERAL INDEX	0.81	0.93	0.95	0.97	1.00	1.04	1.06	1.08	1.10	1.15	1.18	1.21	1.24	1.30	1.33	1.37	1.41
33 HC & E*	0.87	0.97	0.98	1.00	1.00	1.03	1.03	1.06	1.06	1.08	1.09	1.13	1.15	1.20	1.22	1.24	1.29
34 Banking & Insur.	0.87	0.94	0.97	0.99	1.00	1.03	1.05	1.06	1.09	1.13	1.24	1.27	1.29	1.33	1.36	1.38	1.42
35 Education	0.77	0.88	0.91	0.96	1.00	1.05	1.06	1.10	1.13	1.19	1.21	1.24	1.28	1.33	1.38	1.43	1.47
36 Culture & Arts	0.82	0.95	0.95	0.97	1.00	1.00	1.01	1.04	1.06	1.08	1.14	1.17	1.28	1.32	1.35	1.37	1.37
37 Health, etc.**	0.83	0.92	0.92	0.93	1.00	1.05	1.12	1.18	1.23	1.27	1.33	1.39	1.46	1.50	1.53	1.55	1.59
38 Science	0.77	0.97	0.97	0.99	1.00	1.06	1.05	1.05	1.05	1.10	1.14	1.13	1.15	1.23	1.25	1.35	1.41
39 State Admin.	0.87	0.94	0.97	0.99	1.00	1.03	1.05	1.06	1.09	1.13	1.24	1.27	1.29	1.33	1.36	1.38	1.42

* housing-communal and everyday (personal care) services
**health, social services, tourism and sports

TABLE 0.17: SOCIO-CULTURAL SERVICES, SCIENCE AND STATE ADMINISTRATION
   (billion rubles)

	1965	1970	1971	1972	1973	1974	1975	1976	1977	1978	1979	1980	1981	1982	1983	1984	1985
Supporting Data on Employment in Non-Production Service Sectors:																	
40 TOTAL	23.1	28.2	29.3	30.3	31.3	32.7	33.3	33.6	34.5	35.6	36.5	37.3	38.1	38.7	39.0	39.7	40.1
41 HOUSEHOLD SERV.	14.9	18.1	18.7	19.3	19.7	20.5	20.9	21.4	21.9	22.6	23.1	23.5	23.9	24.5	24.8	25.3	25.6
42 HC & E	2.5	3.2	3.3	3.5	3.6	3.8	3.9	4.0	4.1	4.3	4.5	4.6	4.7	4.7	4.8	4.9	5.0
43 Banking & Insur.	0.3	0.4	0.4	0.4	0.4	0.5	0.5	0.5	0.6	0.6	0.6	0.6	0.7	0.7	0.7	0.7	0.7
44 Education	6.6	7.9	8.1	8.3	8.4	8.6	8.8	9.1	9.2	9.5	9.7	9.9	10.0	10.3	10.3	10.6	10.7
45 Culture & Arts	1.0	1.2	1.3	1.3	1.4	1.5	1.5	1.5	1.6	1.7	1.7	1.8	1.8	1.9	1.9	1.9	1.9
46 Health, etc.**	4.6	5.4	5.6	5.8	5.9	6.1	6.2	6.3	6.4	6.4	6.6	6.6	6.7	6.9	7.0	7.1	7.2
47 SCIENCE & ADMIN.	4.1	5.0	5.3	5.5	5.8	6.1	6.2	6.1	6.3	6.5	6.7	6.9	7.1	7.1	7.1	7.2	7.3
48 Science	2.6	3.2	3.4	3.5	3.7	3.9	4.0	3.9	4.0	4.1	4.3	4.4	4.5	4.5	4.5	4.5	4.6
49 Administration	1.5	1.8	1.9	2.0	2.1	2.2	2.2	2.2	2.3	2.4	2.4	2.5	2.6	2.6	2.6	2.7	2.7
Supporting Data on Current Material Purchases by Sector of Use in Current Prices:																	
50 TOTAL	10.0	15.1	16.1	17.2	18.5	19.8	21.1	22.2	23.2	24.8	25.7	27.3	28.7	30.5	32.3	33.7	35.0
51 HOUSEHOLD SERV.	6.8	9.9	10.4	11.0	11.7	12.5	13.5	14.6	15.4	16.4	17.1	17.8	18.6	19.7	20.4	21.0	21.3
52 HC & E	1.5	2.2	2.3	2.4	2.7	2.9	3.1	3.4	3.6	4.0	4.3	4.5	4.7	5.1	5.3	5.5	5.7
53 Education	2.8	4.0	4.2	4.5	4.6	4.9	5.2	5.5	5.8	6.1	6.3	6.5	6.8	7.1	7.4	7.5	7.6
54 Culture & Arts	0.2	0.3	0.4	0.4	0.4	0.5	0.6	0.7	0.7	0.8	0.8	0.8	0.8	0.9	0.9	0.9	0.9
55 Health, etc.	2.3	3.4	3.5	3.7	4.0	4.3	4.7	5.1	5.3	5.5	5.7	6.0	6.3	6.6	6.8	7.0	7.1
56 SCIENCE & ADMIN.	3.3	5.2	5.7	6.2	6.8	7.3	7.5	7.6	7.8	8.4	8.6	9.5	10.1	10.8	11.9	12.7	13.7
57 Science	2.8	4.5	4.9	5.4	6.0	6.4	6.6	6.6	6.7	7.1	7.3	8.1	8.7	9.3	10.3	11.1	12.0
58 Administration	0.5	0.7	0.8	0.8	0.8	0.9	0.9	1.0	1.1	1.3	1.3	1.4	1.4	1.5	1.6	1.6	1.7
Supporting Data on Current Material Purchases by Sector of Use in Constant Prices:																	
59 TOTAL	12.6	16.4	17.2	18.1	19.3	20.1	20.8	21.3	21.9	22.5	22.9	23.5	24.1	24.0	24.8	25.4	26.0
60 HOUSEHOLD SERV.	8.2	10.5	10.9	11.2	11.7	12.2	12.8	13.4	14.0	14.2	14.6	14.7	14.8	14.9	15.0	15.1	15.2
61 HC & E	1.7	2.2	2.3	2.5	2.7	2.9	3.0	3.2	3.4	3.6	3.8	3.8	3.9	4.0	4.1	4.2	4.2
62 Education	3.3	4.3	4.4	4.5	4.6	4.7	4.8	5.0	5.2	5.3	5.4	5.4	5.4	5.4	5.5	5.5	5.5
63 Culture & Arts	0.2	0.3	0.4	0.4	0.4	0.5	0.5	0.6	0.6	0.7	0.7	0.7	0.7	0.7	0.7	0.7	0.7
64 Health, etc.	2.9	3.6	3.7	3.8	4.0	4.2	4.4	4.6	4.7	4.7	4.7	4.8	4.8	4.8	4.8	4.8	4.8
65 SCIENCE & ADMIN.	4.4	5.9	6.3	6.9	7.6	7.8	8.0	7.9	8.0	8.3	8.4	8.9	9.3	9.2	9.8	10.3	10.8
66 Science	3.4	4.6	5.0	5.5	6.0	6.2	6.3	6.2	6.1	6.3	6.4	6.8	7.1	7.0	7.6	8.0	8.5
67 Administration	1.0	1.2	1.3	1.4	1.6	1.7	1.7	1.7	1.9	2.0	2.0	2.1	2.2	2.2	2.2	2.3	2.2
Supporting Data on Capital Outlays on Service Sectors in Constant Prices:																	
68 TOTAL	8.6	10.8	11.7	12.2	12.7	12.4	13.0	12.8	13.3	13.2	13.1	14.0	13.9	13.9	14.3	14.3	14.3
69 HOUSEHOLD SERV.	7.0	8.9	9.4	9.7	10.0	10.0	10.1	9.8	10.1	10.1	10.0	10.3	10.0	10.0	10.4	10.6	10.9
70 HC & E	1.6	2.5	2.7	2.8	3.1	3.1	3.3	3.4	3.6	3.7	3.8	3.9	3.9	3.8	4.2	4.5	4.4
71 Education	3.7	4.0	4.0	4.1	4.5	4.4	4.4	4.3	4.3	4.1	4.1	4.3	4.2	4.2	4.2	3.9	4.2
72 Culture & Arts	0.6	0.7	0.8	0.8	0.8	0.8	0.8	0.7	0.8	0.8	0.7	0.7	0.6	0.6	0.6	0.7	0.8
73 Health, etc.	1.1	1.7	1.9	2.0	1.7	1.6	1.5	1.3	1.3	1.5	1.4	1.4	1.3	1.4	1.3	1.5	1.5
74 SCIENCE & ADMIN.	1.6	1.9	2.2	2.4	2.7	2.5	2.9	3.0	3.2	3.2	3.1	3.7	3.9	3.9	4.0	3.6	3.4
75 Science	1.3	1.5	1.8	2.1	2.3	2.1	2.4	2.5	2.8	2.7	2.7	3.3	3.5	3.5	3.6	3.2	3.0
76 Administration	0.3	0.4	0.4	0.4	0.4	0.4	0.5	0.5	0.4	0.4	0.4	0.4	0.4	0.4	0.4	0.4	0.4

TABLE O.17: SOCIO-CULTURAL SERVICES, SCIENCE AND STATE ADMINISTRATION
(billion rubles)

		1965	1970	1971	1972	1973	1974	1975	1976	1977	1978	1979	1980	1981	1982	1983	1984	1985

Supporting Data on Current Material Purchases by Sector Origin in Current Prices:

		1965	1970	1971	1972	1973	1974	1975	1976	1977	1978	1979	1980	1981	1982	1983	1984	1985
77	HOUSEHOLD SERV.	6.8	9.9	10.4	11.0	11.7	12.6	13.6	14.6	15.4	16.4	17.1	17.8	18.6	19.7	20.4	20.9	21.3
78	Power	0.2	0.3	0.4	0.4	0.4	0.4	0.4	0.4	0.4	0.4	0.5	0.5	0.5	0.7	0.7	0.7	0.7
79	Fuels	0.2	0.3	0.3	0.3	0.3	0.3	0.4	0.5	0.5	0.6	0.7	0.7	0.7	1.0	0.9	0.9	0.8
80	Chemicals	0.2	0.3	0.3	0.4	0.4	0.5	0.6	0.7	0.8	0.8	0.9	1.0	1.0	1.0	1.0	1.1	1.1
81	MBMW	0.1	0.2	0.2	0.2	0.2	0.3	0.3	0.3	0.3	0.3	0.3	0.3	0.3	0.3	0.3	0.3	0.3
82	Wood and Paper	0.1	0.2	0.2	0.2	0.2	0.2	0.3	0.3	0.4	0.5	0.5	0.6	0.6	0.7	0.7	0.7	0.7
83	Constr. Mater.	0.2	0.2	0.2	0.2	0.2	0.3	0.3	0.3	0.3	0.3	0.3	0.3	0.3	0.4	0.4	0.4	0.4
84	N.E.C. Ind.	0.7	1.1	1.2	1.3	1.4	1.5	1.7	1.9	2.0	2.2	2.3	2.4	2.5	2.8	2.9	3.0	3.1
85	Light Industry	1.4	1.8	1.9	2.0	2.1	2.2	2.4	2.6	2.8	3.0	3.1	3.2	3.3	3.4	3.5	3.6	3.6
86	Food Industry	3.0	4.5	4.7	5.0	5.3	5.5	5.7	6.0	6.2	6.4	6.7	6.9	7.2	7.4	7.7	7.9	8.1
87	Agriculture	0.6	0.8	0.9	0.9	1.0	1.1	1.2	1.3	1.4	1.5	1.6	1.6	1.7	1.7	1.9	2.0	2.0
88	Other Prod.	0.1	0.2	0.2	0.2	0.2	0.2	0.3	0.3	0.3	0.3	0.3	0.3	0.4	0.4	0.4	0.4	0.4
89	SCIENCE & ADM.	3.3	5.2	5.7	6.2	6.8	7.3	7.5	7.6	7.9	8.4	8.6	9.5	10.1	10.8	11.8	12.8	13.7
90	Power	0.2	0.3	0.3	0.4	0.4	0.5	0.5	0.5	0.5	0.6	0.6	0.6	0.6	0.9	0.9	0.9	1.0
91	Fuels	0.2	0.3	0.3	0.4	0.4	0.4	0.4	0.5	0.5	0.6	0.6	0.7	0.7	0.9	0.9	1.0	1.0
92	Metallurgy	0.4	0.6	0.7	0.8	0.8	0.9	1.0	1.1	1.1	1.2	1.2	1.2	1.2	1.5	1.5	1.5	1.6
93	Chemicals	0.5	0.6	0.7	0.7	0.8	0.9	0.9	0.9	1.0	1.0	1.0	1.1	1.2	1.2	1.3	1.4	1.5
94	MBMW	1.2	2.2	2.3	2.5	2.9	3.0	3.1	3.0	3.0	3.3	3.5	4.1	4.5	4.2	5.0	5.6	6.0
95	Wood and Paper	0.2	0.3	0.4	0.4	0.4	0.4	0.4	0.4	0.4	0.4	0.4	0.5	0.5	0.5	0.5	0.6	0.7
96	Constr. Mater.	0.2	0.3	0.3	0.3	0.4	0.4	0.4	0.4	0.4	0.4	0.4	0.4	0.4	0.6	0.6	0.6	0.7
97	N.E.C. Ind.	0.2	0.3	0.3	0.4	0.4	0.4	0.5	0.5	0.5	0.5	0.5	0.5	0.6	0.6	0.6	0.7	0.7
98	Light Industry	0.1	0.2	0.2	0.2	0.2	0.2	0.2	0.2	0.3	0.3	0.3	0.3	0.3	0.3	0.4	0.4	0.4
99	Other Prod.	0.1	0.1	0.1	0.1	0.1	0.1	0.1	0.1	0.1	0.1	0.1	0.1	0.1	0.1	0.1	0.1	0.1

Supporting Data on Price Indexes:

		1965	1970	1971	1972	1973	1974	1975	1976	1977	1978	1979	1980	1981	1982	1983	1984	1985
100	Power	0.76	1.01	1.00	1.00	1.00	1.00	1.00	0.97	0.97	0.97	0.97	0.97	0.97	1.20	1.20	1.20	1.20
101	Fuels	0.77	0.97	0.98	0.99	1.00	1.02	1.03	1.05	1.06	1.14	1.17	1.19	1.20	1.59	1.61	1.60	1.63
102	Metallurgy	0.69	0.97	0.98	0.99	1.00	1.01	1.02	1.06	1.08	1.09	1.11	1.13	1.14	1.39	1.40	1.42	1.42
103	Chemicals	0.90	1.01	1.00	1.00	1.00	0.99	1.00	1.02	1.05	1.07	1.12	1.12	1.14	1.17	1.18	1.20	1.23
104	MBMW	0.88	0.98	0.98	0.99	1.00	1.03	1.07	1.08	1.11	1.15	1.17	1.20	1.24	1.27	1.31	1.34	1.38
105	Wood and Paper	0.66	0.91	0.93	0.96	1.00	1.04	1.08	1.13	1.18	1.23	1.28	1.32	1.36	1.58	1.63	1.65	1.70
106	Constr. Mater.	0.85	0.98	0.96	0.98	1.00	1.04	1.06	1.07	1.09	1.10	1.15	1.18	1.19	1.39	1.42	1.43	1.46
107	N.E.C. Ind.	0.88	1.01	1.00	1.00	1.00	1.01	1.01	1.02	1.02	1.04	1.09	1.12	1.15	1.22	1.25	1.28	1.33
108	Light Industry	0.72	0.87	0.89	0.92	1.00	1.05	1.12	1.15	1.19	1.23	1.27	1.31	1.34	1.47	1.50	1.52	1.55
109	Food Industry	0.87	0.95	0.96	0.99	1.00	1.01	1.03	1.04	1.06	1.09	1.11	1.14	1.17	1.21	1.21	1.23	1.25
110	Agriculture	0.83	0.95	0.98	1.04	1.00	1.03	1.10	1.11	1.16	1.15	1.25	1.28	1.36	1.36	1.55	1.62	1.65
111	Other Prod.	0.87	0.92	0.98	0.99	1.00	1.02	1.06	1.06	1.08	1.10	1.14	1.18	1.21	1.29	1.30	1.31	1.36
112	Construction	0.63	0.91	0.95	0.98	1.00	1.01	1.03	1.06	1.07	1.09	1.11	1.13	1.16	1.22	1.25	1.46	1.52
113	Machinery Invest	0.82	0.94	0.96	0.98	1.00	1.05	1.09	1.11	1.16	1.21	1.24	1.28	1.32	1.37	1.43	1.46	1.53

TABLE 0.18: TOTAL USSR DEFENSE BUDGET IN CURRENT AND CONSTANT 1973 PRICES
(billion rubles)

	1965	1970	1971	1972	1973	1974	1975	1976	1977	1978	1979	1980	1981	1982	1983	1984	1985
Current Prices:																	
1 TOTAL OUTLAYS	36.7	51.2	55.7	57.3	59.8	63.7	68.4	72.7	73.6	80.2	83.4	88.2	93.4	104.2	106.6	113.8	120.1
2 Procurement	16.4	22.9	24.3	25.5	26.8	28.2	30.5	32.3	33.8	35.5	37.3	38.9	41.2	45.5	49.2	53.7	58.2
3 O & M	10.7	15.3	16.5	18.0	17.7	19.5	21.1	23.1	23.2	26.4	28.1	29.4	31.9	34.5	33.4	35.9	36.1
4 Military RDT&E	3.8	6.1	6.4	6.9	7.3	7.5	8.1	8.3	8.7	9.2	9.7	10.9	11.4	12.1	12.8	13.2	13.5
5 Military Space	1.1	1.5	1.5	1.6	1.7	1.8	1.9	2.0	2.1	2.2	2.3	2.5	2.5	2.9	3.1	3.3	3.6
6 Defense Constr.	3.7	4.2	5.6	4.0	4.8	5.2	5.2	5.4	4.1	5.2	4.4	4.7	4.7	7.5	6.2	5.8	6.6
7 Pensions	1.0	1.3	1.4	1.4	1.5	1.5	1.6	1.6	1.7	1.7	1.7	1.8	1.8	1.8	1.9	1.9	2.0
8 WEAPONS & SPACE	15.4	20.8	21.8	23.1	24.4	25.6	27.3	28.7	30.0	31.2	32.2	34.0	36.0	40.9	44.1	48.9	53.2
9 Materials	9.0	11.7	12.2	13.0	13.6	14.1	15.0	15.8	16.5	17.1	17.5	18.7	20.1	24.0	26.7	30.9	34.6
10 Total Wages	5.4	7.6	8.0	8.4	8.8	9.3	9.8	10.2	10.6	11.1	11.6	12.0	12.4	12.9	13.2	13.7	14.1
11 Social Security	0.5	0.7	0.7	0.8	0.8	0.8	0.9	0.9	1.0	1.0	1.0	1.1	1.1	1.5	1.6	1.6	1.7
12 Depreciation	0.5	0.8	0.9	1.0	1.2	1.4	1.6	1.8	1.9	2.0	2.1	2.3	2.4	2.5	2.6	2.7	2.8
13 ADMINISTRATION	7.9	11.2	11.4	12.3	12.9	14.3	15.2	15.7	16.1	18.2	18.7	22.2	22.8	23.5	24.2	25.1	26.5
14 Materials	2.0	3.1	3.1	3.4	3.5	4.5	5.0	5.1	5.2	6.7	6.7	9.7	9.8	9.4	9.7	10.1	11.0
15 Food & Uniforms	2.3	3.1	3.2	3.5	3.6	3.8	4.0	4.2	4.3	4.5	4.7	5.1	5.3	5.9	6.0	6.2	6.3
16 Total Wages	3.1	4.2	4.4	4.6	4.9	5.1	5.3	5.5	5.7	5.9	6.2	6.4	6.7	6.9	7.3	7.4	7.6
17 Social Security	0.2	0.3	0.3	0.3	0.3	0.3	0.3	0.3	0.3	0.4	0.4	0.4	0.4	0.5	0.5	0.5	0.5
18 Depreciation	0.3	0.5	0.5	0.5	0.5	0.6	0.6	0.6	0.6	0.7	0.7	0.7	0.7	0.8	0.8	0.9	1.0
19 SERVICES	6.2	9.7	10.2	10.7	11.5	12.3	12.9	13.3	13.9	14.7	15.7	17.0	17.7	18.9	19.9	20.6	21.5
20 Transp. & Comm.	1.2	2.1	2.3	2.4	2.6	2.9	2.9	3.2	3.3	3.5	3.8	4.0	4.1	4.5	4.8	5.0	5.3
21 Housing, etc.	0.2	0.3	0.3	0.3	0.4	0.4	0.5	0.5	0.5	0.6	0.6	0.7	0.7	0.7	0.8	0.8	0.9
22 Education	0.8	1.1	1.1	1.2	1.3	1.4	1.5	1.5	1.6	1.7	1.7	1.8	1.9	2.0	2.0	2.1	2.2
23 Health	0.6	0.8	0.8	0.9	1.0	1.0	1.1	1.1	1.2	1.3	1.3	1.4	1.5	1.6	1.6	1.6	1.7
24 RDT & E	3.3	5.3	5.5	5.7	6.1	6.4	6.7	6.8	7.0	7.5	7.9	8.8	9.2	9.9	10.3	10.7	11.2
25 Central Admin.	0.1	0.2	0.2	0.2	0.2	0.2	0.3	0.3	0.3	0.3	0.3	0.3	0.3	0.4	0.4	0.4	0.4
26 INVESTMENT	2.4	4.0	5.2	5.8	4.7	4.9	6.2	8.0	7.8	9.2	10.7	8.4	10.4	11.7	10.3	11.5	10.2
27 Dual Use M & E	0.4	0.7	0.9	1.0	1.0	1.0	1.3	1.2	1.4	1.4	1.7	1.9	1.9	1.5	1.6	1.6	1.7
28 Cap. Investment	2.0	3.4	3.7	3.8	3.8	4.0	4.7	5.3	5.7	6.2	6.6	6.8	7.1	7.3	8.0	8.0	8.2
29 Weapons Prod.	1.1	1.9	2.0	1.8	1.7	2.0	2.4	2.8	3.0	3.4	3.7	3.5	3.7	3.8	4.1	3.9	4.3
30 Military R&D	0.5	0.8	1.0	1.2	1.2	1.2	1.4	1.6	1.7	1.7	1.8	2.1	2.2	2.3	2.5	2.5	2.3
31 Housing	0.5	0.8	0.8	0.8	0.9	0.9	1.0	1.0	1.1	1.1	1.1	1.2	1.2	1.3	1.4	1.6	1.7
32 Inventories	0.0	0.3	0.2	0.5	0.2	0.5	0.2	0.4	0.7	0.7	2.5	2.7	-0.2	2.9	0.5	0.5	1.1
33 Mater. Supplies	0.0	-0.3	0.4	0.5	-0.4	-0.7	0.0	1.0	0.1	0.9	-0.1	-2.9	1.6	0.0	0.2	1.4	-0.8
Constant Prices:																	
34 TOTAL OUTLAYS	44.1	53.9	57.8	58.4	59.8	61.4	64.6	67.6	67.2	70.1	71.0	73.7	76.1	81.2	81.3	83.4	85.4
35 Procurement	19.1	23.7	25.0	25.9	26.8	27.4	28.6	29.8	30.5	30.8	31.6	32.1	33.0	35.3	37.1	39.2	41.3
36 O & M	13.2	16.4	17.4	18.5	17.7	18.7	20.0	21.4	21.1	22.9	23.8	24.2	25.7	26.5	25.2	26.2	25.6
37 Military RDT&E	4.9	6.2	6.6	6.9	7.3	7.1	7.7	7.9	8.2	8.3	8.5	9.6	9.9	9.8	10.2	9.8	9.6
38 Military Space	1.4	1.5	1.6	1.6	1.7	1.7	1.8	1.9	2.0	2.0	2.0	2.2	2.2	2.3	2.5	2.5	2.6
39 Defense Constr.	4.5	4.7	5.8	4.1	4.8	5.1	5.0	4.9	3.7	4.4	3.5	3.7	3.6	5.4	4.5	3.8	4.3
40 GENERAL INDEX	0.83	0.95	0.96	0.98	1.00	1.04	1.06	1.08	1.10	1.14	1.17	1.20	1.23	1.28	1.31	1.37	1.41
41 MBMW	0.88	0.98	0.98	0.99	1.00	1.03	1.07	1.08	1.11	1.15	1.17	1.20	1.24	1.27	1.31	1.34	1.38
42 RDT & E	0.77	0.97	0.97	0.99	1.00	1.06	1.05	1.05	1.05	1.10	1.14	1.13	1.15	1.23	1.25	1.35	1.41
43 Services	0.81	0.93	0.95	0.97	1.00	1.04	1.06	1.08	1.10	1.15	1.18	1.21	1.24	1.30	1.33	1.37	1.41
44 Construction	0.75	0.89	0.92	0.97	1.00	1.02	1.05	1.09	1.11	1.19	1.27	1.29	1.34	1.42	1.45	1.63	1.67
45 Machinery	0.82	0.94	0.96	0.98	1.00	1.05	1.09	1.11	1.16	1.21	1.24	1.28	1.32	1.37	1.43	1.46	1.53

TABLE 0.19: THE BALANCE OF IMPORTS AND EXPORTS IN 1973 DOMESTIC PRICES BY SECTOR
(billion rubles)

	1965	1970	1971	1972	1973	1974	1975	1976	1977	1978	1979	1980	1981	1982	1983	1984	1985
1 TOTAL REVENUES	10.2	12.5	12.5	16.2	14.9	11.8	18.1	18.2	16.4	17.9	20.9	26.5	31.2	34.3	34.5	34.4	37.3
2 Imports	20.5	29.9	30.5	34.1	34.6	34.7	40.4	41.7	41.8	44.0	47.6	53.9	59.9	63.8	64.2	65.9	67.9
3 Exports	10.2	17.4	18.0	18.0	19.7	22.9	22.3	23.5	25.4	26.1	26.7	27.4	28.8	29.5	29.7	31.5	30.6
4 POWER	0.0	0.1	0.1	0.1	0.1	0.1	0.1	0.2	0.2	0.2	0.2	0.2	0.3	0.4	0.5	0.6	0.6
5 FUELS	-1.5	-2.1	-2.0	-1.9	-2.0	-2.3	-2.6	-3.0	-3.5	-3.6	-3.9	-4.1	-4.1	-4.2	-4.3	-4.4	-4.2
6 Imports	0.2	0.3	0.5	0.8	0.9	0.7	0.8	0.8	0.8	0.8	0.6	0.4	0.4	0.6	0.7	0.8	0.7
7 Exports	1.7	2.4	2.5	2.7	2.9	3.0	3.4	3.8	4.2	4.4	4.4	4.5	4.5	4.8	5.0	5.2	5.0
8 METALS	-0.8	-2.0	-2.1	-2.0	-2.0	-1.5	-1.2	-1.5	-1.4	-1.0	-0.4	-0.3	-0.3	-0.1	-0.6	-0.7	-0.7
9 Imports	0.8	1.2	1.2	1.3	1.4	2.0	2.2	2.0	1.8	2.1	2.7	3.0	3.1	3.3	3.0	2.8	2.8
10 Exports	1.6	3.2	3.3	3.3	3.4	3.5	3.4	3.5	3.3	3.1	3.1	3.3	3.4	3.4	3.5	3.5	3.5
11 CHEMICALS	1.9	2.4	2.5	2.4	2.1	2.5	2.6	2.5	2.6	2.6	3.2	4.0	4.0	3.6	3.7	3.6	4.2
12 Imports	2.2	3.3	3.5	3.4	3.1	3.8	4.0	3.9	4.0	4.1	4.8	5.9	6.3	5.8	6.0	6.3	6.9
13 Exports	0.3	1.0	1.0	1.0	1.1	1.3	1.3	1.4	1.4	1.5	1.6	1.9	2.3	2.2	2.3	2.7	2.7
14 M & E	3.4	2.2	1.6	2.3	0.5	0.0	2.2	2.2	1.2	2.5	1.9	2.5	2.8	4.2	6.2	6.2	7.4
15 Imports	5.7	6.2	5.8	6.8	6.4	6.8	8.9	10.0	10.3	12.4	12.1	12.9	13.7	16.0	18.0	18.6	19.3
16 Exports	2.3	4.0	4.3	4.5	5.9	6.8	6.7	7.8	9.1	9.9	10.2	10.5	10.9	11.8	11.8	12.4	11.9
17 WOOD f.p.	0.0	-0.2	0.0	0.2	0.0	-0.3	0.4	0.2	0.1	0.3	0.3	0.8	1.1	0.6	0.5	0.4	0.6
18 Imports	0.8	1.3	1.4	1.5	1.4	1.3	1.9	1.7	1.6	1.6	1.8	2.4	2.5	2.1	1.9	1.9	2.0
19 Exports	0.8	1.5	1.4	1.3	1.4	1.6	1.5	1.5	1.5	1.3	1.4	1.5	1.4	1.5	1.4	1.4	1.4
20 CONSTR. MATER.	0.0	0.3	0.2	0.2	0.3	0.3	0.2	0.3	0.3	0.3	0.3	0.3	0.3	0.4	0.4	0.5	0.4
21 Imports	0.1	0.4	0.4	0.3	0.4	0.4	0.3	0.4	0.4	0.4	0.4	0.4	0.4	0.5	0.5	0.5	0.5
22 Exports	0.1	0.1	0.1	0.1	0.1	0.1	0.1	0.1	0.1	0.1	0.1	0.1	0.1	0.1	0.1	0.1	0.1
23 N.E.C. INDUSTRY	0.2	0.5	0.5	0.6	0.4	0.4	0.5	0.5	0.5	0.4	0.5	0.5	0.9	1.1	1.2	1.3	1.3
24 Imports	0.3	0.7	0.8	0.9	0.8	0.8	1.0	1.0	1.0	1.0	1.0	1.1	1.4	1.5	1.7	1.8	1.8
25 Exports	0.1	0.2	0.3	0.3	0.4	0.4	0.4	0.5	0.5	0.6	0.6	0.6	0.5	0.5	0.5	0.5	0.5
26 LIGHT INDUSTRY	4.5	8.7	9.1	9.5	9.3	9.5	9.6	9.2	8.9	8.6	8.6	9.9	11.0	12.2	12.5	12.6	14.3
27 Imports	5.7	10.5	11.0	11.6	11.3	11.5	11.8	11.6	11.3	11.1	11.0	12.3	13.6	14.5	14.7	14.9	16.5
28 Exports	1.3	1.8	1.9	2.1	1.9	2.0	2.1	2.4	2.4	2.5	2.4	2.5	2.6	2.3	2.2	2.3	2.1
29 FOOD INDUSTRY	2.1	2.4	2.1	2.6	3.4	2.4	4.2	4.8	6.0	4.8	6.8	9.3	10.4	11.6	10.5	10.0	8.9
30 Imports	3.6	4.6	4.2	4.3	5.1	5.1	6.3	6.5	7.7	6.6	8.7	10.7	12.2	13.3	12.4	12.1	10.9
31 Exports	1.5	2.2	2.1	1.7	1.7	2.7	2.1	1.7	1.7	1.8	1.8	1.5	1.7	1.8	1.9	2.1	2.0
32 AGRICULTURE	0.6	0.4	0.7	2.4	2.9	0.8	2.3	3.3	2.2	3.4	3.8	4.1	5.7	5.5	5.0	5.6	5.9
33 Imports	0.9	1.3	1.6	3.1	3.8	2.1	3.2	3.8	2.8	3.7	4.4	4.6	6.2	5.9	5.3	5.9	6.2
34 Exports	0.4	0.8	0.9	0.7	0.8	1.2	0.8	0.4	0.6	0.3	0.6	0.4	0.5	0.4	0.3	0.3	0.3
35 OTHER PROD.	0.0	0.0	0.0	0.0	-0.1	-0.1	-0.1	-0.1	-0.1	-0.1	-0.1	-0.2	-0.2	-0.1	-0.1	-0.1	-0.1
36 Imports	0.1	0.1	0.1	0.1	0.1	0.1	0.1	0.1	0.2	0.2	0.2	0.2	0.2	0.2	0.2	0.2	0.3
37 Exports	0.1	0.1	0.1	0.1	0.2	0.2	0.2	0.2	0.3	0.3	0.3	0.4	0.4	0.3	0.3	0.4	0.4
38 Import Deflator	0.73	0.89	0.92	0.95	1.00	1.06	1.15	1.19	1.23	1.28	1.30	1.35	1.38	1.42	1.47	1.50	1.56
39 Export Deflator	0.81	0.96	0.96	0.98	1.00	1.02	1.03	1.05	1.07	1.09	1.12	1.14	1.17	1.34	1.37	1.39	1.40

TABLE 0.20: NET EXPORTS IN DOMESTIC AND FOREIGN PRICES

	1965	1970	1971	1972	1973	1974	1975	1976	1977	1978	1979	1980	1981	1982	1983	1984	1985
Net Exports (Balance of Payments) in Billions of Domestic Rubles:																	
1 NET EXPORTS	1.3	2.0	3.6	1.2	3.1	5.2	-3.3	0.7	4.7	4.1	5.2	4.9	4.7	5.3	5.9	6.5	3.6
2 in 1973 prices	2.0	2.5	4.1	1.3	3.1	4.5	-2.5	0.5	3.2	2.6	3.4	3.1	3.0	3.5	3.8	4.1	2.1
3 Trade Balance	0.2	1.3	1.7	-1.4	0.4	2.2	-4.8	-1.2	2.6	0.9	3.2	3.1	2.6	4.2	5.0	5.4	2.0
4 Monetary Income	1.0	0.6	1.9	2.5	2.7	3.0	1.5	1.9	2.1	3.3	2.0	1.7	2.0	1.1	1.0	1.1	1.6
Net Exports (Balance of Payments) in Billions of Gold Rubles:																	
5 NET EXPORTS	1.1	1.4	2.6	1.4	2.5	4.6	-1.2	1.5	5.8	5.2	7.4	7.8	7.9	8.5	9.9	11.0	6.1
6 Trade Balance	0.2	0.9	1.2	-0.6	0.3	2.0	-2.7	-0.7	3.2	1.1	4.5	5.0	4.5	6.8	8.3	9.1	3.4
7 Monetary Income	0.9	0.4	1.3	1.9	2.2	2.6	1.5	2.3	2.6	4.1	2.9	2.8	3.4	1.7	1.6	1.9	2.7
Monetary Income in Billions of U.S. Dollars:																	
8 MONETARY INCOME	0.9	0.4	1.2	1.6	1.6	2.0	1.1	1.7	1.9	2.8	1.9	1.8	2.2	1.1	1.0	1.2	1.7
9 Gold Sales	0.6	0.0	0.8	1.1	0.9	1.2	0.7	1.4	1.6	2.7	1.9	1.6	2.7	1.1	0.8	1.0	1.8
10 in metric tons	60	0.0	60	200	280	230	140	340	340	n.a.	n.a.	n.a.	n.a.	n.a.	n.a.	n.a.	n.a.
11 Net Interest*	0.0	-0.1	-0.1	-0.1	-0.1	-0.1	-0.6	-0.7	-0.8	-1.0	-1.1	-1.2	-1.8	-1.3	-1.1	-1.2	-1.5
12 Invisibles**	0.3	0.5	0.5	0.6	0.8	0.9	1.0	1.0	1.1	1.1	1.1	1.4	1.3	1.3	1.3	1.4	1.4
13 U.S.Dollar/G.R.	0.98	0.92	0.90	0.82	0.74	0.76	0.72	0.75	0.74	0.68	0.65	0.65	0.64	0.64	0.63	0.63	0.62
14 D.R./G.R. Ratio	1.12	1.45	1.40	2.45	1.25	1.12	1.75	1.73	0.81	0.79	0.70	0.63	0.59	0.62	0.60	0.59	0.59

* excluding intra-CEMA transactions
**including transactions with CEMA countries

TABLE 0.21: SECOND ECONOMY BY SECTOR
        (billion rubles)

	1965	1970	1971	1972	1973	1974	1975	1976	1977	1978	1979	1980	1981	1982	1983	1984	1985
1 TOTAL*	5.9	8.8	9.2	10.1	10.6	11.4	12.1	12.9	14.1	14.9	16.2	17.6	18.9	20.6	21.9	23.1	24.6
2 REPAIRS	0.7	1.2	1.2	1.3	1.4	1.5	1.7	1.9	2.1	2.2	2.3	2.5	2.7	2.8	3.1	3.2	3.5
3 Footware**	0.2	0.3	0.3	0.4	0.4	0.4	0.4	0.5	0.5	0.6	0.6	0.7	0.7	0.7	0.8	0.8	0.9
4 Apparel**	0.3	0.4	0.4	0.4	0.5	0.5	0.6	0.6	0.7	0.7	0.7	0.7	0.8	0.8	0.9	1.0	1.1
5 Auto Service	0.0	0.1	0.1	0.1	0.1	0.2	0.2	0.2	0.3	0.3	0.3	0.4	0.4	0.5	0.5	0.6	0.6
6 Electronics	0.0	0.1	0.1	0.1	0.1	0.1	0.1	0.2	0.2	0.2	0.3	0.3	0.3	0.4	0.4	0.4	0.5
7 Flats	0.2	0.3	0.3	0.3	0.3	0.3	0.4	0.4	0.4	0.4	0.4	0.4	0.4	0.4	0.5	0.5	0.5
8 RURAL HOUSING	0.8	1.1	1.2	1.2	1.2	1.3	1.3	1.3	1.4	1.4	1.5	1.6	1.7	1.7	1.8	1.8	1.8
9 Registered	0.7	1.0	1.1	1.1	1.1	1.1	1.2	1.2	1.2	1.2	1.3	1.4	1.4	1.4	1.5	1.5	1.5
10 Unregistered	0.1	0.1	0.1	0.1	0.1	0.1	0.1	0.1	0.2	0.2	0.2	0.2	0.2	0.3	0.3	0.3	0.3
11 SERVICES	3.3	4.7	5.0	5.4	5.8	6.0	6.5	6.8	7.3	7.6	8.1	8.6	9.1	9.6	9.9	10.2	10.6
12 Transportation	0.2	0.3	0.3	0.4	0.4	0.4	0.5	0.5	0.6	0.6	0.7	0.8	0.9	1.0	1.1	1.2	1.3
13 Housing (Rent)	2.1	2.7	2.8	3.0	3.2	3.3	3.4	3.5	3.5	3.6	3.6	3.6	3.6	3.6	3.5	3.5	3.5
14 Registered	0.7	0.9	0.9	1.0	1.1	1.1	1.2	1.2	1.2	1.2	1.2	1.1	1.1	1.1	1.0	1.0	1.0
15 Unregistered	1.4	1.8	1.9	2.0	2.1	2.2	2.2	2.3	2.3	2.4	2.4	2.5	2.5	2.5	2.5	2.5	2.5
16 Communal	0.2	0.4	0.4	0.4	0.5	0.5	0.6	0.7	0.8	0.9	1.0	1.1	1.2	1.3	1.4	1.4	1.5
17 Education	0.4	0.6	0.6	0.7	0.7	0.7	0.8	0.8	0.9	0.9	1.0	1.1	1.2	1.3	1.3	1.4	1.5
18 Culture	0.0	0.1	0.1	0.1	0.1	0.1	0.1	0.1	0.2	0.2	0.2	0.2	0.3	0.3	0.3	0.3	0.3
19 Health	0.4	0.6	0.7	0.8	0.9	1.0	1.1	1.2	1.3	1.4	1.6	1.8	1.9	2.1	2.3	2.4	2.5
20 THEFTS	0.7	1.1	1.2	1.4	1.4	1.7	1.7	1.9	2.4	2.7	3.1	3.6	4.2	5.1	5.9	6.6	7.4
21 Fuels	0.1	0.2	0.2	0.3	0.3	0.4	0.4	0.5	0.7	0.8	1.0	1.2	1.4	1.8	2.2	2.5	2.9
22 Utilities	0.1	0.1	0.1	0.2	0.2	0.2	0.2	0.2	0.3	0.3	0.3	0.4	0.4	0.4	0.4	0.4	0.4
23 Gasoline	0.0	0.1	0.1	0.1	0.1	0.2	0.2	0.3	0.4	0.5	0.7	0.8	1.0	1.4	1.8	2.1	2.5
24 Constr. Mater.	0.3	0.4	0.5	0.5	0.5	0.5	0.5	0.5	0.6	0.6	0.6	0.6	0.7	0.7	0.7	0.7	0.7
25 Other	0.2	0.3	0.3	0.3	0.3	0.4	0.4	0.4	0.4	0.5	0.5	0.6	0.7	0.8	0.8	0.9	0.9
26 TRADE (SCALPING)	0.4	0.7	0.7	0.8	0.8	0.9	0.9	1.0	1.0	1.1	1.2	1.3	1.3	1.4	1.3	1.3	1.2
27 NET TOTAL***	4.9	7.4	7.7	8.5	9.0	9.8	10.4	11.2	12.4	13.2	14.4	15.6	16.8	18.5	19.7	20.9	22.3
28 in 1973 prices	7.0	8.2	8.3	8.8	9.0	9.2	9.4	9.6	9.9	10.1	10.3	10.6	10.8	11.1	11.3	11.5	11.7
29 deflator	0.70	0.90	0.92	0.97	1.00	1.06	1.11	1.16	1.25	1.30	1.40	1.48	1.56	1.67	1.74	1.82	1.90

* excluding production of moonshine
**including illicit manufacturing
***total included in GNP (total less registered new rural housing and stolen construction materials)

UN Classification of Soviet Economic Sectors

Code number	Branches, sub-branches and groups	Scope of branches, sub-branches and groups
1	2	3
01	INDUSTRY	
01.01	Electrical and thermal power industry	Generation, transmission and distribution of electrical and thermal power
01.02	Fuel industry	
01.02.01	Coal industry	Coal mining. Cleaning and briquetting of coal
01.02.02	Coke and coke products industry	Manufacture of coke, semi-coke, tar, coke-oven gas and other coke products
01.02.03	Oil extracting industry	Extraction of oil, gasoline, mineral wax and associated petroleum gas
01.02.04	Gas extracting industry	Extraction and distribution of natural gas (excluding associated gas)
01.02.05	Artificial gas industry	Production and distribution of artificial gas
01.02.06	Extraction of other combustible materials	Extraction of peat and manufacture of peat briquettes

Extraction and briquetting of bituminous (oil) shale |
01.03	Ferrous metallurgy (including ore-mining)	
01.03.01	Mining and processing of ferrous metal ores	Mining, processing and sintering of iron, manganese and chromite ores
01.03.02	Ferrous metallurgy	Production of iron, blast-furnace ferro-alloys, goods made from ferrous metal powders, steel, rolled ferrous metals

Note: The repair and renovation of industrial goods (including the repair of industrial consumer goods) and the manufacture of spare parts, units and components (other than multi-purpose ones for general mechanical engineering) are classified under the appropriate branches of industry.

Code number	Branches, sub-branches and groups	Scope of branches, sub-branches and groups
1	2	3
		Production of steel and iron pipes
		Production of electric ferro-alloys
		Production of sintered metal-ceramic alloys and mixtures
		Production of rolled metal goods, including production of steel electrodes
		Production of forged, pressed and stamped goods from ferrous metals and castings, recovery of ferrous metal scrap
01.04	Non-ferrous metallurgy (including ore-mining)	
01.04.01	Mining and processing of non-ferrous metal ores	Mining and processing of non-ferrous, rare and precious metal ores
01.04.02	Non-ferrous metallurgy	Production of non-ferrous, rare and precious metals and their alloys; rolled stock, semi-manufactures, pipes and rolled products; powders of non-ferrous, rare and precious metals and their alloys (including hard alloys), sintered metal-ceramic alloys and mixtures; recovery of non-ferrous metal scrap
		Production of forged, pressed and stamped goods made from non-ferrous metals and castings

Code number	Branches, sub-branches and groups	Scope of branches, sub-branches and groups
1	2	3
	Mechanical engineering and metal-working industry (01.05, 01.06, 01.07)	
01.05	Mechanical engineering industry a/	
01.05.01	Industry of machines, processing equipment and installations for industry	Production of steam-boilers, turbines, internal combustion engines, steam power engines, nuclear reactors (excluding production of engines for transport vehicles)
		Production of machine tools for metal working using cutting and other metal-working techniques
		Production of machines and equipment for processing metals by hot or cold plastic deformation
		Production of machines, equipment and installations for drilling and operation of oil or gas wells
		Production of machines, equipment and installations for extracting and processing coal, peat and ores
		Production of machines, equipment and installations for producing coke and semi-coke
		Production of machines, processing equipment and installations for ferrous and non-ferrous metallurgy, including foundry equipment
		Production of machines, equipment and installations for welding and metal-coating of products

---

a/ Excluding the electrical and electronic engineering industry (01.06) and the industry of metal structures and other metal products (01.07).

Code number	Branches, sub-branches and groups	Scope of branches, sub-branches and groups
1	2	3
		Production of equipment for manufacturing of electric bulbs
		Production of machines, processing equipment and installations for chemical processes and oil refining
		Production of machines, processing equipment and installations for the preparation of rubber and plastics
		Production of machines and processing equipment for manufacturing pulp, paper and cardboard
		Production of machines and processing equipment for manufacturing building materials and refractories
		Production of machines and processing equipment for logging and woodworking
		Production of machines and processing equipment for manufacturing glassware and ceramics
		Production of machines and processing equipment for the textile, knitwear, wearing apparel, leather, fur and footwear industries
		Production of machines and processing equipment for the food industry
		Production of processing equipment for the mixed-feed industry
		Production of machines, processing equipment and installations for the printing and publishing industry

Code number	Branches, sub-branches and groups	Scope of branches, sub-branches and groups
1	2	3
		Production of goods for general mechanical engineering (lubricating equipment, reducer gears, speed regulators, filtration units, hydraulic and pneumatic drives, hydraulic and pneumatic control devices, etc.)
01.05.02	Precision engineering industry	Production of non-electrical equipment, devices and instruments (control and measuring devices, optical and medical apparatus and instruments, apparatus for scientific research and laboratories, office supplies and calculating machines)
01.05.03	Bearings industry	Production of roller and plain bearings
01.05.04	Industry of machines and equipment for construction and road work	Production of machines and equipment for earth-moving and bed (foundation) laying, rock excavation and tunnel building, manufacture of concrete, mortar and lime, construction and repair of roads and railway lines, assembly of conductors and power-transmission lines, insulation and trimming work, etc.
01.05.05	Industry of tractors, machines and equipment for agriculture and forestry	Production of tractors and self-propelled tractor frames Production of machines and equipment for soil cultivation, sowing, planting, harvesting and threshing and for cleaning, sorting, processing and drying of agricultural products, land-improvement work, livestock raising, vegetable and flower growing, viticulture, forestry (excluding manual farming tools - hoes, scythes, sickles, rakes etc., classified under 01.07, "Industry of metal structures and other metal products")

Code number	Branches, sub-branches and groups	Scope of branches, sub-branches and groups
1	2	3
01.05.06	Industry of means of transport	Production of motor vehicles (passenger cars, buses, trucks, tractor units, prime movers, special-purpose motor vehicles, trailers and semi-trailers, apparatus for converting semi-trailers into trailers), trolley-buses, motor cycles, motor scooters and bicycles
		Production of railway rolling stock (locomotives, railcars, goods and passenger cars, sleeping cars, mail cars, restaurant cars, etc.) and trams
		Building of ships and boats for sea, river and lake navigation
		Production of aircraft (airplanes, helicopters, etc.)
01.05.07	Industry of other machines and equipment	Production of vending machines for beverages, food products, non-food goods, etc.
		Production of machinery for public utilities
		Production of hoisting and hauling equipment, cranes of all types, pumps, compressors, ventilation and air-conditioning equipment and installations, refrigeration equipment and installations (other than electrical household goods), general-purpose heat exchangers, vessels and tanks (reservoirs), metal bins and other general-purpose metal containers, machines and processing equipment for the manufacture of containers and packaging, bench equipment, accessories and installations for processing equipment
		Production of laundering and dry-cleaning equipment

Code number	Branches, sub-branches and groups	Scope of branches, sub-branches and groups
1	2	3
01.06	Electrical and electronic engineering	
01.06.01	Electrical engineering industry	Production of electrical and thermal-electric machines, equipment, apparatus and tools for industrial use
		Production of electrical machines and equipment for household use
		Production of electrical-engineering materials
		Production of electrical welding equipment
		Cable production and production of accumulators
		Production of electrical control and measuring devices, electrical laboratory apparatus and instruments, automation equipment, and calculating machines (other than electronic)
		Production of electrical drives for valves
01.06.02	Electronic engineering industry	Production of equipment and components for radio engineering, television and communications, electronic control and measuring devices, electronic instruments, electronic laboratory devices, electronic automation equipment and computer hardware
		Production of vacuum-manufactured goods (including electric bulbs, discharge lamps) and other electronic goods
		Production of robots and manipulators
		Production of electric metal-ceramic goods

Code number	Branches, sub-branches and groups	Scope of branches, sub-branches and groups
1	2	3
01.07	<u>Industry of metal structures and other metal products</u>	Production of metal structures, metallic consumer goods, metal furniture, metal containers, metal goods, plumbing fixtures and other metal products, including metal hand tools (hoes, scythes, sickles, rakes, etc.) Production of metal structures for mine-rescue equipment Production of industrial and pipeline mountings, goods made from wire (nails, netting)
01.08	<u>Chemical and oil refining industry (including rubber-asbestos industry)</u>	
01.08.01	Mining of mineral raw materials for the chemical industry	Mining and processing of chemical raw materials: potassium and rock salts (including evaporation), sulphur (including sulphur-bearing raw materials - iron pyrites), phosphorites and appetites, boron-bearing raw materials, etc.
01.08.02	Oil refining industry (including processing of gas)	Refining of oil and gasoline (production of benzine, kerosene, diesel fuel, fuel oil, lubricating oils, greases and other products of oil refining, including raw materials for petroleum chemistry) Processing of natural and associated gas Production of synthetic liquid fuel
01.08.03	Basic organic and inorganic chemical industry	Manufacture of basic organic chemical products (hydrocarbons, halogen, hydroxyl and hydrocarbon compounds, hydrocarbon acids, organic anhydrides and compound ethers, organic compounds of nitrogen and sulphur, organic compounds with multiple molecular functions, heterocyclic compounds and

Code number	Branches, sub-branches and groups	Scope of branches, sub-branches and groups
1	2	3
		organic salts, paint thinners and removers, synthetic alcohols, crude carbon-furnace soot, tanning extracts)
		Manufacture of electrocarbon goods
		Manufacture of basic inorganic chemical products (non-metallic chemical elements, oxides of metals and non-metals, inorganic acids and per acids, hydroxides and other inorganic non-metal compounds, other inorganic chemical products)
		Manufacture of basic inorganic salts (salts of halogenic hydroxy acids, sulphides and phosphides, salts of oxy-acids containing sulphur and phosphorus, acid salts of carbon, nitrogen, chromium, manganese, boron, arsenic, titanium, vanadium, molybdenum, tungsten, silicon, aluminum and other chemical elements, and complex salts)
01.08.04	Industry of chemical fertilizers and crop protection agents	Manufacture of simple fertilizers (phosphatic, nitrogeneous, potassic) and mixed fertilizers (consisting of two or three elements)
		Manufacture of chemical crop protection agents - active or synthetic substances (insecticides, fungicides, herbicides, rodenticides, etc.)
01.08.05	Industry of chemical (synthetic and artificial) fibres	Manufacture of polyamide, polyester, polyacrylonitrile, polypropylane, polyvinyl, polyurethane and other synthetic fibres
		Manufacture of staple (viscose, polynose and cuprammonium fibre) and regenerated protein fibre, acetate, triacetate, organzine and other artificial fibres

Code number	Branches, sub-branches and groups	Scope of branches, sub-branches and groups
1	2	3
01.08.06	Industry of synthetic rubber and basic high-molecular products	Manufacture of synthetic rubbers (rubbers based on homopolymers, copolymers, hydrocarbons, copolymers of various monomers), rubbers based on polycondensates, rubber stocks and intermediate products
		Manufacture of basic high-molecular products and intermediate goods (high-molecular products based on natural products, saturated polymers obtained from monomers, hydrocarbons and halogen derivatives, saturated polymers obtained from unsaturated compound ethers and esters, terpines and heterocyclic vinyls, products obtained from polycondensation)
01.08.07	Industry of dyes, pigments, lacquers and paints	Manufacture of organic dyes and pigments (direct dyes; dyes forming ionic or stable bonds; solvent soluble dyes; dyes made from a mixture of different groups; dye-forming couplers for fibre; chemical bleaches)
		Manufacture of inorganic pigments and enamels (artificial pigments and metal based synthetic pigments, phosphorescent pigments, etc., enamels, frit and glaze), lacquers and paints (primers, putties, paints, lacquers, enamels, drying oils, water colours)
01.08.08	Industry of soap manufacture, detergents, perfumes and cosmetics	Manufacture of soap (household and toilet soap, shaving soap, soap paste, powder and flakes, and liquid soap)
		Manufacture of synthetic detergents (active or synthetic substances)
		Manufacture of perfumes and cosmetics
		Manufacture of essential oils and aromatic substances

Code number	Branches, sub-branches and groups	Scope of branches, sub-branches and groups
1	2	3
01.08.09	Industry of drugs and medicines	Manufacture of drugs, including drugs for veterinary use, manufacture of biostimulators Manufacture of disinfectants Manufacture of herbal preparations
01.08.10	Industry of tyres and rubber-asbestos products	Manufacture of tyres and other tyre industry products (regular tyres, cellular and solid, rim bands and detachable bands, balloon tyres) Manufacture of industrial rubber products (conveyor belts, drive belts, gaskets, couplings, bushes, hoses etc.) and industrial asbestos products (sheets, packings, protective textiles and clothing) Manufacture of rubber footwear Manufacture of rubber sheets and mats, rubber sports equipment
01.08.11	Industry of plastic products	Manufacture of pipes, fittings, rods, sheets, mats, foil and intermediate products for industrial components, manufacture of containers, electrical insulation materials with organic binding resins Manufacture of building materials with a polymer base (linoleum, lincrust, etc.) Manufacture of fibreglass materials Manufacture of consumer goods, plastic furniture, protective products made of plastic and other plastic products Manufacture of bottles, jars, cases, trays, boxes, drums, cans, tubes, sacks, bags, and other plastic containers

Code number	Branches, sub-branches and groups	Scope of branches, sub-branches and groups
1	2	3
		Manufacture of synthetic leather
01.08.12	Industry of other chemical products	Manufacture of photosensitive products, magnetic tapes and disks, reagents and high-purity chemicals, binders, catalysts and their vehicles, molecular sieves, ancillary chemical products and chemical products not classified elsewhere
		Manufacture of chemical and biochemical feed additives and other microbiological products
		Manufacture of wood chemistry products
		Manufacture of inks and dyes (printing and duplicating inks, dyes, printing paste)
		Manufacture of magnetic lacquers, organo-silicon compounds, synthetic glues, absorbents
		Manufacture of synthetic diamonds and corundums
01.09	<u>Building materials industry</u>	
01.09.01	Mining and quarrying of non-ore minerals	Mining, quarrying and processing of stone, sand and clay
		Mining and quarrying of asbestos, barite, mica, raw materials for refractories, etc.
01.09.02	Cement industry and industry of other bonding materials	Manufacture of cement and clinker
		Manufacture of lime, gypsum, lime filler and other bonding materials
01.09.03	Industry of concrete and concrete elements	Manufacture of fresh concrete and slurry, prefabricated ferro-concrete elements, articles made of plain concrete and slurries

Code number	Branches, sub-branches and groups	Scope of branches, sub-branches and groups
1	2	3
01.09.04	Industry of refractory materials	Manufacture of moulded refractory goods of all types and makes, refractory bodies and concretes, refractory powders and mortars
01.09.05	Industry of asbestos cement, ceramics, insulation materials and other building materials	Manufacture of asbestos-cement products, basic structural clay products (bricks, tiles for roofing, floors and facing of buildings, glazed tiles, etc.)
		Manufacture of waterproofing, thermal-insulation and sound-proofing materials
		Manufacture of other building materials (asphalt, soft roofing, etc.)
01.10	Logging and woodworking industry	
01.10.01	Logging	Felling and dressing of timber crop, hauling and transportation of timber, cutting and sorting of timber
		Production of wooden articles at logging sites (props, posts, lattice work, lathes, shingles, bark, etc.)
	Woodworking industry (01.10.02-01.10.06)	
01.10.02	Sawn timber industry	Production of sawn timber and sleepers
01.10.03	Veneer and board industry	Manufacture of veneer (for industrial use, laminated or constituted), plywood (for special applications, interior finishing, form work, coaches, aircraft, etc.), panel boards and other laminated wood products
		Manufacture of particle board, fibre board and boards made of other ligno-cellulose materials

Code number	Branches, sub-branches and groups	Scope of branches, sub-branches and groups
1	2	3
01.10.04	Industry of building elements and structures	Manufacture of doors, window frames and various elements for them, shutters and blinds, flooring units, prefabricated elements for various wood structures
		Manufacture of various prefabricated wood structures
01.10.05	Furniture industry	Manufacture of furniture of all types from wood and wood substitute (except furniture made solely of metal and plastic), radio and television cabinets
01.10.06	Industry of other woodworking products	Manufacture of wooden containers, wooden sports and leisure equipment, wooden household articles, matches, prefabricated units, wooden instruments, etc.
		Manufacture of wicker work from materials of plant origin
01.11	Paper and wood-pulp industry	
01.11.01	Industry of wood-pulp, paper and cardboard	Manufacture of chemical pulp (cellulose), semi-chemical pulp (hemicellulose) and mechanical wood pulp, including manufactures from cane, straw and other plants
		Manufacture of paper for writing, drawing, printing and wrapping and for technical and industrial uses, covered processed paper
		Manufacture of cardboard for printing, packaging, technical and industrial uses, covered or processed cardboard, millboard, etc.
01.11.02	Industry of paper and cardboard products	Manufacture of paper and cardboard products
		Manufacture of cardboard boxes, cases and drums, paper bags and sacks

Code number	Branches, sub-branches and groups	Scope of branches, sub-branches and groups
1	2	3
01.12	Glass, porcelain and earthenware industry	
01.12.01	Glass industry	Manufacture of structural and industrial glass
		Manufacture of glass containers.
		Manufacture of everyday goods from glass, glass tableware
		Manufacture of industrial glass products (including optical goods), laboratory glassware
		Manufacture of glass and mineral fibre and goods made from such fibre
01.12.02	Porcelain and earthenware industry	Manufacture of porcelain, semi-porcelain, earthenware and ceramic industrial and sanitary goods, including goods made from fireclay and other materials, porcelain goods for various branches of the national economy
		Manufacture of porcelain, semi-porcelain, earthenware and ceramic goods for domestic, everyday household and decorative purposes
01.13	Textile and knitwear industry	
01.13.01	Textile industry	Primary processing of cotton. Manufacture of cotton and cotton-type yarn and fabrics, manufacture of sewing threads, cotton wool, rope, cord and other cotton-type goods
		Manufacture of scoured wool, wool combings and cord ribbons
		Manufacture of woolen and woolen-type yarn and fabrics

Code number	Branches, sub-branches and groups	Scope of branches, sub-branches and groups
1	2	3
		Manufacture of felt, matting (including goods made from them) and other non-woven woolen textiles
		Manufacture of all types of silken (natural and artificial silk) and silk-type yarn and fabrics
		Manufacture of non-woven silken textiles
		Processing of bast fibres (flax, hemp, jute, etc.)
		Manufacture of yarn, fabrics, packaging materials, rope and cord, industrial and other products made from bast fibres
		Manufacture of carpets and carpet products
		Manufacture of net curtain, lace, braid articles (trimmings, cords, etc.), fabrics for umbrellas, batik-type goods (scarves, kerchiefs, decorative wall hangings, etc.), ribbons
01.13.02	Knitwear goods industry	Manufacture of knitwear from all types of yarn. Manufacture of stockings, socks and gloves
01.14	Wearing-apparel industry	Manufacture of outer clothing and underwear from fabrics, non-woven and knitted cloth
		Manufacture of domestic and everyday clothing articles from fabrics of all types, non-woven textiles and knitwear

Code number	Branches, sub-branches and groups	Scope of branches, sub-branches and groups
1	2	3
01.15	Leather, fur and footwear industry	
01.15.01	Production and processing of leather and fur	Production of natural leather and fur (soft and hard leathers, leather with enriched and semi-enriched fur, dressed leather)
		Manufacture of artificial leather based on natural leathers
01.15.02	Footwear industry	Manufacture of footwear with uppers of leather and leather substitutes and footwear made of textile materials, excluding rubber and plastic footwear
01.15.03	Industry of industrial, protective and saddlery articles	Manufacture of drive belts, workers' and protective gloves, gaskets, goods made of leather or its substitutes for the textile industry, harness, protective articles
		Manufacture of morocco leather goods for personal use, school and office goods and goods for service use, travelling and hunting bags, suitcases, decorative goods made of leather or its substitutes
01.15.04	Industry of wearing apparel made of leather, fur and their substitutes	Manufacture of wearing apparel from leather or its substitutes, leather and fur, or leather, fur and textile materials, artificial fur, polyvinyl-chloride fabrics
01.16	Printing and publishing industry	Production of books, brochures, newspapers, magazines, forms, notebooks and calendars, art reproductions, maps and atlases and other printed goods

Code number	Branches, sub-branches and groups	Scope of branches, sub-branches and groups
1	2	3
01.17	Food industry (including manufacture of beverages and tobacco goods)	
01.17.01	Meat industry	Production of meat, animal fats, products obtained from slaughtering livestock and poultry, production of meat goods and preserved meats
01.17.02	Fish industry	Catching of fish, whales and other marine animals and extraction of products from seas and oceans
		Processing of fish, whales and other marine animals, production of other goods from fish and marine animals (including canning)
01.17.03	Dairy industry	Manufacture of milk for consumption, fresh dairy products, ice cream, animal oil, cheese, sheep's-milk cheese, canned dairy products, processed milk products
01.17.04	Industry of vegetable oils and fats	Manufacture of edible vegetable oil, edible vegetable fats (simple and with additives), industrial vegetable oils, other products made from vegetable oil
01.17.05	Flour-milling industry	Manufacture of flour-milling products, shelled products, friable grain products, grain flakes (excluding production of mixed feeds)
01.17.06	Bread and baking pasta industry	Manufacture of bread and other grain products
		Manufacture of pasta-type goods
01.17.07	Sugar industry	Manufacture of raw sugar, refined sugar and other products from processing of sugar beet and sugar cane
01.17.08	Confectionery industry	Manufacture of confectionery goods and goods from cacao

Code number	Branches, sub-branches and groups	Scope of branches, sub-branches and groups
1	2	3
01.17.09	Industry of fruit and vegetable and fruit preserves	Canning and preserving of vegetables, fruit and berries, production of fresh-frozen and dried vegetables, fruits and berries
		Manufacture of other products from processing of vegetables and fruits
01.17.10	Industry of alcoholic products	Manufacture of wine and viticultural products, cognac, vodka and other alcoholic beverages
		Manufacture of raw ethyl alcohol
		Manufacture of beer, malt brewers' yeast
01.17.11	Industry of non-alcoholic beverages	Manufacture of non-alcoholic beverages, mineral and carbonated waters
01.17.12	Tobacco industry	Manufacture of fermented tobacco and tobacco goods
01.17.13	Refrigeration industry	Refrigeration for cooling and freezing products
		Manufacture of ice
01.17.14	Industry of other food products	Manufacture of concentrated foods, culinary products and other manufactured food products not classified elsewhere
		Roasting of natural coffee, manufacture of products from processing of tea, coffee and eggs
		Manufacture of starch, glucose, dextrin and vinegar
01.18	Other manufacturing industries	
01.18.01	Industry of abrasive goods	Manufacture of granulated abrasives, semi-finished goods and common products from manufacturing abrasives, various goods made from abrasive

Code number	Branches, sub-branches and groups	Scope of branches, sub-branches and groups
1	2	3
		materials. Manufacture of flexible abrasives, abrasive pastes
01.18.02	Industry of coal and graphite goods	Manufacture of goods from coal and graphite (manufacture of electrodes, brush collectors, amorphons, blocks, pastes, rings, tablets and other goods made of graphite and coal)
01.18.03	Mixed feed industry	Manufacture of mixed feeds, fodder yeasts and other types of industrial feeds made from various raw materials
01.18.04	Industry of art goods and other cultural goods	Manufacture of art goods and various articles for cultural purposes, phonograph records, copies of motion pictures and film strips, magnetic and video recordings, musical instruments Manufacture of toys, office and school supplies
01.18.05	Jewelry industry	Manufacture of jewelry from precious metals and stones, processing of natural precious and semi-precious stones, gems and amber Manufacture of costume jewelry
01.18.06	Industry of goods based on by-products of plant and animal husbandry	Manufacture of goods from the processing of by-products of plant and animal husbandry (not classified elsewhere), brooms, brushes and dusters (other than plastic) and other goods
01.18.07	Production of water	Production of water (drilling, purifying and distributing of water) for material and non-material consumption
01.18.08	Mining and quarrying of other non-metal ore materials	Mining and quarrying of precious stones (other than diamonds) and semi-precious stones, gems and amber

Code number	Branches, sub-branches and groups	Scope of branches, sub-branches and groups
1	2	3
01.18.09	Other types of industrial production (not classified elsewhere)	Dry-cleaning, dyeing and laundering of linen; provision of various industrial services
		Other types of industrial production not classified elsewhere
02	CONSTRUCTION	
02.01	Construction and assembly work	Construction of industrial projects, railways, roads and bridges, pipelines, housing, etc.; construction of electricity, water, telephone and sewerage networks; construction of waterworks, irrigation and land-improvment systems; construction for watercourse management (including drilling for water supply); construction of logging sites (forest roads, installations for conveyance and transportation at logging sites, etc.); assembly of equipment; installation and insulation work
		Hiring, assembly and dismantling of construction equipment
		Customer supervision of completion of construction, installation and assembly work
		Work on internal plants, branching and connecting with gas-distribution, heating and water-pipe networks, and rendering of any services of the construction or assembly type
		Capital repair of buildings and structures by construction and repair organizations using contractual or direct-labour methods
02.02	Geological prospecting, geodetic and drilling work	Geological prospecting, drilling and mining work, operational and exploratory drilling for oil and natural gas, geodetic and cartographic work for capital investment projects

Code number	Branches, sub-branches and groups	Scope of branches, sub-branches and groups
1	2	3
02.03	Planning and surveying work	Work of planning and surveying organizations for all types of construction
03	AGRICULTURE	
03.01	Production of agricultural products	
03.01.01	Plant-growing	Cultivation of grain, legume and industrial crops, potatoes, vegetables, cucurbits, fodder crops
		Viticulture, horticulture, cultivation of plants for producing seeds and planting material
		Flower growing and cultivation of ornamental plants and other crops
03.01.02	Livestock-raising	Raising of productive and working livestock (including pedigree livestock) and stock for slaughter and raising of other types of animals (fur farming, rabbit breeding, etc.), poultry, bees, silkworms
		Production of milk, wool, eggs, honey
		Production of by-products obtained from raising cattle, pigs, sheep, goats, horses, poultry, bees
03.01.03	Fish farming	Breeding of fish, aquatic animals, etc., catching of fish in ponds and natural water basins where piscicultural techniques are practised (fry breeding, feeding of fish, fertilization of ponds, etc.)
03.02	Operation of irrigation and land-improvement improvement systems	Operation of irrigation, water-supply and drainage systems, including dams, reservoirs, pumping stations and other water-management installations

Code number	Branches, sub-branches and groups	Scope of branches, sub-branches and groups
	1	2	3
03.03	Agricultural services	Direct servicing of the production process in agriculture by enterprises and organizations engaged in the mechanization of agriculture, variety testing stations, sanitary and veterinary inspectorates, plant protection laboratories, seed control laboratories, veterinary clinics, dispensaries and laboratories, agro-chemical and soil-science laboratories; aviation units providing agricultural services; other establishments providing agricultural services
04	FORESTRY	
04.01	Basic forestry operations	Afforestation, forest regeneration, forest management and protection of forests against pests and disease, protection of wild animals, timber distribution; forest reclamation, forest construction work, protection of forests from fires, forestry seed control stations, forest soil laboratories and other work directly related to afforestation and forest regeneration
04.02	Procurement of wild-growing and non-wood forestry products	Collection of seeds, cuttings, resin, buds, roots, cane; collection of non-wood forestry products: berries, mushrooms, wild medicinal plants
04.03	Hunting and trapping	Hunting and trapping of game, birds, etc. Collection of eggs of wild birds, antlers or horns of deer, reindeer, roe deer, goats, etc.
05	TRANSPORT	
05.01	Railway transport	Transport of goods by public railways Transport of passengers by public railways

Code number	Branches, sub-branches and groups	Scope of branches, sub-branches and groups
1	2	3
		Maintenance of power-supply, signalling and communications centres, civilian buildings and structures (other than residential and municipal), and protective forest plantings along railways
		Maintenance of railway tracks, installations and rolling stock
05.02	Motor transport b/	Public goods transport by road
		Public passenger transport by road
		Maintenance of motor transport vehicles
05.03	River transport	Public goods transport by river
		Public passenger transport by river
		Specific services of river transport and port facilities (pilot service, emergency assistance, rescue service, river agencies, freight operations, operations for the regulation, control and inspection of navigation)
		Maintenance and care of vessels and routes, including canals, locks and related installations
		Timber flotation
05.04	Marine transport	Public goods transport by sea
		Public passenger transport by sea

---

b/   This sub-branch does not include the operation of motor vehicle facilities engaged in town cleansing and refuse collection, or servicing health establishments and communications enterprises. Such operations are classified, respectively under the branches "Housing, public utilities and amenities services", "Health services, social security, physical culture and tourism" and "Communication".

Code number	Branches, sub-branches and groups	Scope of branches, sub-branches and groups
1	2	3
		Maintenance of vessels, installations and structures
		Specific services of marine transport and port facilities (pilot service, emergency assistance, rescue service, diving operations, freightage, control and inspection of ships and shipping lanes)
05.05	Air transport	Public goods transport by air
		Public passenger transport by air
		Communications and radar servicing of flights, operation of airports
		Other air transport operations (takeoff and landing services, lighting, etc.)
05.06	Major pipeline transport	Transport of oil, oil products, gas, etc., by major pipelines
		Other operations for pipeline transport (pumping and compression of oil products, pipeline servicing, inspection and control)
05.07	Urban transport	Urban transport of passengers (trams, buses, trolley-buses, underground trains, taxis)
		Urban transport of goods
05.08	Other forms of transport	Public carriage of goods and passengers by horse-drawn and other forms of transport
05.09	Loading, unloading and dispatching operations	Loading, unloading and trans-shipment from or into transport vehicles, including warehousing and storage of goods in transit. Work of dispatching organizations

Code number	Branches, sub-branches and groups	Scope of branches, sub-branches and groups
1	2	3
05.10	Road services	Maintenance and operation of roads, bridges and road structures (other than those listed in group 09.03.01)
06	COMMUNICATION	Postal operations: delivery of mail, money transfers, parcels
		Telegraphic communications, including technical maintenance of telegraph lines, installations and stations; installation of telegraph equipment and teleprinters
		Remote data transmission
		Servicing of telephone calls by public communications units
		Technical maintenance of telephone lines, installations and exchanges; installing and replacement of telephone
		Wired-radio installation, wired-radio relay service, installing and replacement of radio relay receivers
		Radio broadcasting (transmission of radio programmes, etc.)
		Transmission and relaying of television programmes, etc.
07	TRADE, PROCUREMENTS AND MATERIAL SUPPLY	
07.01	Domestic trade	
07.01.01	Wholesale trade	Marketing of consumer goods by wholesale enterprises and organizations: wholesale depots and stores of pharmaceutical administrations, book suppliers and other wholesale organizations marketing consumer goods

Code number	Branches, sub-branches and groups	Scope of branches, sub-branches and groups
1	2	3
07.01.02	Retail trade	Retail sale of food and non-food goods through specialized units - permanent, seasonal, mobile; small-scale wholesale units selling food and non-food goods; units selling goods on commission; pharmacies, stores for the retail sale of building materials, furniture, oil products and other kinds of fuel
07.02	Public catering	Public catering by specialized enterprises: factory kitchens, restaurants, bars, dining rooms of all kinds, cafes, tea rooms, snack bars, cafeterias, kebab stalls and other public catering enterprises
07.03	Foreign trade	Foreign trade and economic scientific and technical co-operation involving: sales, purchases or goods exchanges, provision of services, planning and execution of work, technical assistance and co-operation, sale and purchase of licences to use patents on inventions or technology Organization of international exhibitions and fairs
07.04	Material supply	Material supply by specialized entities (goods-supply enterprises, depots, including work at state material stockpiles - warehouses, shops, offices whose functions are solely or chiefly those of supply)
07.05	State procurement of agricultural products	Procurement, purchase and forward contracting of agricultural products by specialized enterprises
07.06	Collection of scrap metal and other secondary raw materials	Collection and procurement and initial processing of scrap metal and other secondary raw materials (rubber, plastics, glass containers, etc.)

Code number	Branches, sub-branches and groups	Scope of branches, sub-branches and groups
1	2	3
08	OTHER BRANCHES OF MATERIAL PRODUCTION	
08.01	Technical planning	Work of independent organizations engaged in planning and technical (engineering) servicing of all branches of material production, except the work of organizations engaged in planning construction projects and the work of planning organizations carrying out scientific work
08.02	Mechanized and automated processing of data and information c/	Work of independent calculating-machine offices and public computer centres, and other organizations which provide computer services and whose main work is the production of software or hardware, irrespective of which branches receive their products or services
08.03	River-control operations	River control, flood prevention and nature protection
08.04	Other activities of the material sphere	Publishing houses, editorial offices, press offices, telegraph agencies
		Cinema studios, phonograph recording studios
09	HOUSING, PUBLIC UTILITIES AND AMENITIES	
09.01	Housing services	Housing administrations and residential management offices, including hostels (other than hostels for the aged and disabled)
09.02	Hotel-type services	Operations of hotels, guest houses and hostels, inns, motels, etc.

---

c/ Computer centres serving individual branches are included under the relevant branches.

Code number	Branches, sub-branches and groups	Scope of branches, sub-branches and groups
1	2	3
09.03	Public utilities and amenities	
09.03.01	Public utilities	Street lighting, town cleansing (including refuse collection), maintenance and care of streets, boulevards, squares, creation of urban green zones, water-purification and sewerage system
		Management of cemeteries and crematoria
		Maintenance of urban roads, bridges, viaducts, junctions
		Other public utilities
09.03.02	Public amenities	Photographer's studios, hairdressers, messenger services, information centres, housing exchange offices, hire agencies, typing, translation and other business services, bath houses, shower-rooms, apartment-cleaning and floor-polishing organizations, organizations performing individual errands for the population and other non-productive organizations providing everyday services to the population
10	SCIENCE	
10.01	Scientific research	Research work in basic and applied sciences by academies of science and their branches, and by research institutions attached to academies of science and other agencies
		Experimental design work at research institutions and centres
		Planning and design organizations engaged in scientific work, astronomical observatories, research laboratories, scientific and experimental stations of all kinds, nature reserves, etc.

Code number	Branches, sub-branches and groups	Scope of branches, sub-branches and groups
1	2	3
		Centres systematically engaged in research work on computer technology
10.02	Other scientific work	Scientific and technical information, meteorological and hydro-meteorological stations
		Geological prospecting, geodetic and cartographic work (except the work of organizations engaged in deep exploratory drilling requiring capital investment)
11	EDUCATION	
11.01	Pre-school education	Pre-school education in kindergartens maintained by enterprises, institutions and people's soviets, in special pre-school establishments for mentally and physically handicapped children
11.02	General education	Education in general schools, secondary schools, evening and extramural departments of general schools, preventorium schools, and art and music schools providing secondary education
		Education and training in special general schools for physically or mentally handicapped children (deaf and partially deaf, blind and partially blind, etc.)
11.03	Vocational education	Training of qualified personnel at vocational colleges and schools, including their evening and extramural departments, training of physically and mentally handicapped personnel

328

Code number	Branches, sub-branches and groups	Scope of branches, sub-branches and groups
1	2	3
11.04	Special secondary education	Training of physically and mentally handicapped personnel at technical colleges of all kinds, teacher-training schools, various kinds of colleges and other special secondary training establishments. Having qualifications in special secondary education
11.05	Higher education	Education at universities, and institutes in all disciplines, conservatories, higher schools and other educational establishments (including evening class and extramural) offering qualifications in intermediate and higher education
11.06	Retraining and extension course for qualified personnel	Retraining and extension courses at vocational training centres and other special educational establishments (on an extramural basis) for qualified staff
11.07	Other forms of education	Halls and houses for pioneers and schoolchildren, youth clubs, art and culture clubs, centres for young technicians and naturalists, and other out-of-school institutions for children
12	CULTURE AND ART	
12.01	Cultural establishments	Urban and rural houses of culture, planetariums, trade union and student clubs (including youth centres and clubs) Parks of culture and leisure
12.02	Libraries and museums	Public libraries Museums of natural science, technology, history, ethnography, memorial museums, museums of art, botanical and zoological gardens, permanent exhibitions, workshops restoring cultural monuments, works of art, etc.

Code number	Branches, sub-branches and groups	Scope of branches, sub-branches and groups
1	2	3
12.03	<u>Cinemas, radio and television</u>	Distribution and screening of the cinema film (at stationary and mobile theatres) Production and dissemination of radio, wired-radio and television programmes, etc.
12.04	<u>Theatres and other musical establishments</u>	Drama and puppet theatres, opera houses, musical and operetta houses, song and dance ensembles, circuses, philharmonic and symphony orchestras, folk instrument orchestras, agencies organizing shows and tours by artists, ticket agencies of theatres and musical establishments (except for amateur companies)
13	HEALTH SERVICES, SOCIAL SECURITY, PHYSICAL CULTURE AND TOURISM	
13.01	<u>Health services</u>	Public-health and medical institutions and bodies of all kinds: general hospitals, clinics and other hospital institutions, tuberculosis sanatoriums, preventive-care units, public-health and medical services attached to enterprises and organizations, dispensaries, in-patient and other facilities (tuberculosis, dermatological and venereal, oncological, physiotherapy, local dispensaries, dispensaries attached to enterprises, etc.) tuberculosis prevention centres, health centres attached to enterprises and institutions, local in-patient facilities, rural health centres, out-patient departments (rheumatological, gastro-enterological, etc.), polyclinics, first-aid stations, blood-donor centres, public-health aviation centres, pharmaceutical services attached to medical stations, public-health and epidemiological

Code number	Branches, sub-branches and groups	Scope of branches, sub-branches and groups
1	2	3
		institutions, disinfection stations, public health inspection services, public-hygiene laboratories, anti-malaria stations and posts, public-health education centres, dental co-operatives, and other health institutions
		Sanatoriums and spas
13.02	<u>Children's establishments</u>	Nurseries and mother and child institutions
13.03	<u>Social security</u>	Social security organizations: homes, hostels and workshops for the aged and handicapped, national-assistance canteens, industrial disease and injury commissions and other social-welfare institutions
13.04	<u>Physical culture and sport</u>	Organization of physical culture and sport through sports federations, clubs and societies (attached to enterprises, institutions, co-operatives, etc.)
		Sports centres, stadiums, halls, skating rinks, swimming pools, aquatic sports centres, etc.
13.05	<u>Tourism and leisure</u>	Organization of tourism and leisure and other tourist and leisure services
		Holiday homes of all kinds, students' camps
14	FINANCE, CREDIT AND INSURANCE	National and specialized banks, and financial activities of other bodies in the financial and state insurance system
		Savings banks, state lotteries and games
15	GENERAL GOVERNMENT	Central and local government of state affairs through the relevant organs

Code number	Branches, sub-branches and groups	Scope of branches, sub-branches and groups
1	2	3
		Central legislative organs of state power, central and executive organs of state power (state committees and commissions, ministries, offices, main departments, and departments attached to the Council of Ministers, and other central agencies of the state)
		Customs organs
		State service for quality control, inspection and servicing of measuring instruments and safety engineering equipment, etc.
		Local organs of state power (regions, sub-regions, areas, districts, towns, villages), departments and branches of the executive committees of people's soviets
		Courts and judicial agencies, the notarial service and colleges of lawyers
		Fire fighting and security-guard services
		Protection of public safety and defence
		Central and local administration of consumer and producer co-operatives
		State archives
16	OTHER BRANCHES OF THE NON-MATERIAL SPHERE	Organs of political parties, mass, trade union and other public organizations, societies of creative artists, and cultural and other societies of a non-productive character (for religious worship, etc.). Advertising not included elsewhere, and other forms of activity of the non-material sphere

Bibliography

BIBLIOGRAPHY

Alexeev, M. "Soviet Residential Housing in the GNP Accounts." In CIA, <u>Measuring Soviet GNP: Problems and Solutions</u>. A Conference Report. SOV 90-10038, September 1990.

Aslund, A. "How Small is Soviet National Income?" A Paper presented at the First Biennial RAND-Hoover Symposium on the Defense Sector in the Soviet Economy, Stanford, March 1988.

Becker, A. <u>Soviet National Income, 1958-1964</u>. Berkeley: University of California Press, 1969.

Belkin, V. Market and Non-Market Sytems: Limits to Macroeconomic Comparability." A paper presented at the AEI Conference on the Comparison of U.S. and Soviet Economies. Virginia, April 1990.

_____and A. Geronimus, eds., <u>Model' "Dokhod-Tovary" i Balans Narodnogo Khozyaistva SSSR</u>. Moscow: Nauka, 1978.

Birman, I. <u>Secret Incomes of the Soviet State Budget</u>. The Hague: Martinus Nijhofff Publishers, 1981.

_____. <u>Ekonomika Nedostach</u>. New York: Chalidze Publications, 1983.

_____. "Rosefielde and My Cumulative Disequilibrium Hypothesis: A Comment." <u>Soviet Studies</u>, January 1989.

_____. "The size of Soviet military expenditures: methodological aspects," A Paper presented at the AEI Conference on the Comparison of Soviet and American Economies, Virginia, April 19-22, 1990.

Bor, M. <u>Voprosy Metodologii Planovogo Balansa Narodnogo Khozyaistva</u>. Moscow: Izdatel'stvo AN SSSR, 1960.

_____. <u>Effektivnost' Obshchestvennogo Proizvodstva i Problemy Optimal'nogo Planirovaniya</u>. Moscow: Mysl', 1972.

Boretsky, M. "The Tenability of the CIA Estimates of Soviet Economic Growth," <u>Journal of Comparative Economics</u>, December 1987.

Edwards I., Margaret Hughes and James Noren, "US and USSR: Comparisons of GNP," in U.S. Congress, Joint Economic Committee, <u>Soviet Economy in a Time of Change</u>, Volume I. Washington, D.C.: USGPO, 1979.

Ericson, R., "The Soviet Statistical Debate: Khanin vs. TsSU." A Paper Presented at the RAND-Hoover Conference on the Defense Sector in the Soviet Economy held in March, 1988.

Faltsman, V. <u>Proizvodstvennyi Potentsial SSSR: Voprosy Prognozirovaniya</u>. Moscow: Ekonomika, 1987.

Gaidar, E., "Trudnyi Vybor. Ekonomicheskoe Obozrenie Po Itogam 1989 Goda." <u>Kommunist</u>, N2, 1990.

Gallik, D., B. Kostinsky and V. Treml. <u>Input-Output Structure of the Soviet Economy: 1972</u>. Foreign Economic Report, No.18, Washington, D.C.: U.S. Department of Commerce, 1983.

Glushkov, N., and A. Dryabin, eds., <u>Tsena V Khozyaistvennom Mekhanizme</u>, Moscow: Nauka, 1983.

Goskomstat SSSR. <u>Narodnoe Khozyaistvo SSSR v 1980 Godu</u>. Moscow: Finansy i Statistika, 1981.

_____. Finansy SSSR. Statisticheskiy Sbornik (1987-1988). Moscow, 1990.

_____. Metodika Ischisleniya Normal'nykh, Izbytochnykh Sberezheniy i Neudovletvorennogo Sprosa, Moscow, 1990.

_____. Narodnoe Khozyaistvo SSSR v 19.. Godu. Moscow: Finansy i Statistika, 1970-1990.

_____. Osnovnye Pokazateli Balansa Narodnogo Khozyaistva SSSR i Soyuznykh Respublik, Moscow: IIT, 1990.

_____. Promyshlennost' SSSR. Moscow: Finansy i Statistika, 1988.

_____. "Sotsial'no-Ekonomicheskoe Razvitie SSSR v 1989 godu," Statisticheskiy Press-Bullten' N4. Moscow: GKS SSSR, 1990.

_____. SSSR v Tsyfrakh v 1989 Godu, Moscow: Finansy i Statistika, 1990.

_____. Torgovlya SSSR. Moscow: Finansy i Statistika, 1989.

_____. and Gosplan SSSR, Metodika Ischisleniya Valovogo Natsional'nogo Produkta SSSR. Moscow, 31 March 1988.

Gosplan SSSR, Metodicheskie Ukazaniya k Razrabotke Gosudarstvennykh Planov Razvitiya Narodnogo Khozyaistva SSSR. Moscow: Ekonomika, 1974.

Isayev, B., and A. Terushkin, eds., Svodnyi Material'no-Finansovyi Balans, Moscow: Nauka, 1978.

Ivanov Yu. and B. Ryabushkin, "Integratsiya Balansa Narodnogo Khozyaistva i Sistemy Natsional'nykh Schetov." Vestnik Statistiki, N9, 1989.

_____, B. Ryabushkin and M. Eidel'man, "Ischislenie Valovogo Natsional'nogo Produkta SSSR." Vestnik Statistiki, N7, 1988/

Khanin, G., "Ekonomicheskiy Rost: Al'ternativnaya Otsenka," Kommunist, N17, 1988.

Kheinman S., and D.M. Palterovich, eds., Naucho-Tekhnicheskiy Progress i Struktura Proizvodstva Sredstv Proizvodstva. Moscow: Nauka, 1982, pp. 25-54.

Kirichenko, V. "Ochishchenie Statistiki," EKO, N2, 1990.

_____. "Natsional'nyi Dokhod: Spravedlivost' Raspredeleniya," Ekonomika i Zhizn', N13, March 1990.

_____. "Vernut' Doverie Statistike," Kommunist, N3, 1990.

Kornev, A., "Kak Rastut Tseny Na Machiny i Oborudovanie." Voprosy Ekonomiki, N6 1988.

Koryagina, T. "Uslugi Tenevyei Legal'nye." EKO, N2, 1989.

_____. Platnye Uslugi v SSSR, Moscow: Ekonomika, 1990.

Kuznetsov, Ye., and F. Shirokov, "Naukoyomkie Proizvodstva i Konversiya," Kommunist, N10.

Lee, W. The Estimation of Soviet Defense Expenditures, 1955-1975, An Unconventional Approach. New York: Praeger, 1977.

Lokshin, P., Spros, Proizvodstvo, Torgovlya. Moscow: Ekonomika, 1975.

Lopatin, V., "Armiya i Ekonomika," Voprosy Ekonomiki, N10, 1990.

Martynov, V. "SSSR i SSha po Materialam Mezhdunarodnykh Sopostavleniy OON i SEV (raschoty Goskomstata SSSR)." Vestnik Statistiki, N9, 1990.

Mikhailov, L. "Natsional'nyi Dokhod: Proizvedennyi i Ispol'zovannyi," Ekonomika i Zhizn', N40, 1990.

Ministerstvo Finansov SSSR, Gosudarstvennyi Byudget SSSR v 1981-1985. Moscow: Finansy i Statistika, 1987.

Ministerstvo Vneshnei Torgovli, Vneshnyaya Torgovlya SSSR v 19.. Godu. Moscow: Finansy i Statistika.

Panskov, V., "U Opasnoi Cherty," Ekonomika i Zhizn', N15, 1990.

Pitzer, J. "Alternative Methods of Valuing Soviet Gross National Product." In CIA, Measuring Soviet GNP: Problems and Solutions. A Conference Report. SOV 90-10038, September 1990.

Pogosov, I., "Sotsial'no-Ekonomicheskie Preobrazovaniya i Informatsiya." Ekonomika i Matematicheskie Metody, Vol. XXV, N5, 1989.

Pozharov, A. Ekonomicheskie Osnovy Oboronnogo Mogushchestva Sotsialisticheskogo Gosudarstva. Moscow: Voenizdat, 1981.

Prell, M. "The Role of the Service Sector in Soviet GNP and Productivity Estimates," Journal of Comparative Economics, September 1989.

Rapawy, S. "Labor Force and Employment in the USSR," The U.S. Congress, Joint Economic Committee, Gorbachev's Economic Plans, Study Papers, Volume 1, Washington, D.C.: GPO, 1987.

Rogovsky, E., Rutkovskaya, E., and Polyakov, E. "Kachestvennaya Neodnorodnost' Investitsionnykh Resursov i Metody Yeyo Izmereniya," Ekonomika i Matematicheskie Metody, Vol. 26, N3, 1990.

Rumer, B., "Soviet Estimates of the Rate of Inflation, Soviet Studies, Vol. XLI, N2, April 1989.

Rutgeizer, V. Resursy Razvitiya Neproizvodstvennoi Sfery, Moscow: Mysl', 1975.

Ryabushkin, B., "Statisticheskoe Izuchenie Urovnya Zhizni," Vestnik Statistiki, N4, 1988.

Schroeder, G., "Consumption in the USSR and the U.S.: A Western Perspective." A Paper presented at AEI Conference on the Comparison of U.S. and Soviet Economies. Virginia, April 1990.

_____, and I. Edwards, Consumption in the USSR: An International Comparison. A study prepared for the use of the Joint Economic Committee, U.S. Congress, Washington, D.C.: USGPO, 1981.

Sel'tsovsky, V., "Sovershenstvovanie Analiza Effektivnosti Vneshnei Torgovli," Vestnik Statistiki, N6, 1983.

Semenov, V. Finansovo-Kreditnyi Mekhanizm v Razvitii Sel'skogo Khozyistva, Moscow: Finansy i Statistika, 1983.

_____. Prodovol'stvennaya Programma i Finansy, Moscow: Finansy i Statistika, 1985.

Steinberg, D. "Defense Allocations Under Gorbachev: Analyzing New Data for 1988-1990." In Roy Allison, ed., Military and Strategic Issues in the Contemporary USSR. London: Macmillan, 1991.

_____. "Estimating Total Soviet Military Expenditures: An Alternative Approach Based on Reconstructed Soviet National Accounts." In C.G. Jacobsen, ed. The Soviet Defense Enigma. SIPRI. Oxford: Oxford University Press, 1987.

_____. "Trends in Soviet Military Expenditure." Soviet Studies, vol. 42, no. 4, October 1990.

_____. "Pozvol'te Predstavit' Mezhotraslevoi Balans SSSR za 1988." Eko, 12, 1990.

_____. "The Growth of Soviet GNP and Military Expenditures in 1970-1989: An Alternative Assessment." A Paper presented at the Second Biennial RAND-Hoover Symposium on the Defense Sector in the Soviet Economy, Santa Monica, March 1990.

_____. "Valovoi Natsional'nyi Produkt SSSR v 1987 Godu: Al'ternativnye Otsenki." Eko, N1, 1991.

Sverdlik, Sh. "Vol'nyi Ekonomist Protiv Vsei Korolevskoi Rati." Eko, 12, 1990.

_____. "The Balance of Monetary Turnover in the USSR," A Paper presented at the Stanford-Berkeley Seminar, December 24, 1990.

Treml, V., and B. Kostinsky, "Domestic Value of Foreign Trade: Exports and Imports in the 1972 Input-Output Table," Foreign Economic Report N20, U.S. Department of Commerce, Bureau of the Census, October 1982.

U.S. Central Intelligence Agency. Soviet and US Defense Activities, 1970-1979: A Dollar Cost Comparison. A Research Paper, SR 80-100005, January 1980.

_____. "USSR Measures of Economic Growth and Development, 1950-1980." Studies Prepared for the Use of the Joint Economic Committee of the U.S. Congress. Washington, D.C.: U.S. GVO, 1982.

_____. "A Guide to Monetary Measures of Soviet Defense Activities," A Reference Aid, SOV 87-10069, November 1987.

_____. "The Soviet Economy in 1988: Gorbachev Changes Course." A Report presented to the Subcommittee on National Security Economics of the Joint Economic Committee. April 14, 1989.

_____. Measuring Soviet GNP: Problems and Solutions. A Conference Report. SOV 90-10038, September 1990.

_____. Measures of Soviet Gross National Product in 1982 Prices. Submitted to Joint Economic Committee of the U.S. Congress. Washington, D.C.: USGPO, 1990.

Val'tukh K., and B. Lavrovsky, "Proizvodstvennyi Apparat Strtany: Ispol'zovanie i Rekonstruktsiya," Ekonomika i Organizatsiya Promyshlennogo Proizvodstva, N2, 1986.

Volkov, A., "Finansovyi Krizis v SSSR," A Paper presented at the Stanford-Berkeley Seminar, December 24, 1990.

Wilkinson, D. "Perestroika and the Soviet Defense Sector: Some Implications for NATO," The Paper presented at the IV World Congress for Soviet and East European Studies, Harrogate, England, July 21-26, 1990.

Zalkind, A., ed. Dva Podrazdeleniya Obshchestvennogo Produkta. Moscow: Statistika, 1976.

Zhelnorova, N. "Statistika Stanovitsa Ob'yektivnoi." Argumenty i Fakty, N4, 1990.